4

Informatik-Fachberichte 153

Herausgegeben von W. Brauer
im Auftrag der Gesellschaft für Informatik (GI)

D. Meyer-Ebrecht (Hrsg.)

ASST '87
6. Aachener Symposium für Signaltheorie

Mehrdimensionale Signale und Bildverarbeitung

Aachen, 9.–12. September 1987

Proceedings

Springer-Verlag
Berlin Heidelberg New York
London Paris Tokyo

Herausgeber

Dietrich Meyer-Ebrecht
Lehrstuhl für Meßtechnik
Rheinisch-Westfälische Technische Hochschule
D-5100 Aachen

Veranstalter des Symposiums

Rheinisch-Westfälische Technische Hochschule Aachen
Tagungsbüro: Lehrstuhl für Meßtechnik, RWTH Aachen

Programmkomitee

W. Ameling D. Meyer-Ebrecht (Tagungsleitung)
P. L. Butzer H. Meyr
H.-D. Lüke F. Schreiber

Tagungsorganisation

E. Bücken W. Winkler
W. Krybus J. Vossebürger
M. Schmitz

CR Subject Classifications (1987): C.1, I.4, I.5

ISBN 3-540-18401-5 Springer-Verlag Berlin Heidelberg New York
ISBN 0-387-18401-5 Springer-Verlag New York Berlin Heidelberg

CIP-Kurztitelaufnahme der Deutschen Bibliothek. Mehrdimensionale Signale und Bildverarbeitung: proceedings / 6. Aachener Symposium für Signaltheorie, Aachen, 9.–12. September 1987. D. Meyer-Ebrecht (Ed.). – Berlin; Heidelberg; New York; Tokyo: Springer, 1987.
(Informatik-Fachberichte; 153)
Auf d. Haupttitels. auch: ASST '87
ISBN 3-540-18401-5 (Berlin ...)
ISBN 0-387-18401-5 (New York ...)
NE: Meyer-Ebrecht, Dietrich [Hrsg.]; Aachener Symposium für Signaltheorie «06, 1987»; GT

Printed in Germany

Repro- u. Druckarbeiten: Weihert-Druck GmbH, Darmstadt
Bindearbeiten: Druckhaus Beltz, Hemsbach/Bergstraße
2145/3140-543210

Vorwort

Seit 1975 wird das *Aachener Symposium für Signaltheorie* – zunächst unter dem Namen "Aachener Kolloquium" – in zwei- bis dreijährigem Turnus ausgerichtet. Das Symposium widmet sich schwerpunktmäßig den theoretischen Grundlagen der Signalverarbeitung, soll jedoch die Bezüge zur Anwendung und Technologie nicht außer acht lassen. Initiiert wurde das Symposium 1975 als eine gemeinsame Veranstaltung Aachener Institute der Elektrotechnischen und der Mathematisch-Naturwissenschaftlichen Fakultät, die sich in ihren Forschungsarbeiten mit Signalverarbeitung und deren theoretischen Grundlagen befassen. Das Symposium soll ein "Werkstattgespräch" sein, also über den aktuellen Stand laufender Forschungsarbeiten berichten. Es soll insbesondere ein Forum für die Verständigung zwischen den involvierten Fachrichtungen – Ingenieurwissenschaften, Informatik, Mathematik, Physik etc. – sein. In dieser Rolle hat das Symposium schnell breites Interesse im In- und Ausland gefunden.

Schwerpunktthemen wie *diskrete Signale* (1976), *stochastische Signale* (1979), *Signalverarbeitungsmethoden* (1981), *mathematische Methoden in der Signalverarbeitung* (1984) haben das Symposium jeweils auf aktuelle Forschungsgebiete fokussiert. Stets haben *Bilder* als eine besondere Art komplexer Signale in dem Symposium eine entscheidende Rolle gespielt. Obwohl die digitale Bildverarbeitung bereits ein etabliertes Gebiet ist, erleben wir zur Zeit wieder ein wachsendes internationales Interesse an ihren Methoden, Technologien und Anwendungen. Gründe dafür sind, daß wichtige Schlüsseltechnologien für eine effiziente Verarbeitung, Speicherung und Übertragung der meistens sehr umfangreichen Datensätze digitaler Bilder heute verfügbar sind und damit die Schwelle zum praktischen Einsatz deutlich reduziert ist. Die Vielzahl der bereits im medizinischen und industriellen Anwendungsbereich etablierten Verfahren zeigt dies eindrucksvoll. Mit dem Schwerpunkthema *multidimensionale Signale und Bildverarbeitung* soll das diesjährige Symposium dieser Entwicklung Rechnung tragen.

Für die Unterstützung bei der Vorbereitung und Durchführung dieses Symposiums danke ich den Mitgliedern des Programmkomitees und den Mitarbeitern des Tagungsbüros. Insbesondere war die Organisation des Symposiums nur durch den engagierten Einsatz der Mitarbeiter des Tagungsbüros möglich, die diese Arbeit neben ihren vielfältigen Aufgaben aus Lehre und Forschung geleistet haben. Unserer Hochschule danken wir dafür, daß das Symposium in einer adäquaten Umgebung stattfinden kann.

D. Meyer-Ebrecht

Bisher erschienene Tagungsbände in der Reihe Aachener Symposium für Signaltheorie

Spezielle Probleme der Signaltheorie
Kurzfassungen des 1. Aachener Kolloquiums
20. März - 21. März 1975
Hrsg.: H. D. Lüke, IENT, Aachen, IX + 56 Seiten

Theorie und Anwendung diskreter Signale
Kurzfassungen des 2. Aachener Kolloquiums
7. Okt. - 8 . Okt. 1976
Hrsg.: H. J. Tafel, IND, Aachen, XXII + 258 Seiten

Stochastische Signale
Kurzfassungen des 3. Aachener Kolloquiums
4. Okt. - 6. Okt. 1979
Hrsg.: H. Meyr, ERT, Aachen, 251 Seiten

Theorie und Anwendung der Signalverarbeitung
Kurzfassungen des 4. Aachener Kolloquiums
30. Sept. - 2. Okt. 1981
Hrsg.: H. D. Lüke, IENT, Aachen, 381 Seiten

Mathematische Methoden in der Signalverarbeitung
Tagungsband des 5. Aachener Kolloquiums
26. Sept. - 29. Sept. 1984
Hrsg.: P. L. Butzer, LAMA, Aachen, 416 Seiten

INHALT

Eröffnung 1

A Unifying Framework for Reconstructive Imaging Methods
H. Schomberg 2

Interactive 3D Computer Grafic in Medical Imaging
M. W. Engelhorn 16

Theorie mehrdimensionaler Signale 28

Resolution in Computerized Tomography
F. Natterer 29

Pulssequenzen als Modulationsfunktion beim NMR-Imaging
R. Gebhardt, E. Penner, R. Martin, W. Ameling 37

Reconstruction of Two-Dimensional Signals from Irregularly Spaced Samples
D. H. Mugler, W. Splettstößer 41

Nonuniform Sampling Expansions of Two-Dimensional Bandlimited Signals
P. L. Butzer, G. Hinsen 45

Finite Spherical Analogues of the Whittaker-Shannon Sampling Theorem
J. R. Higgins 49

Ein- und zweidimensionale Signale mit quadratischer Phasenfunktion
H. Eggers 52

Zweidimensionale Höhenschichtenfilterung
E. Paulus 56

Optimization of Three-Dimensional Nonrecursive Digital Walsh Domain Filters
K. Rumatowski 60

Dynamic Scene Analysis and Motion Estimation
T. S. Huang 64

Take a Snapshot of the Ewald Sphere
R. Bamler 73

Best Approximation of Multidimensional Random Signals
M. Rosenblatt-Roth 77

Optimization of Quantizers by the Use of Haar Expansions
J. Sawicki 80

Invariant Description of Pictorial Patterns Via Generalized Auto-Correlation Functions
H. Glünder 84

Mehrkanalige Kleinste-Quadrate-Schätzung autoregressiver Parameter zur Sensorausfallerkennung
W. Ptacek, U. Appel 88

Walsh Series in Polar Coordinates
W. R. Wade 93

Nonuniform Sampling for Multidimensional Signals
F. A. Marvasti 97

Methodik 101

Deutung von Bildfolgen anhand ihrer symbolischen Beschreibung
B. Radig 102

Aufgabenangepaßte Abtastraster zur digitalen Verarbeitung von Infra-Rot-Bildsequenzen
D. Coy 116

Paßpunktfreie Justierung von Bildern
P. Schwarzmann, B. Schorer, M. Griesinger 120

Chrominanzsignalquantisierung unter Berücksichtigung des Beitrags des Chrominanzsignals zur Wiedergabeluminanz
G. Bruck 124

HQ-MAC; Ein Konzept zur schmalbandigen, kompatiblen HQTV-Übertragung
M. Silverberg, W. Boie 128

Projection of the Hough Transform
U. Eckhardt, G. Maderlechner 131

Ein regionenorientiertes Segmentierungsverfahren für texturierte Bildvorlagen
R. Mester, U. Franke 135

Contour Extraction of Objects Corrupted with Shadows and Reflections by Object-Adapted Filtering
R. Schmidt 139

Knowledge Based Pictorial Pattern Recognition
H. Bunke 143

Animated 3D-Model of the Human Heart Based on Echocardiograms
L. Jilin, K. Affeld, M. W. Engelhorn, M. Schartl 155

Mitteln bei 3-D-Rekonstruktionen biologischer Objekte
K. Niemann, D. Graf v. Keyserlingk 159

Animation of Medical Objects Using a Transformation Approach between Two Data Models
L. E. Peters, P. Jensch, W. Ameling 163

Compensation of the Dispersion of Optical Systems with Crude A-Priori Knowledge of the Input Signal
K. Kroschel 168

2-D Vector Representation of Multi-Dimensional Symptom Space for Computer Aided Medical Diagnosis
T. Sekiya, A. Watanabe, M. Saito, M. Kikuchi 172

Automatische Detektion von Bereichen schwacher Rückstreuung in Radar-Abbildungen
K. Behrens, E. Mauer 176

Automatische Detektion von Bereichen einheitlicher Interferenzerscheinungen in Radar-Abbildungen durch Speckle Modellierung
A. Ebert, E. Mauer 181

Bildcodierung 186

Irrelevanzreduktion von Bildinformation durch Unscharffilterung
F. Arp 187

Auf Schätzung basierende Bufferüberwachung für einen echtzeitfähigen Transformationskoder
Y. Du 191

Ein regionenorientiertes Bildkodierungskonzept mit sehr hoher Datenreduktion
U. Franke, R. Mester 195

Multiple Coding of Images without Distortion Accumulation
M. Gilge, W. Guse 199

Codierung von Videosequenzen mit niedriger Datenrate durch Vektor-quantisierung und Bewegungskompensation
B. Hammer, A. v. Brandt 203

Multi-Spectral Data Compression Using an Adaptive Spectral Transform
W. C. Huisman 207

Exakte Bildcodierung und Huffman-Codes
D. Manstetten 212

Hierarchical Image Coding
Th. Wendler 217

Data Compression and Decorrelation in Digital Signal Processing of Random Data
M. Maqusi, I. Makhamreh 229

Segmentierung und Vektorquantisierung von Bildsignalen auf der Grundlage eines Composite Source Modells
H. Bohlmann, P. Meissner 233

Technologie der Bildverarbeitung 238

Natural Basis Functions for Image Analysis
P. E. Danielsson 239

IPAS – Ein Pipeline-Bildverarbeitungssystem
K. Gütschow 255

Contextual Image Processing in MRI-Applications
H. Schwarz 259

Pixel Pipe – A New Bus Concept for a High Speed Image Processor
D. Meyer-Ebrecht,Th. Schilling, W. Winkler 263

Bildverarbeitung mit Datenflußrechnern
J. Baston 265

Distance Transforms with Data Flow Techniques
S. T. Dekker, P. P. Jonker, F. C. A. Groen 269

Structflow – Ein Datenflußrechner zur Verarbeitung strukturierter und kontinuierlicher Daten
P. Nitezki 273

Anwendung und Grenzen kommunizierender paralleler Prozesse in der industriellen Bildverarbeitung
R. Föhr, W. Ameling 278

Ein Entwicklungssystem für Multi-Prozessorsysteme mit Signalprozessoren FUJITSU MB 8764 und MB 87064
H. Eulenberg 282

Ein modulares Multi-Signalprozessorsystem
W. Guse, M. Gilge 286

Echtzeit-Symbolextraktion aus Grauwertbildern
R. Massen, P. Janke, M. Simnacher, J. Rösch 290

Ein universelles Kamera-Simulationsmodell für die Erzeugung von Bewegtbildsequenzen
P. Kauff, S.-C. Chen, R. Schäfer 295

Anwendungen 301

Medical Imaging
W. J. Dallas 302

Display of Hemispheric Local Metabolic Rates from Human Brain
C. Nahmias, M. Loken, E. S. Garnett 310

Comparison of Statistical Methods for the Detection of Contrast Material in Echocardiographic Image Sequences
E. Steinmetz, R. Brennecke, D. Jung, N. Wittlich, R. Erbel, J. Meyer 314

Verbesserte Strukturerkennung durch Bildverarbeitung: Anwendung beim menschlichen Gehirn
J. Wasel, D. Graf v. Keyserlingk 319

Computertomographische Rekonstruktion von Vektorfeldern
B. Siemund 323

Some Special Applications of Image Processing in Medicine
V. A. Pollak 327

Bildverarbeitung in der Parameter-Selektiven Kernspintomographie
T. Tolxdorff, K. Gersonde 337

Anwendungsbeispiele aus der Kunststofftechnik – Qualitätskontrolle und Automatisierung durch Bildverarbeitungssysteme
G. Menges, M. Haupt, K. Borgschulte 341

Pattern Recognition for Earthquake Detection
M. Joswig 347

Navigation of an Airborne Vehicle by Model-Based Image Sequence Processing
R. Schmidt, H. Zinner 351

Images and Image Processing for X-Ray Small Angle Scattering
H. W. Halling, H. G. Haubold 356

A New "Switched Kalman Filter" for 3D-Contourmeasuring Problems with a Laser Diode Range Finder
O. Loffeld 361

Ein Geräuschunterdrückungssystem mit zweidimensionaler Mikrofongruppe und nachgeschalteter adaptiver Wiener Filterung
R. Zelinski 372

The Feature Locating for Fingerprint Recognition
H. Wang, S. Yu, Y. Wu 376

Poster 380

Physiologische Meßsignale und deren Auswertung zur Unterstützung medizinischer Entscheidungen bei Operationen
A. Janitzki 381

Nachtrag zur Theorie 383

Hausdorff Distance and Digital Filters
A. Andreev 384

Autorenindex 388

Eröffnung

A UNIFYING FRAMEWORK FOR RECONSTRUCTIVE IMAGING METHODS

Hermann Schomberg
Philips GmbH, Forschungslaboratorium Hamburg
Vogt-Koelln-Str. 30, D-2000 Hamburg 54

Abstract: Reconstructive imaging methods, such as CT or MRI, produce their images in a two stage process: First, the object to be imaged is probed using some form of physical radiation, and then the image is computed, or reconstructed, from the outcome of the probing. The computation is based on a reconstruction algorithm which tries to invert a mathematical model of the probing. In this paper we inspect and study these building blocks of reconstructive imaging methods in more detail. Special attention is given to an unavoidable property of the reconstruction problem, known as ill-posedness. As a result, we obtain something like a unifying framework for reconstructive imaging methods that helps to better understand, design and estimate these methods. Finally, the framework and its applications are briefly illustrated by way of some examples.

1. INTRODUCTION

Imaging methods produce images of objects. These are some typical examples:

photography,
radar,
ultrasound B-mode imaging,
x-ray projection imaging,
conventional x-ray tomography,
x-ray computed tomography (X-CT),
emission computed tomography,
magnetic resonance imaging (MRI).

Any imaging method is realized by an imaging system. The object to be imaged must belong to a class of objects that are admissible to this imaging method. The information that is required to form its image is obtained by probing the object with some sort of physical radiation (electromagnetic waves, elastic waves, elementary particles). Imaging methods can provide information that is otherwise not or not readily available. Moreover, images are well suited to convey this information to human observers. Imaging methods are widely used in fields like geophysics, astronomy, military and civil surveillance, material testing, and medical diagnostics. Reconstructive imaging methods are a special type of imaging methods. They provide "quantitative" images and are predominantly applied in medical diagnostics. In the above list, the last three examples are reconstructive imaging methods.

With a reconstructive imaging method, the imaging system consists of two subsystems, and the imaging process proceeds in two steps. The first subsystem is a measuring apparatus. It is used for the probing. The probing has the nature of a well controlled physical experiment, and will therefore simply be called the experiment. The second subsystem is a comput system. It computes, or reconstructs, the wanted image of the object from the outcome of the experiment. This reconstruction is based on a reconstruction algorithm which is in turn related to a mathematical model of the experiment. Both the model and the reconstruction algorithm may be seen as mathematical mappings, with the reconstruction algorithm trying to invert the model. Thus, a reconstructive imaging method is characterized by four highly interdependent conceptual building blocks: Two of them, the experiment and the reconstruction, belong to and happen within the physical world; the other two, model and reconstruction algorithm, are their respective counterparts in the abstract world of mathematics.

Many statements can be made and conclusions drwan about these building blocks and their interdependencies without reference to a specific reconstructive imaging method. Some of what could be said will be said in section 2. Particular attention is given to the mathematical building blocks and to an unavoidable property of the reconstruction problem, known as ill-posedness. As a result, we shall obtain something like a unifying framework for reconstructive imaging methods that can help to describe, understand, explain, design, evaluate, and compare reconstructive imaging methods. In section 3, we further illustrate the framework and its applications by inspecting several proven or proposed medical reconstructive imaging methods with respect to their models and the ill-posedness of their reconstruction problems.

2. THE UNIFYING FRAMEWORK

We restrict ourselves to those reconstructive imaging methods whose measuring apparatus has a special volume of space in it or near it which holds the object to be imaged during the probing. This object region is that region whose contents are imaged when the imaging system performs its action. It may be flat or truely three-dimensional. If the object region is flat, the resulting image is two-dimensional, otherwise three-dimensional. Before the probing starts, the object to be imaged is positioned in such a way that its portion of interest lies within the object region. The object should not appreciably move during the subsequent probing. We must distinguish between an "object" in the ordinary sense of the word and the contents of object region during the probing. The latter will be called a positioned object if we wish to emphasize the difference. A single object, in the ordinary sense of the word, can lead to many different positioned objects. (A patient may be positioned in many ways in a CT-scanner.) It is the positioned object that is imaged. There is another region, the <u>radiation region</u>, that surrounds the object region and that is traversed by the radiation used for the probing. The medium that fills the radiation region except for the subregion occupied by the object is typically homogeneous and chosen such that it affects the propagation of the radiation only in a trivial way (e.g. air or water). We further restrict ourselves to those reconstructive imaging methods that image the interior of objects (in the ordinary sense of the word) and that record and process the outcome of the experiment digitally. These restrictions are not essential. Most reconstructive imaging methods are still included, and in particular the ones with medical applications.

We begin with a phenomenological, abstract description of the four building blocks. Most statements have been found by an inspection of existing reconstructive imaging methods. We shall sometimes illustrate an abstract notion using X-CT as a concrete example.

Several general statements about the experiment and the measuring apparatus have already been made. Here are some more: The measuring apparatus always includes detectors for the radiation employed. Usually there are also one or more sources that are used to irradiate the positioned object in the object region. If such sources are absent, then the objects radiate themselves, either naturally or artificially. Often, the experiment consists of a sequence of subexperiments which differ from each other only by the setting of one or a few parameters (taking projections at different angles in case of X-CT). The outcome of the experiment consists of a finite set of digitally coded, physically represented data. If we repeat the experiment on the same positioned object, we obtain the same outcome, except for experimental noise. If we perform the experiment on different positioned objects, then we usually obtain different outcomes.

Next we turn to the model. We need to reference points in space, and to this end think a Cartesian x_1-x_2-x_3 coordinate system attached to the imaging system such that its origin and the central portion of the x_1-x_2 plane lie within the object region. The object region can now be described by a set $A \subset \mathbb{R}^a$ where $a = 2$ if it is flat and $a = 3$ otherwise. The radiation region corresponds to a set $A' \subset \mathbb{R}^a$ where $a' \leq a$. We may regard A as a subset of A', even if $a = 2$ and $a' = 3$.

The model, M, may be formulated as a mapping, or operator, between two normed function spaces. We write

$$M:D(M) \subset X \longrightarrow Y,$$

where D(M) denotes the domain of M. The idea is that the functions in X and Y describe potential positioned objects and potential outcomes of the experiment, resp., and that the action of M describe the action of the experiment.

More precisely, for X we may choose $L_2(A', \mathbb{R}^p)$, the Hilbert space of square integrable functions $f: A' \longrightarrow \mathbb{R}^p$. Such an f is physically interpreted as a macroscopic, possibly space dependent, radiation specific material parameter that is exhibited by all admissible positioned objects and the surrounding medium in the radiation region (the x-ray attenuation coefficient in case of X-CT). The parameter p is small, typically even p = 1. D(M) may be smaller than X, but must still contain enough functions so that any potential positioned object can be described by a function in D(M). Outside A, the functions in D(M) are typically equal to zero or another constant. In many cases, then, it makes more sense to choose X as $L_2(A, \mathbb{R}^p)$ and to regard D(M) as a subset of this X, with the understanding that the functions in X are to be extended by zero (or another constant) on the rest of $\mathbb{R}^a$ if needed. Whether X is considered as a real or complex function space is a matter of mathematical convenience. Sometimes one may also wish to regard functions with values in $\mathbb{R}^2$ as complex valued, etc.

Similarly, Y may be chosen as $L_2(B, \mathbb{R}^q)$, where B is a bounded domain in $\mathbb{R}^b$ and q is a small number. Typically we will have a = b and p = q. Y contains functions that are suited to represent possible outcomes of the experiment. These may be seen as functions of some parameters, expressed in a suitable coordinate system, and B is simply the range of these parameters. Actually, however, we do not observe the outcome for all $\underline{y} \in B$, but only at preselected, finitely many points $\underline{y}_1, \ldots, \underline{y}_n \in B$, forming what we shall call the <u>sampling pattern</u>. We may describe this sampling process by the mapping

$$\text{sam}: Y \longrightarrow Y_{dat},$$

where $Y_{dat} = \mathbb{R}^{nq}$ and $g \in Y$ is assigned to the vector $(g(\underline{y}_1), \ldots, g(\underline{y}_n)) \in Y_{dat}$.

Finally, the action of M is chosen such that if $f \in X$ describes a positioned object, then $Mf \in Y$ (or $\text{sam}Mf \in Y_{dat}$) describes the outcome of the experiment.

Some explanations are in order: Assume that an object has been positioned in the object region and that the experiment has been performed on this object. Let $f^* \in D(M)$ be a function that describes the positioned object. As such a function is not uniquely determined by the object, but only up the unavoidable tolerance of some measuring scheme, we may choose f* to be "well behaved" as well. This f* represents what might be called the "true" image of the object. Let $d \in Y_{dat}$ be a vector that describes the observed outcome of the experiment. (As the outcome is represented digitally, d is well determined.) Then samMf* will approximately equal d. That is, we have

$$\text{samMf}^* = d + e\ , \tag{2.1}$$

for some hopefully small vector $e \in Y_{dat}$. This error term subsumes the imperfections of both the experiment and the model. We may think of e as the sum of two contributions, $e = e^{exp} + e^{mod}$, where e^{exp} denotes the experimental error and e^{mod} the modelling error, resp. Such errors are unavoidable. In practice, f* is unknown, and then e is also unknown. However, we may have a feeling for the typical size of the errors, say, or their low order statistics.

One can make plausible that the action of M can always be written in the form

$$(Mf)(\underline{y}) = \int_{A'} K(\underline{x},\underline{y};f)\,f(\underline{x})\,d\underline{x}, \quad \underline{y} \in B. \tag{2.2}$$

The integrating kernel K is q×p-matrix of "ordinary" functions or δ-functions; if present, the δ-functions serve to reduce the integration over the a'-dimensional domain A' to an integration over lower dimensional subsets in A' (surfaces, lines, points). If K does not depend on f, then M acts linearly. In such a case we may choose D(M) as a linear subspace of X, and M becomes a linear operator. If K does depend of f, then M is nonlinear, though still "quasilinear". The dependence of K on f may be fairly complicated, but still there will be rules that tell us how to evaluate K, given f, and these rules may be computerized. If X was chosen as $L_2(A, \mathbb{R}^p)$, then A' in (2.2) is to be replaced by A.

Formula (2.2) reflects a general superposition principle. To find the model of a given experiment, one needs to know the arrangement of this experiment and deterministic, macroscopic laws of nature that govern the propagation of the radiation through the media of interest. Strictly speaking, the model of a given experiment is not unique, but usually there are only a few sensible choices, and one of them is meant here by "the" model. The size of the modelling error may be used to measure the goodness of the model. It may be necessary to preprocess the raw outcome of the experiment in order to achieve a "nice" model (e.g. taking logs in case of X-CT). In such a case we regard the preprocessed data as the outcome of the experiment.

M reduces to the identity mapping if X = Y and if K is the p×p diagonal matrix of δ-functions $\delta(\underline{x}-\underline{y})$. On principle, it would be nice to have an experiment that could well be modelled by the identity mapping. For then (2.2) would reduce to samf* = d + e, i.e. the outcome were a good sampled version of f* and could thus serve immediately as an image of the object. Such an experiment would have to directly measure the physical parameter represented by f* at selected points in the interior of the object being imaged. Unfortunately, it seems to be difficult to devise such experiments; sensitive point MRI, an early and now obsolete version of MRI, provides one of the rare exceptions. This difficulty is one of the reasons for the existence of reconstructive imaging methods: It is much easier to find experiments whose models involve a true integration (over lines, surfaces etc.), but then a reconstruction becomes nessary if we insist on knowing f*. From now on then, we assume that M does involve a true integration. This has important implications, as we shall see.

To choose X and Y as L_2-spaces is mathematically convenient and physically reasonable. The L_2-norm seems to be the weakest physically meaningful norm. Sometimes, however, it may be mathematically even more convenient and physically still sufficient to choose smaller spaces and stronger norms, e.g. Sobolev spaces and their norms or spaces of continuous functions endowed with the maximum norm. In what follows, all norms will be denoted by $\| \ \|$; the underlying space follows from the context.

A mapping of the form (2.2) is continuous, i.e. $\|Mf_n - Mf\| \to 0$ whenever $f_n, f \in D(M)$ and $\|f_n - f\| \to 0$. Hence, if f is close to f', then Mf is close to Mf'. But M is even "more than continuous": For every $f \in D(M)$ there will be $f'_n \in D(M)$ such that $\|Mf'_n - Mf\| \to 0$ and $\|f'_n - f\| \not\to 0$, i.e. it may happen that Mf is close to Mf' although f is not close to f'. This may be understood by the integrating action of M which damps any oscillations of f when M is applied to f. As a result, Mf is a smoother function than f; M is said to have a <u>smoothing property</u>. Because of the smoothing property, for each $\varepsilon > 0$ and each $g \in R(M)$, the range of M, the set

$$S_\varepsilon(g) := \{f \in D(M) \mid \|Mf - g\| < \varepsilon\}$$

is rather large in the sense that it contains many and rather different (with respect to $\| \ \|$) function, no matter how small $\varepsilon > 0$. If M is linear, $S_\varepsilon(g)$ is even unbounded. The smoothing property is a matter of degree. As a rule of thumb, the degree tends to increase with the "smoothness" of the integrating kernel. Also, integration over surfaces tends to be more smoothening than integration over lines etc. Because of the smoothing property, Mf is no longer an arbitrary function in Y. For this reason, R(M) is a proper subset of Y, i.e. M is not surjective (not "onto"). Usually, however, R(M) is still dense in Y. (Exceptions do exist.) M cannot be made surjective by shrinking Y, because the functions that are needed to describe possible outcomes of the experiment do not, in general, exhibit the type of smoothness owned by the functions in R(M), as they are contaminated by error terms (cf. (2.1)). There seem to be no generally valid rigorous mathematical proofs for the statements made in this paragraph, but they have been verified in many special cases.

A mapping of the form (2.2) may be injective ("one-to-one") or not. If it is injective, then for every $g \in R(M)$ the set

$$S_0(g) := \{f \in D(M) \mid Mf = g\}$$

contains exactly one element. In this case M has an inverse operator $M^{-1}: R(M) \subset Y \rightarrow D(M) \subset X$, which assigns to $g \in R(M)$ the single f in $S_0(g)$. If M is not injective, then $S_0(g)$ contains more than element. If M is also linear, then $S_0(g)$ is even unbounded. Hence, if M is not injective, the above set $S_\varepsilon(g)$, which contains $S_0(g)$, becomes even larger.

If M is injective, the inverse operator is not continuous. This follows from the smoothing property of M: Let $f'_n, f \in D(M)$ be such that $\|Mf'_n - Mf\| \rightarrow 0$ but $\|f'_n - f\| \not\rightarrow 0$, and let $g'_n := Mf'_n$, $g := Mf$. Then $\|g'_n - g\| \rightarrow 0$ but $\|M^{-1}g'_n - M^{-1}g\| \not\rightarrow 0$.

We turn to the reconstruction algorithm. It is an algorithm, of course, but may also be seen as a mapping

$$R: Y_{dat} \rightarrow X_{ima},$$

where $X_{ima} = \mathbb{R}^{mp}$ for some m. The vectors in X_{ima} represent possible images; the components u_i of a vector $u \in X_{ima}$ are interpreted as the values of the image represented by u at an underlying two- or three-dimensional mesh of grid points $\underline{x}_1, \ldots, \underline{x}_n \in A$. X_{ima} may also be seen as a subspace of X. We shall illustrate this for the case p = 1: Choose a set of basis functions $\phi_1, \ldots, \phi_m \in X$ with

$$\phi_i(x_j) = \begin{cases} 1 \text{ if } i = j, \\ 0 \text{ if } i \neq j. \end{cases} \tag{2.3}$$

In addition, the ϕ_i should be nonzero only in the vicinity of $\underline{x}_i$ and easy to evaluate. Good choices are functions that are piecewise constant or linear with respect to each of their variables in each of the "pixels" or "voxels" formed by the $\underline{x}_i$. The set of all functions of the form $\phi = \sum u_i \phi_i$ then forms a linear subspace of X which is isomorphic to $X_{ima} = \mathbb{R}^{mp}$. Because of (2.3) we have even $\phi(\underline{x}_i) = u_i$, i.e. ϕ interpolates the vector u. The case p > 1 can be treated similarly. (In the same manner we may also see $Y_{dat} = \mathbb{R}^{nq}$ as a subspace of Y.)

The action of R is specified by a set of rules that can be computerized and is chosen such that $Rd \in X_{ima}$ is a good approximation to $f^* \in X$, where f^* and d are supposed to have the same meaning as in (2.1). This property is the origin of the attribute "reconstructive". Rd represents what might be called the reconstructed image.

Actually, a reconstructive imaging method can have several reconstruction algorithms. The chosen one is realized by the computer subsystem of the imaging system via a computer program. The computer takes the physical representation of d as input and outputs a physical representation of a vector $u \in X$ which is an approximation to Rd. When a digital computer is used, u differs from Rd only by rounding errors. Hence u is also close to f* and thus indeed a meaningful and even "quantitative" image of the positioned object. We may measure the goodness of the image by $\| f^* - \phi \|$, where ϕ is the interpolated version of u. Reconstruction algorithms tend to be computation intense, and often array processors are used to reduce the computation time. Usually the reconstructed image is displayed on a screen.

So far we have been describing the building blocks of reconstructive imaging methods. Next we try to understand their functioning and their interdependencies. As it turns out, this is best done in terms of the mathematical building blocks.

We begin with the reconstruction algorithm. Plainly enough, there must exist one in the first place. Now, the reconstruction algorithm is to solve the <u>reconstruction problem</u>: compute an approximation $\tilde{f}$ to f* whenever f* represents an admissible positioned object that has been subjected to the experiment. Thus, before we can devise a reconstruction algorithm, we have to study the determinacy of the reconstruction problem.

Let $f^* \in D(M)$ represent an admissible positioned object that has been subjected to the experiment, and let $d \in Y_{dat}$ describe the outcome of the experiment. Then almost all we know we know about f* is expressed in the relation (2.1). In that relation, sam, M, and d are known, while f* and e are not. But we have some vague a priori information about f* and d: f* has a certain physical interpretation and is well behaved; and we may guess something like the maximum size or the statistics of e. What does all this information tell us about f*?

Let us assume we know an upper bound, $\hat{\varepsilon} > 0$, for the size of e. If $\hat{\varepsilon}$ is sufficiently small and the sampling pattern sufficiently dense, then d in (2.1) determines a reasonably good approximation g to Mf*, e.g. by interpolation as described earlier. Conversely, if these conditions are not well satisfied, then Mf* remains rather undetermined by (2.1), let alone f*. So let us assume that $\hat{\varepsilon}$ is sufficiently small and the sampling pattern sufficiently dense. Under these conditions, then, d determines a function $g \in Y$, not necessarily in R(M), such that g is close to Mf*. Using our knowledge of sam, M, $\hat{\varepsilon}$, and our a priori knowledge about f*, we will be able to find a reasonably sharp but still safe upper bound $\varepsilon > 0$ for $\|Mf^* - g\|$. Then (2.1) is essentially equivalent to

$$f^* \in S_\varepsilon(g) := \{f \in D(M) \mid \|Mf - g\| \leq \varepsilon\} . \tag{2.4}$$

The set $S_\varepsilon(g)$ contains those functions that are compatible with the measurements. So far any $f \in S_\varepsilon(g)$ could be the wanted f^*. How large, then, is $S_\varepsilon(g)$?

We claim that $S_\varepsilon(g)$ contains a set of the form $S_{\tilde{\varepsilon}}(\tilde{g})$ where $\tilde{g} \in R(M)$ and $\tilde{\varepsilon} > 0$: Usually we will have $\|Mf^* - g\| < \varepsilon$, and then $\tilde{g} := Mf^*$ will do, if $\tilde{\varepsilon}$ is chosen sufficiently small. Should we happen to have $\|Mf^* - g\| = \varepsilon$, this simply means that our choice of ε was not safe; we quickly make a better choice and then have the usual case. More typically, however, some $\tilde{g}$ close to g and some $\tilde{\varepsilon}$ not much smaller than ε will do, though to prove this rigorously in the general case seems to be difficult.

Now, as we have seen earlier, such a set $S_{\tilde{\varepsilon}}(\tilde{g})$ with $\tilde{g} \in R(M)$ is rather large (even unbounded if M is linear), no matter how small $\tilde{\varepsilon}$. We must thus face the fact that the relations (2.4) or (2.1) leave f^* rather undetermined. This situation is usually paraphrased by saying that the problem of solving the equation $Mf = g$ is <u>ill-posed</u>.

Ill-posedness is defined as the absence of well-posedness. The problem of solving the equation $Mf = g$ would be called well-posed, if M were i) injective; ii) surjective; and iii) M^{-1} continuous. If this were true, we could apply M^{-1} to g, and as Mf^* is close to g, $M^{-1}g$ would be close to $f^* = M^{-1}Mf^*$. One could also show that the set $S_\varepsilon(g_\varepsilon) := \{f \in D(M) \mid \|Mf - g_\varepsilon\| \leq \varepsilon\}$ would be bounded and shrink to $\{f^*\}$ as $\varepsilon \to 0$ and $\|g_\varepsilon - Mf^*\| \to 0$. This would certainly be a favorable situation. But as we have seen, M cannot be surjective, and M^{-1}, if it exists, cannot be continuous.

On the other hand, if M is injective, at least Mf^* does determine f^*, so we are not yet willing to give up. (And indeed, X-CT works!)

Let us assume, then, that M is injective. A hint on how to overcome the problem comes from our heuristic explanation of the smoothing property: We argued that $\|Mf - Mf'\|$ could be small and $\|f - f'\|$ large, provided $f - f'$ was a highly oscillating function. Now, if this could only happen if $f - f'$ was such a queer function, then we could shrink the set $S_\varepsilon(g)$ by the additional constraint that its functions should be "smooth" or "regular" in some reasonable sense.

This idea, known as <u>regularization</u>, can indeed be made to work. Many regularization methods do exist. Typically, such a regularization method involves a smoothness criterion and a parameter α that controls and measures the amount of smoothness imposed. For each error level ε and smoothness parameter α it selects a regularized approximation $f_{\varepsilon,\alpha} \in D(M)$ to f^*. The crucial point is the choice of α, i.e. the amount of smoothing applied. If that amount is too small, then $f_{\varepsilon,\alpha}$ may still be so oscillating that $\|f^* - f_{\varepsilon,\alpha}\|$ is unnecessarily large; and if that amount is too large, $f_{\varepsilon,\alpha}$ may be forced to be so smooth that $\|f^* - f_{\varepsilon,\alpha}\|$ is unnecessarily large again. Finding the right compromise requires a minimum of a priori information about f^*. Some regularization methods can exploit more a priori information, e.g. the statistics of e. How good an approximation to f^* can be determined in this way depends on the regularization method employed and, most strongly, on the error level and on the degree of the smoothing property of M. Generally speaking, at a given error level, the achievable accuracy is noticeably less than with a well-posed problem. In "mildly" ill-posed cases the accuracy may still be satisfactory at realistic error levels, while in "strongly" ill-posed cases f^* remains rather undetermined even at very small error levels and despite the best possible regularization. Again, there seem to be no generally vaild proofs for these statements, but they have been verified in many special cases. Ill-posed problems and regularization methods are treated at length in [1].

The preceding discussion also suggests how we might find a reconstruction algorithm. As with a well-posed problem we devise a method that exploits (2.1) to compute an $\tilde{f}$ such that $M\tilde{f}$ is close to g (or sam$M\tilde{f}$ close to d). In addition, we make sure that $\tilde{f}$ is "regular" in some sense and to an amount that we can control, say by the choice of a parameter. (The latter would not be necessary if the reconstruction problem were well-posed.)

There are two strategies that can be followed to exploit (2.1). The first strategy works when M^{-1} is explicitly known, for which there is a chance only if M is linear. We may then construct a discrete, approximate version of M^{-1} and apply it to d or another vector close to d. At some stage of the procedure, regularization must be built in.

Alternatively, we may fully discretize the semidiscrete relation (2.1), neglecting, say, the error term. Using (2.2) one can make plausible that the resulting system of equations will have the general form

$$A(u)u = b(u), \tag{2.5}$$

where u is the vector of unknowns, b(u) is another vector, and A(u) a matrix. If M is linear, A and b will not depend on u. In any case the set of vectors that approximately satisfy (2.5) is very large and contains rather queer elements, reflecting the ill-posed nature of the original problem. The number of equations and unknowns will be very large. We need a numerical method that computes an approximate and "regular" solution of (2.5). Such methods do exist, see [2] for a review. In practice, however, most of them are ruled out by the sheer size of the system. Among those that are left, an iterative method due to Kaczmarz and its variants are particularly attractive. The Kaczmarz method can be made to work even in the nonlinear case, see [8] for examples. Again, regularization must somehow be built in.

In general, the first strategy leads to algorithms that are computationally faster than those obtained with the second strategy. However, the first strategy fails, if M is nonlinear.

If M is not injective, the situation is certainly worse, as now f* is left undetermined even by Mf*. Put differently, the set $S_\varepsilon(g)$, already large by the smoothing property of M, becomes even larger. Yet we may be able to find an algorithm that computes a regularized $\tilde{f}$ such that $M\tilde{f}$ is close to g; for instance, if M is linear, we may compute a regularized pseudoinverse solution. However, such an $\tilde{f}$ need no longer be close to f*. Whether or not such an $\tilde{f}$ is still worth knowing, or even provides a meaningful image, depends on the circumstances. Section 3 contains both "good" and "bad" examples of non-injective models.

This ends our discussion of the mathematical building blocks. We have identified a number of conditions that should be satisfied, if the reconstruction problem is to be meaningfully solvable in practice: The error level should be very small, the sampling pattern sufficiently dense, and the model injective and only mildly smoothing. Also, we have indicated how to find a reconstruction algorithm, if these conditions are satisfied.

As the mathematical building blocks are closely tied to their physical counterparts, the above conditions have immediate implications for these building blocks as well. Conversely, there are physical, technical and other constraints lying on these other building blocks which hence also influence the mathematical ones. These other constraints, however, will not be discussed here. Many of the constraints are contradictory, and their satisfaction is a matter of degree. The designer of a reconstructive imaging method must choose the building blocks such that a viable compromise results. The more difficult part of the job is to find a practicable experiment and an associated injective and only mildly smoothing model. Once this has been achieved, it is usually easier to find a reconstruction algorithm also and to implement it on a computer. The framework presented in this section can help in this design process.

3. EXAMPLES

To further illustrate the framework and its applications, we now inspect a selected choice of proven and proposed medical reconstructive imaging methods with respect to their models and the determinacy of their reconstruction problems. This is certainly a very restricted view of these methods, but it allows us nevertheless to understand why the proven ones work as well as they do and to estimate the potentials and limitations of the proposed ones.

X-Ray Computed Tomography [5]. We consider the standard two-dimensional version. Its model is the two-dimensional Radon transform, defined by

$$(Rf)(r,\theta) = \int_{-\infty}^{\infty} f(r\underline{\omega}^{\perp}(\theta) + s\underline{\omega}(\theta))ds,$$

where $r \in \mathbb{R}$, $0 \leqslant \theta \leqslant \pi$, and

$$\underline{\omega}^{\perp}(\theta) = (-\sin\theta,\cos\theta),\ \underline{\omega}(\theta) = (\cos\theta,\sin\theta). \tag{3.1}$$

The above formula may also be written as

$$(Rf)(r,\theta) = \int_A \delta(r - \underline{x}\cdot\underline{\omega}^{\perp}(\theta))f(\underline{x})d\underline{x},$$

which is now in accordance with (2.2). Physically, f represents the x-ray attenuation coefficient. The Radon transform is injective and mildly smoothing only [7]. The error level is quite low, and efficient reconstruction algorithms are available.

Magnetic Resonance Imaging [6]. Numerous variants exist. We restrict ourselves to the so-called two-dimensional Fourier methods. All these methods have essentially the same model, namely a bandlimited Fourier transform:

$$(F_B f)(\underline{y}) = \int_A e^{-2\pi i\underline{x}\cdot\underline{y}} f(\underline{x})d\underline{x}, \qquad \underline{y} \in B,$$

where B, the parameter region, is a square or a circle centered at the origin of the $\underline{y}$-plane [9]. Typically, $F_B f$ is sampled on a square or polar grid. Physically, f represents the transverse magnetization of the object which is in turn related to its proton density. At our present level of abstraction, all the methods in the class considered differ merely by the sampling pattern and by the details of the meaning of f. The bandlimited Fourier transform is injective: As f has bounded support, its Fourier transform, Ff, is analytic. Hence Ff is uniquely determined by its values in B, i.e. by $F_B f$. But if Ff is uniquely determined by $F_B f$, then so is f, as F is injective. When passing from f to Ff and then to $F_B f$, only the information contained in the frequencies outside B is lost. Hence the reconstruction problem is very mildly ill-posed. The two-dimensional FFT provides an efficient reconstruction algorithm, if the sampling pattern is a square grid.

Emission Computed Tomography [5]. There are two variants, single photon emission computed tomography (SPECT) and positron emission tomography (PET). With these methods, the objects to be imaged are artificially made to emit γ-rays prior to the imaging process. Such an object is then described by a pair of functions, $f = (f_1, f_2)$, where f_1 denotes the local activity and f_2 the local attenuation coefficient. The model of SPECT is the so-called attenuated Radon transform:

$$(R_{att}f)(r,\theta) = \int_{-\infty}^{\infty} f_1(r\underline{\omega}^{\perp}(\theta) + s\underline{\omega}(\theta))\exp(-\int_{s}^{\infty} f_2(r\underline{\omega}^{\perp}(\theta)+\sigma\underline{\omega}(\theta)d\sigma))ds,$$

where $\underline{\omega}(\theta)$ and $\underline{\omega}^{\perp}(\theta)$ are defined as in (3.1). The model of PET, $\tilde{R}$, is similar, except that the second integral extends from $-\infty$ to ∞. Therefore $\tilde{R}$ simplifies to

$$\tilde{R}f = \exp(-Rf_2)Rf_1 .$$

Both models are nonlinear. Both appear to be not injective, but only mildly smoothing (although rigorous proofs are seemingly not available). Nevertheless, at least f_1 can be reasonably well reconstructed from $R_{att}f$ and $\tilde{R}f$, resp. [7]. In case of PET, Rf_2 can also be determined by an extra experiment so that then f_1 can be obtained by inverting the Radon transform.

Ultrasound Computed Tomography [5]. Again there are various versions. We restrict ourselves to that variant that tries to reconstruct the acoustic refractive index from travel time projections. The natural model is another nonlinear variant of the Radon transform:

$$(R_{US}f)(\underline{y}) = \int_{\gamma[\underline{y};f]} f ds, \quad \underline{y} \in B.$$

Here $\gamma[\underline{y};f]$ is a ray in the sense of geometric acoustics (optics) that connects the emitting and receiving ultrasound transducers at their position characterized by $\underline{y}$. Such a ray depends on its endpoints and on the refractive index, f, in between. If f is not constant, $\gamma[\underline{y};f]$ is a curved line. Replacing $\gamma[\underline{y};f]$ in (3.1) by $\gamma[y;f_0]$ where $f_0 \equiv 1$ reduces R_{US} in essence to the Radon transform. In fact, the Radon transform has been used as a model, but appears to be not accurate enough. If B is properly chosen, R_{US} is injective when it is restricted to functions that are sufficiently smooth and close to a constant. The smoothing property is only mild. A nonlinear extension of the Kaczmarz method can be used as a reconstruction algorithm [8]. This imaging method, with R_{US} as the model, has not yet been tried clinically.

SQUID Imaging [4]. This method is in a very early stage. The idea is to reconstruct the electric current density in the brain from measurements of the magnetic field outside the brain. These fields are extremely weak, but can be detected using SQUIDs (superconducting quantum interference devices). The fields are also quasistatic so that time is merely a parameter. The model is based on the Bio-Savart law:

$$(M_{BS}f)(\underline{y}) = \int_A f(\underline{x}) \times \frac{\underline{y} - \underline{x}}{|\underline{y} - \underline{y}|^3} \, d\underline{x}, \qquad \underline{y} \in B.$$

Here $f = (f_1, f_2, f_3)$ represents the electric current density, and B is a closed two-dimensional surface surrounding the three-dimensional object region A. This model is not injective, which may be proved using an idea from [3]. Also, its smoothing property is rather strong and increases with the distance from B to A. This follows from a spherical multipole expansion of the magnetic field. As a result of all this, the reconstruction problem lacks the determinacy needed for good images. It is hoped that using a substantial amount of a priori information about f may reduce the ambiguity of the reconstruction problem.

REFERENCES

[1] Bertero, M., de Mol, C., Viano, G. A., The Stability of Inverse Problems, in: Inverse Scattering Problems in Optics (Baltes, H. P., ed.), Springer (1980), 161-214.

[2] Bjørck, A., Eldén, L., Methods in Numerical Algebra for Ill-Posed Problems, Rep. LitH-MAT-R-33-1979, Dept. of Mathematics, Linkøping University, Linkøping, Sweden.

[3] Bleistein, N., Cohen, J. K., Nonuniqueness of the Inverse Source Problem in Acoustics and Electromagnetics, J. Mathem. Phys. 18 (1977), 194-201.

[4] Dallas, W. J., Fourier Space Solution to the Magnetostatic Imaging Problem, Applied Optics 24 (1985), 4543-4546.

[5] Kak, A. C., Computerized Tomography with X-Ray, Emission, and Ultrasound Sources, Proc. IEEE 67 (1979), 1245-1272.

[6] Locher, P. R., Proton NMR Tomography, Philips Tech. Rev. 41 (1983/84), 73-88.

[7] Louis, A. K., Natterer, F., Mathematical Problems of Computerized Tomography, Proc. IEEE 71 (1983), 379-389.

[8] Schomberg, H., Nonlinear Image Reconstruction from Projections of Ultrasonic Travel Times and Electric Current Densities, in: Mathematical Aspects of Computerized Tomography (Herman, G. T., Natterer, F., eds.), Springer (1981), 270-291.

[9] Twieg, D. B., The k-Trajectory Formulation of the NMR Imaging Process with Applications in Analysis and Synthesis of Imaging Methods, Med. Phys. 10 (1983), 610-621.

INTERACTIVE 3D COMPUTER GRAPHICS IN MEDICAL IMAGING

Michael W. Engelhorn

Technische Universität Berlin
Institut für Technische Informatik
Franklinstr. 28-29
D-1000 Berlin 10
EARN: engelhor@db0tui11

1. ABSTRACT

We concentrate on the method of representing and displaying surfaces of organs three-dimensionally from various digital imaging modalities. A process is described to perform this 3D representation and display by using Computer Vision modelling and Computer Graphics methods. Depending on the modelling function - which plays a keyrole - methods of segmentation, object surface- or object volume detection, visible surface determination, and rendering are discussed. A variety of these methods has been implemented in experimental systems, some of which are installed in a clinical environment.

2. INTRODUCTION

Within modern clinical environment, a growing number of modern digital imaging modalities (as for example CT, DSA, MRI and Ultrasound) produce a rapidly increasing number of digital images. Traditionally this information is twodimensional although the objects, from which the data are acquired, are naturally threedimensional. To overcome this lack of information, concerning the third dimension, usually a set of 2D-images is acquired which are traditionally viewed as an array of 2D-grey-scale slice images. A radiologist for example has thus to perform the 3D-shape of the organs mentally by stepping through a stack of 2D-images. So it seems naturally to ask for a 3D-representation and display of those images because the human eye (and brain in particular) is accustomed to see a 3D-world and to live in it [12].

No doubt, there are advantages to spur methods of the 3D-representation and display into clinical applications [1, 2, 9, 17, 24, 25] but today, one can find most of them in pure research. To come to a 3D-representation and display of organs, a couple of processing steps has to be performed. This means in particular that out of the array of 2D images the organs have to be recognized. After the generation of the digital images the first processing steps would be a set of image processing, feature extraction and object recognition processes - to be referred herein as Computer Vision-methods [19]. After these processes have taken place, reconstruction and modelling operations have to be performed to generate the internal 3D-representation, which plays a keyrole in said system. The remaining steps are the display and interactive manipulation of the organs by means of 3D Computer Graphics methods. The focus hereon should be on a "near-real-time" and "near-realistic" representation for the display, as recently a couple of applications have used interactive manipulations, such as preoperative surgical planning [3, 23]. As for example an integrated system, such as a Medical Workstation (MWS), is proposed defining the required steps (as shown in Fig. 1) [6, 13, 14].

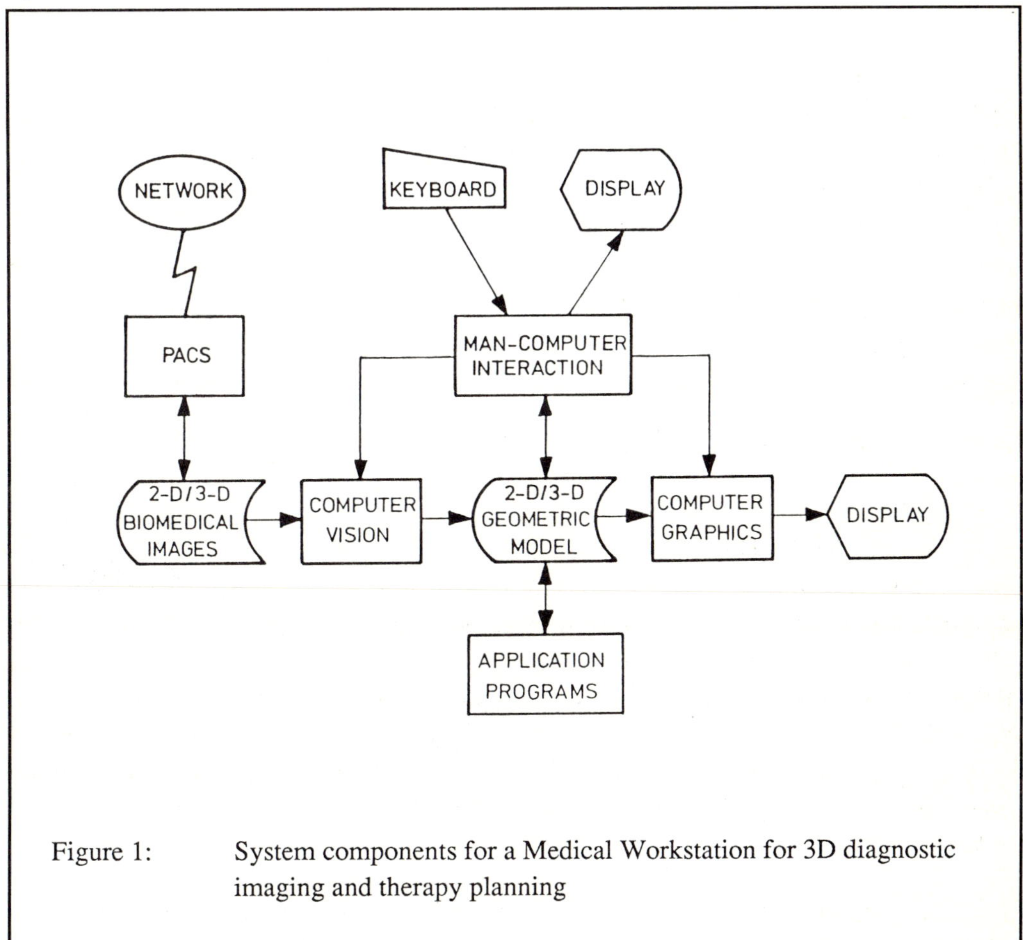

Figure 1: System components for a Medical Workstation for 3D diagnostic imaging and therapy planning

3. *COMPUTER VISION*

Computer Vision algorithms have to be applied when automatic or semi-automatic segmentation of the images is to be performed. They also have to be applied when manual segmentation is hard to perform since low picture quality makes it hard to the observer to follow boundaries of regions of interest. These algorithms can be classified into the traditionally "classical image processing" with "low-level procedures" and the feature extraction and object recognition processing with "high-level procedures" [20]. The low-level procedures, which are problem independent, will operate for

- image enhancement and
- segmentation.

In the case of digital images, coming from image generation systems, we have to deal with distortion, noise, low-contrast image functions, out-of-focus and motion blur which have to be corrected by geometric rectification, noise reduction/suppression, contrast enhancement and filtering as for example by applying the image enhancement functions on these images. After enhancing the images, segmentation of the images can take place by either region based segmentation (e.g. region growing), edge based segmentation (e.g. intensity gradient determination) or contour/boundary segmentation (e.g. contour-following). The segmentation process is greatly dependent on the previous image processing step (e. g. contrast enhancement). The high-level procedures - which are problem independent - will operate on

- feature exraction and
- object recognition.

The feature extraction procedures extract the image primitives of identical characteristics (as for example shape or geometry), which match "rules of similarity" of a model of the objects, and classify them. They depend on the segmentation approach, which has to be performed before. Object recognition attaches "meanings" to components and allows the "naming" of the components. This allows a model based image analysis by matching a description of a model with the interpretation of the image by global consistency with the model description which has prototypical knowledge of the objects.

4. *RECONSTRUCTION AND MODELLING*

4.1 *Geometric modelling*

To represent the 3D-objects, modelling operations have to be performed to generate a computerized internal structure - a representation scheme - of the "reality". Several abstract levels have to be run through until a manageable set is reached. To keep this simple, we focus on geometry only. Various approaches are known to classify different representation schemes [18], which are

- Boundary representation schemes with planar polyhedral surfaces, curved surfaces and functions on the surface,
- Volumetric representation schemes with Spatial enumaration schemes and Constructive Solid Geometry (CSG) and
- Sweep representation schemes.

In the case of boundary representations an object is represented by an approximation of planar, bivariate curved surface patches or as a function of a generic object. Within planar surface patches the geometrical primitives are points, lines or surfaces. Within bivariate surface patches one can find functions representing the local parts of an object such as B-splines. When using volumetric representation schemes, the volume is decomposed into disjunct elements (as for example cubes or parallelepipeds). The spatial enumeration schemes are often used to model organs from digital imaging modalities because of its simple approach as for example the octtree encoding scheme which has a recursive hierarchical structure. The sweep representations schemes represent a surface by moving an arbitrary shaped cross section along a curve in space. A sweeping rule can be defined to transform the shape of the cross section along the curve in space.

"Natural objects" - such as organs - have to be modelled, according to their geometry in order to be more "shape-oriented" and "relational-oriented" in contrast to "technical objects". This demands a previous step when generating a computerized internal structure, which is a "reconstruction oriented" process in contrast to a "construction oriented" method, as in the way technical objects have to be modelled.

4.2 3D-reconstruction

To produce a geometric 3D-model out of a set of 2D-contourlines or pixel areas by a 3D-reconstruction step as for example a surface detection, the object recognition has to be performed first in order to find the contours of the object. This 3D-reconstruction step depends on the model that should be produced and is in its simplest form the "stacking" of 2D-pixel areas to a 3D-voxel volume which can be represented as either a boundary scheme or as a volumetric representation scheme. A more complex method is the determination of the surface of the object by planar polygons out of a set of stacked 2D contour lines. One well-known approach is the "triangulation" method which is described in [8] and in an advanced form in [22].

5. INTERACTIVE 3D COMPUTER GRAPHICS

The keyrole 3D Computer Graphics plays is the one to give the observer a unique and fast impression of a selected portion of a scene which is given by a geometric model. Usually the display area is twodimensional and so the problem persists to avoid ambiguities by displaying threedimensional graphical information on that area. Traditionally within the field of Computer Graphics there are a couple of methods developed to solve the problem, described above. Among these methods there are the following:

- Kinetic effects (as for example dynamic rotation, scaling and translation), and kinetic depth cues,
- perspective projection,
- hidden surface and hidden line elimination,
- illumination of surfaces,
- transparancy and
- surface texturing.

Each 3D displaying technique gives more or less sufficient results for displaying various scenes containing "natural objects". The technique to be chosen for a special scene depends upon what the scene shows and what the observer expects to see. A sufficient solution is sometimes to combine some of the above described techniques.

5.1 *Kinetic effects*

To produce kinetic effects, as for example near real-time rotation, scaling and translation - one unique operation can be performed on the data, selected from the scene: a vector-by-matrix multiplication. With special hardware (matrix multipliers) a large number of polygon-points can be processed several times a second. Performing this operation 25 to 30 times a second on all selected vertices, a real-time-effect of "immediate reaction" of the system is noticed when the line-generation system accepts the amount of data in time. Using homogenious coordinates ([**x**,**y**,**z**,1]), a 4 by 4 transformation matrix and a division operation, also perspective projection can be performed in real-time.

5.2 *Hidden-line- and hidden-surface-elimination*

Assuming "solid" objects, the observer always can view only this portion of an object which faces points towards the viewpoint. With complex objects - and even with simple ones - ambiguities exist of the correct location of the object in space. The hidden-line- resp. hidden-surface-removal process eliminates the portion of the surface which is not seen by the observer [21]. Because the hidden-line/hidden-surface-removal operation is nonlinear in time, it is hard to perfom this operation in real-time. An improvement to this performance can be made by applying a "backface-removal" operation just before the hidden-line/hidden-surface-removal is carried out. The backface-removal is a linear-in-time operation which uses the surface normals to determine which surfaces can be seen by the observer. However, the backface-removal does not solve the hidden-line/hidden-surface-problem completely, but reduces the set of data which has to be processed by the hidden-line/hidden-surface-removal operation and thus gives a better input condition to the nonlinear hidden-line/hidden-surface-removal.

5.3 *Illumination*

After removing the "hidden surfaces" from a scene, the remaining surfaces can be rendered by an illumination process. Based on Lamberts cosine law, a variety of illumination models has been developed for Computer Graphics applications [4, 7, 15]. They range from simple models, which consider only the diffuse portion (1) of the reflected light, to highly complex models, which consider the complex influences on the spectral distribution of the reflected light on surfaces of different materials [11].

$$\mathbf{I_d} = \mathbf{k_d}\, (\vec{\mathbf{N}} \cdot \vec{\mathbf{L}})\, \mathbf{I_s} \qquad (1)$$

with: $\mathbf{I_d}$: intensity of the diffuse reflection,

$\mathbf{k_d}$: reflection konstant of the diffuse reflection ($0 \leq \mathbf{k_d} \leq 1$),

$\vec{\mathbf{N}}$: surface normal,

$\vec{\mathbf{L}}$: vector, pointing towards the light source,

$\mathbf{I_s}$: intensity of the light source.

The simple illumination models do not produce a sufficient impression of the surface of the objects. This depends on the one hand on the fact that the material dependend portion of the reflected light is not considered, and on the other hand that on curved surfaces the transition from one surface patch to the adjacent one is not smooth. Illumination models which allow a smooth surface rendering are for example those developed by Gourauld [5] and Phong [16]. Gourauld's illumination model performs a bilinear interpolation on the intensities between the intensities on the vertices of a surface patch, but does not consider the viewpoint of the observer. Phong's illumination model interpolates linearly the surface normals between the vertices of a surface patch, which is performed for each vertex by averaging the surface normals of the surrounding patches. Then it derives the intensities according to the results of this former computation on the surface normals. This allows to consider the viewpoint of the observer by computing the specular portion of the reflected light also. Models with more than one light source has been developed. As for example Phong's illumination model (2) can be extended to an arbitrary number of light sources, where the intensity of one surface point is performed by summing all intensities of all light sources in the scene.

$$\mathbf{I_p} = [\mathbf{I_d} + \mathbf{k_s}\, (\vec{\mathbf{S}} \cdot \vec{\mathbf{E}})^n]\, \mathbf{I} \qquad (2)$$

with: $\mathbf{I_p}$: intensity of the reflection by Phong's model,

$\mathbf{I_d}$: intensity of the diffuse reflection (1),

$\mathbf{k_s}$: reflection konstant of the specular reflection,

$\vec{\mathbf{E}}$: vector, pointing towards the viewpoint,

$\vec{\mathbf{S}}$: vector, of the (total) reflected beam on a (ideal) surface,

n: specualr concentration (typically 1 .. 200).

A more realistic illumination model is given by the generation of shadowcasting. This is essential for all situations where the light sources are not collinear with the viewpoint to the scene. For this type of illumination, additional "hidden surface elimination" processes - an additional one for each light source - have to be performed before the illumination process can take place.

5.4 Transparency

If objects are enclosed within other objects, resp. objects are surrounding other objects, it is often helpful to change the representation of the surrounding objects from a fully opaque representation to a more or less transparent one. This technique - also described as translucent surfaces - determines in addition to the reflected portion of the intensities a transmitting portion. Dependent of the degree of "translucency" the transmitted and reflected protions of the intensities at a specific surface point are weighted and then summed (3).

$$\mathbf{I}_t = \mathbf{k}\,\mathbf{I}_1 + (1 - \mathbf{k})\,\mathbf{I}_2$$

where k measures the "translucency" resp. (1 - k) the "opaqueness" (0: perfectly translucent, 1: opaque).

More sophisticated models of transparency also consider the refraction of transparent material (e. g. glass), where the refracted beam of the light changes the optical line of sight. These models lead to the so called "raytracing" algorithms, where for each "ray" which reaches the viewers eye (practically each pixel on a rasterdisplay) its intensity is computed by following the ray backwards to its source(s) under the condition that at each surface, which the ray is striking, the ray is broken into three parts:

- diffusely reflected light,
- specularly reflected light and
- transmitted, and therefore refracted light.

For each of the three "subrays" the same procedure has to be applied again. This leads to a recursive algorithm, which generates a (unary to tertiary) tree which represents the reflection - refraction ratio at each surface of the scene for each light source in it as nodes of the tree. Practically the depth of the tree is limited to avoid high, exponential computational costs. This limit also controls the "reality" of the so generated pictures while it limits the way a ray bounced the surfaces.

5.5 Surface texture

The most realistic and natural impression by displaying surfaces of organs can be performed by application of surface textures. This technique rather mappes a texture on a surface than modelling it along with the modelling process, and so far keeps the modelling process simpler. Within the variety of the surface texturing methods, the following are examples [10]:

- Surface textures by varying the surface normals,
- Stochastic surface texturing,
- Texture tiling and
- Syntactic texture generation.

Surface textures by varying the surface normals produces the patterns on the surface by intensity variations of the computation of the reflected light, which depends on the surface normals. The variation of the surface nórmals at the present position - when the illumination process is performed - is computed according to the function F (u,v), where u and v are parameters within the description of the surface. This technique allows the mapping of images on top of the surfaces of objects.

Stochastic surface texturing methods also use the technique, varying the surface normals. But they compute the variation-function F (u,v) by a randomly genereated function (e. g. fBm "fractional Brownian motion"). The random function can be applied as well on one dimension of the surface as on two dimensions on the surface. This function produces a "realistic" impression of irregular textured objects.

The basis of texture tiling is the digitisation and storage of a set of textures and the mapping on surfaces of this textures by referencing them by the polygons. When the polygons are displayed, the textures will also be displayed within the polygon boundaries according to the actual orientation of the polygon in space. This method has the advantage that a set of different textures can be stored in a library and - when needed - fast applied with minimal computation costs.

Using a syntactic method to generate surface textures, a grammar directs the process which texture element has to be placed next at a specific position. A texture pattern herein is composed of a number of windows which are mostly of fixed size, and where each window is represented by a tree of which each node corresponds to a pixel in the window. So far tree-grammars are used to describe the texture pattern.

6. ACKNOWLEDGEMENTS

I am grateful to C. Stüve, who helped to prepare the manuscript.

REFERENCES

[1] B. Bauer, W. Schlegel, R. Boesecke, J. Doll, G. Hartmann, W. J. Lorenz **Three dimensional Planning of Conformation Therapy** Proceedings of the International Symposium on Computer Assisted Radiology, Berlin, June 1985, pp 388-394

[2] H. Becker, J. Rolfes, B. Graul, G. Giebel **Klinische Anwendung von 3-D Bildverfahren in der Computertomographie** Proceedings of the International Symposium on Computer Assisted Radiology, Berlin, July 1987, pp 630-635

[3] F. R. P. Boecker, U. Tiede, K. H. Höhne **Combined Use of Different Algorithms for Interactive Surgical Planning** Proceedings of the International Symposium on Computer Assisted Radiology, Berlin, June 1985, pp 572-577

[4] Klaus D. Bösing **Beleuchtungsmodelle zur realitätsnahen Darstellung dredimensionaler Objekte in der Computer Graphic** Technischer Bericht 85/14 Technische Universität Berlin, Institut für Technische Informatik, Computer Graphics / Computer Vision, Juni 1985

[5] H. Gourauld **Continuous Shading of Curved Surfaces** IEEE Transactions on Computers, Vol. C-20, No. 6, Los alamitos, June 1971, pp 623-629

[6] P. Dev, L. L. Fellingham, S. L. Wood, A. Vassiliadis **A Medical Graphis System for Diagnosis and Surgical Planning** Proceedings of the International Symposium on Computer Assisted Radiology, Berlin, June 1985, pp 602-607

[7] J. D. Foley, A. Van Dam **Fundamentals of Interactive Computer Graphics** Addison Wesley 1982.

[8] H. Fuchs, Z. M. Kedem, S. P. Uselton **Optimal Surface Reconstruction from Planar Contours** Communications of the ACM, Vol. 20, No. 10, 1977, pp 693-702

[9] Gabor T. Herman **Computer Graphics in Radiology** Proceedings of the International Symposium on Computer Assisted Radiology, Berlin, June 1985, pp 540-550

[10] Bernd Knobloch **Methoden der realitätsnahen Darstellung** Technischer Bericht 85/15 Technische Universität Berlin, Institut für Technische Informatik, Computer Graphics / Computer Vision, Juni 1985

[11] R. L. Cook, K. E. Torrance **A Reflectance Model for Computer Graphics** Computer Graphics, Vol. 15, No. 3, ACM SIGGRAPH, New York, August 1981, pp 307-316

[12] **The Third Dimension** Proceedings of the International Symposium on Computer Assisted Radiology, Berlin, June 1985, pp 628-634

[13] Heinz U. Lemke, Klaus Bösing, M. Engelhorn, Dietmar Jackel, Bernd Knobloch, Harald Scharnweber, H. Siegfried Stiehl, Klaus D. Tönnies **3-D Computer Graphic work station for biomedical information modelling and display** Proceedings of SPIE's Medical Imaging Conference, 1-6 February, 1987, Newport Beach, California, U.S.A.

[14] H. Lemke, M. Engelhorn **Work Stations for Computer-Graphic Display in Medical Imaging** Biomedizinische Technik 31, 6 (1986), pp 143-149

[15] William M. Newman, Robert F. Sproull **Principles of Interactive Computer Graphics, Second Edition** McGraw-Hill International Book Company, 1979

[16] Bui Tuong Phong **Illumiantion for Computer Generated Pictures** Communications of the ACM, Vol. 18, No. 6, June 1975, pp 449-455

[17] Michael L. Rhodes, Yu-Ming Azzawi, Eva S. Chu, Alex T. Pang, William V. Glenn, Stephen L. G. Rothman **A Network Solution for Structure Models and Custom Prostheses Manufacturing from CT Data** Proceedings of the International Symposium on Computer Assisted Radiology, Berlin, June 1985, pp 403-412

[18] Harald Scharnweber **Modelling and Representation Principles for Three-Dimensional Natural Objects** Proceedings of the International Symposium on Computer Assisted Radiology, Berlin, July 1987, pp 757-763

[19] H. Siegfried Stiehl **Automatische Verarbeitung und Analyse von kranialen Computer-Tomogrammen** Dr. Ing. Dissertation, Technische Universität Berlin, Fachbereich 20, 1980

[20] H. Siegfried Stiehl **A Framework for Spatial Image Sequence Understanding in Radiology** Technischer Bericht 85/4, Technische Universität Berlin, Institut für Technische Informatik, Computer Graphics / Computer Vision, April 1985

[21] Ivan E. Sutherland, Robert F. Sproull, Robert A. Schumacker **A Characterization of Ten Hidden-Surface Algorithms** Computing Surveys, Vol. 6, No. 1, March 1974, pp 387-441

[22] Klaus D. Toennies **Erzeugung von 3D-Objektrepräsentationen durch Triangulation** Technischer Bericht 86/2, Technische Universität Berlin, Institut für Technische Informatik, Computer Graphics / Computer Vision, Januar 1986

[23] M. W. Vannier, J. L. Marsh, J. O. Warren **Three Dimensional Computer Graphics for Craniofacial Surgical Planning and Evaluation** Computer Graphics (Proc. SIGGRAPH 83), Vol. 17, No. 3, July 1983, pp 263-273

[24] M. W. Vannier, J. L. Marsh, W. G. Totty, L. A. Gilula **Three Dimensional CT Scan Reconstruction at the Mallinckrodt Institute of Radiology** Proceedings of the International Symposium on Computer Assisted Radiology, Berlin, June 1985, pp 578-582

[25] M. W. Vannier, J. L. Marsh, R. H. Knapp, C, J, Offutt **Recent Advances in 3-D CT Scan Reconstruction for Clinical Applications** Proceedings of the International Symposium on Computer Assisted Radiology, Berlin, July 1987, pp 657-661

Theorie mehrdimensionaler Signale

RESOLUTION IN COMPUTERIZED TOMOGRAPHY

F. Natterer
Institut für Numerische und
instrumentelle Mathematik
Einsteinstraße 62
4400 Münster, FRG

1. INTRODUCTION

In computerized tomography one reconstructs a function in $\mathbb{R}^2$ from a finite set of line integrals. The arrangement of the lines is referred to as scanning geometry. In the present paper we shall investigate the possible resolution of various scanning geometries. We say that a scanning geometry has resolution d if functions containing no details of size $\leq d$ can be recovered reliably from the integrals over the lines making up the scanning geometry. By a function containing no detail of size $\leq d$ we mean a function which is (essentially) band-limited with band-width $b = 2\frac{\pi}{d}$, i.e. a function f whose Fourier transform

$$\hat{f}(\xi) = (2\pi)^{-n/2} \int_{\mathbb{R}^n} e^{-ix\cdot\xi} f(x)dx$$

is negligible for $|\xi| \geq b$. Here, n denotes the dimension (in our case n=2) and $x\cdot\xi$ is the dot product. This definition of resolution is quite common in image processing, see e.g. Pratt (78). Note that in the engineering literature a factor of 2π shows up in the exponent in the Fourier integral. This means that our results have to be modified correspondingly when compared to others.

To fix ideas we assume that the function to be reconstructed is concentrated in the unit disk Ω of $\mathbb{R}^2$. The lines are given by their Hesse normal form $x\cdot\theta = s$ with $\theta = (\cos\varphi, \sin\varphi)^T$ a unit vector perpendicular to the line and s the (signed) distance of the line to the origin. φ, s are the coordinates we use for the lines. We consider the following scanning geometries.

1. The parallel geometry.

Here, the lines come in p parallel bunches with 2q+1 lines each (fig. 1a). We thus have the lines

$$\varphi_j = j\Delta\varphi,\quad j=0,\ldots,p-1,\quad \Delta\varphi = \frac{\pi}{p}\,,\quad s_\ell = \ell\Delta s,\ \ell = -q,\ldots,q\,,\ \Delta s = \frac{1}{q}\,.$$

The number of data is essentially 2pq.

2. The parallel interlaced geometry.

It is obtained from the parallel geometry by dropping every second line, retaining only the ℓ's with the parity of j. The number of data is one half of the (full) parallel geometry, i.e. pq.

3. The fan-beam geometry.

Here the lines emanate from p points $a_j = r(\cos\beta_j, \sin\beta_j)^T$ on the circle of radius $r > 1$, $\beta_j = j\Delta\beta$, $\Delta\beta = \frac{2\pi}{p}$, $j=0,\ldots,p-1$. They make an angle $\alpha_\ell = \ell\Delta\alpha$, $\Delta\alpha = \frac{\pi}{2q}$, $\ell = -q,\ldots,q$ with the line joining a_j with the origin (fig. 1b). Since only the lines with $|\alpha_\ell| \leqq \bar{\alpha}(r)$ $= \arcsin(1/r)$ hit the reconstruction region Ω, ℓ actually runs only from $-q_r$ to q_r with $q_r = \lfloor 2q\bar{\alpha}/\pi \rfloor$, the other line integrals being zero. Hence the number of data is essentially $4\bar{\alpha}pq/\pi$. In φ - s - coordinates, the lines are given by

$$\varphi = \beta_j + \alpha_\ell - \pi/2 \quad , \quad s = r\sin\alpha_\ell \;.$$

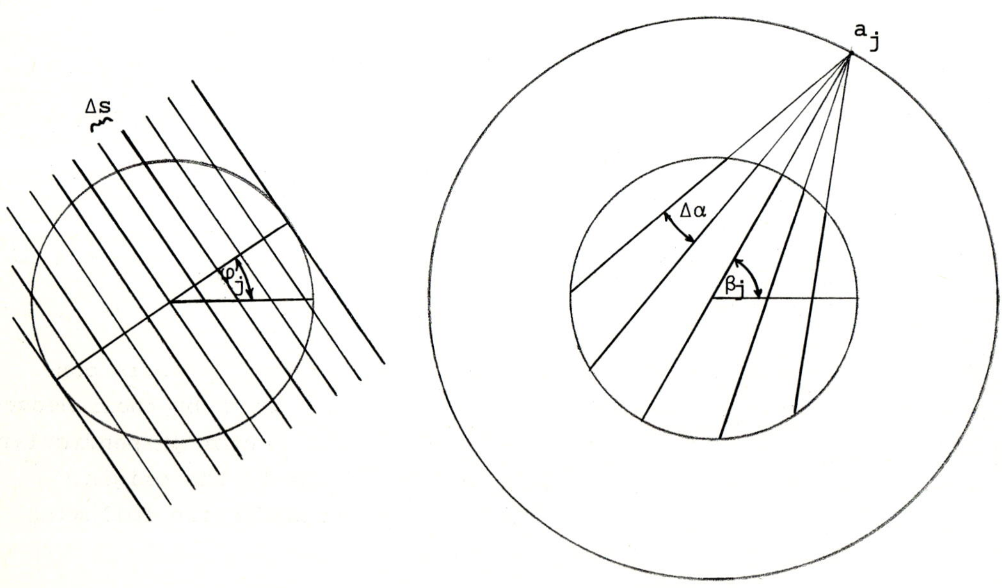

Figure 1: Scanning geometries. a(left) Parallel scanning. b(right) Fan-beam scanning, $r = 2$.

4. The PET (or ring) geometry.

This is a special case of the fan-beam geometry with $r = 1$, $\Delta\alpha = \frac{1}{2}\Delta\beta$, i.e. $p = 2q$. It consists of the $p(p-1)/2$ lines joining p points uniformly distributed on the boundary of Ω.

2. THE BASIC SAMPLING THEOREMS FOR THE RADON TRANSFORM

Let

$$(Rf)(\varphi,s) = \int f(s\Theta + t\Theta^{\perp})dt$$

$$\Theta = (\cos\varphi, \sin\varphi)^T, \quad \Theta^{\perp} = (-\sin\varphi, \cos\varphi)^T$$

be the Radon transform of the function f. Each scanning geometry amounts to sampling Rf on a certain grid on $[0,\pi] \times [-1,+1]$. For fan-beam scanning it is more convenient to use the function

$$g(\beta,\alpha) = Rf(\beta + \alpha - \pi/2 \ , \ r \sin\alpha)$$

on $[0,2\pi] \times [-\frac{\pi}{2}, \frac{\pi}{2}]$.

A grid in $\mathbb{R}^2$ is defined by a real non-singular matrix W, the grid points being $W\ell$ with ℓ any integer vector.

Very loosely, the sampling theorems for the Radon transform (see Natterer (86), Theorems III.3.1-2) can be rephrased as follows.

<u>Theorem 1.</u> Let f have (essential) band-width $b >> 1$. For $0 < \vartheta < 1$ let (see fig. 2a)

$$K = \{(k,\sigma): |\sigma| < b, \ |k| < \max(|\sigma|/\vartheta \ , \ (1/\vartheta - 1)b), \ k \text{ integer}\} .$$

Let W be a real 2×2 -matrix such that the sets $K + 2\pi \ (W^{-1})^T\ell$, ℓ an integer vector, are mutually disjoint. If $Rf(W\ell) = 0$ for all integer vectors ℓ, then Rf is negligible.

<u>Theorem 2.</u> Let f have (essential) band-width $b >> 1$. For $0 < \vartheta < 1$ let (see fig. 2b)

$$K = \{(k,2m): |k| < b/\vartheta , \quad |k-2m| < rb/\vartheta$$
$$|k| \leq \max \quad (|k-2m| , (1/\vartheta - 1)/rb) \quad ,$$
$$k , m \quad \text{integer}\} .$$

Let W be a real 2×2 -matrix such that the sets $K + 2\pi(W^{-1})^T$, ℓ an integer vector, are mutually disjoint. If $g(W\ell) = 0$ for all integer vectors ℓ, then g is negligible.

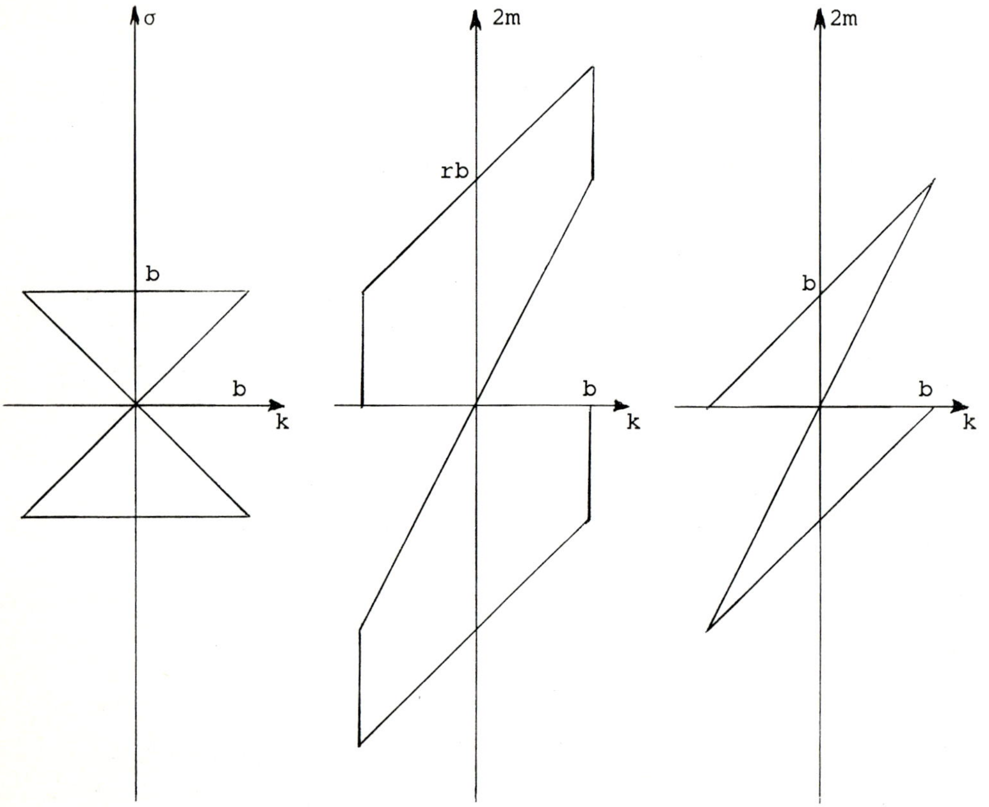

Figure 2: The set K in theorem 1, 2. a(left) Parallel geometry. b(middle) Fan-beam geometry with $r = 2$. c(right) PET geometry (case $r = 1$ of b). For simplicity we have taken $\vartheta = 1$ in all drawings.

What is meant by negligible is explained in the reference given above. For our purpose it means that Rf, hence f, is essentially uniquely determined by its values on the respective grids.

The proofs for Theorem 1, 2 can be divided into two parts: First we show that the supports of the functions Rf, g are essentially given by the sets K. Then we apply a multidimensional version of Shannon's sampling theorem as given by Petersen and Middleton (62). In its simplest form it reads as follows.

Let g be a function on $\mathbb{R}^2$ such that $\hat{g} \in L_1(\mathbb{R}^2)$. Let g vanish on the grid $W\ell$, ℓ an integer, and assume the sets $K + 2\pi(W^{-1})^T\ell$ to be mutually disjoint. Then,

$$|g(x)| \leqq \frac{1}{\pi} \int_{\xi \notin K} |g(\xi)|d\xi \quad .$$

3. APPLICATION TO SPECIFIC SCANNING GEOMETRIES

We now apply the theorems of the preceeding section to the scanning geometries in the introduction. The best results are obtained for ϑ close to 1. In order to avoid cumbersome discussions we therefore put henceforth $\vartheta = 1$, sacrificing a bit of mathematical rigour in favour of simplicity.

1. The parallel geometry.

We put in Theorem 1

$$W = \begin{pmatrix} \pi/p & 0 \\ 0 & 1/q \end{pmatrix} \quad , \quad 2\pi(W^{-1})^T = \begin{pmatrix} 2p & 0 \\ 0 & 2\pi q \end{pmatrix} .$$

From fig. 2a we see that the sets $K + 2\pi(W^{-1})^T$ are mutually disjoint if $b \leqq p$ and $b \leqq \pi q$. These conditions have been already found by Bracewell (56). They suggest to put $p = \pi q$ (approximately, since p,q are integers), i.e. $\Delta s = \Delta\varphi$. Under this optimality relation, the number of data is essentially $2pq = (2/\pi)b^2$.

2. The parallel interlaced geometry.

Here we use Theorem 1 with

$$W = \frac{\pi}{p}\begin{pmatrix} 2 & -1 \\ 0 & 1 \end{pmatrix} \quad , \quad 2\pi(W^{-1})^T = p\begin{pmatrix} 1 & 0 \\ 1 & 2 \end{pmatrix} \quad .$$

The grid generated by W is identical to the grid of the interlaced geometry if the optimality condition $p = \pi q$ is observed. Fig. 2a tells us that the sets $K + 2\pi(W^{-1})^T \ell$ are mutually disjoint if $p \geqq b$. This is the sampling condition for the parallel interlaced scanning geometry which has been found by Lindgren and Rattey (81), see also Cormack (78). It needs only one half of the data, i.e. $(1/\pi)b^2$, of the (full) parallel geometry for the same resolution.

3. The fan-beam geometry.

Here we put

$$W = \pi\begin{pmatrix} 2/p & 0 \\ 0 & 1/(2q) \end{pmatrix}, \quad 2\pi(W^{-1})^T = \begin{pmatrix} p & 0 \\ 0 & 4q \end{pmatrix}$$

in Theorem 2. From fig. 2b we see that the sets $K + 2\pi(W^{-1})^T \ell$ are non-overlapping for $p \geqq 2b$, $q \geqq rb/2$. These are the sampling conditions for the fan-beam geometry. For the minimal values of p, q we have $p = 4q/r$, i.e. $\Delta\alpha = r\Delta\beta$. If p, q are chosen in this way, the number of data is $(4r\bar{\alpha}(r)/\pi)b^2$. For the comparison with other results (see e.g. Joseph and Schulz (80), Rattey and Lindgren (81)) one has to keep in mind that g has to be sampled only for α_ℓ with $|\ell| \leqq q_r$.

4. The PET geometry.

For $r = 1$, the set K is sketched in fig. 1c. We see that now a denser non-overlapping covering of the plane is possible by putting

$$2\pi(W^{-1})^T = p\begin{pmatrix} 1 & 0 \\ 0 & 2 \end{pmatrix}, \quad \text{i.e.} \quad W = \frac{\pi}{p}\begin{pmatrix} 2 & 0 \\ 0 & 1 \end{pmatrix}$$

with $p \geqq b$. W defines the PET geometry. It needs only one half of the points a_j of the general fan-beam geometry, the number of data being only $\frac{1}{2}b^2$.

The following table surveys our findings. It contains the scanning geometries with resolution $2\pi/b$, arranged by increasing number of data.

Scanning geometry	Sampling conditions	Optimality relation	Number of data
parallel interlaced	$p > b,\ q > b/\pi$	$\Delta\varphi = \frac{1}{2}\,\Delta s$	$\frac{1}{\pi}\,b^2$
PET	$p > b$	$\Delta\beta = 2\,\Delta\alpha$	$\frac{1}{2}\,b^2$
parallel (full)	$p > b,\ q > b/\pi$	$\Delta\varphi = \Delta s$	$\frac{2}{\pi}\,b^2$
Fan - beam	$p>2b,\ q>br/2$	$\Delta\beta = r\Delta\alpha$	$\frac{4}{\pi}\ \ldots\ 2b^2$

We see that the fan-beam geometry, which is the most widely used one, is the least efficient one, judged from the number of data required. However it is possible to increase the necessary detector spacing by a factor of 2 by using a detector set-off of $\Delta\alpha/4$, i.e. by putting $\alpha_\ell = \ell\Delta\alpha + \Delta\alpha/4$, see Schwierz, Härer and Wiesent (81).

For $b = 8$, the various line patterns have been sketched in fig. 3. One can see with the naked eye that the parallel interlaced geometry and the PET geometry lead to a significantly more uniform distribution of the lines then the (full) parallel geometry and the fan-beam geometry whose lines concentrate along concentrical circles.

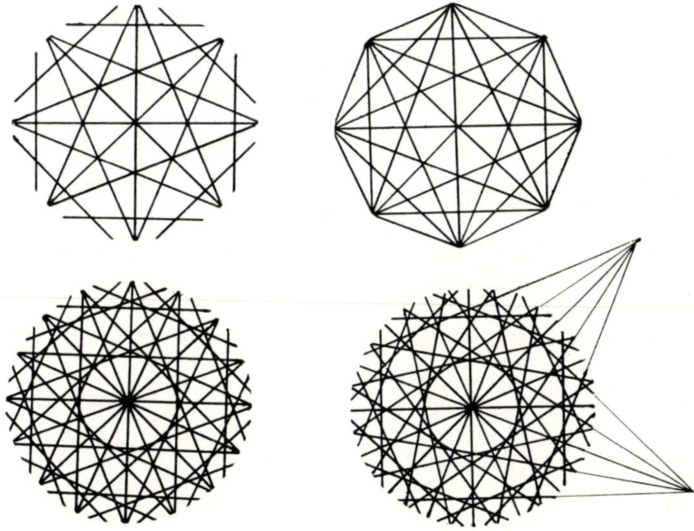

Figure 3: Scanning geometries with the same resolution ($b = 8$) by increasing number of data. a(top left) Parallel interlaced. b(to right) PET. c(bottom left) (Full) parallel. d(bottom right) Fan-Beam, $r = 2$.

4. CONCLUSION

The resolution of some scanning geometries in CT has been determined in the framework of (essentially) band-limited functions. We did not discuss which reconstruction algorithm actually achieve this resolution. For the interlaced parallel geometry and for the PET geometry, such algorithms have been found in Kruse (86). Nothing is known about optimal algorithms for the fan-beam geometry. However there is some numerical evidence that the usual filtered backprojection algorithm does not give the optimal resolution.

REFERENCES

Bracewell, R.N. and Riddle, A.C. (1956). Strip integration in radio astronomy. Aus. J. Phys., 9, 198-217.

Cormack, A.M. (1978). Sampling the Radon transform with beams of finite width. Phys. Med. Biol., 23, 1141-1148.

Joseph, P.M. and Schulz, R.A (1980). View sampling requirements in fan beam computed tomography. Med. Phys., 7, 692-702.

Kruse, H. (1986). Die Auflösung von Rekonstruktionsalgorithmen in der Computer-Tomographie. Dissertation, Fachbereich Mathematik, Universität Münster.

Lindgren, A.G. and Rattey, P.A. (1981). The inverse discrete Radon transform with applications to tomographic imaging using projection data. Advances in Electrics and Electron Physics, 56, 359-410.

Natterer, F. (1986). The Mathematics of Computerized Tomography. Wiley -Teubner.

Pratt, W.K. (1978). Digital Image Processing. Wiley.

Schwierz, G., Härer, W. and Wiesent, K. (1981). Sampling and Discretization Problems in X-ray-CT, in Herman, G.T. and Natterer, F. (eds.), Mathematical aspects of Computerized Tomography. Proceedings, Oberwolfach 1980. Springer.

PULSSEQUENZEN ALS MODULATIONSFUNKTION BEIM NMR-IMAGING

R. Gebhardt, E. Penner, R. Martin, W. Ameling
Lehrstuhl für Allgemeine Elektrotechnik und Datenverarbeitungssysteme
RWTH Aachen, D 5100 Aachen

Einleitung und Zusammenfassung

Das NMR-Imaging ist zwischenzeitlich zu einem bedeutsamen Hilfsmittel in der medizinischen Diagnostik geworden. Von grundsätzlicher Bedeutung beim NMR-Imaging ist die Fragestellung, wie ein Spinsystem (z. B. Wasserstoff-Protonen) 'geeignet' angeregt werden muß, damit die Systemantwort 'optimal' im Hinblick auf die gewünschte Auswertung genutzt werden kann.

Am Beispiel des Hadamard-Imaging wird gezeigt, daß die Wirkung von Pulssequenzen als Modulationsfunktion für die HF-Anregung im Prinzip zu den gleichen Ergebnissen führt wie das Puls-Fourierverfahren. Die Anforderungen, die an die Experimentdurchführung bzw. die Meßtechnik gestellt werden, unterscheiden sich in beiden Fällen jedoch erheblich. Die nachfolgend dargestellten Ergebnisse von Simulationsrechnungen zeigen u. a., daß das Hadamard-Imaging äußerst sensibel auf einen Gleichspannungsoffset im Meßsignal reagiert. Selbst ein Offset, der innerhalb des Rauschpegels liegt, führt zu einem Einbruch im Frequenzspektrum, der die von Bernardo (1) beschriebenen Artefakte erklären kann.

Anregung von Spinsystemen

Das Phänomen der Kernspinresonanz (<u>N</u>uclear <u>M</u>agnetic <u>R</u>esonance) wird durch die Bloch'sche Gleichung (2) dargestellt. Sie beschreibt das dynamische Verhalten der Kernmagnetisierung unter dem Einfluß von außen angelegter magnetischer (statische, inhomogene, hochfrequente) Felder. Für den Fall einer sehr kurzen HF-Anregung mit einer Einhüllenden der Form $\mathrm{rect}(t/\tau)$ ist diese Gleichung analytisch lösbar.

Der Scheitelwert der HF-Anregung und die Dauer τ des Pulses sind ein Maß für den Winkel, um den der Magnetisierungsvektor aus seiner Ausgangslage in die Richtung der x-y-Ebene gedreht wird. Ein 90°-Puls, wie er beim Puls-Fourierverfahren benutzt wird, dreht die Magnetisierung um 90°. Dies erzeugt ein maximales Meßsignal.

Die Fouriertransformierte der Einhüllenden ist bekanntlich proportional $\mathrm{si}(\omega\tau/2)$. Damit bei eingeschaltetem Gradienten (lineare Inhomogenität des statischen Feldes) alle Spinsysteme gleichartig angeregt werden, muß die Bedingung $\omega\tau/2 \ll \pi$ erfüllt sein.

Beim Hadamard-Verfahren werden periodische Pulssequenzen als Modulationsfunktion für die Anregung der Probe benutzt. Damit die Wirkung der einzelnen Pulse superponierbar bleibt, dürfen hierbei pro Puls nur kleine Drehwinkel erzeugt werden. Die Reihenfolge der Scheitelwerte der Pulssequenz entspricht den Werten einer m-Sequenz in der (1,-1)-Darstellung.

Der für die Bildgebung wesentliche Unterschied tritt im Frequenzspektrum der Pulssequenzen zutage. Im Ursprung ergibt sich ein Einbruch um den Faktor 1/N (N: Periode der Sequenz).

Simulationsmodell

Als mögliche Ursachen für den in (1) angegebenen Artefakt werden in Betracht gezogen:

1. Nichtlineares Verhalten der Spinsysteme (zu großer Winkel),
2. Rauscheinflüsse,
3. Gleichspannungsoffset.

Es wird von einem homogenen, linearen Objekt konstanter Spindichte, welches in 1024 Elemente zerlegt ist, ausgegangen. Die Beiträge jedes dieser Elemente zum Meßsignal und ihre Überlagerung wird berechnet und einer inversen Hadamard-, sowie einer Fourier-Transformation zugeführt. Die dem Experiment zugrunde liegenden Parameter werden mit Ausnahme des Abtastzeitpunktes, der hier idealerweise direkt nach dem HF-Puls angenommen wird, (1) entnommen. Zur Untersuchung des nichtlinearen Verhaltens werden mehrere Rechnungen durchgeführt, die sich lediglich in den Scheitelwerten der HF-Pulse unterscheiden. Für die Betrachtung von 2. und 3. werden die Abtastwerte mit einem gaußverteilten, weißen Rauschen sowie einem Gleichspannungsoffset beaufschlagt.

Simulationsergebnisse

Aus den Simulationsrechnungen ergibt sich, daß die in (1) geäußerte Vermutung, das nichtlineare Verhalten der Spinsysteme sei die Ursache für den Artefakt, nicht stichhaltig ist. Die Nichtlinearitäten verursachen eine Verzerrung des abgebildeten Objekts sowie zusätzliche Beiträge, die im Bild außerhalb des Objekts erscheinen. Weiterhin nimmt die Amplitude des Meßsignals - im Vergleich zu der mit einem einzelnen HF-Impuls gleicher Größe erzielbaren Amplitude - beim Verlassen des linearen Bereiches stark ab. In Bild 1 sind die Amplitudenabnahme sowie die Verzerrungen am Beispiel eines 5°-Impulses zu sehen; die Amplitudenabnahme als Funktion des Impulswinkels ist in Bild 2 dargestellt.

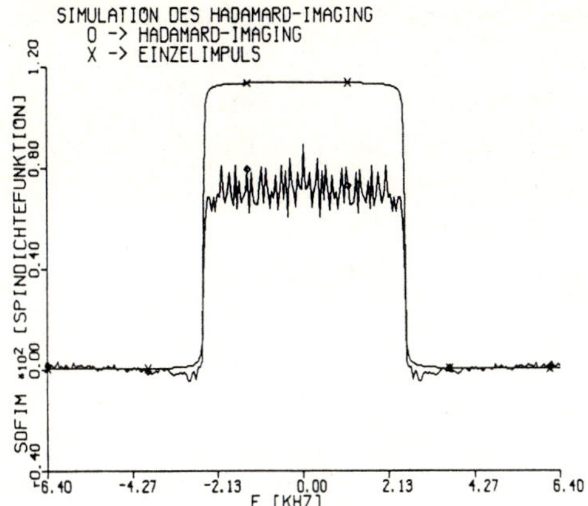

Bild 1: Einzelpulsanregung und Hadamard-Imaging bei Verwendung von 5°-Pulsen

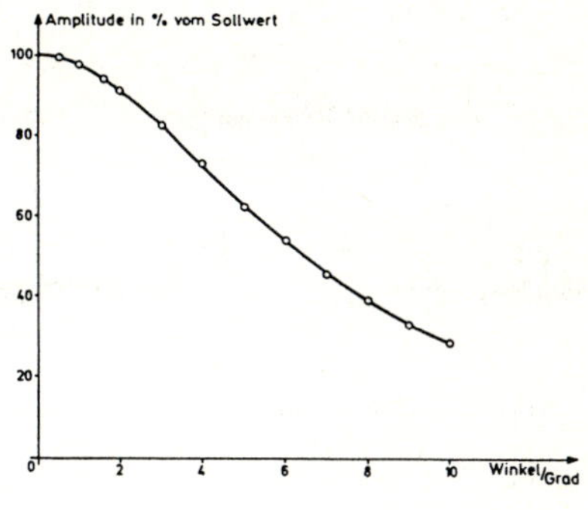

Bild 2: Amplitudenabnahme

Als linear kann das Verhalten bis zu einer Maximalauslenkung der Spins von etwa 30° betrachtet werden (3); dies entspricht bei den hier verwendeten 18 Sequenzen mit je 255 Pulsen einem Winkel von 1,2°. Bei dem in (1) angegebenen Winkel von 1,62° sind die durch Nichtlinearitäten verursachten Einflüsse noch gering, wie Bild 3 zeigt.

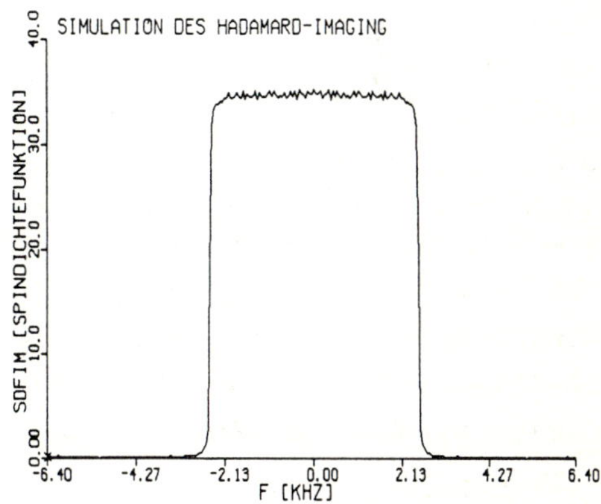

Bild 3: Hadamard-Imaging bei Verwendung von 1,62°-Pulsen

Der im Frequenzspektrum der Pulssequenz auftretende Einbruch muß durch die inverse Hadamard-Transformation kompensiert werden. Daher reagiert das Hadamard-Imaging in der Bildmitte besonders sensibel auf Rauscheinflüsse. Die Simulation ergibt, daß die durch Rauschen verursachten Störspitzen in der Bildmitte wesentlich stärker hervortreten können, als in anderen Bereichen. In Bild 4 wird dieses an fünf übereinander geplotteten Bildern deutlich. Da eine perfekte Sequenz (4) keinen Einbruch im Spektrum aufweist, ließe sich dieser Effekt durch ihre Verwendung eleminieren.

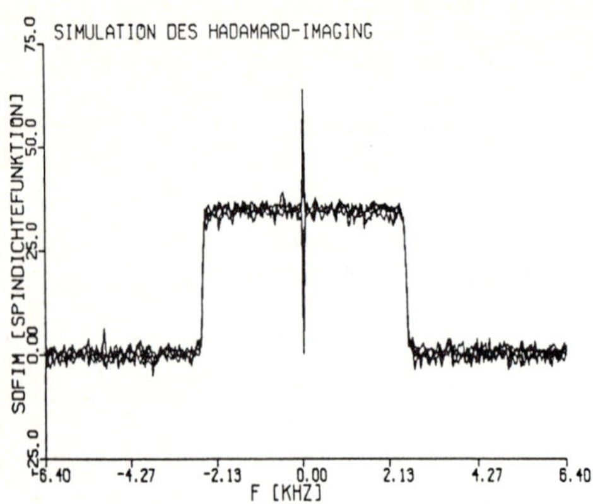

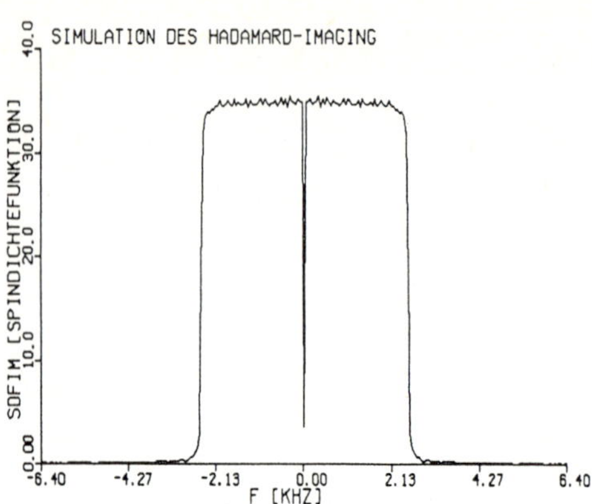

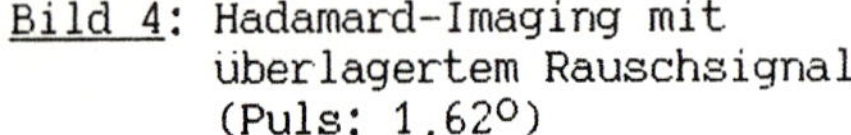
Bild 4: Hadamard-Imaging mit überlagertem Rauschsignal (Puls: 1,62°)

Bild 5: Hadamard-Imaging mit überlagertem Spannungsoffset (Puls: 1,62°)

Als weitere Ursache für eine Störung in der Bildmitte kommt ein Gleichspannungsoffset der Meßsignale in Betracht, der z. B. durch mangelhaften Empfängerabgleich oder unvollständige Basislinienkorrektur des FID's verursacht werden kann. Wie die Simulation zeigt, kann ein Offset, der zwei Größenordnungen kleiner als das Meßsignal ist, bereits zu einem starken Abfall in der Bildmitte führen (siehe Bild 5).

Damit kommen als Ursachen für Artefakte sowohl Rauscheinflüsse, als auch Abgleichfehler in Betracht. Ausgeschlossen werden kann der Einfluß des nichtlinearen Verhaltens der Spinsysteme.

Anmerkung

Die Simulationsrechnungen wurden mit Hilfe des Simulationssystems ACSL auf einem IBM-PC-AT durchgeführt.

Literatur:

1. M. Bernado Jr., D. Chaudhuri, X.-R. Liu and P. C. Lauterbur; Proceedings, Fifth Annual Meeting of the Society of Magnetic Resonance in Medicine, S. 944, Montreal, 1986.

2. F. Bloch; Phys. Rev. 70, p. 460, 1946.

3. D. I. Hoult; I. Mag. Res. Vol. 35, S. 69, 1979.

4. H. D. Lüke; Frequenz 40, S. 215, 1986.

RECONSTRUCTION OF TWO-DIMENSIONAL SIGNALS
FROM IRREGULARLY SPACED SAMPLES

Dale H. Mugler
Santa Clara University
California USA

and

Wolfgang Splettstößer
Entwicklungszentrum für Mikroelektronik
Siemens, Düsseldorf 31

There are many occasions when it is desirable to reconstruct a two-dimensional (2-D) signal from a set of irregularly spaced samples. In [2] the authors mention applications to image processing, where irregularly spaced samples result from sampling related to motion compensation of time-varying imagery, as well as applications to computer graphics, geology, and more.

The reconstruction of a 2-D signal from irregularly spaced samples differs from the corresponding theory for 1-D signals due to the wide variety of sampling distributions possible in 2-D. Methods used to transform an irregular sample set to a regular one become much more complicated in a higher dimension, for example. In [3], results and problems were discussed concerning the generalization of transformation techniques from 1-D to 2-D. This paper is concerned primarily with the amount of variation that may be allowed from a periodic sample set in order to achieve error-free signal reconstruction.

In 1-D, a band-limited finite-energy signal f(t) is one whose Fourier transform $F(w) = (1/\sqrt{2\pi}) \int f(t)e^{-iwt}\,dt$ vanishes outside a finite interval, $(-\pi W,\pi W)$. The inverse relation gives

$$f(t) = \frac{1}{\sqrt{2\pi}} \int_{-\pi W}^{\pi W} F(w)e^{iwt}\,dw, \quad \text{for } F \in L^2(-\pi W,\pi W).$$

Reconstruction of the signal from its periodic samples $f(nT)$ may be done by the Whittaker-Kotel'nikov-Shannon sampling theorem whenever $T \le 1/W$. Reconstruction from <u>irregularly</u> spaced samples is connected closely with the theory of nonharmonic Fourier series, which have the form $\sum c_n e^{i\lambda_n t}$, $-\pi \le t \le \pi$, where the numbers λ_n $(n=0, \pm 1, \ldots)$ are not all integers. In an earlier paper [9], the second-named author noted an extension of the result of Paley & Wiener (see also [8]), which allows for error-free reconstruction whenever $\{\lambda_n\}$ is a sequence of <u>complex</u> constants which do not vary too much from strict periodic samples. In particular, if we suppose that the irregular samples are at $\{\lambda_n T\}$, then the reconstruction is possible whenever $|\lambda_n - n| \le \log 2/\pi \simeq 0.22$. This contrasts with earlier work of Levinson, who showed that the reconstruction is possible for a real sequence $\{\lambda_n\}$ for which $|\lambda_n - n| \le D < 1/4$, see also [4]. A recent description of this can be found in [10] .

Even with irregularly spaced samples, past samples alone are enough to insure error-free recovery. The authors [6] noted that prediction of f(t) is possible for a set of samples $\{f(t_n)\}$ if $|t-t_n|<nT$ for $n\geq 1$ and $TW < \log 2/\pi$. Further, classical Newton series provide the reconstruction formula. Building on the work of Levinson, Beutler [1] obtained a wide class of results including one that assures that the past need only be sampled at an average rate greater than the Nyquist rate.

In 2-D, a matrix notation was developed in [5] to describe periodic sampling that provides a striking analogy to the 1-D case. If $\underline{t}$ represents a column vector, a band-limited signal is defined to be a function $f(\underline{t})$ with a 2-D Fourier transform $F(\underline{w}) = (1/2\pi)\int f(\underline{t})\exp(-i\underline{wt})d\underline{t}$, whose support is contained in a finite region W of "Fourier" space. That is, $F(\underline{w})=0$ if $\underline{w} \notin W$.

A sampling lattice in 2-D is the set of integer linear combinations of two linearly independent column vectors, $\{\underline{v}_1, \underline{v}_2\}$. That is, sample points are at $\underline{t}_{\underline{n}} = V\underline{n}$, where $V = (\underline{v}_1, \underline{v}_2)$ is a "sampling matrix" having basis vectors as columns and $\underline{n}$ is a vector with integer entries. A given lattice does not have a unique basis, but the value of the determinant $|\det V|$ is unique. It is further noted in [5] that this determinant is physically the reciprocal of the sampling rate, so that it is the 2-D analog of the T value in our 1-D sampling. A reciprocal lattice in Fourier space is defined as the set of points, $U\underline{n}$, where U is the matrix such that $U^TV = 2\pi I$.

A rectangular lattice may be the method most often used in practice, and a rectangular lattice corresponds to a sampling matrix $V = \begin{pmatrix} T_1 & 0 \\ 0 & T_2 \end{pmatrix}$. The initial paper in this area, [7], showed that the most efficient lattice for some signals is not rectangular but is hexagonal, for which an appropriate sampling matrix is $V = \begin{pmatrix} T_1/2 & T_1/2 \\ T_2 & -T_2 \end{pmatrix}$. This matrix generates a lattice in which alternate rows are positioned identically and odd-numbered rows are shifted by one-half sample interval with respect to the even-numbered rows.

The condition for signal reconstruction under this rather general formulation of a sampling lattice is that the functions $F(\underline{w}-U\underline{n})$, for all integer vectors $\underline{n}$, do not possess overlapping regions of support. In terms of the representation

$$(1) \qquad f(\underline{t}) = \frac{1}{2\pi} \int_W F(\underline{w})\exp(i\underline{wt})d\underline{w},$$

this means that one must consider the regions formed by translating W by any integer linear combination of $\underline{u}_1$, $\underline{u}_2$ vectors from the matrix U. These regions must not intersect. The formula for the reconstruction in this case of uniform periodic sampling is $f(\underline{t}) = \sum f(V\underline{n})\,\varphi(\underline{t}-V\underline{n})$, where $\varphi(\underline{t}) = (1/2\pi) \int_W \exp(i\underline{wt})d\underline{w}$. The representation (1) also allows one to conclude that for samples at arbitrary points $\underline{t}_{\underline{n}}$,

$$|f(\underline{t}) - \sum c_{\underline{n}} f(\underline{t}_{\underline{n}})| = \left|\frac{1}{2\pi} \int_W F(\underline{w}) \left\{\exp(i\underline{w}\underline{t}) - \sum c_{\underline{n}} \exp(i\underline{w}\underline{t}_{\underline{n}})\right\} d\underline{w}\right| \tag{2}$$

$$\|\exp(i\underline{w}\underline{t}) - \sum c_{\underline{n}} \exp(i\underline{w}\underline{t}_{\underline{n}})\| \; \|F(\underline{w})/2\pi\|$$

where $\| \; \|$ denotes the norm of $L^2(W)$.

In analogy with the irregular spacing of samples in 1-D, we suppose that sample points in 2-D are allowed to stray from points on the lattice $V\underline{n}$ to points at $\underline{t}_{\underline{n}} = V\lambda_{\underline{n}}$, where $\lambda_{\underline{n}}$ is a vector with components $\lambda_{n_1}, \lambda_{n_2}$. This means that each point in the irregular spacing is associated with a point $V\underline{n}$ on the periodic lattice. Throughout the following, we measure the amount of deviation by considering the relations

$$|\lambda_{n_1} - n_1| \leq L_1 \quad \text{and} \quad |\lambda_{n_2} - n_2| \leq L_2. \tag{3}$$

Inequality (2) shows that, similar to the situation in 1-D, sampling at these points $\{\underline{t}_{\underline{n}}\}$ will allow for signal recovery, if the function $\exp(i\underline{w}\underline{t})$ has an $L^2(W)$-convergent expansion over the (non-harmonic) set of exponentials $\exp(i\underline{w}\underline{t}_{\underline{n}})$. It is the particular form of these exponentials and the 2-D integral over W that allows us to simplify this completeness argument.

It is enough to consider a rectangular lattice and a spectrum that is contained within the rectangle Rect: $|w_1| \leq \pi/T_1, |w_2| \leq \pi/T_2$. We suppose that $V = \begin{pmatrix} a & c \\ b & d \end{pmatrix}$ and that spectrum W has no self-intersections in the Fourier space under action by the matrix $U = \frac{1}{K}\begin{pmatrix} d & -b \\ -c & a \end{pmatrix}$, ($k = \det V \neq 0$). Making the transformation $kw_1 = T_1 dw_1' - T_2 bw_2'$, $kw_2 = -T_1 cw_1' + T_2 aw_2'$ maps the vectors $\underline{u}_1$ and $\underline{u}_2$ onto vectors $\underline{u}_1'$, $\underline{u}_2'$ that generate a rectangular lattice in $\underline{w}'$ space. In the integral of (2), the power on the exponent, wt becomes $\underline{w}'\underline{t}'$ under this transformation to $\underline{w}'$, with $kt_1' = T_1(dt_1 - ct_2)$, $kt_2' = T_2(-bt_1 + at_2)$. But this transformation from $\underline{t}$ to $\underline{t}'$ is the same transformation that maps the general sampling matrix V to the rectangular matrix with basis $\underline{v}_1' = (T_1, 0)$, $\underline{v}_2' = (0, T_2)$. The image of W is certainly contained in the rectangular region Rect in $\underline{w}'$-space defined above. Finally, it is not difficult to show that $\underline{v}_1$ maps to $\underline{v}_1'$ in $\underline{t}'$-space, so that our measure (3) of variance from the uniform lattice is not dependent on the use of a rectangular lattice.

Finally, signal reconstruction is possible whenever

$$|\lambda_{n_1} - n_1| + |\lambda_{n_2} - n_2| \leq \log 2/\pi. \tag{4}$$

This result depends on (2) and the discussion above, in conjunction with results presented in [8,p.110] concerning perturbation of orthonormal bases in a Hilbert space. In our case $e_{\underline{n}} = e_{\underline{n}}(\underline{w}) = K\exp(i\underline{w}V\underline{n})$ is an orthonormal system in $H = L^2(\mathrm{Rect})$, for the rectangular $\overline{V}$ matrix above and $K = \sqrt{T_1 T_2}/(2\pi)$. Define $x_{\underline{n}} = K\exp(i\underline{w}\underline{t}_{\underline{n}})$ with $\underline{t}_{\underline{n}} = V\lambda_{\underline{n}}$. The representation of $\exp(i\underline{w}\underline{t})$ in terms of sums of $x_{\underline{n}}$ as noted above, follows from the theorem in [8] by showing that $x_{\underline{n}}$ is a Bessel system in H. The proof from which condition (4) follows has to be omitted here.

The condition in (4) may be applied to the general sampling lattice to make a nice analogy with the 1-D theory. For a general lattice, (4) implies that signal reconstruction is possible whenever samples lie in a parallelogram symmetric about the regular lattice point, with sides parallel to the generating vectors $\underline{v}_1 + \underline{v}_2$ and $\underline{v}_1 - \underline{v}_2$. The total area of one such parallelogram is $2C^2|\det V|$, with $C \leq \log 2/\pi$. This corresponds well to the 1-D case, where irregular samples that lie in an interval about the regular sample point lie in an interval of length no more than 2CT, with similar bounds on the constant C.

References:

1. F.J. Beutler, Error-free recovery of signals from irregularly spaced samples. SIAM Review 8 (1966) 328-335.

2. D.S. Chen and J.P. Allebach, Analysis of error in reconstruction of two-dimensional signals from irregularly spaced samples. IEEE Trans. Acoust. Speech, Signal Processing ASSP-35 (1987) 173-180.

3. J.J. Clark, M.R. Palmer and P.D. Lawrence, A transformation method for the reconstruction of functions from nonuniformly spaced samples. IEEE Trans. Acoust. Speech, Signal Processing ASSP-33 (1985) 1151-1165.

4. J.R. Higgins, A sampling theorem for irregularly spaced sample points. IEEE Trans. Inform. Theory IT-22 (1976) 621-622.

5. R.M. Mersereau and T.C. Speake, The processing of periodically sampled multidimensional signals. IEEE Trans. Acoust. Speech, Signal Processing ASSP-31 (1983) 188-194.

6. D.H. Mugler and W. Splettstößer, Difference methods for the prediction of band-limited signals. SIAM J. Appl. Math. 46 (1986) 930-941.

7. D.P. Petersen and D. Middleton, Sampling and reconstruction of wave-number-limited functions in n-dimensions. Inform. Control 5 (1962) 279-323.

8. H.S. Shapiro, Topics in Approximation Theory. Lecture Notes in Mathematics 187, Springer-Verlag, Berlin, 1971.

9. W. Splettstößer, Unregelmäßige Abtastung determinierter und zufälliger Signale. Kolloquium, DFG-Schwerpunktprogramm Digitale Signalverarbeitung, (1981) 1-4.

10. R.M. Young, An Introduction to Nonharmonic Fourier Series. Academic Press, New York, 1980.

NONUNIFORM SAMPLING EXPANSIONS OF TWO-DIMENSIONAL BANDLIMITED SIGNALS

P.L. Butzer, G. Hinsen
Lehrstuhl A für Mathematatik
RWTH Aachen
Templergraben 55, D-5100 Aachen

1. Introduction

Let $L^2(\mathbb{R})$ ($L^2(\mathbb{R}^2)$, respectively) be the space of all square integrable functions on $\mathbb{R}$ ($\mathbb{R})^2$, i.e.,

$$\|f\|^2_{L^2(\mathbb{R})} := \frac{1}{\sqrt{2\pi}} \int_{-\infty}^{\infty} |f(x)|^2 dx < \infty \quad \left(\|f\|^2_{L^2(\mathbb{R}^2)} := \frac{1}{2\pi} \int_{\mathbb{R}^2} |f(x,y)|^2 d(x,y) < \infty \right).$$

The $L^2(\mathbb{R})$ ($L^2(\mathbb{R}^2)$) Fourier transform is defined by

$$f^\wedge(u) := \underset{\rho\to\infty}{\mathrm{l.i.m.}} \frac{1}{\sqrt{2\pi}} \int_{-\rho}^{\rho} f(x) e^{-iux} dx \quad \left(f^\wedge(u,v) := \underset{\rho\to\infty}{\mathrm{l.i.m.}} \frac{1}{2\pi} \int_{-\rho}^{\rho}\int_{-\rho}^{\rho} f(x,y) e^{-i(ux+vy)} dxdy\right).$$

For $\sigma \geq 0$ let B^2_σ denote the class of all functions $f \in L^2(\mathbb{R})$ which are bandlimited to $[-\sigma,\sigma]$, i.e., which are representable as $f(x) = (2\pi)^{-1/2} \int_{-\sigma}^{\sigma} f^\wedge(v) e^{ivx} dv$, $x \in \mathbb{R}$ Analogously, if $E \subset \mathbb{R}^2$ is bounded and symmetric with respect to the origin, then $f \in L^2(\mathbb{R}^2)$ is called bandlimited to E, in notation $f \in B^2_E$, if

$$f(x,y) = \frac{1}{2\pi} \int_E f^\wedge(u,v)\, e^{i(ux+vy)} d(u,v) \quad ((x,y) \in \mathbb{R}^2). \tag{1.1}$$

A famous theorem of Paley and Wiener states that each function $f \in B^2_\sigma$ is the restriction to $\mathbb{R}$ of an entire function of exponential type σ, i.e., $|f(z)| \leq C \cdot e^{\sigma|\mathrm{Im} z|}$, $z \in \mathbb{C}$. If a function $f(x,y)$ belongs to B^2_E where E is the rectangle $[-\sigma_1,\sigma_1] \times [-\sigma_2,\sigma_2]$, then $f(x,y_0) \in B^2_{\sigma_1}$ for each fixed $y_0 \in \mathbb{R}$ and $f(x_0,y) \in B^2_{\sigma_2}$ for each fixed $x_0 \in \mathbb{R}$ (cf. [4], Chapter 3). Bandlimited functions can be reconstructed from equidistantly spaced samples. Indeed,

Theorem A. (Whittaker-Shannon-Kotel'nikov) If $f \in B^2_{\pi W}$, *then*

$$f(t) = \sum_{n=-\infty}^{\infty} f\left(\frac{n}{W}\right) \frac{\sin \pi(Wt-n)}{\pi(Wt-n)} \qquad (t \in \mathbb{R}).$$

The assumption that the sampling points are equidistant may be weakend (cf. [1]).

Theorem B. (Higgins ([2], 1976)) Let $(t_n)_{n\in\mathbb{Z}}$ *be a sequence of reals with* $|t_n-n| \leq D < 1/4$ *for all* $n\in\mathbb{Z}$, *and let*

$$G(t) := G((t_n);t) = (t-t_o)\prod_{n=1}^{\infty}(1-t/t_n)(1-t/t_{-n}). \tag{1.2}$$

Then, for each $f\in B^2_\pi$,

$$f(t) = \sum_{n=-\infty}^{\infty} f(t_n)\frac{G(t)}{G'(t_n)(t-t_n)} \quad (t\in\mathbb{R}) . \tag{1.3}$$

Theorem C. (Yen ([5], 1956)) Let $t_n = \tau_{n \bmod K} + [n/K]K$ *with* $0\leq\tau_o<\tau_1<\ldots<\tau_{K-1}<K$. *Then (1.3) holds for all* $f\in B^2_\pi$, *the function* $G(t)$ *of (1.2) equaling* $c_o \cdot \Pi_{j=o}^{K-1}\sin\frac{\pi}{K}(t-\tau_j)$ *for some constant* $c_o\neq 0$.

2. A two-dimensional sampling theorem

Theorem. Let P *be a parallelogram centered at the origin, and let* T_A, *with* $T_A(x,y) := (x,y)A$ *for some* $A\in\mathbb{R}^{2\times 2}$ *(= set of all real 2×2 matrices), be one of the linear transforms that map the square* $[-\pi,\pi]\times[-\pi,\pi]$ *onto* P. *Define* $H_D := \{(t_n)_{n\in\mathbb{Z}};\ |t_n-n|\leq D\}$, *and* $Y_{K_o,\delta} := \{(t_n)_{n\in\mathbb{Z}};\ t_n = \tau_{n \bmod K} + [n/K]K$ *for some* $K\leq K_o$ *with* $0\leq\tau_o<\tau_1<\ldots\tau_{K-1}<K$ *and* $\min\{K+\tau_o-\tau_{K-1};\ (\tau_{i+1}-\tau_i),\ 0\leq i\leq K-2\}\geq\delta\}$. *If there exist* $D<1/8$, $\delta>0$ *and* $K_o<\infty$ *such that each of the sequences* $(u_k)_{k\in\mathbb{Z}}$, $(v_{k,l})_{l\in\mathbb{Z}}$, $k\in\mathbb{Z}$, *belongs to* $H_D\cup Y_{K_o,\delta}$, *then there holds for all* $f\in B^2_p$,

$$f(x,y) = \lim_{n\to\infty}\sum_{|k|,|l|\leq n} f(x_{kl},y_{kl})\cdot\frac{G((u_m);u(x,y))}{G'((u_m);u_k)(u(x,y)-u_k)}\,\frac{G((v_{km});v(x,y))}{G'((v_{km});v_{kl})(v(x,y)-v_{kl})}$$

$$((x,y)\in\mathbb{R}^2), \tag{2.1}$$

with $G((t_m);z)$ *defined by (1.2),* $(u(x,y),v(x,y)) = T_{A^t}(x,y)$ *and* $(x_{kl},y_{kl}) = T_{(A^{-1})^t}(u_k,v_{kl})$. *The double series converges absolutely.*

Sketch of Proof. Firstly, $f(x,y) = f(T_{(A^{-1})^t}(T_{A^t}(x,y))) = (f\circ T_{(A^{-1})^t})(u,v)$. By formula (1.1) and since $(f\circ T_A)^\wedge(w,z) = (1/|\det A|)\, f^\wedge(T_{(A^{-1})^t}(w,z))$, $(f\circ T_{(A^{-1})^t})(u,v)$ is bandlimited to $[-\pi,\pi]\times[-\pi,\pi]$. Thus it is possible to apply Theorem B or C to $f\circ T_{(A^{-1})^t}$ as a function of u, obtaining the series $f(x,y) = \sum_{k=-\infty}^{\infty}(f\circ T_{(A^{-1})^t})(u_k,v)G((u_m);u)(G'((u_m);u_k)(u-u_k))^{-1}$. Now one may use Theorem B and C again to expand each term $(f\circ T_{(A^{-1})^t})(u_k,v)$ separately into a sampling series (this time as a function of v). This yields (2.1) at first as an iterated series. Then one shows that this series is absolutely convergent and may be arbitrarily rearranged. By Hölder's inequality one separates the sampling values from the fundamental functions. The resulting series both have finite sums: the fundamental functions are majorized in view of inequalities due to Levinson ([3], 55-57), and the sampling values are estimated with the help of

Lemma: *Let* $g\in B^2_{[-\sigma_1,\sigma_1]\times[-\sigma_2,\sigma_2]}$, *and assume that, more generally than in the Theorem,* $(u_k)_{k\in\mathbb{Z}}$, $(v_{k,l})_{l\in\mathbb{Z}}$ *are strictly increasing sequences with* $0<\delta_1\leqslant u_{k+1}-u_k\leqslant L_1<\infty$, $k\in\mathbb{Z}$ *and* $0<\delta_2\leqslant v_{k,l+1}-v_{k,l}\leqslant L_2<\infty$ *for* $k,l\in\mathbb{Z}$. *Then*

$$\left(\sum_{k=-\infty}^{\infty}\sum_{l=-\infty}^{\infty}|g(u_k,v_{kl})|^2\right)^{1/2}\leqslant\frac{(1+L_1\sigma_1)(1+L_2\sigma_2)}{\sqrt{\delta_1\delta_2}}\sqrt{2\pi}\,\|g\|_{L^2(\mathbb{R}^2)}.$$

This result is a generalization of Theorem 3.3.2 of Nikol'skii ([4], p. 124).

3. Examples

a) Let $P=[-a,a]\times[-b,b]$, $A=\begin{pmatrix}a/\pi & 0\\ 0 & b/\pi\end{pmatrix}$, and $u_k=k$, $v_{kl}=l$, $k,l\in\mathbb{Z}$, (i.e., equidistant). Then for all $f\in B^2_P$ there holds the usual "rectangular sampling" theorem

$$f(x,y)=\lim_{n\to\infty}\sum_{|k|,|l|\leqslant n}f\left(\frac{\pi}{a}k,\frac{\pi}{b}l\right)\frac{\sin(ax-k\pi)}{ax-k\pi}\cdot\frac{\sin(by-l\pi)}{by-l\pi}.$$

b) Let P be the parallelogram with the corners (5,3),(-1,1), (-5,-3) and (1,-1), $A=\pi^{-1}\begin{pmatrix}3 & 1\\ 2 & 2\end{pmatrix}$, and again $u_k=k$, $v_{kl}=l$ for $k,l\in\mathbb{Z}$. Then, for all $f\in B^2_P$,

$$f(x,y)=\lim_{n\to\infty}\sum_{|k|,|l|\leqslant n}f\left(k\frac{\pi}{2}-l\frac{\pi}{4},-k\frac{\pi}{2}+l\frac{3\pi}{4}\right)\frac{\cos(-x+y+(k-l)\pi)-\cos(5x-3y-(k+l)\pi)}{4(3x+y-k\pi)(2x+2y-l\pi)}.$$

c) Let $P=[-2\pi/\sqrt{3},2\pi/\sqrt{3}]\times[-2\pi/3,2\pi/3]$, $A=\begin{pmatrix}2/\sqrt{3} & 0\\ 0 & 2/3\end{pmatrix}$, $u_k=k$, $k\in\mathbb{Z}$ and $v_{kl}=l$ if k and l are both even or odd or $v_{kl}=l+1/3$, otherwise. Connecting the sampling points of the resulting expansion suitably gives a hexagonal pattern. The fundamental functions may easily be calculated using Theorem C.

References:

[1] Butzer, P.L. - Hinsen, G.: Reconstruction of bounded signals from pseudo-periodic, irregularly spaced samples. (to appear)

[2] Higgins, J.R.: A sampling theorem for irregularly spaced sample points. IEEE Trans. Inform. Theory IT-22 (1976), 621-622.

[3] Levinson, N.: Gap and Density Theorems. New York: AMS, Colloq. Publ. Vol. XXVI, 1940.

[4] Nikol'skii, S.M.: Approximation of Functions of Several Variables and Imbedding Theorems. Berlin-Heidelberg-New York: Springer-Verlag, 1975.

[5] Yen, J.L.: On nonuniform sampling of bandwidth-limited signals. IRE Trans. Circuit Theory CT-3 (1956), 251-257.

FINITE SPHERICAL ANALOGUES OF THE WHITTAKER-SHANNON SAMPLING THEOREM

J. R. Higgins

Department of Science, Cambridgeshire College of Arts and Technology
Cambridge, England.

Reconstruction from samples for signals having finite Fourier expansions have been discussed recently by J.L.Brown [1], in the setting of Hilbert space with reproducing kernel. This setting generalises to other truncated expansions; however the interplay between the existence and nature of explicit sampling formulae on the one hand, and the location of the sample points on the other, can be a delicate matter.

I will report first on the theoretical background - briefly since little more is involved here than the theory of finite dimensional inner product spaces. Then I will go on to report on current investigations into the special case where functions are defined on the two-dimensional domain $\mathbb{S}_2$, the surface of the unit ball in $\mathbb{R}_3$, rather than on a circle as in [1]. Consequently the truncated expansions involve surface spherical harmonics.

The spherical case is interesting because it gives rise to a diversity of phenomena. These include the presence of orthogonal, partially orthogonal, or biorthogonal sampling representations, which appear to demand sometimes more, sometimes less, in the way of regularity, or "fair coverage" of the sphere, by the sample points.

On the practical side, these spherical sampling formulae would reconstruct globally defined phenomena such as sea-level temperature or pressure, or ground contours on our planet or on other heavenly bodies, from samples provided that they could be modelled with sufficient accuracy by truncated expansions in spherical harmonics.

It is a pleasure to thank my colleague Boz Kempski for his generous interest in this project. He has contributed several computer programs which we feel are essential to alleviate the heavy algebra involved in searching for sampling formulae on the sphere.

Finite sampling

The sampling representations to be reported on take one or other of the forms

$$f(t) = \sum_{n\in I} f(s_n)\, k(s_n,t), \qquad (1)$$

or

$$f(t) = \sum_{n\in I} f(s_n)\, \psi_n(t). \qquad (2)$$

Here, $\{s_n\}$ is a set of points belonging to E, the domain of f, k is a reproducing kernel, $\{\psi_n\}$ is biorthogonal to $\{k(s_n,t)\}$, and f has a truncated expansion in a set $\{\varphi_n\}$ which has an apprpriate completeness and orthogonality property; the truncation being to N terms, where N = card I.

Clearly (1) bears a formal similarity to the well known Whittaker-Shannon sampling series. When the points $\{s_n\}$ are such that

$$k(s_n,s_m) = \delta_{nm} \qquad (3)$$

(1) is an orthogonal expansion; $\{s_n\}$ will always be required to be such that $\{k(s_n,t)\}$ possesses a completeness property.

Spherical examples

We now take E to be the unit sphere, let T be a point on the sphere haveing the usual spherical coordinates (φ,ϑ) of polar co-latitude, and longitude, and suppose f to have a suitably truncated expansion in spherical harmonics. For each natural number N there is an unambiguous truncation containing $(N + 1)^2$ terms; one now seeks a set of $(N + 1)^2$ points on the sphere such that the orthogonality criterion (3) is satisfied. When N = 1, the "tetrahedral" points

$$T_1\colon (0,0);\quad T_2\colon(\varphi_o,0);\quad T_3\colon(\varphi_o,\tfrac{2\pi}{3});\quad T_4\colon(\varphi_o,\tfrac{4\pi}{3})$$

where $\varphi_o = \arccos(-\tfrac{1}{3})$, are shown to provide the unique solution. The 4 - point version of (1) is

$$\begin{aligned} f(T) = \frac{1}{4\pi}\{&f(T_1)(3\cos\varphi + 1) + f(T_2)(-\cos\varphi + 2\sqrt{2}\sin\varphi\cos\vartheta + 1)\\ &+ f(T_3)(-\cos\varphi - \sqrt{2}\sin\varphi\cos\vartheta + \sqrt{2}\sqrt{3}\sin\varphi\sin\vartheta + 1)\\ &+ f(T_4)(-\cos\varphi - \sqrt{2}\sin\varphi\cos\vartheta - \sqrt{2}\sqrt{3}\sin\varphi\sin\vartheta + 1)\}. \end{aligned}$$

A computer was then programed to make a similar search in the cases N = 2 through 15. It was found that there are no sets of points satisfying (3) in these cases. At the same time, it is possible in all these cases to choose points which produce partial orthogonalities within $\{k(s_n,t)\}$. These point sets show remarkable patterns of regularity on the spherical surface, and a modified version of (1) is true with linear combinations of samples for the coefficients.

Biorthogonal sampling formulae are being investigated, also with the help of the computer. For example, if the six "octohedral" points $O_1:(0,0)$; $O_2:(\frac{\pi}{2},0)$; $O_3:(\frac{\pi}{2},\frac{\pi}{2})$; $O_4:(\frac{\pi}{2},\pi)$; $O_5:(\frac{\pi}{2},\frac{3\pi}{2})$; $O_6:(\pi,0)$ are prescribed as sample points, there is only one out of a possible ten 6 - term truncations of the expansion in spherical harmonics for which a corresponding biorthogonal sampling representation holds. It is

$$\begin{aligned} f(T) &= f(O_1)\ \tfrac{1}{2} \cos\varphi(\cos\varphi + 1) \\ &+ f(O_2)\ \tfrac{1}{4}(1 + 2\sin\varphi\cos\vartheta - \cos^2\varphi + \sin^2\varphi\cos 2\vartheta) \\ &+ f(O_3)\ \tfrac{1}{4}(1 + 2\sin\varphi\cos\vartheta - \cos^2\varphi - \sin^2\varphi\cos 2\vartheta) \\ &+ f(O_4)\ \tfrac{1}{4}(1 - 2\sin\varphi\cos\vartheta - \cos^2\varphi + \sin^2\varphi\cos 2\vartheta) \\ &+ f(O_5)\ \tfrac{1}{4}(1 - 2\sin\varphi\cos\vartheta - \cos^2\varphi - \sin^2\varphi\cos 2\vartheta) \\ &+ f(O_6)\ \tfrac{1}{2} \cos\varphi(\cos\varphi - 1). \end{aligned}$$

Reference

[1] J.L.Brown, "An RKHS analysis of sampling theorems for harmonic - limited signals", IEEE Trans. ASSP - 33, No. 2, 1985, p. 437-440.

Ein- und zweidimensionale Signale mit quadratischer Phasenfunktion.

von

Helmuth Eggers, Forschungsinstitut Ulm
AEG Aktiengesellschaft, Sedanstraße 10, 7900 Ulm

1.) Einführung:

Ein- und zweidimensionale, komplexe und diskrete Signale s(m,n) mit guten Autokorrelationseigenschaften finden in der Radar- und Kommunikationstechnik, in der Medizintechnik und in vielen weiteren Gebieten Anwendung [EGG86]. Deshalb hat sich die Synthese solcher Signale als ein breites Arbeitsgebiet herausgebildet. Es gibt allerdings kein systematisches Verfahren, daß die allgemeine Synthese dieser Signale gestattet. In den bekannten Syntheseverfahren beschränkt man sich deshalb auf die Betrachtung eines bestimmten Signaltyps.

Die vorliegende Arbeit folgt diesem Ansatz; sie betrachtet ausschliesslich ein- und zweidimensionale Signale, die durch die folgende Gleichung (1) definiert sind:

$$s(m,n) = \begin{cases} e^{jp(m,n)} & m=0(1)M-1,\ n=0(1)N-1 \\ 0 & \text{sonst} \end{cases} \qquad (1)$$

p(m,n) Phasenfunktion mit quadratischem Verlauf

Es muß also nur noch untersucht werden, wie die quadratische Phasenfunktion p(m,n) zu gestalten ist, um Signale mit möglichst guten Autokorrelationseigenschaften zu erhalten.

Jedem Signal nach (1) läßt sich eine Autokorrelationsfunktion (AKF) zuordnen, die in der üblichen Weise definiert ist (z.B. [EGG86]). Die AKF hat ihren Hauptwert fuer die Verschiebung k=l=0, dieser ist gleichzeitig der Maximalwert und entspricht der Signalenergie. Alle anderen AKF-Werte werden als Nebenwerte bezeichnet. Sie stellen die (Selbst-)Störung dar, die vom Signal selbst hervorgerufen wird und sollten deshalb möglichst klein sein. Als Maß für die Güte eines Signals bezüglich der Nebenwerte seiner AKF hat sich der Merit-Faktor (MF) durchgesetzt [MOE86,GOL71,EGG86]:

$$MF = \frac{\text{Energie im Hauptwert der AKF}}{\text{Energie in allen Nebenwerten}} \qquad (2)$$

Gute Autokorrelationseigenschaften werden demnach durch hohe Werte des Merit-Faktors angezeigt.

2.) Eindimensionale Signale:

In diesem Fall lautet die allgemeine quadratische Phasenfunktion:

$$p(m) = 2\pi(am^2 + bm + c) \qquad (3)$$

Es soll nun untersucht werden, ob es einen Satz von Phasenfaktoren a,b,c gibt, der auf besonders hohe Merit-Faktoren führt. Dazu ist ein handlicher Ausdruck für die Nebenwerte der AKF erforderlich, der mit vertretbarem Aufwand das Auffinden optimaler Phasenfaktoren erlaubt. Die Darstellung des Signals (1) wird in die Formel für die AKF eingesetzt, ebenso die allgemeine quadratische Phasenfunktion (3). Nach dem Ausklammern konstanter Terme verbleibt die Summe über eine endliche geometrische Reihe, die nach der bekannten Formel berechnet werden kann. Daraus folgt das Betragsquadrat der AKF:

$$|akf_{ss}(k)|^2 = \frac{\sin^2(2\pi ak(N-|k|))}{\sin^2(2\pi ak)} \qquad (4)$$

Die Energie in den Nebenwerten der AKF ergibt sich dann als Summation für alle Verschiebungen k (außer k=0) über die Betragsquadrate (4). Die Phasenfaktoren b und c spielen keine Rolle für den Merit-Faktor des Signals, sie können deshalb so gewählt werden, daß sich bei einer Quantisierung der Phasenfunktion möglichst geringe Abweichungen ergeben. Die günstigen Korrelationseigenschaften übertragen sich dann bis zu einem gewissen Grad auch auf das Signal mit der quantisierten Phasenfunktion.

In der folgenden Tabelle 1 werden die Phasen- und Merit-Faktoren verschiedene Signale aufgelistet.

Tabelle 1: Phasen- und Meritfaktoren für eindimensionale Signale

N	a_{max}	MF_{1D}	N	a_{max}	MF_{1D}
10	0,0539194	4,74904	100	0,0050248	15,9840
20	0,0256449	7,68315	125	0,0040015	17,7118
30	0,0168843	8,46772	150	0,0033414	19,1760
40	0,0126573	10,3815	175	0,0028722	20,8988
50	0,0100765	11,0006	200	0,0025062	22,4081
60	0,0084027	12,5274	225	0,0022227	23,6730
70	0,0071812	13,0517	250	0,0020029	24,7882
80	0,0062888	14,3594	275	0,0018243	26,1434
90	0,0055785	14,8218	300	0,0016694	27,3647

Man kann Tabelle 1 entnehmen, daß der optimale Phasenfaktor für eindimensionale Signale mit geringen Abweichungen $1/(2\cdot M)$ ist. Stellt man sich vor, daß das Signal s(m) durch Abtastung mit der Abtastperiode 1 aus einem Signal s(t) der Dauer N mit linearer Frequenzmodulation [LEK82] entstanden ist, so ergibt sich eine anschauliche Erklärung für diesen Wert des Phasenfaktors a: Es ist derjenige Wert, bei dem die Abtastung gerade mit der 'Nyquist-Rate' erfolgt. Unterhalb dieses Wertes ergibt sich eine schlechte AKF durch den mangelnden Frequenzhub der linearen Frequenzmodulation; oberhalb verschlechtert sich die AKF durch stark ansteigendes Aliasing.

Vergleicht man die Werte aus Tabelle 1 mit den in der Literatur [SCHO71, GOL72, MOE86, LEK82] genannten Werten, so fällt auf, daß für die binären Signale in der Literatur der Wert $a=1/(4 \cdot M)$ angegeben wird, was durch eine eigene Untersuchung bestätigt wurde. Eine anschauliche Erklärung dafür ergibt sich, wenn das Binärsignal als Produkt zweier Signale mit (im Falle von Binärsignalen) nahezu gleicher Bandbreite verstanden wird. Das eine dieser Signale enthält als Phasenmodulation die analoge Phasenfunktion mit $a=1/(4 \cdot M)$, das andere den Quantisierungsfehler auf ganzzahlige Werte von π. Da sich die Spektren dieser Signale im Frequenzbereich falten, ergibt sich nahezu die doppelte Bandbreite für das Signal mit quantisierter Phasenfunktion und demzufolge auch nur der halbe Wert für den optimalen Phasenfaktor.

3.) Zweidimensionale Signale:

Für zweidimensionale Signale gelten im Prinzip die gleichen Überlegungen wie für eindimensionale Signale. Die Phasenfunktion lautet hier:

$$p(m,n) = 2\pi[\ am^2 + bmn + cn^2 + dm + en + f] \qquad (5)$$

Mit der gleichen Vorgehensweise wie im Eindimensionalen ergibt sich ein der Gl.(4) ähnlicher Ausdruck für das Betragsquadrat AKF. Auch hier können die Phasenfaktoren d,e,f wieder zur Verringerung des Fehlers bei Quantiserungsvorgängen eingesetzt werden. Durch numerische Optimierung ergeben sich die in Tabelle 2 aufgeführten Phasenfaktoren.

Tabelle 2: Phasen- und Meritfaktoren für zweidimensionale Signale

M	N	a_{max}	b_{max}	c_{max}	MF_{2D}
5	5	0,0000000	0,1946576	0,0000000	6,5345461
5	10	0,2151584	0,1938142	0,0166636	3,1686441
5	15	0,0149151	0,1964608	0,0091717	3,6827131
10	10	0,0000000	0,0981257	0,0000000	10,1399800
10	15	0,0459056	0,0985564	0,0147047	3,8294672
10	20	0,0391816	0,0920977	0,0202241	3,5597105

Tabelle 2: (Fortsetzung)

M	N	a_{max}	b_{max}	c_{max}	MF_{2D}
15	15	0,0000000	0,0657451	0,0000000	13,6169015
20	20	0,0000000	0,0494520	0,0000000	16,8548634

Für quadratische Signale ergibt sich a=c=0 und b=1/N. Die so konstruierten Signale sind Transformationsmatrizen einer diskreten Fourier-Transformation der Länge N. Signale dieser Art sind schon lange für ihre guten Korrelationseigenschaften bekannt [FRA63,LEK81] und werden in der Radartechnik z.B. unter der Bezeichnung Frank-Codes verwendet.

Bei Signalen mit ungleichen Kantenlängen ergeben sich auch für die Phasenfaktoren a und c Werte ungleich Null. Auffällig sind jedoch die wesentlich geringern Merit-Faktoren.

4.) Zusammenfassung:

Es werden ein- und zweidimensionale komplexe und diskrete Signale mit quadratischer Phasenfunktion untersucht. Durch numerische Optimierung konnten Phasenfunktionen gefunden werden, die hohe Merit-Faktoren ergeben. Es werden anschauliche Deutungen für die beobachteten Werte der Phasenfaktoren gegeben.

5.) Literatur:

[MOE86] Moeser,M.: Ein Konstruktionsverfahren für binäre Folgen mit kleinen Seitenkeulen der antizyklischen Autokorrelierten. ntz Archiv, Bd.8, Heft 7, 1986

[SCH70] Schroeder, M. R.: Synthesis of Low-Peak-Factor Signals and Binary Sequences With Low Autocorrelation IEEE Trans. on Information Theory, Vol.IT-16, 1970, pp.85-89

[GOL72] GOLAY,M.J.E.: A Class of Finite Binary Sequences With Alternate Autocorrelation Values Equal to Zero IEEE Tr. on Inform.Theory, Vol.IT-18,No.3,May 1972,pp.449-450

[FRA63] Frank,R.L.: Polyphase Codes with Good Nonperiodic Correlation Properties IEEE Tr. on Inform. Theory, January 1963, pp.43-45

[LEK81] Lewis,B.L.;Kretschmer,F.F.Jr.: A New Class of Polyphase Pulse Compression Codes and Techniques IEEE Tr. on Aerospace and Electronic Systems, Vol.AES-17, May 1981, pp.364-371

[LEK82] Lewis,B.L.; Kretschmer,F.F.Jr.: Linear Frequency Modulation Derived Polyphase Pulse Compression Codes IEEE Tr. on Aerospace and Electronic Systems, Vol.AES-18, September 1982, pp.637-641

[EGG86] Eggers, H.: Synthese zweidimensionaler Folgen mit guten Autokorrelationseigenschaften Dissertation, RWTH Aachen, 1986

ZWEIDIMENSIONALE HÖHENSCHICHTENFILTERUNG

Erwin Paulus
Institut für Nachrichtentechnik, TU Braunschweig
Schleinitzstr. 23, D-3300 Braunschweig, Fed. Rep. Germany

Einführung

Ein zweidimensionales Signal ist im allgemeinen eine eindeutige Funktion von zwei unabhängigien Variablen und läßt sich als Fläche im dreidimensionalen Raum auffassen. Aus Gründen der Anschaulichkeit sollen die beiden unabhängigien Variablen Ortskoordinaten bedeuten. Bei diskreten Ortskoordinaten sei der Einfachheit halber angenommen, daß jedes Koordinatenpaar den Mittelpunkt eines quadratischen Flächenelementes festlegt und daß der zugehörige Funktionswert nicht nur im Mittelpunkt sondern überall innerhalb des Flächenelementes gilt. Bei quantisierten (diskreten) Funktionswerten ist die Funktion stückweise konstant und die genannte Fläche bildet gleichsam eine aus übereinanderliegenden Schichten aufgebaute "Terassenlandschaft", wobei es für jede "Höhe" - d.h. für jedes Quantisierungsniveau - eine Schicht gibt. Jede Höhenschicht läßt sich als zweidimensionales binäres Signal (Binärbild) darstellen, wobei ein "schwarzes" Flächenelement anzeigt, daß der zugehörige Funktionswert über oder bei dem betreffenden Quantisierungsniveau liegt, während bei einem "weißen" Flächenelement der Funktionswert unter dem Quantisierungsniveau bleibt. Der Stapel der übereinanderliegenden Höhenschichten, läßt sich als ein dreidimensionales binäres Signal mit einem zusammenhängenden schwarzen und einen ebenfalls zusammenhängenden weißen Volumenbereich auffassen. Die Trennfläche zwischen den beiden Bereichen repräsentiert das quantisierte zweidimensionale Signal.

Allgemeiner gesprochen läßt sich ein quantisiertes M-dimensionales Signal immer als (M+1)-dimensionales binäres Signal darstellen. In [1] sind Filter für M-dimensionale quantisierte Signale beschrieben, die als verschiebungsinvariante nichtrekursive Filter für (M+1)-dimensionale binäre Signale arbeiten. In [2] ist gezeigt, daß sich insbesondere Medianfilter für quantisierte M-dimensionale Signale als nichtrekursive verschiebungsinvariante Filter für binäre (M+1)-dimensionale Signale realisieren lassen: Durch binäre Medianfilterung jeder Höhenschicht ergeben sich die Höhenschichten des mediangefilterten Signals.

Im folgenden wird der nichtrekursiven Medianfilterung noch einige Aufmerksamkeit gewidmet, vor allem aber wird ein neuer spezieller Algorithmus zur Höhenschichtenfilterung vorgestellt. Alle Überlegungen stützen sich weitgehend auf das Konzept der "Ursignale" (root signals), das aus der Literatur als wirksames Mittel zur deterministischen Beschreibung von Medianfiltern bekannt ist [3], [4]. Unter den Ursignalen eines Filters sollen diejenigen Signale verstanden werden, die durch das Filter nicht verändert werden können. Die Kenntnis dieser invarianten Signale oder ihrer wesentlichen Merkmale ist insbesondere bei solchen Filtern von Interesse, die bei wiederholter Anwendung auf ein zunächst beliebiges Signal schließlich ein Ursignal liefern müssen. Aus der Literatur [3], [4] ist bekannt, daß das bei Medianfiltern mit Sicherheit zutrifft, wenn das Signal von endlicher Ausdehnung ist, d.h. außerhalb eines bestimmten Bereichs verschwindet oder konstant bleibt. Für alle hier betrachteten Verfahren der Filterung gilt, daß bei einem Ursignal jede einzelne Höhenschicht ein binäres Ursignal ist. Daher sind jeweils die binären Ursignale von besonderem Interesse.

Medianfilterung

Das Ausgangssignal g(m,n) eines zweidimensionalen Medianfilters ist bei jedem Koordinatenpaar m,n gleich dem Medianwert unter denjenigen Werten des Eingangssignals, die innerhalb eines "Fensters" mit dem Zentrum m,n liegen. Wenn das Eingangssignal f(m,n) nur innerhalb eines endlichen Bildfeldes gegeben ist, ist g(m,n) nur innerhalb eines Ausschnittes davon definiert. Die Verkleinerung des Bildfeldes ergibt sich daraus, daß g(m,n) nur dann bestimmt werden kann, wenn das Fenster vollständig innerhalb des ursprünglichen Bildfeldes liegt. Die Vielfalt der Ursignale hängt sehr stark von der Form und der Größe des Fensters ab. Bei quadratischem Fenster gibt es (außer bei der Größe 3x3) in einem binären Ursignal keine zusammenhängenden schwarzen oder weißen Bereiche mit geschlossenen Konturen, d.h. jede Trennlinie zwischen schwarz und weiß läßt sich nur über den Bildrand schließen. Außerdem können solche Trennlinien keinen beliebigen, sondern nur einen im wesentlichen geraden Verlauf haben. Bei anderen Fensterformen kann es innerhalb des Bildfeldes in sich geschlossene Konturen geben, Form und Größe der möglichen umschlossenen Gebiete hängen dabei stark von der Form der Größe des Fensters ab und ihre Vielfalt ist immer sehr stark eingeschränkt.

Die separable Medianfilterung besteht aus zwei nacheinander ausgeführten Medianfilterungen, wobei in einem Fall das Fenster nur in m-Richtung (Zeilenrichtung), im anderen nur in n-Richtung (Spaltenrichtung) ausgedehnt ist. Das Ergebnis der Filterung ist im allgemeinen davon abhängig, in welcher Reihenfolge die beiden Schritte ausgeführt werden. Die Ursignale sind allerdings für beide Varianten dieselben; sie sind dadurch gekennzeichnet, daß in jeder Höhenschicht jede Zeile und jede Spalte ein binäres Ursignal eines eindimensionalen Medianfilters ist. In der einzelnen

Höhenschicht kann es innerhalb des Bildfeldes geschlossene Konturen geben, wobei die Abmessungen der eingeschlossenen Fläche sowohl in Zeilen- als auch in Spaltenrichtung über einer bestimmten Mindestgröße liegen müssen. Die Vielfalt möglicher Konturverläufe ist wiederum sehr stark eingeschränkt, geschlossene Konturen sind umso eher konvex, je kleiner die umschlossene Fläche ist.

Nivellierungsalgorithmus

Eine weitaus größere Vielfalt an Ursignalen als bei Medianfiltern ergibt sich, wenn der folgende Algorithmus zur Filterung verwendet wird:

Für jede "Größe" $1 \leq S \leq S_{max}$

1) Schwärzen aller extremen weißen Objekte der Größe S
2) Weißen aller etremen schwarzen Objekte der Größe S
3) Wiederholung ab 1) so lange es noch etreme Objekte der Größe S gibt

Unter einem "Objekt" ist dabei jede in einer beliebigen Höhenschicht liegende zusammenhängende weiße oder schwarze Fläche zu verstehen, die von ihrer schwarzen bzw. weißen Umgebung durch eine geschlossene Kontur getrennt ist. Ein schwarzes Objekt ist dann ein extemes Objekt, wenn es zu jedem seiner Bildelemente in der darüberliegenden Höhenschicht eine weißes Bildelement gibt. Dementsprechend muß es zu jedem Bildelement eines extremen weißen Objektes in der darunterliegenden Höhenschicht ein schwarzes Bildelement geben. Für die Größe S sind verschiedene Festlegungen denkbar. Naheliegend ist es, unter der Größe eines Objekts z.B. seine Fläche oder seinen Umfang zu verstehen. Im eindimensionalen Fall (M=1) ist S notwendigerweise die Länge des Objekts. Eine Verkleinerung des Bildfeldes ergibt sich aus der Forderung, daß an jeder Stelle ein Objekt der Größe S_{max} vollständig innerhalb des ursprünglichen Bildfeldes Platz haben soll.

Der Algorithmus ergibt immer bereits nach einmaliger Anwendung ein Ursignal. Der Schritt 1) bedeutet ein Abtragen der Maxima und Schritt 2) ein Auffüllen der Minima von f(m,n). Als Folge von 3) ergeben sich schließlich Plateaus, die größer als S_{max} sind. Im eindimensionalen Fall (M=1) sind die Ursignale genau dieselben wie bei einem Medianfilter mit einem Fenster der Länge $2S_{max}+1$. Im zweidimensionalen Fall gibt es in jeder Höhenschicht eines Ursignals nur Objekte, die größer als S_{max} sind. Im übrigen aber ist die Form der Objekte beliebig - im Gegensatz zu den Medianfiltern, bei denen es immer Einschränkungen für die Konturverläufe gibt (siehe oben).

Signalanalyse

Sei f(m,n) ein gegebenes Signal von dem bekannt ist, daß es sich aus zwei Anteilen zusammensetzt, von denen der eine sicher ein Ursignal eines gegebenen Filters und der andere sicher kein solches Ursignal ist:

$$f(m,n) = f_0(m,n) + e(m,n)$$

Die Aufgabe soll darin bestehen, das Ursignal f (m,n) zu finden. Durch - nötigenfalls mehrmalige - Anwendung des gegebenen Filters geht aus f(m,n) ein Ursignal $g_0(m,n)$ hervor. Im Idealfall ist

$$g_0(m,n) \equiv f_0(m,n)$$

Bei den hier betrachteten Filtern zeigt die Erfahrung, daß der Idealfall nur unter sehr speziellen Voraussetzungen über e(m,n) erreicht werden kann. Eine gute Übereinstimmung zwischen $g_0(m,n)$ und $f_0(m,n)$ ergibt sich aber allem Anschein nach schon unter der weitaus weniger einschränkenden Voraussetzung, daß die Filterung von e(m,n) allein zu einem Ursignal $g_{e0}(m,n)$ = const führt.

Bildsignale sind häufig aus zwei Anteilen zusammengesetzt, von denen der eine stückweise glatt ist. Die Trennlinien zwischen den einzelnen glatten Stücken können dabei auf Objektkonturen, Texturgrenzen, Schattengrenzen und dgl. zurückgehen. Solche stückweise glatten Anteile lassen sich in vielen Fällen hinreichend genau durch Ursignale von Höhenschichtenfiltern annähern. Dabei kommt es vor allem darauf an, die Trennlinien möglichst genau anzunähern, während die glatten Verläufe zwischen der Trennlinien weitaus weniger genau zu sein brauchen. Daraus folgt, das echte zweidimensionale Medianfilter mit quadratischem Fenster am wenigsten und der Nivellierungsalgorithmus am besten für die Analyse von Bildsignalen geeignet ist.

Literatur

[1] PRESTON, JR., K.: Ξ-Filters. IEEE Trans. ASSP-31 (1983), 861-876.

[2] FITCH, J.P., E.J. COYLE, and N.C. Gallagher, JR.: Median Filtering by Threshold Decomposition. IEEE Trans. ASSP-32 (1984), 1183-1188.

[3] NODES, Th.A., and N.C. GALLAGHER: Median Filter: Some Modifications and Their Properties. IEEE Trans. ASSP-30 (1982), 739-746.

[4] HUANG, T.S.: Two-Dimensional Digital Signal Processing II. New York: Springer, 1981.

OPTIMIZATION OF THREE-DIMENSIONAL NONRECURSIVE DIGITAL WALSH DOMAIN FILTERS

Karol Rumatowski
Institute of Electronics and Telecommunication,
Technical University of Poznan, ul. Piotrowo 3a,
60-965 Poznan, Poland

1. Introduction

As known, the three-dimensional (3-D) nonrecursive digital filtering is a computational process by which a 3-D digital input signal (sequence of samples) is transformed into a 3-D digital output signal. The output samples of such a filter are determined as weighted sums of past and present input samples. This type filtering process can be implemented by other transformations than the well-known Fourier transformation. Among many transformations, special role plays the 3-D Walsh transformation, which is based on the 3-D Walsh functions [1]. The 3-D Walsh functions are two-valued functions, only taking the two values +1 or -1. Therefore, the 3-D Walsh transform can be computed much faster than the 3-D Fourier transform.

In the paper the principle of the 3-D nonrecursive digital filtering by the use of the 3-D discrete Walsh transform (DWT) is described. Particularly the optimum Walsh filter problem with 3-D discrete data is considered. The optimal 3-D Walsh domain filter based upon some modified minimum-mean-square-error criterion is determined.

2. Three-dimensional nonrecursive digital Walsh filtering process

The block-diagram of the 3-D digital Walsh filtering process is shown in Fig.1.

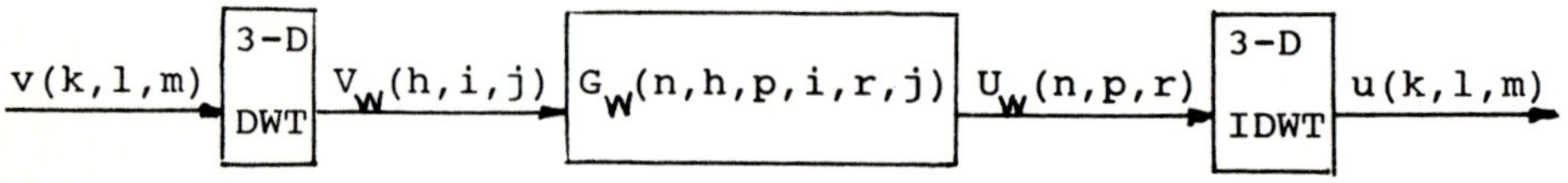

$$v(k,l,m) \rightarrow \boxed{\text{3-D DWT}} \xrightarrow{V_W(h,i,j)} \boxed{G_W(n,h,p,i,r,j)} \xrightarrow{U_W(n,p,r)} \boxed{\text{3-D IDWT}} \rightarrow u(k,l,m)$$

Fig.1 The general block-diagram of the 3-D digital filtering in the Walsh domain

The 3-D Walsh filtering operation is given by the following formula

$$U_W(n,p,r)=\sum_{h=0}^{N-1}\sum_{i=0}^{N-1}\sum_{j=0}^{N-1} G_W(n,h,p,i,r,j)V_W(h,i,j) \quad (1)$$

where

$G_W(n,h,p,i,r,j)$ is the filter weighting function in the Walsh domain,

$$V\ (h,i,j)=\frac{1}{N^3}\sum_{k=0}^{N-1}\sum_{l=0}^{N-1}\sum_{m=0}^{N-1} v(k,l,m)\,wal(h,m)\,wal(i,l)\,wal(j,k) \quad (2)$$

denotes the 3-D DWT of the N input data v(k,l,m), and $U_W(n,p,r)$ is the 3-D Walsh spectrum of the filtered signal u(k,l,m), which is determined by the 3-D inverse discrete Walsh transform (IDWT), i.e.,

$$u(k,l,m)=\sum_{n=0}^{N-1}\sum_{p=0}^{N-1}\sum_{r=0}^{N-1} U_W(n,p,r)\,wal(m,n)\,wal(l,p)\,wal(k,r) \quad (3)$$

It is possible to significantly reduce the computational efforts assuming that the component $U_W(h,i,j)$ of the Walsh output spectrum only depends on the $V_W(h,i,j)$-component of the Walsh input spectrum. Then the filtering effect is obtained by multiplying the 3-D DWT of the 3-D input data with attenuation coefficients $G_W(h,i,j)$ which represent the sequency response of the filter. It denotes that the filtering operation is given by the following formula

$$U_W(h,i,j)=G_W(h,i,j)V_W(h,i,j) \quad (4)$$

The block-diagram of this filtering process type is shown in Fig.2.

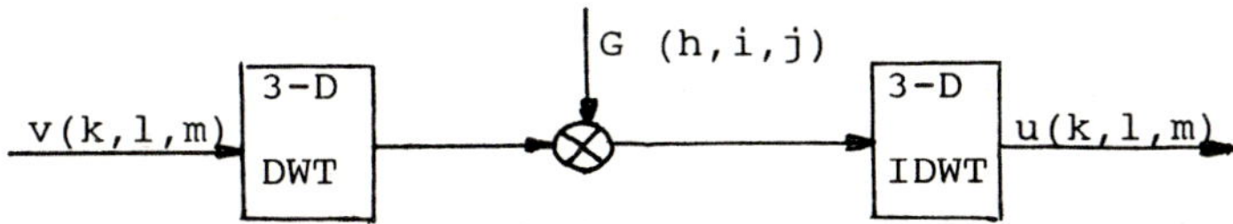

Fig.2. 3-D Walsh filtering process

In such case some errors of the filtering are to be expected. However, if the properties of the input signal are known, the above mentioned errors can be minimized.

3. The three-dimensional optimal nonrecursive digital Walsh filter

Let the discrete input signal $v(k,l,m)=u(k,l,m)+z(k,l,m)$ $(k,l,m=0,1,\dots,N-1;\ N=2^q,\ q\text{-integer})$ be a discrete stationary random process. The task of the to extract the signal $u(k,l,m)$. Two possible cases are considered.

Case (I). The signal $u(k,l,m)$ and noise $z(k,l,m)$ appear on the input of the filter.

Case (II). The signal $u(k,l,m)$ does not appear on the input of the filter.

Then the error of the filtering process can be defined as

$$E[e\ ]=E\{\frac{1}{N^3}\{\sum_{k=0}^{N-1}\sum_{l=0}^{N-1}\sum_{m=0}^{N-1}[u_{I}(k,l,m)-u(k,l,m)]^2+$$

$$+\sum_{k=0}^{N-1}\sum_{l=0}^{N-1}\sum_{m=0}^{N-1}u^2_{II}(k,l,m)\}) \tag{5}$$

where the two components correspond to the above mentioned cases and E denotes the expectation operator.

According to Parsevals theorem, the error (5) can also be expressed by the following equation:

$$E[e^2]=E\{\sum_{h=0}^{N-1}\sum_{i=0}^{N-1}\sum_{j=0}^{N-1}[U_{WI}(h,i,j)-U_{W}(h,i,j)]^2+$$

$$+\sum_{h=0}^{N-1}\sum_{i=0}^{N-1}\sum_{j=0}^{N-1}U^2_{WII}(h,i,j)\} \tag{6}$$

Taking into account that

$$U_{WI}(h,i,j)=G_W(h,i,j)V_W(h,i,j) \tag{7}$$

$$U_{WII}(h,i,j)=G_W(h,i,j)Z_W(h,i,j) \tag{8}$$

the error can be minimized with respect to the attenuation coefficients

$G_W(h,i,j)$.

Finally, we obtain

$$G_W(h,i,j)=\frac{E[U_W(h,i,j)V_W(h,i,j)]}{E[V_W^2(h,i,j)]+E[Z_W^2(h,i,j)]} \tag{9}$$

where

$E[U_W(h,i,j)V_W(h,i,j)]$ represents the Walsh crosspower spectral density function,

$E[V_W^2(h,i,j)]$ and $E[Z_W^2(h,i,j)]$ represent the Walsh power spectral density functions.

It can be seen that if the Walsh power densities spectra $E[U_W(h,i,j)]$ and $E[Z_W(h,i,j)]$ do not overlap then $G_W(h,i,j)=0$ for h,i,j such that $E[Z_W(h,i,j)]=0$ and

$$G\ (h,i,j)=\frac{E[U_W(h,i,j)]}{E[V_W(h,i,J)]}$$

for such h,i,j that $E[U_W(h,i,j)]=0$.

4. Conclusions

The 3-D DWT enable us to realize simplified 3-D digital filter. The optimum filter can be determined on the basis of some modified mean-square-error criterion. The quality of 3-D Walsh domain filters depends on the sequency power density spectrum of the 3-D input signal as well as on the sequency response of the 3-D filter.

References

[1] H.F.Harmuth, Transmission of Information by Orthogonal Functions, Springer, Berlin/Heidelberg/New York, 1972.

[2] M.Maqusi, Applied Walsh Analysis, Heyden and Son, London, 1981.

[3] K.Rumatowski, "Walsh transform applied to digital filtering", Signal Processing 10, Elsevier Science Publishers B.V. (North Holland), 1986, pp.253-263.

DYNAMIC SCENE ANALYSIS AND MOTION ESTIMATION

T.S. Huang
Coordinated Science Laboratory
University of Illinois
1101 W. Springfield Ave.
Urbana, IL 61801, USA

ABSTRACT

We give a brief overview of dynamic scene analysis and then concentrate on one of its major difficulties: motion/structure estimation from image sequences. It is seen that though typical dynamic scenes contain multiple moving objects which may not be rigid, even the simplest problem of motion estimation, that of two-view motion analysis of a single rigid object has not been satisfactorily resolved. Much remains to be done.

I. INTRODUCTION

A major problem in computer vision is the analysis and understanding of dynamic scenes [1]. The goal is to build an autonomous system which will look at a dynamic scene and come up with a description of the unfolding events. Such a system can conveniently be thought of as comprising two modules. The first module extracts from the observed raw data (e.g., an image sequence taken by a TV camera), low and intermediate level features such as 3-D shapes and motion parameters of objects in the scene. Then the second module arrives at a symbolic description of the dynamic scene by high-level reasoning based on the low/intermediate level features as well as other a priori information about the scene.

In Section II, we give some examples of dynamic scene analysis systems and point out that one of the major obstacles in constructing such a system lies in the first module. Specifically, the problem of estimating 3-D motion and structure remains a challenge to researchers in computer vision.

In section III, we give an overview of one of the approaches to motion estimation - using feature correspondences, and summarize some of the difficulties.

II. DYNAMIC SCENE ANALYSIS

II.A. Examples of Complete 2-D Systems

One can find complete dynamic scene-analysis systems in the literature in the biomedical area. Two excellent examples are Ref. 2, which describes a rule-based system for characterizing blood cell motion, and Ref. 3, which describes a system for analyzing the motion of left-ventricle walls. In both cases the "scenes" are basically 2-D in nature, and therefore the task of the low/intermediate-level module is greatly simplified.

II.B. Examples of Partial 3-D Systems

For truly 3-D scenes a complete dynamic scene-analysis system is hard to construct. The main problem is that the low/intermediate-level features the high-level module needs for its reasoning may be very difficult, if not impossible, to extract from the raw data. In fact, the low/intermediate-level module will probably need help from high-level reasoning to improve its performance. Some impressive examples of high-level modules are Refs. 4, 5, and 6. Reference 5 describes a system that observes traffic scenes and produces natural-language descriptions of them. In particular, the system will recognize and verbalize interesting occurrences (events) in the scene -- e.g., one car is overtaking another. Reference 6 describes an expert system for event identification. The applications considered are simple assembly-line tasks. However, in both systems the low/intermediate-level features needed by the high-level modules are furnished at least in part by human operators.

II.C. Event Description and Identification

To convey the flavor of typical high-level modules, the results in Ref. 6 will now be briefly described.

This work presents a formal mechanism for specifying the steady and time-varying characteristics of a large class of physical events. The rules which embody these specifications are then used to identify occurrences of particular events in a changing scene, starting from low-level data such as the positions and orientations of objects as they vary over time. In this manner, a high-level description of changes occurring in a scene is formed.

This mechanism has been implemented as a package which runs in the Interlisp environment on a VAX 11/780 computer. A knowledge base has been constructed for the application of this system in a simplified assembly-line context, including robot arms, bolts, nuts and other simple objects, and events such as transportation of objects in different manners, stacking objects, fastening objects together with bolts, and so forth.

The process of recognizing physical events occurs at such a low level in humans that much of its inherent complexity is hidden from view. When this process is duplicated computationally, this complexity becomes apparent. Some of the finer points which must be dealt with include the following:

(a) Time-related parts of an event. Many events contain parts or sub-events which may or may not be required to fit together in a particular manner in time. All instances of these events must nevertheless be recognized.

(b) Parts of wholes. If a hand is seen to be grasping an object, for example, it is necessary to have some idea of whether or not that object is part of a larger object. If so, then the event should be taken in a more general sense of the hand grasping the larger object, and so forth.

(c) Avoiding redundancy. If a simple event is found to occur (for example, a hand carrying an object) then if later a higher level event is found which contains the first event as a component (for example, stacking that object on another object) the first event should be excluded from a final description of changes in the scene, as a statement that the latter event has occurred is sufficient.

Figure 1 illustrates the sequence of actions for a particular example involving a robot arm and two blocks. The robot hand first makes a false grasping motion at BLOCK1 followed by an actual grasping of the object, a lifting of the object and a setting down of the object on top of BLOCK2.

Input to the implemented system, however, is provided at a level which is considerably less refined than the verbal description above. The input file contains only data which falls into one of the following three categories.

(1) Static attributes and relationships of objects. This includes dimensions of objects and relationships of parts of an object to the whole (e.g. the parts of the robot hand). All composite objects are represented using constructive geometry based on simple objects such as spheres, cylinders and blocks.

(2) Time-varying positions and orientations of objects. These are input as graphs of values over time. For positions, each object is traced along the "world" coordinate axes "X," "Y" and "Z." For orientations, each object is associated with zero to three "object axes." Spheres have no object axes, cylinders have one. Components relating vectors along these axes to the world axes "X," "Y" and "Z," comprise the input for orientations of objects.

(3) Simple spatial relationships of the objects. For this example these include notions of "betweenness" (one object directly between two others), touching (or more precisely, very close proximity), and support (when one object touches a part of another which faces downwards).

Given input of the form described above, the resulting description of changes occurring in the scene as produced by the implemented package is as follows:

```
(HAND1 IS PICKING BLOCK1 UP FROM TIME 0 TO TIME 6)
(HAND1 IS GRASPING BLOCK1 FROM TIME 2 TO TIME 4)
(HAND1 IS HOLDING BLOCK1 FROM TIME 4 TO TIME 6)
(HAND1 IS STACKING BLOCK1 ON BLOCK2 FROM TIME 6 TO TIME 9)
(HAND1 IS HOLDING BLOCK1 FROM TIME 9 TO TIME 10)
```

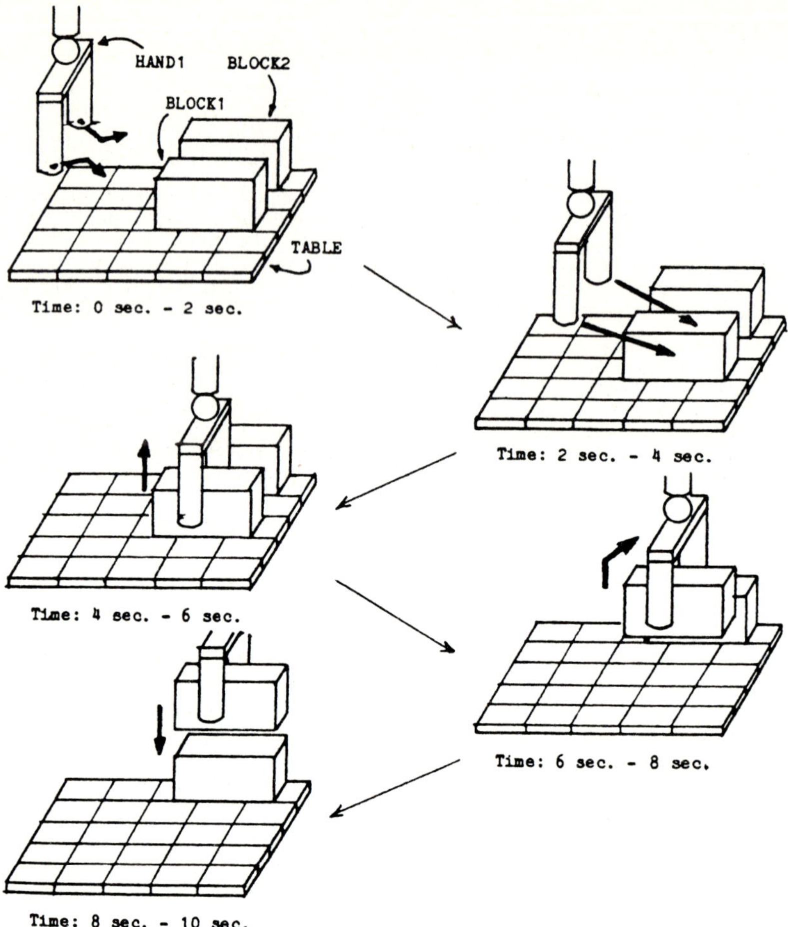

Figure 1. Sequence of actions for the example.

(This high-level description should be contrasted with the low-level description in the input file, which is some 550 lines long.)

A number of observations can be made about this resultant description. First of all, the implemented package has correctly identified the major events occurring in the example. Of equal importance however, it has also attached reasonable boundaries in time to these events. Secondly, although the system must identify many lower-level events in the course of finding the high-level events, only the top-most events are included in the final description. For instance, HAND1 is holding BLOCK1 from time 4 to time 10, but for those times at which a higher-level event occurs which explains the holding (e.g. stacking), the event of holding is not mentioned.

Without going into details, we mention very briefly that the implemented system can be seen as the integration of three parts: a knowledge base of rules for determining the occurrences of particular events, an interpreter for evaluating these rules, and a supervisory program which forms a high-level description of changes in a scene from the input low-level description by a process of successive refinement.

It should be emphasized that this work does not consider the very difficult problem of how to extract the input information needed for the implemented system from raw image data.

III. MOTION ESTIMATION

III.A. Two-View Motion Estimation

Generally, a scene contains a number of objects moving differently. To detect and tract these objects and to estimate and predict the motion of each of them are formidable tasks indeed. In fact, to date even the following simplest motion estimation problem has not been solved satisfactorily: Given two time-sequential images (perspective views) of a single rigid object, estimate its equivalent rotation and translation from the first time instant to the second. In this section, we shall review results pertaining to this "two-view" motion estimation problem, and point out some of the difficulties.

Several approaches to motion estimation have been proposed and studied [7]. Here we shall look at only one of them: the approach of using feature correspondences. We first state the problem more precisely.

The basic geometry of the problem is sketched in Fig. 2. The object-space coordinates are denoted by lowercase letters, and the image-space coordinates by uppercase letters. Let the two perspective views (central projections) be taken at t_1 and t_2, respectively, and $t_1 < t_2$. The coordinates at t_2 are primed, and the coordinates at t_1 are unprimed. Specifically, consider a particular physical point P on the surface of a rigid body in the scene. Let (x,y,z) be the object-space coordinates of P at time t_1, (x', y', z') the object-space coordinates of P at time t_2, (X,Y) the image-space coordinates of P at time t_1, (X' , Y') the image-space coordinates of P at time t_2, and

$$\Delta X \triangleq X' - X \ , \ \Delta Y \triangleq Y' - Y \tag{1}$$

the image-space shifts (or displacements) of P from t_1 to t_2.

It is well known from kinematics that the object coordinates of P at time instants t_1 and t_2 are related by

$$\begin{bmatrix} x' \\ y' \\ z' \end{bmatrix} = R \begin{bmatrix} x \\ y \\ z \end{bmatrix} + T = \begin{bmatrix} r_{11} & r_{12} & r_{13} \\ r_{21} & r_{22} & r_{23} \\ r_{31} & r_{32} & r_{33} \end{bmatrix} \begin{bmatrix} x \\ y \\ z \end{bmatrix} + \begin{bmatrix} \Delta x \\ \Delta y \\ \Delta z \end{bmatrix} \tag{2}$$

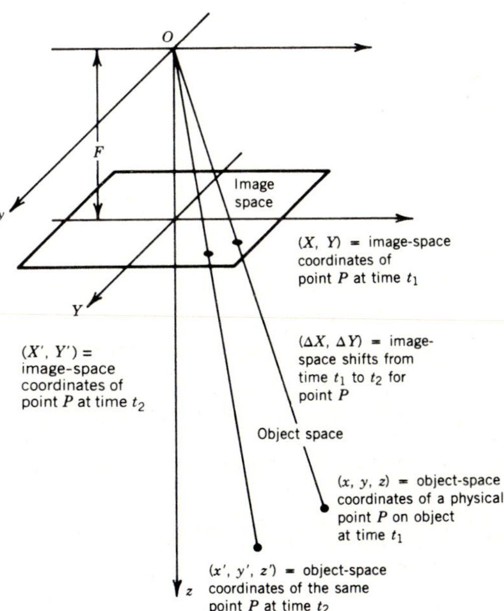

Figure 2. Basic geometry for motion analysis.

where R represents a rotation and T a translation. To make the representation unique the rotation is specified around an axis passing through the origin of our coordinate system. Let $n = (n_1, n_2, n_3)$ be a unit vector along the axis of rotation, and θ be the angle of rotation from t_1 to t_2. Then the elements of R can be expressed in terms of n_1, n_2, n_3, and θ. Since $n_1^2 + n_2^2 + n_3^2 = 1$, there are six motion parameters we have to determine: n_1, n_2, θ; $\Delta x, \Delta y, \Delta z$. However, from the two perspective views, it is impossible to determine the magnitude of the translation, i.e., if the object size and position as well as the translation are scaled by the same factor, one gets exactly the same two image frames. One can therefore determine the translation to only within a scale factor.

To summarize, our problem is: Given two image frames at t_1 and t_2, find the motion parameters T (to within a scale factor) and R. As shown below, the equations relating the motion parameters to the image-point coordinates inevitably involve the ranges (z-coordinates) of the object points. Therefore, in determining the motion parameters, one also determine the ranges of the observed objects points. It will be seen that the translation vector T and the object point ranges can be determined to within a positive global scale factor. The value of this scale factor could be found if the magnitude of T or the absolute range of any observed object point is known.

III.B. Using Point Correspondences

Consider a two-stage method to solve the posed problem. In the first stage, one finds point correspondences in the two perspective views (images). A point correspondence is a pair of image coordinates (X_i, Y_i), (X_i', Y_i') which are images at t_1 and t_2, respectively, of the same physical point on the object. Then, in the second stage, one determines the motion parameters from these image coordinates by solving a set of equations.

III.B.1. Finding point correspondences

In order to be able to find point correspondences, the images must contain points that are distinctive in some sense. For example, images of man-made objects often contain sharp corners which are relatively easy to extract [8]. More generally, image points where the local gray-level variations (defined in some way) are maximum can be used [9].

In any case, in each of the two images a large number of distinctive points are extracted. Then one tries to match the two point patterns in the two images using spatial structures of the patterns [10]. The matching will be successful only if the amount of rotation (θ) is relatively small (so that the perspective distortion is small). For example, in Ref. 8, good matching results are obtained if $\theta < 5°$. This restriction may be relaxed if there is some a priori information about the object [11].

It can be readily appreciated that this first stage is by far the more difficult of the two stages in the method. No algorithm exists which works well for a large variety of images. For many images, no existing algorithms work.

III.B.2. Nonlinear algorithm

From Fig. 2 and Eq. (2), we get

$$X' = \frac{(r_{11}X + r_{12}Y + r_{13})z + \Delta x}{(r_{31}X + r_{32}Y + r_{33})z + \Delta z}$$

$$Y' = \frac{(r_{21}X + r_{22}Y + r_{23})z + \Delta y}{(r_{31}X + r_{32}Y + r_{33})z + \Delta z} \tag{3}$$

where r_{ij}'s can be expressed in terms of n_1, n_2, n_3, and θ. Eliminating z from Eq. (3), we get

$$(\Delta x - X'\Delta z)\left\{Y'(r_{31}X + r_{32}Y + r_{33}) - (r_{21}X + r_{22}Y + r_{23})\right\}$$

$$= (\Delta y - Y'\Delta z)\left\{X'(r_{31}X + r_{32}Y + r_{33}) - (r_{11}X + r_{12}Y + r_{13})\right\} \tag{4}$$

Equation (4) is nonlinear in the 6 unknowns: Δx, Δy, Δz; n_1, n_2, θ. Also, it is homogeneous in Δx, Δy, Δz. Therefore, as mentioned earlier, one can only hope to find T to within a scale factor. After finding T (to within a scale factor) and R, one can find z_i for each observed point to within the same scale factor using Eq. (3).

To fix ideas, let the translation sought after be the unit translation vector

$$\hat{T} = (\hat{\Delta x}, \hat{\Delta y}, \hat{\Delta z}) \triangleq \frac{1}{\sqrt{\Delta x^2 + \Delta y^2 + \Delta z^2}} T \tag{5}$$

Then, Eq. (4) can be considered as a nonlinear equation in the five unknowns: $\hat{\Delta x}$, $\hat{\Delta y}$; n_1, n_2, θ . Thus, with 5-point correspondence, there are five equations with five unknowns. Well-known iterative techniques can then be used to find solutions. In practice, because of noise in the image data, one tries to find more than 5-point correspondences and seek a least-squares solution.

III.B.3. Linear algorithm

It turns out that by introduction of appropriate intermediate variables (which are functions of the motion parameters), Eq. (5) becomes linear [12,13]. If we define

$$E = \begin{bmatrix} e_1 & e_2 & e_3 \\ e_4 & e_5 & e_6 \\ e_7 & e_8 & e_9 \end{bmatrix} = GR \tag{6}$$

where

$$G = \begin{bmatrix} 0 & -\hat{\Delta z} & \hat{\Delta y} \\ \hat{\Delta z} & 0 & -\hat{\Delta x} \\ -\hat{\Delta y} & \hat{\Delta x} & 0 \end{bmatrix} \text{(skew symmetric)} \tag{7}$$

$\hat{T} = (\hat{\Delta x}, \hat{\Delta y}, \hat{\Delta z})$ is the unit translation vector defined in Eq. (5), and R is the orthonormal rotation matrix. Then Eq. (4) becomes

$$[X' \; Y' \; 1] \, E \begin{bmatrix} X \\ Y \\ 1 \end{bmatrix} = 0 \tag{8}$$

which is linear and homogeneous in the nine new unknowns: $e_1, e_2, \ldots, e_9$.

The algorithm consists of two steps:

Step 1 - From 8 or more point correspondences determine E to within an unknown scale factor k.

Step 2 - Decompose kE to obtain R and $\hat{T}$

Step 1 is relatively simple; it amounts to finding the least-squares solution of a set of linear equations (8). Step 2 is more complicated, and will not be discussed here. The reader is referred to Ref. 7 for several algorithms. It can be shown [14] that, except for degenerate cases, 8 or more point correspondences yield a unique solution for R and T

III.C. Using Straight Line Correspondence

In the presence of image noise and/or due to the spatial sampling, the coordinates of feature points cannot be determined accurately. This may make the estimation of motion parameters unreliable. Usually, it is easier to detect and determine the location of straight edges than feature points. Therefore, the question arises: Can one estimate 3-D motion parameters by using straight line correspondences?

III.C.1. Finding line correspondences

Images of man-made objects often contain straight edges. These straight edges can be detected using edge point detectors (such as the Sobel operator) followed by Hough transform. One can first detect straight edges in both image frames and then uses structural information to match the two straight-line patterns. The algorithm of Cheng and Huang [15] can be used to do the matching if the motion from t_1 and t_2 is small. Alternative algorithms include Faugeras, et al. [16].

III.C.2. Nonlinear algorithm (over 3 views)

By a straight-line correspondence over two frames, one knows the equations in the image plane at t_1 and t_2 of a 3-D line on the object:

$$\text{At } t_1: \quad \alpha X + \beta Y = 1 \tag{9}$$

$$\text{At } t_2: \quad \alpha' X + \beta' Y = 1 \tag{10}$$

where $(\alpha,\beta) \longleftrightarrow (\alpha',\beta')$. Note that one does not assume any point correspondences on the two lines. Unfortunately, a little reflection convinces one that no matter how many straight-line correspondences we know over two frames, it is impossible to determine R and T uniquely.

With straight-line correspondences over three image frames (at $t_1 < t_2 < t_3$), it is possible to determine the motion rotations R_{12}, T_{12} (from t_1 to t_2) and R_{23}, T_{23} (from t_2 to t_3). An equation involving R_{12} and R_{23} can be obtained. Let the equations in the image plane at t_1, t_2, and t_3, of a 3-D straight-line be given by Eqs. (9), (10) and

$$\text{At } t_3: \quad \alpha'' X + \beta'' Y = 1 \tag{11}$$

Then it can be shown that [17]

$$q' \cdot (R_{12} q \times R_{23}^{-1} q'') = 0 \tag{12}$$

where

$$q = (\alpha, \beta, -1)$$
$$q' = (\alpha', \beta', -1), \text{and}$$
$$q'' = (\alpha'', \beta'', -1)$$

Here a 3-element array is considered as either a vector or a column matrix from context.

Equation (12) is nonlinear in the six unknown motion parameters (three from each rotation matrix). With 6 or more straight-line correspondences over 3 views, we can solve these nonlinear equations by least-squares. It is easy to show that once the rotations are found, the translation vectors can be obtained by solving linear equations.

II.C.3. Linear algorithm (over 3 views)

Linear algorithms exist which require 13 or more straight line correspondences over 3 views [18,19].

Similar to the point case, a set of 27 intermediate variables are defined which are elements of:

$$F \triangleq T_{13} R_{12}^{(1)t} = R_{13}^{(1)} T_{12}^{t} \tag{13a}$$

$$G \triangleq T_{13} R_{12}^{(2)t} - R_{13}^{(2)} T_{12}^{t} \tag{13b}$$

$$H \triangleq T_{13} R_{12}^{(3)t} - R_{13}^{(3)} T_{12}^{t} \tag{13c}$$

where $R_{ij}^{(k)}$ denotes the kth column of R_{ij} and the superscript "t" denotes matrix transposition. These intermeidate unknowns can be obtained to within a global scale factor from solving linear equations. Then, the rotation and translation parameters are determined from them. The readers are referred to Refs. 18 and 19 for details.

IV. CONCLUDING REMARKS

One of the most difficult part of dynamic scene analysis is motion/structure estimation. The problems of multiple moving objects [20,21] motion prediction [22], and nonrigid objects [23] have been studied only slightly. Even the simplest problem of two-view (and three-view) motion estimation of a single rigid object has not been resolved satisfactorily, mainly because of two interrelated factors:

(i) For many real-world images, it is extremely difficult to extract and locate features (points, lines) reliably and accurately, and to match them over two or three views.

(ii) Existing algorithms for determining the motion parameters from feature correspondences are sensitive to errors in the image coordinates of the features. The dependence of estimation errors on feature configurations and motion parameter values are not well understood [24].

Acknowledgment. This work was supported by the National Science Foundation Grant IRI-8605400.

References

[1] T. S. Huang (ed.), *Image Sequence Processing and Dynamic Scene Analysis*, Springer-Verlag: Heidelberg, 1983.

[2] M. D. Levine, P. B. Nobel, and Y. M. Youssef, "A rule-based system for characterizing blood cell motion," in Ref. 1.

[3] J. K. Tsotses, J. Mylopoulos, H. D. Corvey, and S. W. Zucker, "A framework for visual motion understanding," *IEEE Trans. PAMI*, vol. 2, no. 6, pp. 563-573, Nov. 1980.

[4] J. O'Rourke and N. Badler, "Model-based image analysis of human motion using constraint propagation," *IEEE Trans. PAMI 2*, 522-536 (1980).

[5] B. Neumann, "Natural language description of time-varying scenes," *Bericht nr. 105, FBI-HH-B-105/84*, Aug. 1984, Fachberich Informatik, Univ. Hamburg, W. Germany.

[6] G. C. Borchardt, "A computer model for the representation and identification of physical events," Tech. Rep. T-142, Coordinated Science Laboratory, Univ. of Illinois, Urbana, IL, May 1984.

[7] T.S. Huang, Motion Analysis, in "Artificial Intelligence Encylopedia," Wiley, 1987, pp. 620-632.

[8] J. Q. Fang and T. S. Huang, "A corner finding algorithm for image analysis and registration," *Proc. AAAI-82*, Pittsburgh, PA, Aug. 18-20, 1982, pp. 46-49.

[9] H. P. Moravec, "Obstacle avoidance and navigation in the real world by a seeing robot rover," Ph. D. dissertation, Stanford Univ., Sept. 1980.

[10] J. Q. Fang and T. S. Huang, "Some experiments on estimating the 3-D motion parameters of a rigid body from two consecutive image frames," *IEEE Trans. on PAMI*, vol. 6, no. 5, pp. 547-555, Sept. 1984.

[11] W. K. Gu, J. Y. Yang, and T. S. Huang, "Matching perspective views of a 3-D object using composite circuits" *Proc. 7th ICPR*, July 30-Aug. 2, 1984.

[12] H. C. Longuet-Higgins, "A computer program for reconstructing a scene from two projections," *Nature*, vol. 293, pp. 133-135, Sept. 1981.

[13] R. Y. Tsai and T. S. Huang, "Uniqueness and estimation of 3-D motion parameters of rigid bodies with curved surfaces," *IEEE Trans. PAMI*, vol. 6, no.1, pp. 13-27, Jan. 1984.

[14] H. C. Longuet-Higgins, "The reconstruction of a scene from two projections-configurations that defeat the 8-point algorithm," *Proc. 1st Conf. Artificial Intelligence Applications*, Dec. 5-7, 1984, Denver, CO, pp. 395-397.

[15] J. K. Cheng and T. S. Huang, "Image registration by matching relational structures," *Pattern Recognition*, vol. 17, no. 1, pp. 149-160, 1984.

[16] O.D. Faugeras, F. Lustman, and G. Toscasi, Motion and structure from motion from points and lines, *Proc. 1st. Int. Conf. Computer Vision*, June 8-11, 1987, London, England.

[17] Y.C. Liu and T.S. Huang, "Estimation of rigid body motion using straight-line correspondences," *Proceedings of IEEE Workshop on Motion: Representation and Analysis*, May 7-9, 1986, Kiawah Island, SC, pp. 47-52.

[18] M.E. Spaetsakis and J. Aloimonos, Closed form solution to the structure from motion problem form line correspondences, Tech. Rept. CAR-TR-374, Center for Automation Research, Univ. of Maryland, March 1987.

[19] Y.C. Liu and T.S. Huang, A linear algorithm for motion estimation using straight line correspondences, Technical Note ISP-309, Coordinated Science Laboratory, University of Illinois at Urbana-Champaign, April 15, 1987.

[20] H.H. Chen and T.S. Huang, Multiple object motion determination by matching 3-D points, to appear in Pattern Recognition.

[21] H.H. Chen and T.S. Huang, "Multiple object motion estimation by matching 3-D line segments," Technical Note ISP-120, December 15, 1986; Coordinated Science Laboratory, University of Illinois at Urbana-Champaign.

[22] T.S. Huang, J. Weng, and N. Ahuja, 3-D motion from Image sequences: modeling, understanding, and prediction, *Proceedings of IEEE Workshop on Motion: Representation and Analysis*, May 7-9, 1986, Kiawah Island, SC, pp. 125-130.

[23] S. S. Chen, "Shape and correspondence of nonrigid objects," in "Advances in Computer Vision and Image Processing, Vol. 3" ed. by T.S. Huang, JAI Press, 1987.

[24] J. Weng, T.S. Huang, and N. Ahuja, Error analysis of linear algorithm for motion estimation, in *Proc. 1st Int. Conf. Computer Vision*, June 8-11, 1987, London, England.

Take a Snapshot of the Ewald Sphere

Richard Bamler
Lehrstuhl für Nachrichtentechnik, Technische Universität München
Arcisstr. 21, 8000 München 2, Fed. Rep. of Germany

Abstract

It is shown that the Fraunhofer diffraction pattern of a tilted aperture screen can be interpreted as a "monocular view" of the Ewald sphere. Two of those patterns recorded at opposite tilting angles are used to give a stereoscopic impression of the Ewald sphere. The approximations involved are discussed.

1 Introduction

The Ewald sphere plays an important role in forward scattering (e.g. crystallography [1,2], ultrasonic tomography [3,4], microscopy [5,6]), back scattering (e.g. microwave imaging), and three-dimensional image display (e.g. holography [7]). In his fundamental paper [8] Wolf has shown that it acts as a transfer function for scatterer reconstruction within the first Born approximation.

2 The Diffraction Pattern of a Tilted Aperture

The display of the Ewald sphere will be achieved by two stereoscopic views of its inner surface. We will use the phenomenon of distortion of the Fraunhofer diffraction pattern introduced by tilting the diffracting aperture.

In Fig. 1(a) the apparatus under discussion is sketched. It is a common coherent-optical Fourier transformer. The output field $u_{2d}(x,y)$ in the back focal plane $(z = 2d)$ is a scaled version of the two-dimensional Fourier transform of the input field $u_o(x,y)$ in the front focal plane $(z = 0)$:

$$u_{2d}(x,y) \sim U_o(\lambda d\, f_x, \lambda d\, f_y) \qquad \text{(Capital letters denote Fourier transforms.)}$$

Usually the field $u_o(x,y)$ in the **input plane** is generated by illuminating a transperancy with a collimated laser beam e^{jkz}. In our apparatus, however, this transparency is tilted by an angle α with respect to the input plane. We will refer to the plane defined by the position of the transparency as the **aperture plane**. The experimental findings are the following:

The tilt of the aperture plane introduces geometric distortions in the displayed "spectrum" at $z=2d$ in such a way that straight lines become ellipses [9,10]. Here we will give a Fourier space explanation of this phenomenon. Then the relation of this experiment to the Ewald sphere will become obvious.

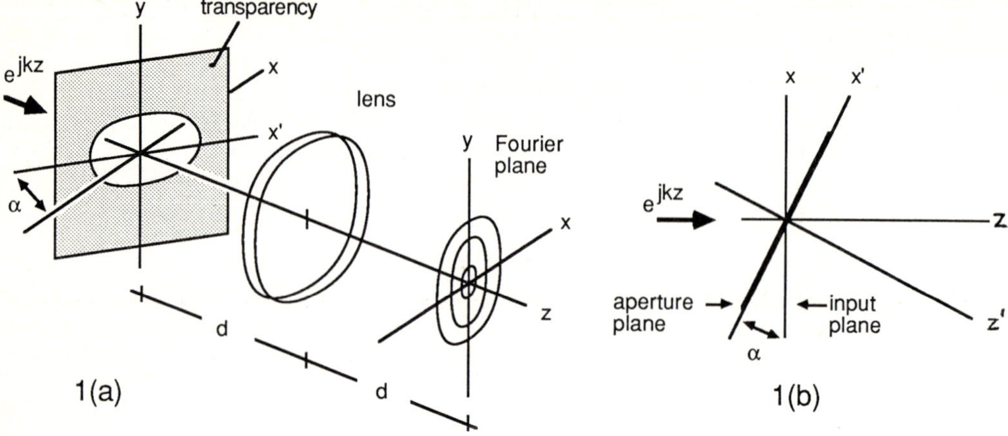

The Fourier transformer performs its operation on an **assumed** field in its input plane. Thus we need only 'trace back' the actual diffracted field to this input plane. The transparancy is characterized by an amplitude transparence function $m(x',y)$. Using the coordinate system of Fig. 1(b), the field in the aperture plane is

$$u'_o(x',y) = [\, m(x',y)\, e^{jkz}\,]\big|_{z'=0} = m(x',y)\, e^{jk_{x'}x} \qquad \text{with} \quad k_{x'} = (2\pi/\lambda)\sin\alpha$$

The propagation of the field u'_o is adequately described by the angular spectrum method (for $z' \geq 0$):

$$u'_{z'}(x',y) = \iint U'_o(f_{x'},f_y)\; e^{j2\pi z'\sqrt{1/\lambda^2 - (f_{x'}^2 + f_y^2)}}\; e^{j2\pi(x'f_{x'}+yf_y)}\; df_{x'}\, df_y$$

Obviously the factor $S'_{z'}(f_{x'},f_y) = e^{j2\pi z'\sqrt{1/\lambda^2 - (f_{x'}^2 + f_y^2)}}$ acts as a two-dimensional linear transfer function for wave propagation between parallel planes [10,11,12]. Thus $u'_{z'}$ is related to u'_o by a two-dimensional convolution:

$$u'_{z'}(x',y) = u'_o(x',y) \overset{x'\;y}{**} s'_{z'}(x',y)$$

with $s'_{z'}$ being the Fourier retransform of $S'_{z'}$. However, the field in a plane parallel to the aperture plane is of no interest to the experiment described, since we want to trace back the field to the input plane which is not parallel to the aperture plane. For that purpose we formally expand the above convolution to three dimensions and allow arbitrary values of z'. Then the input field is given by

$$u_o(x,y) = u'(x',y,z')\big|_{z=0} = [(\, u'_o(x',y)\,\delta(z')\,) \overset{x'\;y\;z'}{***} s'(x',y,z')]\,\big|_{z=0} \tag{1}$$

The former index z' is now treated as a variable. To describe this convolution in Fourier space the three-dimensional Fourier transform $S'(f_{x'},f_y,f_{z'})$ of $s'(x',y,z')$ is needed. This is obtained by Fourier transforming $S'_{z'}(f_{x'},f_y)$ with respect to z'. The shifting theorem of Fourier calculus yields:

$$S'(f_{x'},f_y,f_{z'}) = \delta\!\left(f_{z'} - \sqrt{1/\lambda^2 - (f_{x'}^2 + f_y^2)}\,\right) \qquad \text{(Evanescent waves have been discarded.)}$$

This transfer function obviously occupies the Ewald hemisphere ($f_{x'}^2+f_y^2+f_{z'}^2 = 1/\lambda^2$ and $f_{z'} \geq 0$) a section of which is drawn in Fig. 2b.

Now we can sequentially transform eq.(1) into Fourier space (see Fig. 2):

(a) The spectrum of $u'_o(x',y)\ \delta(z') = m(x',y)\ \delta(z')\ e^{j2\pi x' \sin\alpha / \lambda}$ is constant in $f_{z'}$ - direction. It can be derived from the original spectrum $M(f_{x'},f_y)$ of the transparency via the shifting theorem: $U'_o(f_{x'},f_y) = M(f_{x'} - \sin\alpha / \lambda,\ f_y)$

(b) The three-dimensional convolution in eq. (2) means multiplication of $U'_o(f_{x'},f_y)$ with the hemisphere $S'(f_{x'},f_y,f_{z'})$. The product is the intersection of both spectra.

(c) Taking the field only for $z = 0$ means projection in Fourier space onto the f_x - f_y - plane: $U_o(f_x,f_y) = \int M(f_{x'} - \sin\alpha / \lambda,\ f_y)\ \ S'(f_{x'},f_y,f_{z'})\ df_z$

From this construction rule the spectral distortions become obvious. They may be derived geometrically from Fig. 2.

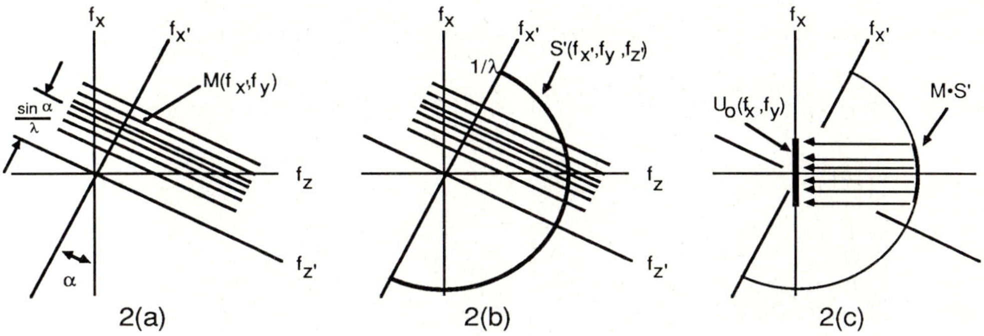

3 Interpretation and Experimental Results

The derived relation between M and U_o can be given a useful interpretation: We imagine the Ewald sphere S' to be the dome of a planetarium. The 'floor' of this planetarium be the f_x - f_y - plane. Using this metaphor steps (a) and (b) of Fig. 2 mean projecting a pattern in the dome. This pattern is the two-dimensional spectrum of the input transparency in the apparatus of Fig. 1. The projection angle is α and the centre of the pattern (spetral DC-term) is in the zenith of the dome. Viewed from the location of the projector $(f_{x'},f_y,f_{z'}) = (\sin\alpha / \lambda,0,0)$ or $(f_x,f_y,f_z) = (\tan\alpha / \lambda,0,0)$ the pattern looks undistorted. However, step (c) claims a backprojection onto the floor, i.e. the observer stands exactly underneath the zenith (again parallel projection instead of central perspective is assumed). This causes the said distortions. For a stereoscopic view of the Ewald sphere (the dome) we really need two different views of the projected pattern. Unfortunatelly these cannot be obtained by our apparatus. Instead we use two projections with angles α and $-\alpha$, and record each backprojected pattern as a monocular observer underneath the zenith would see it. This can be done by using two tilts of the aperture plane α and $-\alpha$, or (with real-valued input transparencies and thus symmetric spectra) simply take two shots of the distorted spectrum and later flip one of them over for viewing. Fig. 3 shows the experimental results, stereo pairs of two different spectra. The input transparencies were a star-pattern and a point-raster, respectively. The angle $\alpha = 60°$ has been chosen to give a somewhat exaggerated impression of the curvature. The reader is encouraged to try visual fusion of each of the stereo pairs.

Fig. 3

Literature

[1] Ewald, P.P., Annalen der Physik, 49 (1), 1-38, 1916.
[2] Ewald, P.P., Annalen der Physik 49 (2), 117-143, 1916.
[3] Mueller, R.K., Kaveh, M., Wade, G., Proc. IEEE 67 (4), 567-587, 1979.
[4] Devaney, A.J., Ultrasonic Imaging 4, 336-350, 1982.
[5] Streibl, N., Optik, 66 (4), 341-354, 1984.
[6] Streibl, N., J. Opt.Soc.Am.A., 2 (2), 121-127, 1985.
[7] Lohmann, A.W., Optik, 51 (2), 105-117, 1978.
[8] Wolf, E., Optics Communications, 1 (4), 153-156, 1969.
[9] Patorski, K., Optica Acta, 30 (5),673-679,1983.
[10] Bollmann, H., Insitut für Nachrichtentechnik ,Techn. Univ. München (internal report), initiated by: H. Platzer, H. Glünder,1984.
[11] Goodman, J.W., In: Introduction to Fourier Optics, MacGraw-Hill,New York, 1968.
[12] Gaskill, J.D., In: Linear Systems, Fourier Transforms, and Optics, Wiley , New York, 1978.
[13] Bracewell, R.N., In: The Fourier Transform and its Applications, McGraw-Hill, New York, 1965.

BEST APPROXIMATION OF MULTIDIMENSIONAL RANDOM SIGNALS

Millu Rosenblatt-Roth
Center for Automation Research, University of Maryland
College Park, MD 20742, USA

1. Generalities on random fields.

Let: E^m be the set of sites $t = \{t^{(i)}, 1 \leq i \leq m\}$ with $t^{(i)}$ integers; τ, S finite, $\tau \subset S \subset E^m$; ξ_t a random variable taking values in the measurable space (X_t, Σ_t) with probability measure P_{ξ_t}, $x_t \in X_t$, $T_t \in \Sigma_t$; $\xi(\tau) = \{\tau_t, t \in \tau\}$, $(X(\tau), \Sigma(\tau))$ the product of (X_t, Σ_t), $t \in \tau$; $\xi = \xi(S)$, a random field with values in (X, Σ) with probability P_ξ.

We use the known concepts of: (a) environment (system of neighborhoods) $\hat{S} = \{\hat{S}_t \subset S - t, t \in S\}$ [4], (b) Markov random field with characteristic kernel $\hat{p}(T_t \mid x(\hat{S}_t)$, $\hat{p} = \{\hat{p}_t, t \in S\}$. For other concepts and notation see [5].

2. The Markov random mesh.

Let K be an order in S, $N = \{S_t, t \in S\}$ <u>a partial environment</u>, $S_t(K) = \{t'; t' \leq t(K)\}$ with sites in K-order. The order K is <u>adapted</u> to N if $S_t \subset S_t(K)$, $t \in S$.

<u>Definition 1.</u> The random field ξ is a <u>Markov random mesh</u> if there exists a partial environment N and an order K adapted to it, so that for all $t \in S$, the random variables $\xi(S_t(K)-S_t)$, $\xi(S_t)$, ξ_t form a Markov chain.

We remark that Markov chains of any order are Markov random meshes on $S \subset E^1$.

The expressions $p_t = p_t(T_t \mid x(S_t))$ representing the conditional probabilities of $\xi_t \in T_t$ under condition $\xi(S_t) = x(S_t)$, are the partial characteristic kernels of the Markov random mesh considered and $p = \{p_t, t \in S\}$. For given N, let: $A_t = \{t'; t \in S_{t'}\}$. $B_t = \cup \{S_{t'}; t' \in A_t\}$, $R_t = t \cup A_t$, $\hat{S}_t = S_t \cup A_t \cup (B_t - t)$.

<u>Theorem 1.</u> Let ξ be a Markov random mesh with N, p. Then (a) $P_\xi(dx)$ is a product of $p_t(dx_t \mid x(S_t))$ for all $t \in S$; (b) p_t is a marginal of P_ξ, $t \in S$; (c) $P_\xi(dx_t \mid x(S-t))$ is the ratio of the product of $p_{t'}(dx_{t'} \mid x(S_{t'}))$, $t' \in R_t$ to its integral in P_{ξ_t} over X_t; (d) ξ is a Markov random field with environment $\hat{S}$ and $\hat{p}$ given by (c) above.

Let v be the cardinality of S and $S^+(K) = S_v(K)$. If $t \in S$ is posted at the site with natural number $t^+ (1 \le t^+ \le v)$, let $f(t) = t^+$ be the one-to-one, order preserving mapping established by it between S and S^+. Endowing corresponding t, t^+ with the same structures, the mapping extends to random fields, so that $f(\xi) = \xi^+(K)$ with invariance of cardinality of sets and of the values of partial (and complete) characteristic kernels.

Theorem 2. The random field ξ is a Markov random mesh with N, p iff there exists an adapted order K such that $\xi^+(K)$ is a Markov random mesh with $N^+(K)$, $p^+(K)$. The Markov random mesh ξ with N, p is a Markov random field with $\hat{S}$, $\hat{p}$ iff there exists an adapted order K, such that the Markov random mesh $\xi^+(K)$ with $S^+(K)$, $p^+(K)$ is a Markov random field with $\hat{S}^+$, p^+.

3. Best approximation of random fields with Markov random meshes.

Let: $P = P_\xi \in L$, for some random field ξ; $P^{(N)}$ the probability measure of a Markov random mesh with N and with $p_t(T_t \mid x(S_t))$, $t \in S$ given by the conditional marginal probability measures of $\xi_t \in T_t$ under condition $\xi(S_t) = x(S_t)$, calculated from P. The relation $P^{(N)} = (A(N))P$ defines the projector $A(N)$ on $L(N) = (A(N))L$, the totality of Markov random meshes with N. Let $I_{A(N)}(P) = h(P:P^{(N)})$, $I^*_{A(N)}(P) = h(P^{(N)}:P)$ be the amount of information, resp. conjugate information corresponding to the class L(N) of Markov random meshes with N, determined by $P \in L$.

Theorem 3. If $P \in L$, then $I_{A(N)}(P) \ge 0$, $I^*_{A(N)}(P) \ge 0$ with equalities iff $P \in L(N)$.

Theorem 4. In order that $I_{A(N)}(P)$, $I^*_{A(N)}(P)$ be finite, it is necessary that $P << [A(N)]P$ resp. $[A(N)]P << P$; in this case $I_{A(N)}(P) = h(P:[A(N)]P)$, $I^*_{A(N)}(P) = h([A(N)]P:P)$.

Lemma. A(N) is a completely regular projector.

Theorem 5.

$$\min\{h(P:Q);\ Q \in L(N)\} = I_{A(N)}(P)$$

and this minimum is reached iff $Q = [A(N)]P$.

Let $\overline{P}(t)$ be the Fourier transform of $P \in L$ and $\|\cdot\|$ the total variation of a completely additive function.

Theorem 6.

$$\sup\{|\overline{P}(t) - (\overline{[A(N)]P})(t)|^2; t\} \le \|P - [A(N)]P\|^2 \le$$

$$\leq 2 \min\{I_{A(N)}(P),\ I^{*}_{A(N)}(P)\}\ .$$

Let $N_i = \{S_t^{(i)},\ t \in S\}$, $i = 0,1,2$, with $S_t^{(0)} = \phi(t \in S)$, and let $A(N_i)$ be the projectors on $L(N_i)$ $(i = 0,1,2)$. Then $A_0 = A(N_0)$ is the independent projector and $I_{A(N_0)}(P) = I_0(P)$ the amount of independent information determined by P. In the set of projectors A(N), let us introduce a partial order such that $A(N_2) < A(N_1)$ if $L(N_2) \subset L(N_1)$.

Proposition. If $S_t^{(1)} \subset S_t^{(2)}$, $t \in S$, then (a) $A_0 < A(N_1) < A(N_2)$, (b) $0 \leq I_{A(N_2)}(P) \leq I_{A(N_1)}(P) \leq I_0(P)$.

The proofs of Theorems 1,2 are straightforward, and those of Theorems 3-6 follow from [5].

References

[1] K. Abend, T. J. Harley, L. N. Kanal, Classification of binary random patterns, IEEE Trans. Inf. Th., v. IT.11, No. 4, 1965, pp. 538-543.

[2] J. Besag, Spatial interaction and the statistical analysis of lattice systems, J. Roy. Stat. Soc., Ser. B, v. 36, N. 2, 1974, pp. 192-236.

[3] I. M. Gelfand, A. M. Yaglom, Calculation of the amount of information about a random function contained in another such function, Usp. mat. nauk, 12, 1957, pp. 3-52.

[4] J. Moussouris, Gibbs and Markov random systems with constraints, J. Stat. Phys., Vol. 10, No. 1, 1974, pp. 11-33.

[5] M. Rosenblatt-Roth, Sur la meilleure approximation des mesures de probabilité, C. r. Acad. Sci. Paris, v. 304, No. 12, 1987, pp. 343-346.

OPTIMIZATION OF QUANTIZERS BY THE USE OF HAAR EXPANSIONS

Janusz Sawicki
Technical University of Poznań
Institute of Electronics and Telecommunications
ul.Piotrowo 3A 60965 Poznań - Poland

1. Introduction

Quantizers are usually specified by the partition of the input signals and the centroids of the output signals, or, alternatively, by step-like functions determining the relation between input and output signals. The characteristic parameters or functions are optimized with respect to the distortion being the result of quantization. The optimization methods for the case of noise-free input signals, as well as for the case of noisy signals, described by their probability density functions, were already elaborated and presented by many authors [1, 2, 3, 4].

An optimization method, using Haar expansions of the above mentioned step-like functions, is presented in the paper. The quality functional, after certain transformations, is minimized with respect to the coefficients of the Haar expansion.

2. Quantization of noise-free signals

The criterion of optimality is the minimum of the distortion defined by the mean square error which equals the difference between the input signal x and its quantized output $Q(x)$ - Fig.1

$$D = E\left[(x - Q(x))^2\right] = \int_X (x - Q(x))^2 f_x(x)\, dx \tag{2.1}$$

where $f_x(x)$ denotes the probability density function of the input signal x ,

X the range of values taken by x .

In order to determine the optimal step-like shape of $Q(x)$, minimizing the functional D , let us introduce a new variable $y=F(x)$ defined by the equation

$$\frac{dy}{dx} = f_x(x) \tag{2.2}$$

The integral will take the form

$$D = \int_Y \left[F^{-1}(y) - Q\left(F^{-1}(y)\right)\right]^2 dy =$$

$$= \int_Y \left[F^{-1}(y) - Q^*(y)\right]^2 dy \tag{2.3}$$

where $Q^*(y) = Q\left(F^{-1}(y)\right)$,

Y - is the interval of y values taken by the new variable and corresponding to X.

A suboptimal solution of the above problem can be obtained if the step-like function $Q^*(y)$ will be replaced by its Fourier series with respect to Haar functions i.e.

$$Q^*(y) = \sum_{i=1}^{M} a_i \operatorname{haar}(i, y) , \tag{2.4}$$

where $\operatorname{haar}(i, y)$ - the i-th order Haar function ,

a_i - the unknown Fourier coefficient.

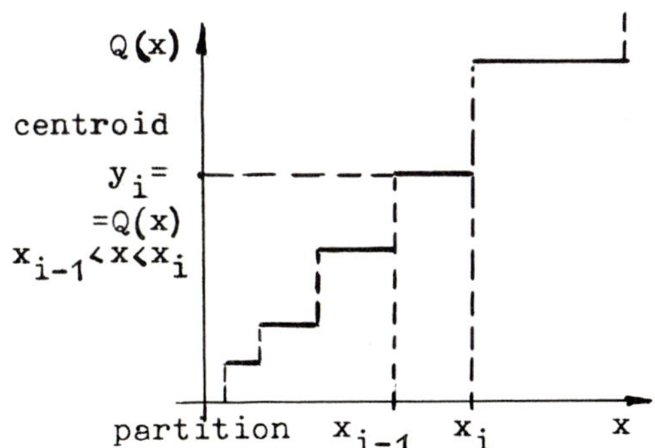

Figure 1: The $Q(x)$ vs. x diagram of the quantizer

The values of a_i's can be found from the well known conditions concerning partial derivatives

$$\frac{\partial D}{\partial a_j} = 0 \qquad j = 1,\ldots,M \quad . \tag{2.5}$$

It can be seen that

$$\frac{\partial D}{\partial a_j} = \frac{\partial}{\partial a_j} \int_Y \left(F^{-1}(y) - \sum_{i=1}^{M} a_i \operatorname{haar}(i, y)\right)^2 dy =$$

$$= \int_Y 2\left[F^{-1}(y) - \sum_{i=1}^{M} a_i \; haar(i, y)\right] haar(j, y)\, dy \qquad (2.6)$$

and

$$a_j = \frac{1}{A} \int_Y F^{-1}(y)\; haar(j, y)\, dy \qquad \text{for} \quad j = 1,\dots,M \qquad (2.7)$$

where

$$A = \int_Y \left[haar(j, y)\right]^2 dy \;.$$

It must be noted that the jump points /discontinuities/ of the above used Haar expansion are uniformly distributed along the y-axis if only $M = 2^k$ /k - integer/ - see definition of Haar functions. For $k{:=}k+1$ the number of jumps increases twice; each Haar component of the added subset results in one, two or three discontinuities.

For a given number N $/N < M/$ of intervals into which the input signal is quantized, it seems possible to optimize the shape of $Q^*(y)$ by choosing N largest coefficients among all calculated a_i's. The intervals between subsequent partitions are quantized in this case, of course; the least difference $x_i - x_{i-1}$ results from the shape of the highest order Haar function and equals the half of the subinterval where non-zero values are taken by this function. The least difference $x_i - x_{i-1}$ can be chosen however, as small as necessary, by increasing the value of M. In practical cases it would be probably limited by purely technical reasons.

The suboptimal quantizer is specified finally by

$$Q_{so}(x) = Q_{so}\left(F^{-1}(y)\right) = \sum_{\substack{i=1 \\ i \neq i_1, i_2, \dots, i_k}}^{M} a_i \; haar\left(i, F^{-1}(y)\right) = \qquad (2.8)$$

$$= \sum_{i=1}^{M} a_i \; haar(i, x) \;,$$

where $i_1, i_2, \dots, i_k$ denote orders of Haar functions which are eliminated in the optimization procedure /it holds of course $M=N+k$/.

3. Quantization of noisy signals

Let us assume the input signal u of the quantizer to be the sum of the intelligence signal x and the noise z

$$u = x + z \quad .$$

In this case the criterion of quality of the quantizing process is the expected value of the squared difference between x and the quantized output $Q(u)$, according to the formula

$$D = E\left[(x - Q(u))^2\right] = \int_U \int_X (x - Q(u))^2 f_{xu}(x, u)\, dx\, du \qquad 3.1$$

where $f_{xu}(x, u)$ - is the joint probability density function of x and u ,

U, X - sets of admissible values of the variables x and u .

In order to optimize the shape of $Q(u)$ let us introduce a new variable $y = F(x, u)$ resulting from the differential equation

$$\frac{dy}{dx} = f_{xu}(x, u) \; .$$

The criterion D will take the form

$$D = \int_U \int_Y \left[F^{-1}(y, u) - Q(u)\right]^2 dy\, du \quad , \qquad 3.2$$

where $F^{-1}(y, u)$ - denotes the inverse function of $F(x, u)$ with respect to x ,

Y - is the interval of y values.

As it was shown above, a suboptimal solution, having the form of a Fourier series with respect to Haar functions, approximating $Q(u)$ can be obtained. It has a similar form as given in Equ.2.8. The coefficients result from the formula

$$a_j = \frac{1}{AB} \int_U \int_Y F^{-1}(y, u)\, \mathrm{haar}(j, u)\, dy\, du \qquad \text{for } j=1,\ldots,M \qquad (3.3)$$

where

$$A = \int_U \left[\mathrm{haar}(i, u)\right]^2 du \quad ,$$

B - the interval of y values.

The quantizer is optimized in a similar way as described in Section 2.

References

1. Gersho A., On the Structure of Vector Quantization, IEEE Trans. on Inf.Theory, IT-28,pp.157-166, March 1982
2. Kurtenbach A.J., Wintz P.A., Quantizing for Noisy Channels, IEEE Trans.on Comm.Techn., COM-A,No 2, pp.291-302, April 1969
3. Lloyd S.P., Least Squares Quantization in PCM, IEEE Trans.on Inf. Theory, IT-28, pp.128-137, March 1982
4. Kroschel K., On Optimal Quantization of Noisy Signals, Signal Processing III: Theories and Applications, I.T.Young et al. /editors/, Elsevier Science Publishers B.V./North-Holland/, EURASIP 1986

INVARIANT DESCRIPTION OF PICTORIAL PATTERNS VIA GENERALIZED AUTO-CORRELATION FUNCTIONS

Helmut Glünder
Lehrstuhl für Nachrichtentechnik, Technische Universität München
Arcisstr. 21, D-8000 München 2, FRG

Abstract. A systematic approach to geometrically invariant pattern description is proposed. It is based on the definition of transformation invariants. The generalized auto-correlation function is introduced as a signal representation from which such invariant descriptors can be derived. Descriptors remaining invariant under all similarity transformations are briefly discussed.

Introduction

In contrast to coding (redundancy removal), pattern description implies a considerable loss of information (irrelevance) and thus in general does not permit unequivocal signal reconstruction. Pattern descriptions are essentially determined by specific tasks and pattern repertoires. Especially pictorial patterns represent enormous amounts of information and thus allow for large varieties of descriptions. This manifold, however, is reduced by quite a pragmatic demand: the invariance of pattern descriptions under geometric transformations. Although it is intuitively clear that increased invariance causes decreased richness of description, only few systematic and practical approaches to the *principal limits* of invariant pattern description are known. The following considerations are based on the definition of transformation invariants and thus are independent from both the purpose of descriptions (except for the claim for invariance itself) and the patterns to be described. Consequently, the resulting descriptions cannot be based on any kind of prototype detection, e.g., on *pattern elements*. In fact, they express general but nonetheless obvious *pattern properties*; mainly various types of symmetry (cf. Radig and Schlieder 1984), i.e., categories outside the reach of template matching techniques, including integral transformations.

Limits of invariant pattern description

The limits of invariant pattern description are given by the invariants of geometric transformations. These are those properties of pattern functions that remain unaltered under certain transformations; for example:

GROUP OF TRANSFORMATIONS	INVARIANTS
• projective	cross-ratio (length double-ratio)
• affine	+ area ratio, length ratio, parallelism
• similarity	+ angle
• congruence	+ area, length

Therefore, invariants represent the categories of invariant pattern description. But how can they be dealt with in practice? – Optimal invariant pattern description, with respect to certain transformations, means to specify the degree to which invariant properties are present in a pattern. Testing this for a single type of transformation is easy: The pattern is transformed and, for every value of the transformation parameter, the distance to the untransformed pattern is determined according to a suitable measure. The resulting comparison function, by definition is invariant under the applied transformation (inherent invariance). In general, this does not hold for several types of transformations, e.g., groups. In these cases the multiple-invariant intersection of the sets of invariants must be determined.

Generalized auto-correlation function

A mathematical formulation of the aforementioned principle is introduced by the generalized auto-correlation function (ACF) (cf. Glünder et al. 1984, and Strube 1985).

$$c(\mathfrak{z}_1 \ldots \mathfrak{z}_m) = \int \bigvee_{i=1}^{m} [\, b(\mathcal{T}_i\, \boldsymbol{x})\,]\, d\boldsymbol{x}, \qquad \text{for} \quad b(\boldsymbol{x}) \geq 0 \tag{1}$$

with $\bigvee_{i=1}^{m} [\ldots]$, symbolizing a mathematical operation on m terms; e.g., multiplication: $\prod_{i=1}^{m} [\ldots]$;

and $\mathcal{T}_i = \mathcal{T}(\mathfrak{z}_i)$, indicating geometric transformations, characterized by the ν-dimensional vector $\mathfrak{z}_i = (\mathfrak{z}_{i1} \ldots \mathfrak{z}_{im})^T$, and acting on the n-dimensional argument $\boldsymbol{x}$ of a pattern function $b(\boldsymbol{x})$;

According to the last section, the extraction of invariant properties from pictorial patterns (n=2) is based on comparisons between a pattern and its transformed versions (m=2). Furthermore, the following investigations are confined to multiplicative comparisons, and to the group of similarity transformations (ν=5). Therefore, equation (1) can be written in the following way

$$c_{sim}(\xi,\eta,s,\psi,\sigma) = \iint b(x,y) \cdot b\{[r/s \cdot \cos(\varphi+\psi)]+\xi,\ [\sigma \cdot r/s \cdot \sin(\varphi+\psi)]+\eta\}\, dxdy, \tag{2}$$

with $r = \sqrt{x^2 + y^2}$ and $\varphi = \arctan(y/x)$; where $\sigma = -1$ leads to reflections at the x-axis.

The thus defined (second order) similarity-ACF *contains* invariant information about the shape (in the mathematical sense) of patterns. Invariance can be expected with regard to all similarity transformations; i.e., to changes of size (s), sense (σ), as well as shift (ξ,η) and rotational (ψ) position of a pattern.

Similarity-invariant descriptors

The complete similarity-ACF is *not invariant at all.* However, three subspaces are invariant under the group of similarity transformations if one gets rid of the constraints imposed by the fixed points of the rotations, scale changes, and reflections. The most pronounced (primary) descriptors are obtained via maximum detection with respect to the limiting parameters.

$$C_s(s) = \max_{\xi,\eta} \{c_{sim}(\xi,\eta,s,\psi=0^\circ,\sigma=+1)\} \qquad \text{with } 0 < s \leq 1, \text{ or } \quad 1 \leq s < \infty$$

$$C_\psi(\psi) = \max_{\xi,\eta} \{c_{sim}(\xi,\eta,s=1,\psi,\sigma=+1)\} \qquad \text{with } 0^\circ \leq \psi \leq 180^\circ$$

$$C_\sigma(\sigma) = \max_{\xi,\eta,\psi} \{c_{sim}(\xi,\eta,s=1,\psi,\sigma)\} \qquad \text{with } \sigma = \pm 1 \quad \text{and } 0^\circ \leq \psi < 360^\circ$$

The descriptor C_s indicates the maximum degree of size-similarity of a pattern as a function of scale. Hence, patterns consisting of straight lines that join or intersect at a common point result in C_s=const. Contrarily, any fraction of a circle line leads to a sudden decrease of this function for $s \neq 1$. The descriptor C_ψ reveals the maximum angular bindings of a pattern; thus, a perfect circle is the only pattern with C_ψ=const. Patterns of at least one perfect axial symmetry are identified by $C_\sigma(-1)=C_\sigma(+1)$. In all the other cases the coefficient $C_\sigma(-1)$ expresses the maximum degree of glide-reflection symmetry.

Compound descriptors show restricted invariance, for example:

$$C_1(\psi,\sigma) = \max_{\xi,\eta} \{c_{sim}(\xi,\eta,s=1,\psi,\sigma)\} \qquad \text{with } 0^\circ \leq \psi < 360^\circ \qquad \text{(rotation variant)}$$

$$C_2(s,\psi) = \max_{\xi,\eta} \{c_{sim}(\xi,\eta,s,\psi,\sigma=+1)\} \qquad \text{with } 0 < s \leq 1 \text{ and } 0^\circ \leq \psi < 360^\circ, \text{ or } 0 < s < \infty \text{ and } 0^\circ \leq \psi < 180^\circ. \qquad \text{(sense variant)}$$

Each value of all these descriptor functions reflects the maximum degree of pattern binding for the optimal shift position(s). However, no information is obtained about its relation to values for other positions. Consequentely, more of the invariant information is revealed, if maximum detection is accompanied by other, e.g., statistical methods, such as analysis of variance, etc.

Related descriptors

There are no *straightforward* descriptors invariantly indicating shift congruence of a pattern. Doyle (1962) showed, however, that such descriptors can be derived from the translational (classical) ACF $c_t(\xi,\eta)$. For this purpose, pattern b in equation (2) is replaced by its ACF c_t, which is a centred and central-symmetric function. Thus, the shift transformations are obsolete and the ranges of the rotation angles are halved. The three resulting (secondary) descriptors express inner bindings of the translational ACF, with respect to size, angle and sense. Although seemingly similar to the primary descriptors, they are of significantly reduced descriptive power and less obvious, but well-suited to supplement the primary descriptors.
Another quite important approach uses the fixed point function f_2, instead of pattern b in equation (2).

$$f_2(x_f,y_f) = \iint c_{f2}'(x_f,y_f,s,\psi)\, ds d\psi$$

The integrand c_{f2}' is a binary function which is derived from the similarity-ACF. It contains the 'optimal' fixed point coordinates (x_f,y_f), i.e., those shift positions that lead to the maxima of descriptor C_2. The resulting descriptors express highly abstract pattern properties, such as compactness, regularity, etc. Besides this, the fixed point that appears most frequently can serve for a final pattern-coherent description.

Concluding remarks

The generalized second order ACF represents a reasonable basis and a tool for the creation of a family of invariant pattern descriptors. For the case of similarity-invariance, indications are given for a systematic approach to the limits of pattern description, as defined by the group-invariants. There are two fundamentally different strategies to compute the generalized ACF: the 'explicit' one, where geometric transformations are actually performed, and the 'implicit' one, which is based on generalized dipole moments. The examples shown on the next page were computed, following an explicit technique, by correlating the pattern with its discretely transformed versions. Correlation and geometric transformations were performed by a new kind of incoherent-optical analog-correlator, maximum detection by digital evaluation. The implicit computation, based on comparisons between every two points of suitably chosen pairs of pixels, is advantageous for serial processing (cf. Glünder and Kramer 1986) but also for the implementation on special purpose parallel computing networks (Glünder 1986).

References

Doyle W (1962) Operations useful for similarity-invariant pattern recognition. J ACM **9**:259-267

Glünder H (1986) Neural computation of inner geometric pattern relations. Biol Cybern **55**:239-151

Glünder H, Kramer T (1986) Description of planar patterns by invariant features – an attempt towards the explanation of visual pattern recognition. In: Guiho G (ed) Proc of 8th ICPR. IEEE Comp Soc Press, Washington DC, pp 1090-1093

Glünder H, Gerhard A, Platzer H, Hofer-Alfeis J (1984) A geometrical-transformation-invariant pattern recognition concept incorporating elementary properties of neural circuits. In: Wein M (ed) Proceedings of the 7th International Conference on Pattern Recognition (ICPR). IEEE Comp Soc Press, Washington DC, pp 1376-1379

Radig B, Schlieder C (1984) RS-automorphisms and symmetrical objects. In: Wein M (ed) Proceedings of the 7th International Conference on Pattern Recognition (ICPR). IEEE Comp Soc Press, Washington DC, pp 1138-1140

Strube HW (1985) A generalization of correlation functions and the Wiener-Khinchin theorem. Signal Process **8**:63-74

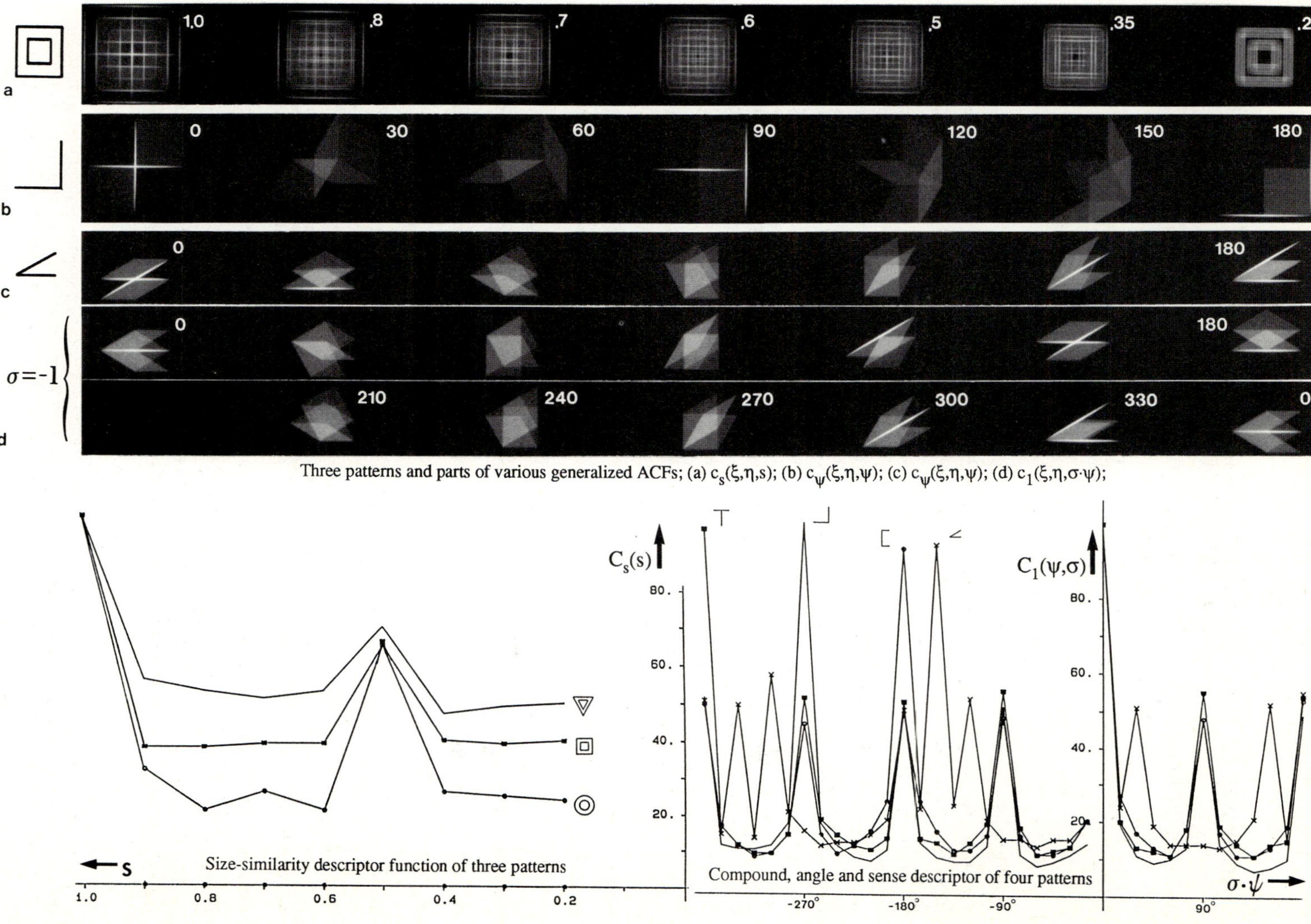

Three patterns and parts of various generalized ACFs; (a) $c_s(\xi,\eta,s)$; (b) $c_\psi(\xi,\eta,\psi)$; (c) $c_\psi(\xi,\eta,\psi)$; (d) $c_1(\xi,\eta,\sigma\cdot\psi)$;

Size-similarity descriptor function of three patterns

Compound, angle and sense descriptor of four patterns

MEHRKANALIGE KLEINSTE-QUADRATE-SCHÄTZUNG AUTOREGRESSIVER PARAMETER ZUR SENSORAUSFALLERKENNUNG

W. Ptacek, U. Appel
Institut für Mathematik und Datenverarbeitung
Fakultät für Elektrotechnik
Universität der Bundeswehr München
D-8014 Neubiberg, FRG

Übersicht:

In der Automatisierungstechnik wird in einem zu überwachenden Prozeß häufig mit mehreren Sensoren an unterschiedlichen Stellen gemessen. Die so erhaltenen Meßsignale sind zwar nicht identisch, aber statistisch voneinander abhängig. Unter Ausnutzung dieser sog. "analytischen Redundanz" werden mehrkanalige Verfahren vorgestellt, die sprunghafte Änderungen des Leistungsspektrums von Meßsignalen - wie sie oftmals typisch sind für den Ausfall oder die plötzliche Änderung von Eigenschaften eines Systems - erkennen. Dabei werden sowohl Änderungen in dem gemessenen Prozeß (gemeinsamer Sprung in allen Kanälen) als auch Ausfälle und gegebenenfalls "Ausreißer" einzelner Sensoren (Sprung in einem einzigen Kanal) erkannt und voneinander unterschieden. Neben einem q-kanaligen multivariaten Ansatz, der bei Vernachlässigung des Rechenaufwandes diese Aufgabe prinzipiell löst, werden insbesondere sinnvolle und effiziente Vereinfachungen dieses allgemeinen Ansatzes angegeben. Die wesentlichen Elemente dieser mehrkanaligen Verfahren sowie deren Eigenschaften werden erläutert.

1. Einleitung

In der Vergangenheit sind zahlreiche Verfahren zur Ausfallerkennung in dynamischen Systemen entwickelt worden. Einen ausgezeichneten Überblick darüber findet man in dem Aufsatz von Willsky /4/ . Alle diese analytischen Ansätze setzen jedoch die a priori Kenntnis der Systemparameter voraus. Im Gegensatz dazu ist die einzige Voraussetzung für das hier vorgestellte Verfahren die Modellierbarkeit des Meßsystems durch das in Bild 1 dargestellte Blockschaltbild. Ein dominierender Prozeß, dargestellt durch farbiges Rauschen, wird von mehreren Sensoren mit unbekannter Übertragungsfunktion überwacht. Die auf diese Weise erhaltenen Multi-Sensor-Signale sind zwar nicht identisch, aber statistisch voneinander abhängig, da sie von ein- und demselben Prozeß abgeleitet sind. Unter Ausnutzung der in der Kovarianzmatrix der Sensorsignale enthaltenen "analytischen Redundanz" lassen sich diejenigen Abweichungen von der gegenwärtigen Korrelation erkennen, die durch einen Sensorausfall hervorgerufen werden.

Die hohe Detektionsgüte des hier beschriebenen Ausfallerkennungssystems wird durch die Dekorrelation der Meßsignale erreicht. Dabei wird jedes Sensorsignal durch eine Linearkombination seiner vergangenen Abtastwerte und den Abtastwerten der anderen Meßkanäle vorhergesagt. Die Prädiktionsfehler - häufig auch Residualsignale genannt - enthalten somit die nicht vorhersagbaren (dekorrelierten) Anteile des jeweiligen Meßsignals. Während sprunghafte Veränderungen in den Signaleigenschaften aller Meßkanäle unterdrückt werden, treten in der Kovarianzmatrix der Residualsignale Ausfälle und gegebenenfalls "Ausreißer" in einem einzelnen Kanal deutlicher hervor. Anhand dieser sprunghaften Veränderungen in den Residualenergien, die durch einen maximum-likelihood-Ansatz /1/ optimal detektiert werden können, läßt sich ein Sensorausfall selbst bei zeitvariablen Prozessen eindeutig erkennen.

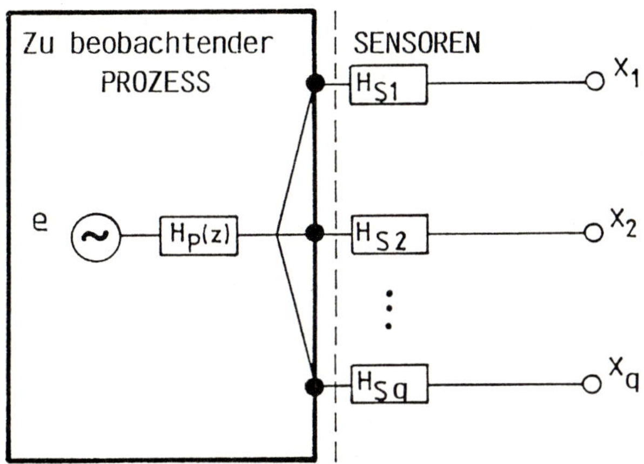

Bild 1: Modell eines q-kanaligen Meßaufbaus.

2. Allgemeiner q-kanaliger multivarianter Ansatz

Das Prinzip des neuen Ausfallerkennungssystems ist es also, unter Benutzung eines AR-Parameterschätzalgorithmus' ein inverses Filter für die Sensorsignale zu bestimmen, um anhand der Kovarianzmatrix der Filterausgangssignale (Residualsignale) eine Ausfallentscheidung zu treffen. Durch den folgenden, nichtrekursiven Ansatz zur AR-Parameterschätzung lassen sich die wesentlichen Eigenschaften des q-kanaligen multivariaten Verfahrens aufzeigen.

Die Erzeugung des Residualsignals für den Kanal i beruht auf der Vorhersage des Meßsignals $x_i(n)$ durch eine gewichtete Mittelung der vergangenen Abtastwerte $x_j(n-m)$ (mit $1 \leq m \leq p$, $1 \leq j \leq q$) sowie der <u>aktuellen Signalwerte</u> $x_j(n)$ (mit $1 \leq j \leq q$, $j \neq i$) der übrigen Kanäle.

Damit ergeben sich der Schätzwert $\underline{\hat{x}}(n)$ und der Prädiktionsfehler $\underline{e}(n)$ zu:

$$\underline{\hat{x}}(n) = \sum_{m=0}^{p} \underline{x}(n-m) \bullet \underset{\sim}{A}_m(n) \qquad \text{mit } \underline{x}(n) = \left[x_1(n) \; \ldots \; x_q(n)\right] \tag{2.1}$$

$$\underline{e}(n) = \underline{x}(n) - \underline{\hat{x}}(n) \qquad \text{mit } \underline{e}(n) = \left[e_1(n) \; \ldots \; e_q(n)\right] \tag{2.2}$$

Die Koeffizientenmatrizen $\underset{\sim}{A}_m(n)$ (für $1 \leq m \leq p$) sind allgemeine Matrizen der Dimension $q \times q$. Lediglich die Matrix $\underset{\sim}{A}_0(n)$ weist als Besonderheit eine mit Nullen besetzte Hauptdiagonale auf. Die Minimierung des quadratischen Fehlers über einen Beobachtungszeitraum von v Abtastwerten führt auf das folgende System von Normalgleichungen zur Bestimmung der Koeffizientenblockmatrix $\underset{\approx}{A}(n)$ zum Zeitpunkt n.

$$\underset{\approx}{A}(n) = \underset{\approx}{C}(n)^{-1} \bullet \underset{\approx}{B}(n) \qquad \text{mit } \underset{\approx}{A}(n) = \left[\underset{\sim}{A}_0^T(n) \; \ldots \; \underset{\sim}{A}_p^T(n) \right]^T \tag{2.3}$$

Kovarianzblockmatrix:

$$\underset{\approx}{C}(n) = \begin{bmatrix} \underset{\sim}{C}_{0,0}(n) & \ldots & \underset{\sim}{C}_{0,p}(n) \\ \vdots & & \vdots \\ \underset{\sim}{C}_{p,0}(n) & \ldots & \underset{\sim}{C}_{p,p}(n) \end{bmatrix}$$

$$c_{l,k}^{(i,j)} = \sum_{t=n-v+1}^{n} x_i(t-l) \bullet x_j(t-k)$$

$$\underset{\approx}{B}(n) = \begin{bmatrix} \underset{\sim}{C}_{0,0}(n) \\ \vdots \\ \underset{\sim}{C}_{p,0}(n) \end{bmatrix} \qquad \begin{matrix} 0 \leq l,k \leq p \\ 1 \leq i,j \leq q \end{matrix}$$

Wie aus der Gleichung (2.3) hervorgeht, wird zur Dekorrelation der Meßsignale die maximal verfügbare Information der Kovarianzblockmatrix verwendet. Damit stellt dieser Ansatz eine wesentliche Erweiterung zur vektorautoregressiven Parameterschätzung /3/ dar, die keinen Gebrauch von den ersten q Spalten und Zeilen von $\underset{\approx}{C}(n)$ macht. Doch gerade die aktuellen Abtastwerte von Multi-Sensor-Signalen zeigen eine starke statistische Bindung, die bei ihrer Berücksichtigung zu einer erheblichen Verringerung der Prädiktionsfehlerenergie führt und somit sprunghafte Veränderungen deutlicher hervortreten läßt. Neben den guten Detektionseigenschaften zeichnet sich das neue Verfahren vor allem durch die Tatsache aus, daß für das oben beschriebene Minimierungsproblem ein numerisch gut konditionierter (matrixinversionsfreier) rekursiver Algorithmus angegeben werden kann.

3. $\binom{q}{2}$ -Verarbeitung

Zunächst ist bei einer zweikanaligen Verarbeitungsweise eine eindeutige Unterscheidung von Änderungen im gemessenen Prozeß und Ausfällen einzelner Sensoren prinzipiell erschwert. So kann der Ausfall eines Sensors neben einem Anstieg des Prädiktionsfehlers im zugehörigen Kanal auch dazu führen, daß das ungestörte Meßsignal schlechter vorhergesagt wird, und die Residualenergie auch in diesem Kanal ansteigt; denn im Gegensatz zum q-kanaligen Verfahren existieren hier keine weiteren Meßsignale, die diesen Informationsverlust kompensieren können. Damit ist anhand der Kovarianzmatrix der Residualsignale allein keine eindeutige Unterscheidung zwischen einem Sensorausfall und einem sprunghaften Anstieg der Energie im zu beobachtenden Prozeß möglich. Erst durch Kombination der q Sensoren zu je zwei ($\binom{q}{2}$) erreicht man, daß bei einem Sensorausfall immer eine "ungestörte" Zweier-Kombination existiert, anhand deren der defekte Sensor eindeutig identifiziert werden kann. Damit lassen sich durch paarweises Zusammenfassen der Sensorausgangssignale und anschließender zweikanaliger Weiterverarbeitung unter Beibehaltung der guten Detektionseigenschaften des multivariaten Ansatzes, aber unter Umgehung dessen hohen Aufwandes, zwei robuste Verfahren zur mehrkanaligen Sensorausfallerkennung angeben:

1) Das bivariate Verfahren ist der numerisch besonders einfach zu handhabende, zweikanalige Sonderfall des multivariaten Parameterschätzalgorithmus' und verwendet die vollständige Kovarianzmatrix von Meßsignalpaaren. Die Eigenschaften dieses Verfahrens sind mit dem allgemeinen Ansatz vergleichbar.

2) Die Verarbeitung als zweidimensionaler Verbundprozeß (Joint process) ist die am weitestgehende Vereinfachung. Bei diesem Ansatz wird ein Schätzwert des Signals $x_2(n)$ (Joint-Zweig) durch eine gewichtete Linearkombination der Abtastwerte $x_1(n-m)$ $(0 \leq m \leq p-1)$ berechnet.

$$\hat{x}_2(n) = \sum_{m=0}^{p-1} x_1(n-m) \cdot d_m(n) \tag{3.1}$$

$$e(n) = x_2(n) - \hat{x}_2(n) \tag{3.2}$$

Ebenso wie in (2.3) führt auch hier die Minimierung der Residualenergie innerhalb eines Beobachtungszeitraums von v Abtastwerten auf ein System von Normalgleichungen zur Bestimmung der Filterkoeffizienten d_m.

$$\underline{d}(n) = \underset{\sim}{c}^{-1} \cdot \underline{b} \qquad \text{mit } c_{l,k} = \sum_{t=n-v+1}^{n} x_1(t-l) \cdot x_1(t-k) \tag{3.3}$$

mit

$$\underline{d}(n) = \left[d_1 \; \dots \; d_p\right] \qquad b_l = \sum_{t=n-v+1}^{n} x_1(t-l) \cdot x_2(n)$$

$$0 \le l, k \le p-1$$

Damit unterdrückt das Filter jene Komponente im Signal $x_2(n)$ bestmöglich im Sinne eines Kleinste-Quadrate-Approximationkriteriums, welche mit dem Signal $x_1(n-m)$ (für $0 \le m \le p-1$) korreliert sind. Im Gegensatz zum bivariaten Ansatz wird jedoch zur Dekorrelation von $x_2(n)$ kein Gebrauch von den vergangenen Abtastwerten $x_2(n-m)$ (für $1 \le m \le p-1$) gemacht. Sprunghafte Veränderungen der Signaleigenschaften im Joint-Zweig bilden sich zwar ebenfalls in einem abrupten Anstieg der Residualenergie ab, während "Gemeinsames" unterdrückt wird, doch bleibt die Detektionsgüte hinter dem bivariaten Verfahren zurück. Auf der Grundlage des zweidimensionalen Verbundprozesses wurde ein aufwandsgünstiger, rekursiver Algorithmus für die Erkennung von Parametersprüngen bei mehrkanaliger Messung entwickelt und ist in /2/ ausführlich beschrieben.

4. Zusammenfassung

In diesem Aufsatz wurden drei Ansätze für die Entwicklung eines Ausfallerkennungssystems angegeben. Das einzige a priori Wissen, das in diesen Methoden Verwendung findet, ist die Modellierbarkeit des Meßsystems mittels eines Blockschaltbildes gemäß Bild 1, dessen Systemparameter unbekannt sind. Für alle Ansätze existieren rekursive und numerisch gut konditionierte Algorithmen. Die Verfahren wurden anhand zahlreicher Simulationsläufe mit Multi-Sensor-Signalen getestet, die die Vibrationen eines Motor-Getriebe-Systems aufnahmen, und zeigten die erwarteten guten Detektionseigenschaften.

5. Literatur

/1/ Appel U., Brandt A. v. (1983). Adaptive sequential segmentation of piecewise stationary time series.
Information Sciences, 29, pp. 27-56

/2/ Appel U., Ptacek W. (1986). Sensor fault detection by means of joint ladder estimation.
Proc. EUSIPCO-86, The Hague, pp. 1009-1012

/3/ Ptacek W., Appel U. (1986). Detection of sensor faults by means of multivariate calculation methods.
Proc. 2nd IFAC Workshop on Adaptive Systems in Control and Signal Processing, Lund, Sweden, pp. 155-160

/4/ Willsky A. S. (1976). A survey of design methods for failure detection in dynamic systems.
Automatica, 12, pp. 601-611

Walsh Series in Polar Coordinates

WILLIAM R. WADE

Mathematics Department
University of Tennessee
Knoxville, TN 37996-1300, U. S. A.

Walsh functions $\Psi_0, \Psi_1, \ldots$ have many applications to information theory (see [**3**]), especially to problems of pattern recognition and image enhancement. The Walsh system shares many properties with other orthonormal systems but is distinguished by the fact that each Ψ_j is locally constant, takes on only the values ± 1, and the intervals of constancy shrink uniformly to points as $j \to \infty$.

For two-dimensional problems the hybrid double Walsh system

$$\Psi_k(x)\Psi_j(y) \tag{1}$$

is often used. This approach is not entirely satisfactory. First, the supports of the functions (1) fill the unit square. Thus for problems which are circular in nature (e.g., data from sonar, air traffic control, or large tropical depressions) there is much waste storing data from corners which are both unwanted and unneeded. Secondly, intervals of constancy of (1) do not shrink uniformly to points as $k+j \to \infty$. And third of all, ordering the double Walsh system is problematic and delicate. For example, when considering convergence of double Walsh-Fourier series should one use rectangular sums, circular sums, or sums based on some other polygonal region? The difficulty in answering this question is illustrated by how little is known about convergence by any of these methods. Moreover, when the answers are known they are not always what one wants. For example, there exist $f \epsilon L^p$, $1 < p < 2$, whose double Walsh-Fourier series diverge a.e. when summed over any regular polygonal region [**4**].

We introduce polar Walsh functions on the unit disc Δ_0 which eliminate these difficulties. First, set

$$w_0 \equiv 1$$

and

$$w_1(r,\theta) = \begin{cases} +1 & 0 \le \theta < \pi \\ -1 & \pi \le \theta < 2\pi, \end{cases}$$

$0 \le r < 1$. For each $j \epsilon [2^\ell, 2^{\ell+1})$, $\ell = 1, 2, \ldots$ define sets $\Delta_j^{(i)}$, $i = 1, 2$, by the following process. If $\ell = 2m$ is even, write j uniquely as

$$j = 2^{2m} + p2^m + q$$

where $0 \leq p < 2 \;\;, 0 \leq q < 2$ and set

$$\Delta_j^{(1)} = \left\{(r,\theta) : \sqrt{\frac{p}{2^m}} \leq r < \sqrt{\frac{p+1}{2^m}},\ \frac{q\pi}{2^{m-1}} \leq \theta < \frac{(q+\frac{1}{2})\pi}{2^{m-1}}\right\},$$

$$\Delta_j^{(2)} = \left\{(r,\theta) : \sqrt{\frac{p}{2^m}} \leq r < \sqrt{\frac{p+1}{2^m}},\ \frac{(q+\frac{1}{2})\pi}{2^{m-1}} \leq \theta < \frac{(q+1)\pi}{2^{m-1}}\right\}.$$

If $\ell = 2m+1$ is odd, write j uniquely as

$$j = 2^{2m+1} + p2^{m+1} + q$$

where $0 \leq p < 2^m$, $0 \leq q < 2^{m+1}$ and set

$$\Delta_j^{(1)} = \left\{(r,\theta) : \sqrt{\frac{p}{2^m}} \leq r < \sqrt{\frac{p+\frac{1}{2}}{2^m}},\ \frac{q\pi}{2^m} \leq \theta < \frac{(q+1)\pi}{2^m}\right\},$$

$$\Delta_j^{(2)} = \left\{(r:\theta) : \sqrt{\frac{p+\frac{1}{2}}{2^m}} \leq r < \sqrt{\frac{p+1}{2^m}},\ \frac{q\pi}{2^m} \leq \theta < \frac{(q+1)\pi}{2^m}\right\}.$$

Then $\left\{\Delta_j^{(1)} : i = 1,2,\ j = 2^\ell, 2^\ell+1, \ldots, 2^{\ell+1}-1\right\}$ forms a partition of the interior of Δ_0 for each $\ell \geq 1$.

Set $\rho_0 \equiv w_1$. For $\ell = 1,2,\ldots.$ Define ρ_ℓ on the interior of Δ_0 by

$$\rho_\ell(r,\theta) = \begin{cases} +1 & (r,\theta)\epsilon\Delta_j^{(1)} \\ -1 & (r,\theta)\epsilon\Delta_j^{(2)} \end{cases}$$

for $2^\ell \leq j < 2^{\ell+1}$. For each integer $k \geq 2$ define the polar Walsh function w_k by

$$w_k(r,\theta) = \rho_{\ell_1}(r,\theta)\cdots\rho_{\ell_\nu}(r,\theta)$$

where k is written uniquely as

$$k = 2^{\ell_1} + \cdots + 2^{\ell_\nu}$$

and $\ell_1 > \ell_2 > \cdots > \ell_\nu \geq 0$. Finally, extend each polar Walsh function to the closed unit disc by

$$w_k(1,\theta) = \lim_{r\uparrow 1} w_k(r,\theta)$$

$0 \leq \theta < 2\pi$, $k = 0,1,\ldots.$

One can show that the polar Walsh system is complete and orthonormal on $L^2(\Delta_0)$ if two dimensional Lebesque measure is normalized by π. Thus for each f integrable on Δ_0 define polar Walsh-Fourier coefficients by

$$\hat{f}(j) = \frac{1}{\pi}\int_0^{2\pi}\int_0^1 f(r,\theta)w_j(r,\theta)r\,dr\,d\theta,$$

$j = 0,1,\ldots.$ Since the w_j's are constant on the sets $\Delta_j^{(i)}$, $i = 1,2$, and the area of these sets diminishes to zero as $j \to \infty$, it is not difficult to see that the Riemann-Lebesque lemma holds for polar Walsh-Fourier coefficients, i.e., $\hat{f}(j) \to 0$ as $j \to \infty$. On the other hand, this property nearly characterizes polar Walsh-Fourier coefficients.

THEOREM 1. *Suppose f is integrable on the unit disc Δ_0 and $S = \sum a_j w_j$ is a polar Walsh series whose coefficients satisfy $a_j \to 0$ as $j \to \infty$. If the partial sums of S satisfy $S_n \to f$ a.e. and*

$$\limsup_{n \to \infty} |S_n(r, \theta)| < \infty \tag{2}$$

for all $(r, \theta) \epsilon \Delta_0 \setminus \bigcup_{k=1}^{\infty} w_k$, where $w_1, w_2, \ldots$ is a countable collection of embedded arcs in Δ_0, then

$$a_j = \hat{f}(j)$$

for $j = 0, 1, \ldots$.

Let $S(f)$ represent the polar Walsh series of f, i.e.,

$$S(f) = \sum_{j=0}^{\infty} \hat{f}(j) w_j .$$

Then one can show the following:

THEOREM 2. *Let f be integrable on Δ_0. If $f \epsilon L^p(\Delta_0)$ for some $1 \leq p < \infty$ then $S_{2^n}(f)$ converges a.e. and in L^p norm to f. If f is continous on Δ_0 then $S_{2^n}(f)$ converges uniformly to f on Δ_0.*

Polar Walsh series can even be used to approximate non-integrable functions:

THEOREM 3. *If f is measurable and a.e. finite on Δ_0 then there is a polar Walsh series whose 2^nth partial sums converge to f a.e..*

Clearly, polar Walsh functions take on only the values ± 1, are constant on sets of the form $\Delta_j^{(i)}$, $i = 1, 2$, and these sets shrink uniformly to points. The support of each polar Walsh function is precisely the unit disc. And, polar Walsh functions have a natural ordering which leads to good convergence and uniqueness properties as indicated in Theorems 1, 2, and 3. Hence this system offers a replacement for the double Walsh system. Moreover, the same techniques are easily adapted to other geometric configurations including the unit square. Thus we offer a system which may compete favorably with the double Walsh system even in square geometries.

Detailed proofs of these results will not be given. They can be constructed by modifying techniques introduced for Haar systems in [**1**] and [**2**]. A Fast Polar Walsh Transform is also possible. An exposition will appear later.

References

[1] G. E. Albert and W. R. Wade, *Haar systems for compact geometries*, (preprint).

[2] R. F. Gundy, *Martingale theory and pointwise convergence of certain orthogonal series*, Trans. Amer. Math. Soc. **124** (1966), 228–248.

[3] H. F. Harmuth, "Transmission of Information by Orthogonal Functions," Springer-Verlag, Berlin, 1972.

[4] D. C. Harris, *A.e. divergence of multiple Walsh-Fourier series*, Proc. Amer. Math. Soc. (in print).

NONUNIFORM SAMPLING FOR MULTIDIMENSIONAL SIGNALS

Farokh A. Marvasti
Illinois Institute of Technology
Chicago, Illinois

We will discuss the nonuniform samples of periodic signals first and then analyse any type of multidimensional signal. A bandlimited periodic signal is represented by a finite number of Fourier coefficients and hence can be represented by the same number of nonuniform samples per period. Therefore in general, random sampling at any rate is sufficient to retrieve the signal; this is because the random samples are distinct and occur at different epochs of time.

For nonuniform sampling of bandlimited signals, Gaarder[1] has tried to extend the work of Yen[2] and Kahn[3] et al. to periodic nonuniform samples of multidimensional signals. However, since Yen's and Kahn et al.'s are already so complex, their extension to multidimensions is of little more than theoretical interest. Gaarder uses the theorems on uniform sampling to derive the frequency spectrum of the periodic nonuniform samples. However, since the constituent uniform samples are less than the Nyquist rate, the frequency domain is an aliased version of the original signal. Using this method, Gaarder argues that if nonuniform samples are at a higher rate, it is possible to retrieve the multidimensional signal from this kind of nonuniform samples, a fact which defeats the whole purpose of nonuniform sampling in practice.

The extension of Lagrange interpolation to multidimensional nonuniform samples has not been reported in the literature. One problem is that for 1-D signals, the interpolation function is derived from the product of zero instances at the nonuniform samples while for the multidimensional signals, the Osgood product containing zero contours should be used. This is why sampling using closed contours is a more natural extension of 1-D sampling than using isolated points. In general, using the same argument as the 1-D sampling case, a multidimensional signal is uniquely represented by a set of irregular contour (or isolated) samples if only if no bandlimited signal can be found having zero-crossings at the sampling contours (or nonuniform) points). This theorem shows that nonuniform sampling is more restrictive than contour sampling; i.e., there is less probability that a set of nonuniform samples can uniquely a multidimensional signal. A sufficient condition for a sampling set is that the average density of nonuniform samples (or contour samples) be higher than the Nyquist rate. This condtion is also valid for stochastic multidimensional signals with random sampling.

Stark[4] has considered the sampling theorem for 2-D signals in polar coordinates. His first theorem is the extension of Bessel interpolation to 2-D signals,i.e.,

If we assume that $f(r,\theta)$ is periodic in the θ direction and bandlimited, then this special nonuniform sampling interpolation is unique and is given by

$$f(r,\theta) = \sum_{i=1}^{\infty}\sum_{k=0}^{2K} f(\frac{z_{0i}}{a}, \frac{2\pi k}{2K+1})\, \phi_{0i}(r)\, (2K+1)^{-1} + 2(2K+1)^{-1}$$
$$X \sum_{n=1}^{K}\sum_{i=1}^{\infty}\sum_{k=0}^{2K} f(\frac{z_{ni}}{a}, \frac{2\pi k}{2K+1})\, \phi_{ni}(r) \;.\; \cos[\, n(\theta - \frac{2\pi k}{2K+1})\,]$$

where K, as a measure of bandwidth, is the number of Fourier series coefficients of $f(r,\theta)$ expanded in θ.

If the signal is not bandlimited but has a finite region of support in the nth order Hankel transform, the interpolation is

$$f(r,\theta) = \sum_{k=0}^{2K}\sum_{i=1}^{\infty} f(\frac{z_{ni}}{a}, \frac{2\pi k}{2K+1})\, \phi_{ni}(r) \;.\; \frac{\sin[\frac{2K+1}{2}(\theta - \frac{2\pi k}{2K+1})]}{\sin[\frac{1}{2}(\theta - \frac{2\pi k}{2K+1})]}$$

For nonbandlimited signals, the uniqueness theorems still hold. Clark[5] et al. has extended the transformation of Papoulis [6] to 2-D signals. The uniform sampling theorem for 2-D signals is

$$f(\vec{x}) = \sum_{\{\vec{x}_s\}} f(\vec{x}_s)\, \frac{J_1[2\pi B(\vec{x} - \vec{x}_s)]}{2\sqrt{3}B|\vec{x} - \vec{x}_s|}$$

where the sampling set $\vec{x}_s$ is

$$\vec{x}_s = \{\vec{x} = n\vec{v}_1 + m\vec{v}_2, \quad n,m = 0,\pm 1,\pm 2, \ldots, \vec{v}_1 \neq k\vec{v}_2\}$$

The parameters J_1 and B in the above equation are respectively the first order Bessel function of the first kind and the equivalent bandwidth of 2D Fourier transform. Under the coordinate transformation

$$\vec{\xi} = \vec{\gamma}(\vec{x})$$

i.e.,

$$f(\vec{x}) = h(\vec{\xi}) = h\,[\vec{\gamma}(\vec{x})\,]$$

The result is

$$f(\vec{x}) = \sum_{\{\vec{x}_s\}} f(\vec{x}_s) \frac{J_1 [2\pi B (\vec{\gamma}(\vec{x}) - \vec{\gamma}(\vec{x}_s))]}{2\sqrt{3}\, B \,|\, \vec{\gamma}(\vec{x}) - \vec{\gamma}(\vec{x}_s) \,|}$$

f(x) is nonbandlimted if g(ξ) is to be bandlimited. Clark et al[5] has considered an approximate interpolation when f(x) is bandlimited and g(ξ) is approximately assumed to be bandlimited.

The spectrum of nonuniform samples of multidimensional signals can also be generalised from the spectrum of 1-D signals.[7] One method of analysis is to consider the nonuniform samples as the extension of FM zero crossings to multidimensional signals. Each nonuniform sample can be represented as

$$\delta [\vec{t} - n\vec{T} - t_o x(\vec{t})$$

where δ is the multidimensional delta function and x is a bandlimited signal with zero crossings passing through $\vec{t}_n$. In other words $\vec{t}_n$ are the crossings of x(t) and sawtooth planes. The Fourier series expansion of the sum of the above delta functions yields the frequency spectrum of the nonuniform samples (see reference [6]).

BIBLIOGRAPHY AND COMMENTS

1 N. T. Gaarder," A note on multidimensional sampling theorem," Proc. IEEE, vol. 60, pp.247-248, Feb. 1972. (The author tries to generalize periodic nonuniform sampling of Yen [158] to multidimensional signals. However, due to complexity of the equations, he considers a trivial, impractical case).

2 J. L. Yen, "On nonuniform sampling of bandwidth-limited signals," IRE Trans. on Circuit Theory, vol. CT-3, pp.251-257, Dec. 1956.

3 R. E. Kahn and B. Liu, "Sampling representations and optimum reconstruction of signals," IEEE Trans. Inform. Th., vol.IT-11, pp.339-347, July 1965. (The paper is on uniform sampling and optimum reconstruction methods. Their model is similar to that of Papoulis [173] for periodic nonuniform samples.)

4 H. Stark," Sampling theorems in polar coordinates," J. Opt. Soc. Am., vol.69, pp.1519-1525, Nov.1979.

5 J. J. Clark, M. R. Palmer, P. D. Lawrence," A transformation method for the reconstruction of functions from nonuniformly spaced samples," IEEE Trans. Acoustic, Speech, and Signal Processing, vol.ASSP-33, no.4, Oct. 1985. (The authors try to extend Papoulis' method [151] to general nonuniform sampling. However, it turns out that this generalization is valid only for a certain class of nonbandlimited signals and not valid for bandlimited signals. They have also tried to generalize the same concept to multidimensional signals with some applications to radio astronomy.)

6 A. Papoulis, *Signal Analysis*, New York: McGraw-Hill, 1977, pp.194-195. (Papoulis uses his Generalized Sampling Theorem to derive the periodic nonuniform sampling formula of [158]. References [167]-[169] consider the sensitivity to noise of this Generalized Sampling Theorem.)

7 F. Marvasti," Error free reconstruction of a general bandlimited signal from FM zero-crossings," A proprietary report prepared for Ampex Corp., Redwood City, CA., July 18, 1985.

--, "The reconstruction of a signal from the zero crossings of an FM signal," Trans. IECE Japan, vol.E68, no. 10, Oct. 1985.

Methodik

Deutung von Bildfolgen anhand ihrer symbolischen Beschreibung

Bernd Radig

Institut für Informatik der Technischen Universität München
Arcisstraße 21, 8000 München 2

Zusammenfassung

Deutung von Einzelbildern und Bildfolgen abstrahiert vom Bildsignal. Dazu wird die ikonische Darstellung in eine symbolische Beschreibung überführt. Zusätzlich zur Objektidentifikation in Einzelbildern ergibt sich die Aufgabe, korrespondierende Bildsymbole innerhalb der Folge zu bestimmen. Die Kenntnis der Korrespondenz wird benutzt, um Tiefeninformation zu gewinnen (Stereobilder), dreidimensionale Körper zu rekonstruieren (Schichtaufnahmen) oder Bewegung zu ermitteln (Zeitfolgen). Diese Aufgaben werfen grundsätzliche Probleme auf:

- Welche Symbole werden benutzt, wie werden sie repräsentiert und wie aus dem Bildsignal extrahiert.
- Wie werden Objekte, Konfigurationen von Objekten, Szenenbeschreibungen repräsentiert und gewonnen.
- Welches Wissen wird benötigt, um die gewünschte Information (Identität, Handlung, ...) zu rekonstruieren.

Als Lösungsvorschlag für die Basisprobleme wird die Verwendung von relationalen Strukturen diskutiert. Sie repräsentieren Bildsymbole, Bildobjekte und Szenenbeschreibungen. Hierarchisch aufgebaute Strukturen modellieren Objekt- und Szenenprototypen. Strukturen werden zur Identifikation und Korrespondenzermittlung miteinander verglichen, indem mathematische Abbildungen zwischen ihnen ermittelt werden, die wiederum als relationale Strukturen repräsentierbar sind.

Weitere Probleme der Bildinterpretation sind in diesem formalen Rahmen nicht lösbar. So muß Wissen über die Entstehung der Bilder, die Zuverlässigkeit und Adäquatheit der symbolischen Beschreibung und über das Interpretationsziel geeignet formuliert werden. Welche Systemarchitektur geeignet ist, welche Repräsentationsformalismen adäquat sind und nach welchen Kriterien sie beurteilt werden, wie die Aufgabenbeschreibung (das Interpretationsziel) in den Systementwurf einfließt und wie ein solcher Entwurf effizient in ein effizientes Programmsystem umgesetzt wird, sind noch weitgehend offene Fragen.

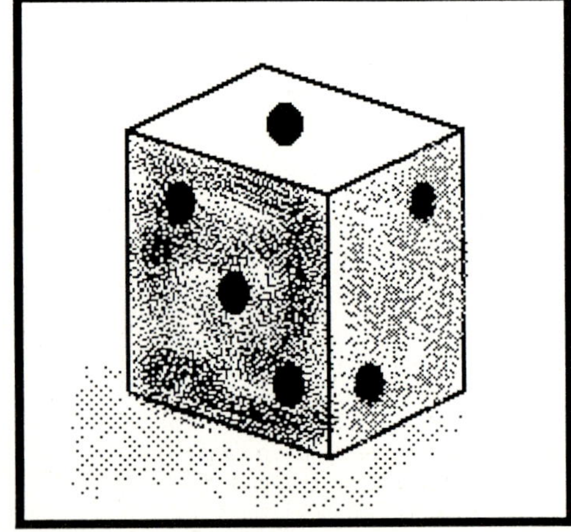

Abbildung 1: Stereobildpaar eines Würfels zur Illustration der Abstraktionsschritte bei der Bildinterpretation

Abstraktion

Die Interpretation von Einzelbildern und Bildfolgen kann man als eine geordnete Sequenz von Abstraktionsschritten ansehen (eine stilisierte Stereoaufnahme eines Würfels, siehe Abb.1 oben, illustriert die Schritte):

Transformationsebene
Transformationen im Signalraum, etwa

- Wechsel des Koordinatensystems im Farbraum,
- Rauschverminderung,
- Kontrastverbesserung,
- Differentiation, etc.
  [Ballard, Brown 82]

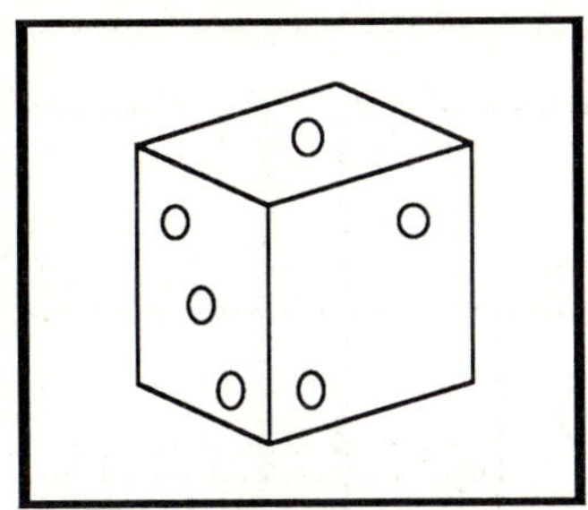

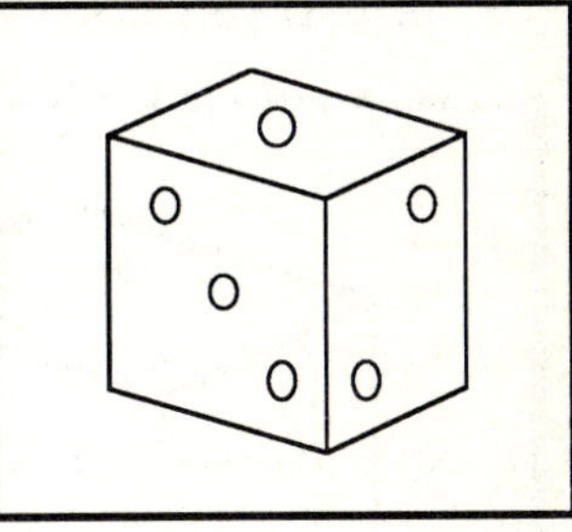

Abbildung 2: Durch approximierte Differentation gewonnene Kantenzüge

Symbolebene
Übergang in den Symbolraum, etwa

- Konturverfolgung,
- Bereichsverschmelzung,
- Formapproximation,
- Texturbeschreibung, etc.
  [Ballard, Brown 82]

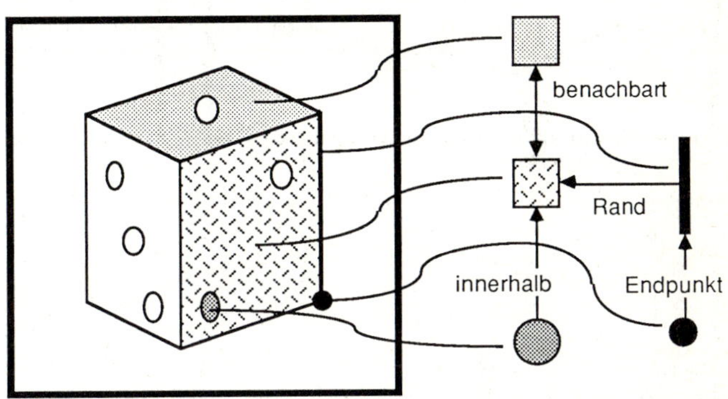

Abbildung 3
Durch Segmentation gewonnene symbolische Beschreibung. Ausschnitt aus der als Graph veranschaulichten relationalen Struktur.

2D-Objektebene
Objektsynthese (zweidimensional), etwa

- Diskrete Relaxation [Kitchen, Rosenfeld 79],
- Konsistente Markierung [Haralick, Shapiro 79,80],
- ε-Homomorphismen [Shapiro, Haralick 81],
- Produktionensysteme [Stein 83, Riseman, Hanson 85, Venable et al. 85],
- Subgraph Isomorphismen [Cheng, Huang 81], etc.

Abbildung 4
Ausschnitt aus der Beschreibung eines zweidimensionalen Modells eines Quaders mit drei Flächen und Geradenstücke der gemeinsamen Ränder. Knoten repräsentieren die elementaren Bildsymbole, Pfeile die gerichtete Randrelation.

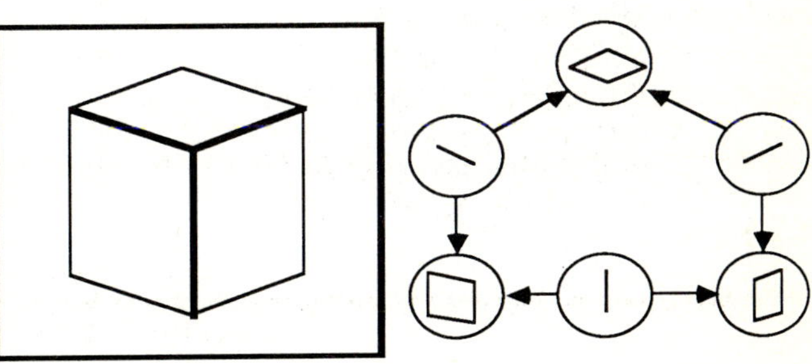

Korrespondenzebene

Korrespondenz zwischen den Bildern, etwa

- Optischer Fluß [Horn, Shunck 81, Neumann 83, Heeger 86, Wohn 86, Waxman, Wohn 86],
- Verschiebungsvektoren an markanten Punkten [Nagel 83],
- Vergleich symbolischer Beschreibungen [Radig 82, Bunke 85],
- Bewegungskompensation [May, Wolf 83, Huang et al. 86], etc.

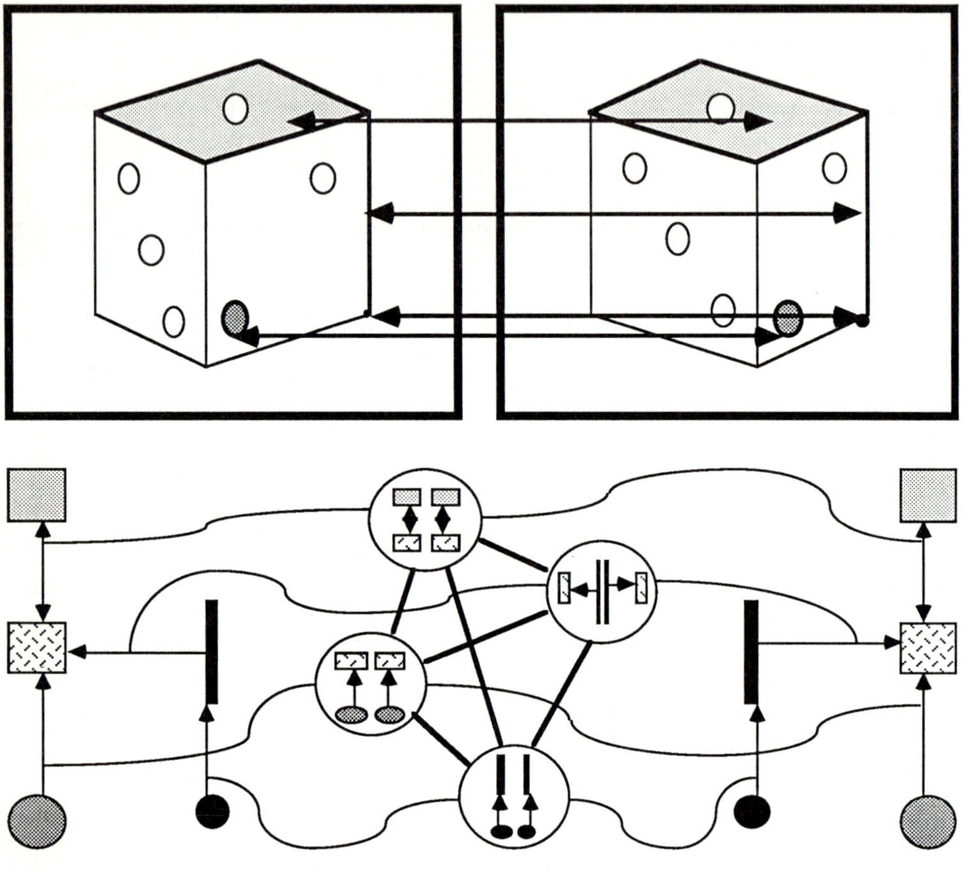

Abbildung 5: Korrespondenz über Strukturvergleich; links und rechts Ausschnitte aus relationaler Beschreibung, in der Mitte Kompatibilitätsgraph. Knoten sind Paare von Relationstupeln, Kanten verbinden verträgliche Zuordnungen; oben die Visualisierung der Zuordnung für den dargestellten Ausschnitt.

3D-Objektebene

Objektsynthese (dreidimensional), etwa

- Tiefenschätzung [Huang, Tsai 81],
- Oberflächenrekonstruktion [Westphal 85],
- Volumenbestimmung [Martin, Aggarwal 83],
- Objektidentifikation [Brooks 81, Besl, Jain 85, Freytag et al. 86, Grebner 86, Adorni et al. 85], etc.

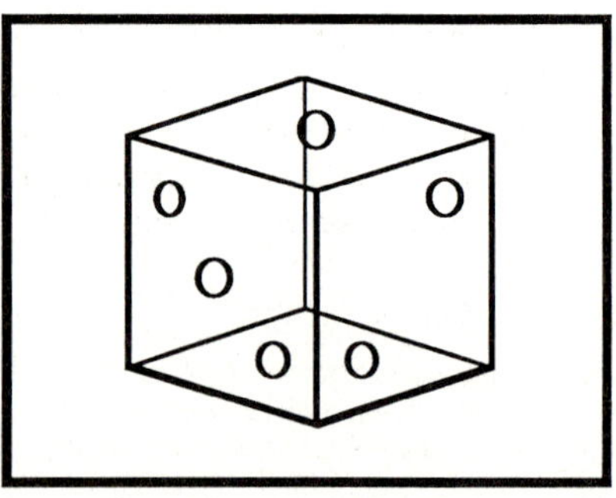

Abbildung 6
Veranschaulichung eines rechnerinternen dreidimensionalen Drahtmodells des Würfels.

Interpretationsebene
Bildfolgeninterpretation, etwa

- Objektbeschreibung [Asada, Tsuji 83],
- Szenenbeschreibung [Nagel 84],
- Handlungsbeschreibung [Neumann, Novak 83], etc.

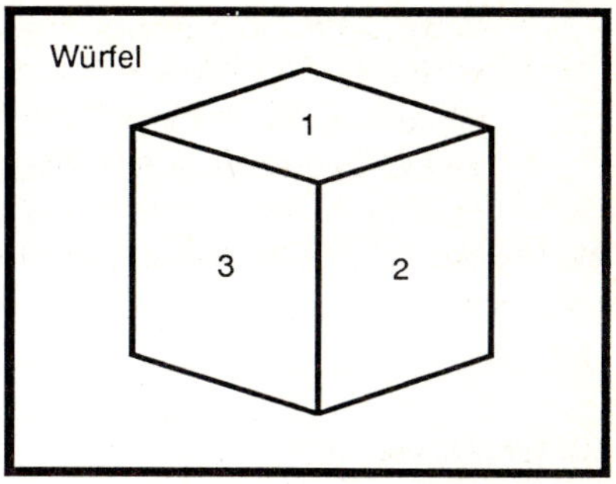

Würfel
1
3
2

Abbildung 7
Interpretation des würfelförmigen Quaders als Würfel und der dunklen Punkte als Augenzahl

In jedem dieser Schritte kann explizit repräsentierbares Wissen herangezogen werden, um den jeweiligen Abstraktionsvorgang zu steuern [Niemann 84], [Nagel 85], [Haton 87], [Niemann, Bunke 87]. Auf diese Weise kann

- a priori Wissen kontrolliert eingebracht werden und
- in späteren Stufen gewonnene Erkenntnis in frühere Stufen zurückgeführt werden.

Eine typische Vorgehensweise ist das Planen von Detailuntersuchungen im Bild [Kelly 71] oder in einer Folge auf Grund der bei einer Grobanalyse gewonnenen Ergebnisse. Die Extrapolation von Objektorten aus den Bewegungszuständen der Objekte oder das Aufschieben von Tiefenschätzungen bis zu dem Bild einer Folge, in der das interessierende Objekt in einer genügend allgemeinen und ungestörten Sicht zu beobachten ist [Asada et al. 84], sind andere Formen der Wissensrückführung. Während bei der Planung unter Umständen eine Abstraktionsebene nicht verlassen wird, erfolgt die Rückkopplung in letzteren Beispielen über mehrere Ebenen.

Je nach Repräsentationsform der bei jedem Abstraktionsvorgang erarbeiteten Ergebnisse und der Kontrollstrukturen für die Auswahl der Operatoren, die auf die Zwischenergebnisse wirken, ist die Wissensrückführung in nahezu jedem Anwendungsfall individuell gelöst [Nagao 84]. Ein Beispiel ist die Steuerung der Parameter zur Extraktion von Oberflächen sich bewegender Gegenstände in einer Bildfolge, bei der die Merkmale der Oberflächen aus vorangegangenen Bildern und die Bewegungsextrapolation der Objekte ausgenutzt wird [Radig 81, Shneier et al. 86]. Auf der 2D-Objektebene und der Korrespondenzebene, auf denen eine formale Vereinheitlichung der Repräsentationsform und der Operationen möglich ist, erscheint der Versuch am ehesten erfolgversprechend, die Metaebene der Wissensrückführung ebenfalls von der speziellen Anwendungssituation zu entkoppeln [Radig 85].

Als Repräsentationsform können auf den Ebenen 2D-Objektsynthese und Korrespondenz relationale Beschreibungen für die rechnerinternen Modelle gewählt werden [Radig 82, Shapiro 82, Connell, Brady 87]. Die auf der Symbolebene erzeugten Symbole werden durch Merkmalsrelationen spezifiziert. Ihre gegenseitigen, geometrischen, topologischen, funktionalen und weitere Beziehungen werden ebenfalls durch Relationen ausgedrückt. Die Gruppierung von Relationstupeln verschiedener Art zu Relationengebilden ermöglicht eine einheitliche Darstellung wesentlicher, bei der Bildfolgeninterpretation benötigter Strukturen:

- symbolische Repräsentation des Bildinhaltes,
- Korrespondenzbeziehung zwischen einander entsprechenden Strukturen in aufeinanderfolgenden Bildern,
- Beschreibung von Prototypen für zu identifizierende Objekte,
- Inkarnationen der Prototypen in der beobachteten Szene.

Die Beziehungen zwischen diesen Strukturen können als mathematische Abbildungen zwischen Relationengebilden (R-Morphismen) ausgedrückt und berechnet werden [Radig 84]. Solche Beziehungen betreffen

- die mehrstufige Organisation komplexer Prototypen aus einfacheren Komponenten [Sielaff 86],
- die Identifikation von Elementen der Prototypenbeschreibung mit zugeordneten Elementen der Bildbeschreibung über die Inkarnationen,
- die Korrespondenz von Elementen der Beschreibung aufeinanderfolgender Bilder.

Zur Unterstützung des Speicherns und Wiederauffindens von Relationengebilden können speziell ausgebildete, relationale Datenbanksysteme eingesetzt werden [Benn, Radig 84, Benn 86].

Relationalstruktur

Zur Berechnung von R-Morphismen zur Konstruktion von Inkarnationen und zum Ermitteln der Korrespondenzbeziehungen läßt sich ein einheitliches Verfahren formulieren. Es geht von der Vorstellung aus, daß Relationstupel aus zwei Bereichen - Bild-Prototyp für die Inkarnation beziehungsweise Bild-Bild für die Korrespondenz - einander zugeordnet, also Paare gebildet werden müssen (s. Abb.5). Von den vielen möglichen Paaren sollen nur solche ausgewählt werden, die in ihren Merkmalen gut übereinstimmen und bei denen die Beziehungen, die auf der einen Seite gelten, möglichst vollständig auch auf der anderen Seite erfüllt sind. Formal kann dieses Auswahlproblem auf das wohlbekannte Problem zurückgeführt werden, die maximalen Cliquen, das sind die größten vollständig verbundenen Teilgraphen, eines Graphen zu finden [Ambler et al. 75]. Dabei werden solche Paare von Relationstupeln als Knoten in einem Kompatibilitätsgraphen aufgenommen, die in ihren Merkmalen genügend gut übereinstimmen. Zwei Knoten werden durch Kanten miteinander verbunden, wenn dadurch eine eindeutige Abbildung zwischen den Relationstupeln und den in ihnen auftretenden Symbolen erhalten bleibt. Die umfangreichsten Cliquen in dem Kompatibilitätsgraphen bezeichnen dann die besten - strukturerhaltenden und merkmalsverträglichen - Entsprechungen zwischen den beiden zu vergleichenden Beschreibungen [Sakane, Kasvand 82], [Bolles, Cain 82]. Abweichungen von Bild zu Bild oder Unterschiede zwischen dem Bildobjekt und seinem Prototyp werden toleriert.

Zwei Begriffe tauchen bei der Bewertung des Vergleiches auf, nämlich merkmalsverträglich und strukturgetreu. Was darunter zu verstehen ist, läßt sich im Einzelfall formal präzise angeben. Wie jedoch Ähnlichkeitsmaßstäbe aus Inhalt und Handlung einer in der Bildfolge beobachteten Szene abgeleitet werden können, entzieht sich bisher einer allgemeingültigen Lösung. Zwei Beispiele mögen diese Behauptung illustrieren.

- Als ein Typ von Bildsymbolen werden Bereiche von einem Segmentationsverfahren geliefert. Eines ihrer Merkmale sei der mittlere Grauwert. Schreibt etwa ein Prototyp für eine Fläche den Grauwert G_p vor und es wird der Grauwert G_b beobachtet, so ist ein zwischen 0 und 1 normiertes Maß für die Merkmalsverträglichkeit $1 - |G_p - G_b| / (G_p + G_b) \geq \theta$. Für $\theta = 1$ wird eine exakte Übereinstimmung gefordert, bei $\theta = 0$ wird jeder beobachtete Grauwert akzeptiert. Der Einstellung des θ - Wertes müßte eigentlich eine vollständige Modellierung der Beleuchtungsverhältnisse in den zu beobachtenden Szenen, der Oberflächenorientierungen und der Reflektanzeigenschaften der realen Oberfläche zu Grunde liegen.
- Ein Segmentationsverfahren liefert Bereiche und sie berandende Konturlinien als Bildsymbole. Randrelationen und Nachbarschaftsrelationen beschreiben die Topologie der Oberfläche eines Objektes. Falls ein Teil der Oberfläche eines Objektes durch einen Bewegungsvorgang verdeckt wird, führt der Strukturvergleich bei der Korrespondenzanalyse zu einer unvollständigen Übereinstimmung, da Relationstupel fehlen. Kann jedoch das Bewegungsverhalten vorhergesehen oder extrapoliert werden, so sollte die nur teilweise Übereinstimmung von Bild zu Bild bei Nichtabbildung der erwarteten Relationstupel nicht nachteilig bewertet werden.

Die Formulierung des Strukturvergleichs als das Berechnen von R-Morphismen zwischen Relationengebilden hat den Vorteil, daß die Berechnung durch explizit repräsentierte Parameter gesteuert wird. Die Merkmalsverträglichkeit wird über Schwellenwerte definiert, die Strukturähnlichkeit über die gewichteten Beiträge der geforderten

Relationstupel zur Abbildung. Genau und nur in diesen Faktoren lassen sich Objektinkarnation und Korrespondenzermittlung beeinflussen.

Formal läßt sich der Strukturvergleich folgendermaßen definieren:

Ein Relationengebilde G = (M,R) besteht aus einer Liste von Relationen $R = (R_1, R_2, \ldots, R_n)$, die auf einer geeigneten Trägermenge M definiert sind. Die Relation R_i habe k_i Komponenten. R heißt Relationalstruktur auf M, das n-Tupel $(k_1, k_2, \ldots, k_n)$ heißt Stellenverteilung der Relationalstruktur. Die Aufgabe des Strukturvergleiches zweier Relationengebilde G und G' besteht nun darin, eine Abbildung zwischen beiden Gebilden zu konstruieren, die gewünschte Eigenschaften aufweist. Sie soll die beiden Strukturen möglichst vollständig aufeinander abbilden (der Idealfall einer eineindeutigen Abbildung (Isomorphismus) läßt sich bei Realweltbildern kaum erreichen), so daß Strukturunterschiede sowie Abweichungen in den Attributwerten minimal sind. Die Abbildung ist eine zweistellige Relation aus Tupeln von R_i und R_i': $A = \{(r_{ip}, r'_{iq}), \ldots\}$, mit i=1...n, wobei p und q verträgliche Tupel aus r_i bzw. aus r_i' sind.

Es ist sinnvoll, die Tupel aus den Relationen der Gebilde durch Symbole eindeutig zu kennzeichnen. Diese Symbole werden als Referenz in anderen Tupeln verwendet. Somit besteht jedes Relationselement aus einem es selbst kennzeichnenden Symbol, aus Referenzsymbolen und aus Komponenten, die Werte von Attributen angeben. Als Beispiel für ein so aufgebautes Relationengebilde, das ausreicht, um aus Geradenstücken bestehende Strichzeichnungen wie den Quader aus Abb.4 zu repräsentieren, werden nur zwei Relationen, PUNKT und GERADE benötigt: PUNKT = $\{(p_i, x_i, y_i), \ldots\}$ und GERADE = $\{(g_i, p_k, p_j), \ldots\}$. Dabei sind die x- und y-Komponente Attributwerte, nämlich die Koordinaten im Bild, p identifiziert den individuellen Punkt, g die individuelle Gerade und die beiden p-Komponenten in der GERADE-Relation sind Referenzen auf die beiden Punkte, zwischen denen sich das Geradenstück erstreckt.

Damit besteht ein Tupel eines Relationengebildes typischerweise aus Symbolen s und Attributwerten w: $r = (s, \ldots, w \ldots)$. Eine Abbildung $A = \{((s, \ldots, w, \ldots), (s', \ldots, w', \ldots)), \ldots\}$ zwischen zwei Relationengebilden G und G' soll dann strukturerhaltend heißen, wenn die durch sie induzierte Abbildung $S = \{(s, s'), \ldots\}$ zwischen den Symbolen eineindeutig ist. Zulässige Abweichungen in den Attributwerten lassen sich dadurch kontrollieren, daß eine geeignete Funktion $\Theta(r, r') = \Theta(w, \ldots, w', \ldots) \rightarrow [0,1]$ die Ähnlichkeit der Attributwerte feststellt und gefordert wird, daß $\Theta(r, r') \geq \theta$ gilt, wobei $0 \leq \theta \leq 1$ein vorgebbarer Schwellenwert ist. Ein so normiertes Maß für die Merkmalsverträglichkeit für Grauwerte, $1 - |G_p - G_b| / (G_p + G_b) \geq \theta$, war oben als Beispiel angegeben. Die Ähnlichkeitsfunktion Θ wird je nach Relation unterschiedlich zu wählen sein. Außerdem läßt sich vereinbaren, daß nur Tupel aus entsprechenden Relationen aufeinander abgebildet werden, sodaß der Strukturvergleich vermöge A, Θ und θ in disjunkte Teile zerlegbar ist, nämlich $A_i = \{(r_{ik}, r'_{il}), \ldots\}$ mit $r_{ik} \in R_i$ und $r'_{il} \in R'_i$ sowie $\Theta_i(r, r') \geq \theta_i$, i=1...n.

Die Konstruktion der gewünschten strukturerhaltenden und merkmalsverträglichen Abbildung, der R-Morphismen ist im Wesen einfach:

1. Für alle Relationen R_i und R'_i, i=1...n werden
   1.1 alle Zuordnungen $z = (r_{ik}, r'_{il})$ gebildet, k=1...$|R_i|$ und l=1...$|R'_i|$;
   1.2 aus der Menge werden alle Zuordnungen eliminiert, für die $\Theta_i(r, r') < \theta_i$ gilt.
2. Die verbliebenen Zuordnungen werden auf Eineindeutigkeit der induzierten Symbolabbildung untersucht. Dazu werden Paare von Zuordnungen gebildet, für die die Symbolabbildung eineindeutig ist.
   $k = ((r_{ik}, r'_{il}), (r_{is}, r'_{it})) = (((s_k, \ldots), (s'_l, \ldots)), ((s_s, \ldots), (s'_t, \ldots))) \in K \Leftrightarrow S = \{(s_k, s'_l), (s_s, s'_t), \ldots\}$ eineindeutig.
3. Aus den Potenzmengen von K werden die maximalen gewählt, für die die induzierte Relation S eineindeutig ist. Sie darin enthaltenen Zuordnungen (r_{ik}, r'_{il}) repräsentieren die besten strukturerhaltenden und merkmalverträglichen R-Morphismen.

Der Konstruktionsprozeß läßt sich als Problem veranschaulichen, in einem Graphen maximale Cliquen, also total verbundene Subgraphen, zu finden. Die Knoten z des Graphen sind merkmalsverträgliche Zuordnungen, seine Kanten k verbinden im Sinne der Symbolabbildung kompatible Zuordnungen. Der Graph in Abb.5 (unten, Mitte) veranschaulicht einen Ausschnitt eines solchen Kompatibilitätsgraphen.

Ein einfaches Beispiel illustriere den Vorgang. Die Abb.8 zeigt zwei der Würfelkanten aus dem einführenden Beispiel in einem Stereobild. Das Geradenstück G1 im linken Bild hat die Endpunkte P1 und P2 mit den Koordinaten (5,5) bzw. (8,3). Die Geradenstücke im rechten Bild sind mit H und ihre Endpunkte mit Q bezeichnet. G1 ... P1 ... H1 ... Q3 sind die Symbole und dienen als Referenzen in den beiden Relationengebilden, die unter Verwendung der schon eingeführten Relationen PUNKT und GERADE aufgebaut werden. Damit haben die Relationen die Elemente

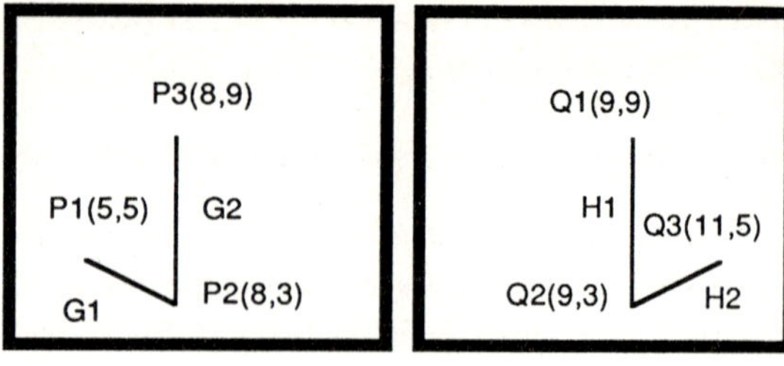

Abbildung 8: Zwei Kanten aus Stereo-Würfelbild mit Bezeichern für Punkte und Geraden sowie Punktkoordinaten

PUNKT = { (P1,5,5), (P2,8,3), (P3,8,9) }
PUNKT' = { (Q1,9,9), (Q2,9,3), (Q3,11,5) }
GERADE = { (G1,P1,P2), (G2,P2,P3) }
GERADE' = { (H1,Q2,Q1), (H2,Q3,Q2) }

Schritt 1.1 der Abbildungskonstruktion liefert 9 Zuordnungen zp_1=((P1,...),(Q1,...)), zp_2=((P1,...),(Q2,...)) bis zp_9=((P3,...),(Q3,...)) für die PUNKT-Relation sowie 4 Zuordnungen zg_1=((G1,...),(H1,...)) bis zg_4=((G2,...),(H2,...)) für die GERADE-Relation. Für Schritt 1.2 muß eine Ähnlichkeitsfunktion festgelegt werden, in diesem Fall nur für die PUNKT-Relation. Eine einfache Abstandsfunktion reicht zur Illustration: $\Theta_P = 1 / (1+|X-X'|+|Y-Y'|)$, wobei X die erste und Y die zweite Koordinate angibt. Mit θ_P = 1/6 bleiben die Zuordnungen übrig, die als Knoten im Kompatibilitätsgraphen der Abb.9 erscheinen. Im 2. Schritt wird die paarweise Kompatibilität festgestellt. So ist beispielsweise der Knoten zp_5=((P2,...),(Q2,...)) nicht mit dem Knoten zp_6=((P2,...),(Q3,...)) kompatibel, da hier das Symbol P2 einmal auf Q2, das andere mal auf Q3 abgebildet wird. Der 3. Schritt, maximale Cliquen zu suchen, ist ein in der Graphentheorie wohlbekanntes Problem, für das eine große Anzahl von Algorithmen existiert. Vier Cliquen existieren in dem Beispielgraphen. Eine besteht aus einem isolierten Knoten. Sie bedeutet, daß für den Fall der Abbildung der Geraden G1 auf die Gerade H1 keine weiteren Zuordnungen möglich sind. Eine weitere Clique ordnet den Punkt P2 dem Punkt Q3 und den Punkt P3 dem Punkt Q1 zu. Eine Gerade zwischen den Punkten P2 und P3 existiert, aber zwischen Q1 und Q3 nicht. Die dritte Clique sagt aus, daß mit einer Abbildung von P2 auf Q3 noch die Zuordnung verträglich ist, die G2 mit H2 identifiziert. Die größte der Cliquen, dick ausgezogen gezeichnet in Abb.9, repräsentiert die Zuordnung der beiden senkrechten Geradenstücke, also der drei beteiligten Relationstupel, sowie von der Struktur her die noch mögliche Abbildung der Geraden G1 auf H2.

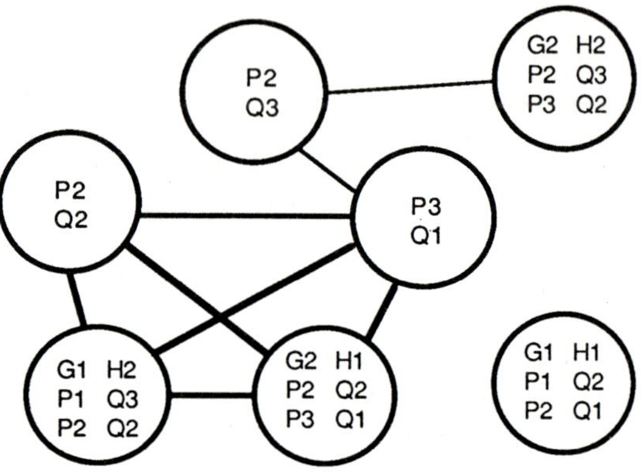

Abbildung 9: Verträgliche Zuordnungen im Kompatibilitätsgraphen mit größter, maximaler Clique (dick ausgezogen)

Obwohl für das Problem der Cliquenidentifikation keine obere Komplexitätsschranke angegeben werden kann, zeigt die Erfahrung aus Bildbeispielen, daß der Zeitaufwand des gesamten Verfahrens grob mit der dritten Potenz der Zahl der im Kompatibilitätsgraphen generierten Knoten steigt. Durch eine hierarchische Strukturierung der Relationengebilde [Sielaff 86] läßt sich der Aufwand verringern. Sielaff versucht die automatische Hierarchisierung von bildbeschreibenden Relationalstrukturen durch eine Zielfunktion zu steuern, die den Vergleichsaufwand minimiert. Eine Implementation des Verfahrenskerns in PROLOG ist sehr kompakt, da sich Relationen durch PROLOG-Prädikate und die Kompatibilitätsprüfung durch Regeln elegant ausdrücken lassen. Die Konstruktion einer eineindeutigen Symbolabbildung läßt sich als Unifikationsschritt, beschränkt auf einfache Grundsubstitutionen, auffassen. Die Bedeutung möglicher Verallgemeinerungen müßte noch untersucht werden.

Das Ergebnis des Strukturvergleiches sind typischerweise mehrere, strukturerhaltende und merkmalsverträgliche Abbildungen. In dem gegebenen Beispiel ist es nicht schwer, den "besten" R-Morphismus auszuwählen, in diesem Fall die umfangreichste. Für recht komplexe Strukturen ist die Bewertung nicht mehr durch Augenschein zu leisten. Enthält ein Bild symmetrische Figuren, können sogar eine größere Anzahl gleichwertiger Abbildungen auftreten, deren Isomorphie erst einmal zu analysieren ist [Radig, Schlieder 85]. Ergebnisse können je nach Einsatz des Strukturvergleichs unterschiedliche Bedeutung haben. Stammen etwa beide Strukturen aus einem Stereobildpaar, so identifizieren die Zuordnungen korrespondierende Bildsymbole. An den Strukturvergleich läßt sich dann ein Verfahren zur Tiefenschätzung anschließen. Werden Prototypen für Objekte als Relationengebilde formuliert, so identifiziert der Vergleich mit einer Bildbeschreibung diese Objekte in der Aufnahme. Über die gefundenen Zuordnungen lassen sich Inkarnationen der Prototypen aufbauen, nämlich Relationengebilde, die die gefundene Ausprägung des Prototyps im Bild beschreiben.

Wissensgestützte Interpretation

Der Ansatz, das Problem der Korrespondenz von Bildbeschreibungen und der Identifikation von Objekten auf die Konstruktion von Abbildungen mit vorgebbaren Eigenschaften zwischen relationalen Strukturen zurückzuführen, erlaubt es, ein vielseitig einsetzbares, formal gut definierbares Werkzeug bereitzustellen. Gerade die Möglichkeit, eine wichtige Teilaufgabe der Interpretation von Einzelbildern und Bildfolgen als abgeschlossen zu betrachten, öffnet den Blick für die sich bei diesem Ansatz ergebenden Fragen.

Die Vorteile des Ansatzes bestehen darin, daß die Komplexität des Strukturvergleiches sich durch die Übertragung auf das Cliquenproblem in Graphen mit analysierbaren Eigenschaften besser abschätzbar wird. Zwei Gruppen von explizit, unabhängig vom formalen Verfahren bereitzustellenden Parametern steuern den Vergleichsvorgang. Es sind das die Ähnlichkeitsfunktionen Θ und die Schwellenwerte θ, mit deren Hilfe die Merkmalsverträglichkeit von Tupelpaaren festgestellt wird. Wie die Parameter gewählt werden, ist, im Gegensatz zu manchen früheren Ansätzen [Jacobus et al. 80], eben nicht Teil des Verfahrens, sondern frei vorgebbar. Damit ergibt sich das Problem, aus dem intendierten Anwendungsbereich des Interpretationssystems, in das der Strukturvergleich eingebettet ist, Gesetzmäßigkeiten abzuleiten, nach denen die Parameter zu bestimmen sind. Für dieses Problem ist bisher eine allgemeingültige Lösung genauso wenig in Sicht wie für die Frage nach der Bewertung des Vergleichsergebnisses. Einige einfache Möglichkeiten liegen nahe, beispielsweise die die Merkmalsverträglichkeit kennzeichnenden Werte geeignet zu kombinieren und zu normieren sowie weiterhin die Zahl der gefundenen Zuordnungen zwischen Relationstupeln ins Verhältnis zur Zahl der im Idealfalle erreichbaren Zuordnungen zu setzen. Sofort liegt dann auch nahe, Merkmale und Relationen gemäß ihrer Bedeutung oder der Zuverlässigkeit ihrer Ermittlung mit Gewichtsfaktoren zu versehen und diese in die Bewertung einzubeziehen. Auch hier ist das methodische Vorgehen, nach der solche Gewichte und Kombinationsvorschriften aus dem Anwendungsfall abzuleiten sind, weitgehend ungeklärt. Es ist offen, ob sich, von Spezialfällen abgesehen, die Bedeutung der Ergebnisse des Strukturvergleichs überhaupt durch so ein einfaches Formelwerk einfangen läßt.

Ein erster Ansatz für eine flexible Lösung von Teilproblemen in den Bereichen der Parameterisierung und Bewertung des Strukturvergleiches liegt in der Verwendung wissensbasierter Methoden. Damit kann das Justieren von Parameterwerten, die Auswahl von Kombinationsverfahren, das Festlegen von Gewichten oder die Bestimmung von Ähnlichkeitsfunktionen problemnäher formuliert werden. Ist beispielsweise in einer Bildfolge, die ein sich bewegendes Objekt aufzeichnet, aufgrund des Vergleiches der ersten beiden Bilder zu ermitteln, in welche Richtung sich das Objekt mit welcher Geschwindigkeit bewegt, so läßt sich der Ort des Objektes im dritten Bild extrapolieren. Dabei können Kenntnisse über die physische Natur des Objektes verwendet werden, etwa um seine Trägheit abzuschätzen. Die Abschätzung ermöglicht es, beispielsweise die Schwellenwerte für die Verträglichkeit von Punktkoordinaten genauer einzustellen. Bewegt sich das Objekt von der Kamera fort, so lassen die Kenntnis der Objektivbrennweite und die Berücksichtigung der Gesetze der Perspektive eine exaktere Einstellung der Verträglichkeitsschwellen für Flächengrößen zu. Verdeckt ein sich bewegendes Objekt ein anderes, so ist von vornherein klar, daß der Strukturvergleich ein unvollkommenes Resultat abliefern muß. Trotzdem sollte das Ergebnis nicht in der Bewertung herabgesetzt werden, da ein besseres aufgrund der physikalischen Gesetze nicht zu erwarten ist. Werden nach der Beobachtung mehrerer Bilder die Kenntisse über das Bewegungsverhalten zuverlässiger, so kann auch das Justieren des Vergleichs und der Bewertung präzisiert werden. In der Abb.10, die zwei Schnappschüsse einer nach links rollenden Kugel zeigt, wird das (gemusterte) Objekt im Hintergrund verdeckt. In dem Bild ist die symbolische Beschreibung visualisiert, die Geradenstücke, Punkte und Flächen verwendet. Die Abb.11 stellt das Ergebnis des Strukturvergleiches dar. Die Kantenstücke, die einander zugeordnet werden konnten, sind in den beiden Bildern gezeichnet. Wie zu erwarten war, ist der rechte Vorsprung des Hintergrundobjektes nicht im Vergleichsergebnis enthalten. Im linken Teilbild von Abb.10 grenzt die waagerechte Unterkante des Vorsprunges an die senkrechte Kante. Im rechten Teilbild hat sich zwischen beide ein schräges Kantenstück zusätzlich eingeschoben. Dadurch sind die waagerechte und die senkrechte Kante kürzer geworden und haben keinen gemeinsamen Endpunkt mehr.

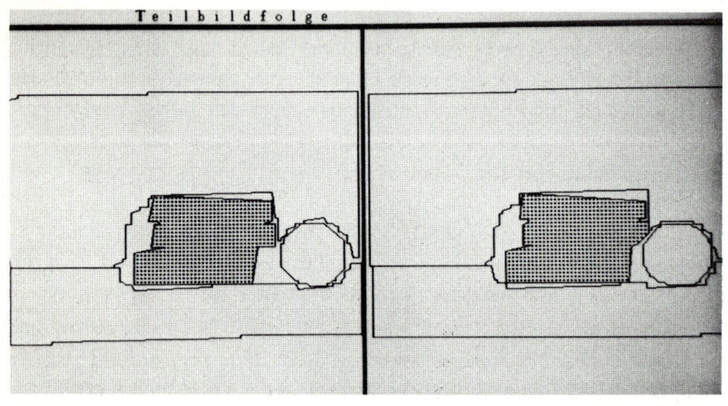

Abbildung 10
Kantenbilder von zwei Aufnahmen einer Bildfolge, auf der eine Kugel von rechts nach links rollt und das mit Punktschraffur gefüllte Hintergrundobjekt verdeckt.

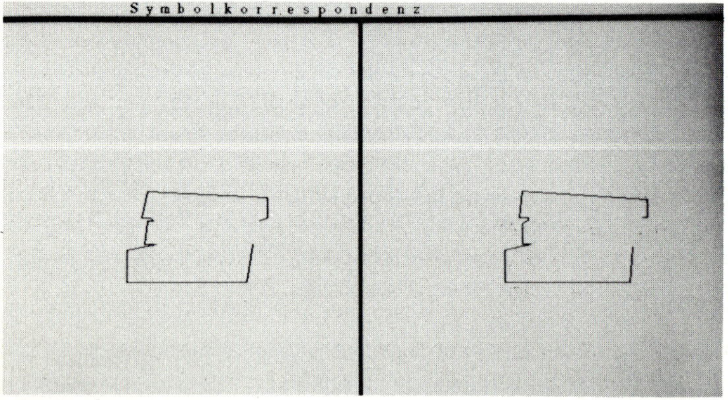

Abbildung 11
Ergebnis des Strukturvergleiches; die einander zugeordneten Geradenstücke des Hintergrundobjektes sind dargestellt.

Eine Möglichkeit, die Anpassung von Vergleich und Bewertung zu formalisieren, besteht darin, aus Bildern und der Information über die zeitliche Entwicklung Hypothesenobjekte zu konstruieren, die regelgesteuert dynamisch modifiziert werden und die aktuelle Erwartung über das nächste Bild ausdrücken. Die Hypothesenobjekte dienen als Vorgabe für die Korrespondenzermittlung enger als es das ihnen zugrundeliegende Bild ermöglicht, ausgestattet mit Toleranzvorgaben. Die Hypothesengenerierung läßt sich natürlich auch auf Objektkonfigurationen und ganze Szenenaragements erweitern. Sie ist auch nicht auf Zeitfolgen eingeschränkt, sondern kann auch beispielsweise für Schichtaufnahmen in der Medizin eingesetzt werden, wobei etwa typische Anatomien als Richtschnur für die Extrapolation dienen können. Der Aufbau eines experimentellen Bildinterpretationssystems, mit

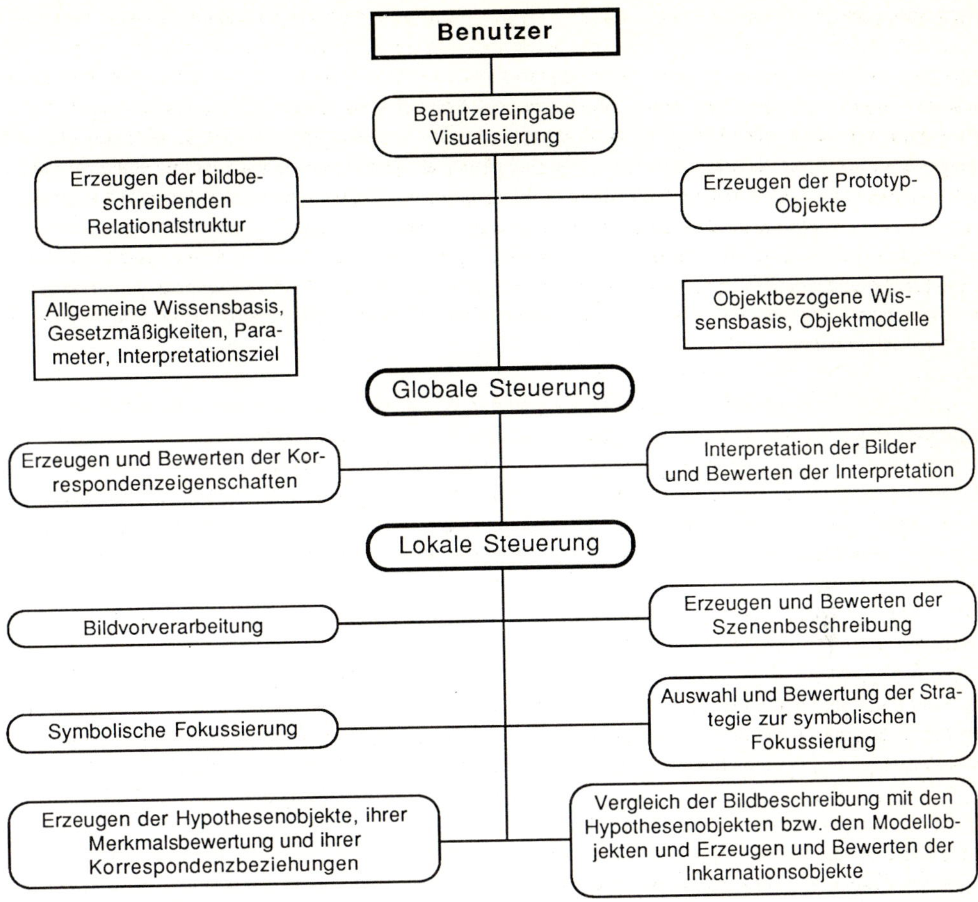

Abbildung 12: Struktur eines Experimentiersystems zur Erprobung wissensbasierter Techniken bei der Steuerung und Bewertung des Vergleichs relationaler Strukturen

dem wir zur Zeit eine wissensbasierte Steuerung des Strukturvergleiches untersuchen, ist in Abb.12 dargestellt. Davon sind, mit Unterstützung der Deutschen Forschungsgemeinschaft, bisher der Strukturvergleich und die Visualisierungskomponente fertiggestellt. Für die Bildvorverarbeitung, hier hauptsächlich Segmentation, wurde das System PSIWAG von der Gesellschaft für Strahlen und Umweltforschung importiert [Eckstein, Pöppl 86]. Der Baustein »Erzeugen der Hypothesenobjekte ...« führt die Extrapolation durch, wobei er Information über die Entwicklung der Szene vom Modul »Erzeugen und Bewerten der Szenenbeschreibung« erhält und Regeln aus den beiden Wissensbasen über Objekte und über allgemeine Gesetzmäßigkeiten bezieht. Die »Symbolische Fokussierung« dient zu Anfang hauptsächlich dazu, die Experimente auf interessante Bildausschnitte zu konzentrieren. Später soll sie in der Lage sein, Anforderungen wie "Verfolge nur bewegte, zweibeinige Objekte" in die Vernachlässigung irrelevanter Strukturteile umzusetzen. Während die »Lokale Steuerung« die Verarbeitungsschritte kontrolliert, die den Vergleich zweier Bilder betreffen, wird durch die »Globale Steuerung« die Auswertung der gesamten Bildfolge bzw. des aktuell bearbeiteten Abschnitts gelenkt. Die Module »Erzeugen der Skizze« und »Erzeugen der Prototyp-Objekte« unterstützen den Entwurf der Relationengebilde für die Bildrepräsentation und für die als Prototypen dienenden Objektmodelle. Das Modul »Benutzereingabe, Visualisierung« wickelt den Dialog ab. Dabei ist die Frage noch ubefriedigend beantwortet, wie nicht direkt visualisierbare Information, beispielsweise die geometrische Veranschaulichung von Schwellenwerten, d.h. ihrer Bedeutung, dargestellt werden kann.

Weiterentwicklung

Schon der Entwurf eines oben skizzierten Systems wirft mehr Frage auf, als durch den Betrieb des Systems beantwortet werden können. Gerade wenn es darum geht, Entwurfsentscheidungen festzulegen und zu implementieren, wird bewußt, wie willkürlich Entscheidungsparameter als frei betrachtet werden, weil eben die Konsequenzen einer Wahl nicht überschaubar sind. Wenn explizites Wissen verwendende Systeme einen wirklichen Fortschritt im Gebiet der Deutung von Bildern einleiten sollen, so muß die grundsätzliche Flexibilität solcher Systeme durch weitgehend automatisierbare Verfahren für ihren methodischen Entwurf nutzbar gemacht werden. Das Umsetzen der Aufgabenstellung für ein Sichtsystem in Entscheidungen insbesondere auf der Ebene der Konzeptualisierung ist eher durch Intuition als durch Systematik gekennzeichnet. Die heuristische Festlegung von Systembausteinen erstreckt sich bis auf die Auswahl von elementaren Bildverarbeitungsoperatoren auf der Transformationsebene.

Ein Ansatz, um überhaupt zu präzisen Fragen vorzustoßen, könnte darin bestehen, von einer Beschreibung der Aufgabenstellung, also des Interpretationszieles, ausgehend Entwurfsentscheidungen zu treffen. Der vom Allgemeinen zum Speziellen fortschreitende Spezifikationsvorgang wird, wenn er überhaupt eingehalten wird, typischerweise von ausgedehnten Experimentierphasen unterbrochen, die die Realisierbarkeit von Teilkonzepten klären sollen. Aus dieser Beobachtung ergibt sich die Forderung zu untersuchen, wie solche Experimentierphasen besser unterstützt werden können. Ein Unterstützungssystem dafür muß Experimente protokollieren, Empfehlungen für eine systematische Variation von Parametern geben, Alternativen zu Entscheidungen anbieten und vor allem die Ergebnisse objektivieren und verwalten, um sie später auch für andere Aufgaben zur Verfügung zu haben.

Einige der Experimente dienen zur Überprüfung, ob und mit welcher Qualtität auf einer Abstraktionsebene benötigte Information zur Verfügung gestellt werden kann. Wie beispielsweise die Robustheit von Operatoren auf der Transformations- und Symbolebene meßbar ist und wie sie sich auf die Güte des Gesamtsystems auswirkt, ist vielfach beim Entwurf kaum entscheidbar. Die Antwort ergibt sich erst beim Einsatz des Systems. Demgemäß besteht eine weitere Fragestellung darin, die Qualitätsanforderungen an Informationsstrukturen formal zu spezifizieren, um die Eignung von Abstraktionsschritten schon beim Entwurf überprüfen zu können. Experimente sollten ihre Ergebnisse in einer hier verwendbaren Beschreibung abliefern. Beispielsweise erfordert die in den Beispielen zur relationalen Beschreibung verwendete GERADE-Relation, daß das Segmentationsverfahren Objektkanten durch Geradenstücke approximiert. Iterative Approximationsverfahren verwenden ein Fehlermaß zum Abbruch. Das Fehlermaß hat natürlich einen direkten (leider bisher nicht exakt untersuchten) Einfluß auf die Wahl der Schwellenwerte und der Ähnlichkeitsfunktion für die Geradenmerkmale bzw. in dem verwendeten Beispiel auf die Streuung der Koordinaten der Geradenendpunkte. Der Einfluß erstreckt sich aber noch bis hin zur Bewertung der vom Strukturvergleich erzeugten R-Morphismen. Bei genauer Betrachtung der Abb.10 läßt sich erkennen, daß die Approximationspolygone für die Kugel in beiden Bildern recht unterschiedlich ausfallen. Als Konsequenz sind keine strukturisomorphen Abbildungen zwischen beiden Kugelpolygonen möglich. Die entstehenden R-Morphismen weisen unterschiedliche Zuordnungslücken auf und gaukeln unterschiedliche Rotationswinkel der Kugel vor. Das letztere ist irrelevant für die Bewertung, das erstere wirft die Frage auf, ob die verwendete Relationalstruktur für das Bildmaterial überhaupt adäquat ist. Hier ist die Antwort negativ.

Um das Experimentieren mit symbolischen Beschreibungsformen, also mit Relationalstrukturen, zu erleichtern, wären Methoden hochwillkommen, die eine Änderung in den Relationen weitgehend automatisch in eine Anpassung der vor dem Strukturvergleich liegenden Verarbeitungsschritte umsetzen. Wird etwa ein Kreis als neues Symbol zur besseren Beschreibung der Kugel in Abb.10 eingebracht, so sollten in den Segmentationsprozeß automatisch Verfahren zur Kreisapproximation einbezogen werden. Hier ergeben sich schon auf der technische Ebene Probleme. Beispielsweise ist unser Experimentalsystem teilweise in PROLOG und teilweise in C implementiert. Einführen und Ändern neuer Relationen wird durch Einfügen oder Ändern von PROLOG-Prädikaten sehr vereinfacht. Für die automatische Anpassung der prozedural formulierten Systemmodule an die neu definierte Relationalstruktur gibt es bisher noch kein befriedigendes Vorgehen. Wenn etwa die PUNKT-Relation um eine Komponente erweitert werden soll, die die Bildnummer angibt, so sind mühsame Schnittstellenanpassungen zu leisten. Aber nicht nur die technische Seite bereitet Schwierigkeiten, die noch am ehesten zu beheben wären.

Mit der Änderung der Repräsentationsform für Objekte und Szenen ist darüberhinaus die Frage zu stellen, welche weiteren Wissenseinheiten von der Änderung betroffen sind. Eine automatische Nachführung sollte nicht nur die formale Struktur betreffen, sondern auch semantische und pragmatische Aspekte berücksichtigen. Wird beispielsweise als neue Relation die Beziehung »links von« für Objektbeschreibungen verwendet, so ist an der relationalen Beschreibung von Objekten nicht mehr feststellbar, ob sie links-rechts spiegelsymmetrisch sind. (Falls Symmetrieuntersuchungen an Werkstücken das Einsatzziel des Sichtsystems sind, so ist diese Änderung verfehlt.) Aber auch die formale Nachführung ist nicht mühelos. Prototypische Beschreibungen für Objekte können sehr umfangreich werden, wenn man von Spielzeugwelten absieht, und werden meist manuell erstellt werden müssen. Eine Bibliothek von Prototypen irrtumsfrei an eine Änderung der Repräsentationsform anzupassen und die Umstellung zu überprüfen, ist eine aufwendige Maßnahme. Das automatische Erstellen von prototypischen Objektbeschreibungen aus Bildbeispielen, technischen Zeichnungen oder CAD-Daten könnte hier eine Erleichterung schaffen. Automatische Lernverfahren sind jedoch noch nicht genügend weit entwickelt, um routinemäßig eingesetzt werden zu können. Connell und Brady haben den Winstonschen Ansatz des Lernens an visuell präsentierten Beispielen jüngst wieder aufgegriffen [Connell, Brady 87], [Winston 75]. Wenn man die zwölf Jahre überschaut, die zwischen beiden Veröffentlichungen liegen, wird offensichtlich, wie dringend die Intensivierung der Forschung auf diesem Gebiet ist.

Nicht die Technik der Wissensrepräsentation, der Inferenz, der Effizienzsteigerung oder der systemtechnischen Einbettung sollte in der Grundlagenforschung über wissensbasierte Sichtsysteme im Vordergrund stehen, sondern die Fragen nach den Quellen des anzuwendenden Wissens, nach den Methoden der Übertragung des Wissens in ein Sichtsystem hinein, nach Verfahren zur Überprüfung des aktuellen Wissensbestandes auf seine Qualität und Eignung für die angestrebte Problemlösung sowie damit verbunden nach der systematischen Anpassung des Wissensbestandes an modifizierte und neue Aufgabenstellungen. Fortschritte in den genannten Bereichen werden die Qualität, die Verbreitung und das Nutzen/Kosten-Verhältnis wissenbasierter Sichtsysteme verbessern.

Literaturhinweise:

A. Ambler et al., A Versatile System for Computer Controlled Assembly, *Artif. Intell.* **6** (1975), S.129-156

G. Adorni, M. Di Manzo, E. Trucco, Partial Occlusions in Scene Analysis: a Structural Approach, *Proc. of the IASTED International Symposium Robotics and Automation*, Lugano 1985, S.36-39

M. Asada, S. Tsuji, Represantation of Three-dimensional Motion in Dynamic Scenes, *Comp. Vision, Graphics, Image Proc.* **21** (1983), S.118-144

M. Asada et al., Analysis of Three-dimensional Motions in Blocks World, *Pattern Recognition* **17** (1984), S.57-71

D.H. Ballard, C.M. Brown, *Computer Vision*, Prentice Hall, Englewood Cliffs 1982

W. Benn, B. Radig, Retrieval of Relational Structures for Image Sequences, *Proc. Intern. Conf. Very Large Data Bases*, Singapur, August 1984, S.533-536

W. Benn, *Dynamische nicht-normalisierte Relationen und symbolische Bildbeschreibung*, Informatik-Fachberichte 128, Springer-Verlag, Berlin Heidelberg 1986

P. Besl, R. Jain, Three-dimensional Object Recognition, *Comput. Surveys* **17** (1985), S.75-145

R. Bolles, R. Cain, Recognizing and Locating Partially Visible Objects, *Int. J. Robotics Research*, **Vol. 1**, No. 3 (1982), S.57-82

R. Brooks, Symbolic Rreasoning among 3-D Mmodels and 2-D Images, *Artif. Intell.* **17** (1981), S.285-348

H. Bunke, *Modellgesteuerte Bildanalyse*, Teubner, Stuttgart 1985

J. Cheng, T. Huang, A Subgraph Isomorphism Algorithm Using Resolution, *Pattern Recognition* **13** (1981), S.371-379

J. Connell, M. Brady, Generating and Generalizing Models of Visual Objects, *Artif. Intell.* **31** (1987), S.159-183

W. Eckstein, S.J. Pöppl, Konzept einer universellen Programmiersprache für Bildverarbeitungsanwendungen, *8. DAGM-Symposium*, Informatik-Fachberichte 125, Springer-Verlag, Berlin Heidelberg 1986, S.170-175

W. Eckstein, S.J. Pöppl, PSIWAG - A Language for Logic Programming in Image Analysis, *Proc. 8th International Conf. Pattern Recognition*, Paris 1986, S.1117-1121

R. Freytag, W. Haettich, H. Wandres, Development Tools for a Model Directed Workpiece Recognition System, *Pattern recognition* **19** (1986), S.267-278

K. Grebner, Model-based Analysis of Industrial Scenes, *Proceedings CVPR `86*; IEEE Computer Society Conference on Computer Vision and Pattern Recognition, Florida (1986), S.28-33

R. Haralick, L. Shapiro, The Consistent Labeling Problem, *IEEE Trans. Pattern Anal. Mach. Intell.*, Part I **PAMI-1** (1979), S.173-184, Part II **PAMI-2** (1980), S.193-203

J.-P. Haton (Hrsg.), *Fundamentals in Computer Understanding: Speech and Vision*, Cambridge University Press, Cambridge 1987

D. Heeger, Models for Motion Perception, Proposal for a Ph. D. dissertation, University of Pennsylvania, Philadelphia 1986

B. Horn, B. Shunck, Determining Optical Flow, *Artif. Intell.* **17** (1981), S.185-203

T. Huang, R. Tsai, Image Sequence Analysis: Motion Estimation, in *Image sequence analysis*, Hrsg .: T. Huang, Springer-Verlag, Berlin Heidelberg 1981, S.1-18

T. Huang, J. Weng, N. Ahuja, 3-D Motion from Image Sequences: Modeling, Understanding and Prediction, *Proc. of the Workshop on Motion: Representation and analysis*, Charleston 1986, IEEE Comput. Soc. Press 1986, S.125-130

C. Jacobus et al., Detection and Analysis of Matching Graphs of Intermediate-level Primitives, *IEEE Trans. Pattern Anal. Mach. Intell.* **PAMI-2** (1980), S.495-510

M. Kelly, Edge Detection in Pictures by Computer Using Planning, *Mach. Intell.* **6** (1971), S.739-409

L. Kitchen, A. Rosenfeld, Discrete Relaxation for Matching Relational Structures, *IEEE Trans. Syst. Man. Cyb.* **SMC-9** (1979), S.869-874

W. Martin, J. Aggarwal, Volumetric Description of Objects from Multiple Views, *IEEE Trans. Pattern Anal. Mach. Intell.* **PAMI-5** (1983), S.150-158

F. May, W. Wolf, Trennung bewegter Objekte im bewegten Umfeld einer Bildszene durch mehrstufige Bewegungskompensation, *5. DAGM-Symposium*, Karlsruhe (1983), S.54-59

M. Nagao, Control Strategies in Pattern Analysis, *Pattern Recognition* **17** (1984), S.45-56

H.-H. Nagel, Displacement Vectors Derived from Second Order Intensity Variations in Image Sequences, *Comp. Vision, Graphics, Image Proc.* **21** (1983), S.85-117

H.-H. Nagel, Spatio-temporal Modeling Based on Image Sequences, *Proc. Int. Symp. Image Proc. Applications*, Tokyo 1984

H.-H. Nagel, Wissensgestützte Ansätze beim maschinellen Sehen: Helfen sie in der Praxis?, in *Wissensbasierte Systeme - GI-Kongreß 1985*, Informatik-Fachberichte 112, Hrsg.: W. Brauer, B. Radig, Springer-Verlag, Berlin Heidelberg 1985, S.170-198

B. Neumann, Optical flow, Mitteilung IFI-HH-M-108 des Fachbereichs Informatik der Universität Hamburg (1983)

B. Neumann, H.-J. Novak, Event Models for Recognition and Natural Language Description of Events in Real World Image Sequences, *Proc. Int. Joint Conf. Artif. Intell.* Karlsruhe (1983), S.724-726

H. Niemann, Mustererkennung - eine einführende Übersicht, *Handbuch der modernen Datenverarbeitung*, Vol. **21**, No. 115 (1984), S.3-22

H. Niemann, H. Bunke, *Künstliche Intelligenz in Bild- und Sprachanalyse*, Teubner, Stuttgart 1987

B. Radig, Inferential Region Extraction in TV-sequences, *Proc. 7th Int. Joint Conf. Artif. Intell.*, Vancouver (1981), S.719-721

B. Radig, Symbolische Beschreibung von Bildfolgen I: Relationengebilde und Morphismen, Bericht IFI-HH-B-90 des Fachbereichs Informatik der Universität Hamburg (1982)

B. Radig, Image Sequence Analysis Using Relational Structures, *Pattern Recognition* **17** (1984), S.161-167

B. Radig, Bildverstehen und Künstliche Intelligenz, *Proc. German Workshop on Artificial Intelligence*, Sept. 1984 in Wingst, Springer-Verlag, Berlin Heidelberg 1985, Informatik-Fachberichte 103, S.88-108

B. Radig, Chr. Schlieder, Modellierung symmetrischer Werkstücke, *Robotersysteme* **1** (1985), S. 35-42

E. Riseman, A. Hanson, A Methodology for the Development of General Knowledge-based Vision Systems, in *Wissensbasierte Systeme - GI-Kongreß 1985*, Informatik-Fachberichte 112, Hrsg.: W. Brauer, B. Radig, Springer-Verlag, Berlin Heidelberg 1985, S.257-288

S. Sakane, T. Kasvand, Segment-Matching of Shapes based on Clique Detection, Report ERB-946, National Research Council Canada (1982)

A. Sanfeliu, K. Fu, A Distance Measure Between Attributed Relational Graphs for Pattern Recognition, *Proc. 6th Int. Conf. Pattern Recognition*, München (1982), S.162-168

Ch. Sielaff, Hierarchical Decomposition and Synthesis of Relational Descriptions - the Modelgraph, *Proc. 8th International Conf. Pattern Recognition*, Paris (1986), S.1207-1209

L. Shapiro, R. Haralick, Structural Description and Inexact Matching, *IEEE Trans. Pattern Anal. Mach. Intell.* **PAMI-3** (1981), S.504-519

L. Shapiro, Organization of Relational Models, *Proc. 6th Int. Conf. Pattern Recognition*, München (1982), S.360-365

M. Shneier, R. Lumia, E. Kent, Model-based Strategies for High-level Robot Vision, *Computer Vision, Graphics, and Image Processing* **33** (1986), S.293-306

G. Stein, Automatische Strukturanalyse von Bildsignalen aufgrund rechnerinterner Modelle aus lokalen Formmerkmalen, *5. DAGM-Symposium*, Karlsruhe (1983), S.319-324

S. Venable, D. Richter, M. Wiedemann, A Rule-based System for Improving on Image Segmentation, *2nd Conference on Artificial Intelligence Applications*, Florida, IEEE Computer Society (1985), S.94-99

A. Waxman, K. Wohn, Image Flow Theory: A Framework for 3-D Inference from Time-varying Imagery, Preprint, to be published in *Advances in Computer Vision*

H. Westphal, Dreidimensionale Modellierung konkaver bewegter Objekte und Untersuchungen zur Ausnutzung von Helligkeitsveränderungen zur Formbestimmung, Dissertation, Universität Hamburg 1985

P.H. Winston, Learning Structure Descriptions from Examples, in *The Psychology of Computer Vision*, Hrsg.: P.H. Winston, McGraw-Hill, New York 1975, S.157-209

K. Wohn, A Contour-based Approach to Image Flow, Doctoral dissertation, University of Maryland 1986

Aufgabenangepaßte Abtastraster zur digitalen Verarbeitung von Infra-Rot-Bildsequenzen

von Dieter Coy, Wedel*

1. Einleitung

Die weiter fortschreitende Miniaturisierung elektronischer Komponenten eröffnet der digitalen Echtzeit-Bildverarbeitung immer mehr Möglichkeiten. Dies gilt für die Bildqualitätssteigerung, die Bildcodierung, die quantitative Analyse in der Mikroskopie, die Röntgenbildanalyse und die multispektrale Erderkundung. In allen Fällen ist die Bildverarbeitung stark von der Bildaufnahmetechnik beeinflußt, und in diesem Beitrag sollen die diesbezüglichen Besonderheiten bei der Verarbeitung von IR-Bildern (Infra-Rot-Bildern) dargestellt werden.

2. Bildübertragung und Bildverarbeitung

In Bild 1 sind ein konventionelles Bildübertragungssystem und ein um eine Bildverarbeitung erweitertes Bildverarbeitungssystem gegenübergestellt. Der Vergleich erlaubt folgende Aussagen:

a) Das Bild am Eingang des Bildverarbeitungsblocks und das Bild auf dem Monitor-Schirm unterscheiden sich in der "Qualität". Z. B. kann ein unterabgetastetes Bild am Eingang des Bildverarbeitungsblocks störendes Aliasing enthalten und trotzdem eine gute Reproduktion des Bildes auf dem Monitor-Schirm erlauben, bedingt durch Tiefpaßfilterungen in Kanal, Elektronik und Monitor.

b) Zur Lösung der gestellten Bildverarbeitungsaufgabe muß das hereinkommende Bild aufgabenspezifisch optimal aufbereitet werden, d. h. der optisch/elektrische Komplex muß als Ganzes optimiert werden, also inklusive Szenen-Eigenschaften, technolgischen Randbedingungen und Aspekten der Datenfusion. Für verschiedene Aufgabenstellungen wird es somit verschiedene Lösungen geben.

c) Die Anzahl der Zeilen und Spalten im Bild muß am Anfang und am Ende des Übertragungssystems nicht dieselbe sein. Falls z. B. die Anzahl Zeilen speziell für die Bildverarbeitung optimiert wird und dem Betrachter dieselbe Qualität angeboten werden soll, ist eine höhere Zeilenzahl im Monitor notwendig. Diese Technik vermeidet die Tiefpaßfilterung durch die Monitor-Modulationsübertragungsfunktion.

Um diese Aussagen zu erläutern, wird der optisch/elektrische Block detaillierter behandelt. Dabei wird davon ausgegangen, daß der Block in guter Näherung ein lineares Bildaufnahmesystem bezüglich der IR-Intensität darstellt und daß seine Gesamtübertragungsfunktion in drei Faktoren zerlegt werden kann:

1) die Übertragungsfunktion des abbildenden optischen Systems
2) das Sensorelement als Meßblende und
3) den elektrischen Datenkanal

* Dr.-Ing. Dieter Coy, AEG, Industriestraße 29, D-2000 Wedel, FRG

3. Zum optisch/elektrischen Block

3.1 Zur Übertragungsfunktion der Optik

Für die Impulsantwort g_P und die Linienantwort g_L gilt unter der Annahme von Beugungsbegrenzung und inkohärenter Beleuchtung

$$g_P \sim \left(\frac{J_1(\rho)}{\rho}\right)^2 \quad \text{bzw.} \quad g_L \sim \left(\frac{\sin(x)}{x}\right)^2$$

mit der 1. Nullstelle bei $\rho_1 = 1,22\pi$ bzw, $x_1 = \pi$.
Dies ist in Bild 2a mit langen bzw. kurzen Strichen skizziert.

3.2 Abtastung im IR-Gerät (Scanning)

Der IR-Sensor besteht in vielen Konstruktionen aus einer Spalte von 120 bis 288 Detektorelementen, die sich in einem Vakuum befinden und gekühlt werden. Mittels eines schwenkbaren Spiegels wird diese Sensorspalte scheinbar über das Bild bewegt und die Szene somit abgetastet. Nach einem Schwenk von links nach rechts wird der Spiegel gekippt, so daß die Sensorelemente im Rücklauf die Szene genau zwischen den bereits gelesenen Zeilen abtasten (siehe Pfeile in Bild 2b). sämtliche Zeilen eines Halbbildes werden gleichzeitig abgetastet und mittels Multiplexer in ein serielles Signal umgewandelt.

Jedes Sensorelement hat dabei die Wirkung einer Meßblende. Bei ideal punktförmiger Meßblende können 2 benachbarte Impulsantworten aufgelöst werden, wenn das Maximum der zweiten in die erste Nullstelle der ersten fällt. Diese entspricht dem Abtasttheorem nach Shannon und führt zu einem Abtastabstand von $\rho_1/2$ (rotationssymmetrisch) bzw. $x_1/2$ (kartesisch) /1/. Aus praktischen Gründen (mechanische Herstellung, S/N-Verhältnis etc.) wird die Fläche der Sensorelemente jedoch in der Größenordnung des Streukreises der Optik gewählt. Eine möglichst geringe Störung des Nutzfrequenzbereiches bei möglichst kleinem Aliasing wird durch kreisrunde Sensorelemente erreicht, deren Durchmesser zwischen dem 2,44-fachen und 2,0-fachen der Abtastschrittweite liegt /2/. Diese kreisrunden Sensorflächen besitzen auf den Hauptachsen ein kleineres Aliasing als die quadratischen Sensorflächen und sind ihnen darum vorzuziehen /2/. Die gemeinsame Tiefpaßfilterung von Optik und Sensor gestattet eine Vergrößerung des Abstandes auf ρ_1 bzw. x_1.

3.3 Der elektrische Datenkanal

Er beeinflußt das Sensorsignal nur parallel zu den Abtastzeilen. Der Datenkanal wird als Tiefpaß ausgebildet, um das breitbandige Rauschen beim nachfolgenden Digitalisieren entsprechend den Forderungen des Abtasttheorems zu begrenzen.

4. Aufgabenspezifische Abtastraster

4.1 Abtastung mit Offsetraster

Im Hinblick auf die Aussage b) wird für die Aufgabenstellung "Beobachtung natürlicher Szenen" zweckmäßigerweise die Tatsache mitberücksichtigt, daß vertikale und horizontale Konturen in Bildern sehr viel häufiger auftreten als diagonale /3/, und daß der Mensch diagonale Strukturen schlechter wahrnimmt /4/. Der Abtastabstand in horizontaler und vertikaler Richtung wird durch die mit der Sensorfläche gefilterten Linienantworten zu x_1 festgelegt. Im orthogonalen Raster ist jedoch die Abtastauflösung diagonal um den Faktor $\sqrt{2}$ höher als horizontal oder vertikal. Durch Weglassen jedes 2. Bildpunktes ergibt sich ein Offsetraster (o in Bild 2), bei dem das Abtasttheorem streng nur noch horizontal und vertikal erfüllt ist /5/. Wegen des seltenen Auftretens diagonaler Strukturen ist das Aliasing in diagonaler Richtung ebenfalls selten, so daß sich bei tolerierbarem Aliasing eine um den Faktor 2 reduzierte Datenmenge und damit eine erhebliche Erleichterung für die Bildverarbeitung ergibt. Eine vergleichbare Beobachtungsaufgabe stellt die Vermessung von linienförmigen Strukturen (z. B. Rissen) in industriellen Szenen dar.

4.2 Abtastung mit Hexagonalraster

Im Hinblick auf die Aussage b) wird für die Aufgabenstellung "Analyse punktförmiger Objekte" die Einführung einer Rotationssymmetrie vorteilhaft sein. Das Raster, mit dem dies näherungsweise erfüllt werden kann, ist das Hexagonalraster. Für dieses Raster sind alle Methoden der linearen Signaltheorie für die ikonische Bildverarbeitung anwendbar. Zusätzlich ist die Datenmenge im Hexagonalraster immer noch um ca. 13,4 % geringer, verglichen mit dem Orthogonalraster/6/. Das Hexagonalraster ist in Bild 2 durch x gekennzeichnet mit einem Abtaststand von ρ_1. Für dieses Raster sind jedoch kreisförmige (sechseckige) Sensorelemente vorzuziehen.

4.3 Restprobleme

Es muß darauf hingewiesen werden, daß das Halbbildverfahren zu einem Bewegungsaliasing führt, da die in beiden Halbbildern aufgezeichneten Bewegungsphasen bei einer Vollbilddarstellung falsch zusammengefügt werden /5/. Dies gilt insbesondere für das Offsetraster, das ruhende Bildvorlagen voraussetzt. Eine progressive Abtastung mit künstlicher Zerlegung in zwei Halbbilder ist für die Bildverarbeitung optimal, derzeit jedoch aus sensortechnologischen Randbedingungen noch nicht realisierbar. Für geringe Bewegungen ist eine Lösung für die Rekonstruktion von Objekten mit unterschiedlichen Bewegungsphasen in /7/ angegeben worden. Für große Bewegungen ist bei hinreichend großen Objektflächen mit Korrelationstechniken eine Zuordnung und Analyse möglich /8/. Die in Bild 2 mit rationalen Zahlen angegebenen Werte der Systemantworten und der Abtastabstände dienen der einfachen Darstellung, genauere Werte, etwa die beste Größe für die Sensorelemente, ergeben sich aus einem Kompromiß von Detailforderungen.

5. Zusammenfassung

Ausgehend von einem Bildübertragungssystem mit Bildverarbeitung, wird verdeutlicht, daß der Aliasingfehler unterschiedlich für Bildverarbeitung und Beobachter auftritt. Der optisch/elektrische Block wird daraufhin kurz vorgestellt und für zwei unterschiedliche Aufgabenstellungen hinsichtlich des Aliasingfehlers durch Abtastraster konfiguriert. Bei entsprechend angepaßten Sensorflächen können diese Abtastraster gebildet werden durch Umschalten der Taktphase für die Spalten um 180° nach jeder Scanphase.

/1/ Schröder, G.: Technische Optik, Vogel-Verlag, Würzburg, 1980

/2/ Linge, H.: Experimentelle Bestimmung der Übertragungsfunktion von Scanningmikroskopphotometern, Diss. Uni Göttingen, 1981

/3/ Keskes, N., etal: Statistical Study of Edges in TV-Pictures, IEEE Trans. on Communications, Vol.COM 27, Aug.1979, pp.1239-1246

/4/ Apelle, S.: The Oblique Effect in Man and Animals, Psychological Bulletin 1972, Vol. 78, no. 4, pp. 266

/5/ Wendland, B.: Zur Theorie der Bildabtastung, ntz Archiv, Bd. 4 (1982), S. 393 ff

/6/ Merserau, R.: The Processing of Hexagonally Sampled Two-Dimens. Signals, Proc. of the IEEE, Vol.67, No.6, June 1979, pp.930-949

/7/ Coy, D.: Zur Rekonstruktion von Fernsehvollbildsequenzen aus Halbbildsequenzen bei geringer Bewegung, Aachener Kolloquien 1984

/8/ Coy, D. etal: Extraktion von Lageinformationen aus TV-Sequenzen ..., VDI-Bericht "Automatisierte Meßsysteme", 1985

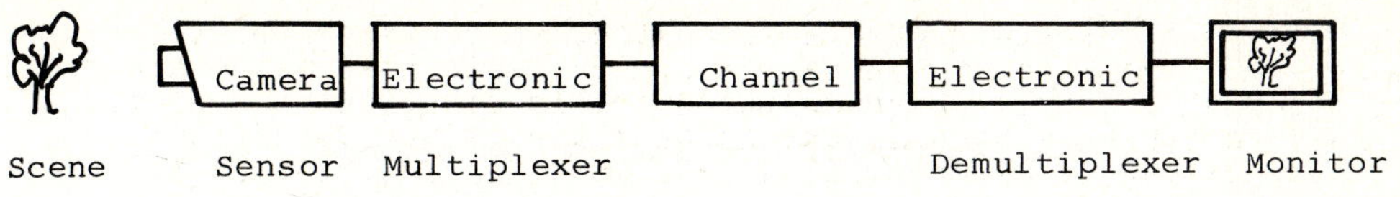

a) Conventional Image Transmission System

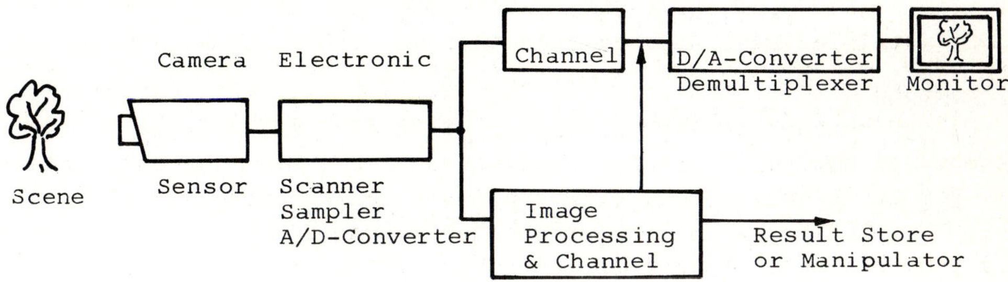

b) Image Processing within a Transmission System

Bild 1 Bilduebertragungssysteme

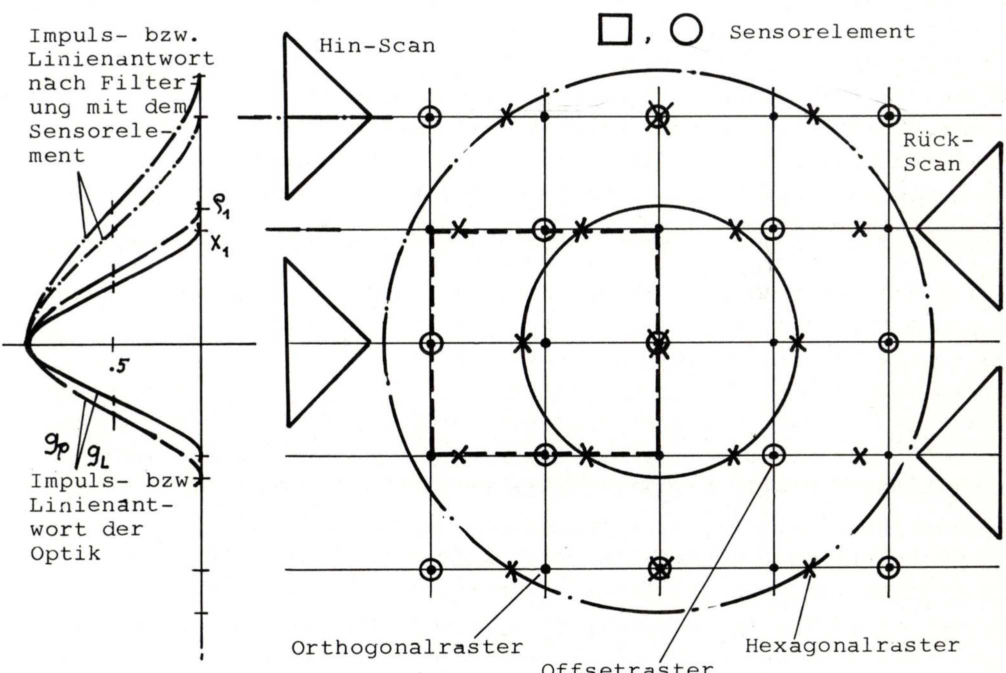

Bild 2 Systemantworten und Abtastraster

PASSPUNKTFREIE JUSTIERUNG VON BILDERN

P. Schwarzmann, B. Schorer, M.Griesinger*
Inst. für Physikalische Elektronik, Prof. W.H. Bloss, Univ. Stuttgart
Pfaffenwaldring 47, 7000 Stuttgart 80
*Daimler Benz AG, Postfach 600202, 7000 Stuttgart 60

Einleitung

Das Aufsuchen korrespondierender Bildteile in Bildpaaren ist von Bedeutung sowohl in der Objekterkennung, als auch bei der Justierung von Bildern gegenüber einer Musterlage. Wird eine solche Justierung über Korrelationsverfahren vorgenommen, trägt der gesamte Bildbereich zum Ergebnis bei, wohingegen Paßpunktverfahren sich nur auf sehr kleine Bildbereiche stützen müssen.

Korrelationsverfahren werden erheblich behindert, wenn bei der Objekterkennung zwischen dem Musterbild und dem aktuellen Bild unterschiedliche Beleuchtungsverhältnisse oder unterschiedliche Sensoreigenschaften der Kamera vorliegen. Für die Justierung von Bildern völlig unterschiedlichen Bildinhaltes wie z.B. Vorder- und Rückseite einer Leiterbahnplatine oder unterschiedliche Masken in der Halbleiterfertigung, ist man sogar allein auf die Justierung über eng begrenzte Paßpunkte angewiesen.

Für die Behandlung von beleuchtungs- und sensorunterschiedlichen Bildern mit Korrelationsverfahren ist aus der Literatur /TOM83/ ein Verfahren bekannt. Eine Erweiterung dieses Verfahrens zur gegenseitigen Justierung auch im Bildinhalt völlig verschiedener Bilder über ein Korrelationsverfahren wird in diesem Beitrag vorgestellt.

Zum besseren Verständnis sei das Verfahren /TOM83/ kurz skizziert.

Die beiden zur Deckung zu bringenden Bilder werden mit der Fouriertransformation in den Ortsfrequenzraum transformiert. Dort wird das Betragsspektrum auf einen konstanten Wert gesetzt; das Phasenspektrum bleibt unverändert (Phase-only-Verfahren). Die so gefilterten Bilder werden in den Ortsraum zurücktransformiert und miteinander korreliert. Das Ergebnis dieser Korrelation zeigt für den Ort der genauen Paßlage ein sehr steiles und lokal eng begrenztes Maximum.

Die beschriebene Bildfilterung besitzt bezüglich üblich auftretender Bildinhalte als erwünschte Eigenschaften:

- Gewöhnlich besitzen Bilder eine Betragscharakteristik im Frequenzraum, die für hohe Frequenzen abfällt; durch Konstantsetzen des Betrages erreicht man daher einen Hochpaßeffekt, d.h. Strukturen werden hervorgehoben und damit in einer anschließenden Korrelation überbewertet.
- Die gefilterten Bilder werden 'rauschähnlicher' (konstanter Betrag - Whitening); damit kann für die Paßlage

ein scharfes Maximum der Korrelationsfunktion erwartet werden, da echtes Rauschen als Korrelationsfunktion eine Deltafunktion besitzt.

- Translationsempfindliche Teile sind in der Phase codiert und bleiben daher nach der Filterung erhalten im Gegensatz zum Betragsspektrum, das translationsinvariant ist.

Die Erweiterung des dargestellten Verfahrens auf die Justierung auch völlig unterschiedlicher Bilder wird im folgenden beschrieben.

Erweiterte Methode

Das Ziel, 2 beliebige Bilder immer wieder in eine einmal vorgegebene Paßlage (Musterlage) zueinander zu bringen, wird wie folgt erreicht:

Für die korrekte Paßlage -unabhängig vom Inhalt der beiden Bilder- wird als Korrelationsfunktion von Filterungen der Originalbilder eine Deltafunktion mit Maximum für den Ort der korrekten Paßlage erzwungen.

Dies geschieht dadurch, daß im Frequenzraum für die Korrelationsfunktion in Betrag und Phase das Spektrum einer Deltafunktion hergestellt wird.

Es seien $m_1(x,y)$ und $m_2(x,y)$
Muster für 2 zur Deckung zu bringende beliebige Bilder und
$g_1(x,y)$ und $g_2(x,y)$
zu m_1 bzw. m_2 ähnliche aktuelle Bilder, die entsprechend dem Mustersatz m_1, m_2 zur Deckung zu bringen sind.

Die zugehörigen Fouriertransformierten seien die komplexen Funktionen $M_1(u,v)$, $M_2(u,v)$, $G_1(u,v)$, $G_2(u,v)$

Die aktuellen Bilder g_1 und g_2 werden im Fourierraum folgenden Filterungen unterzogen:

$$H_1 = \frac{G_1}{M_1} \qquad H_2 = \frac{G_2}{M_2}$$

Die Korrelationsfunktion $k(x,y)$ von h_1 und h_2 stellt sich im Fourierraum dann wie folgt dar:

$$K(u,v) = \frac{G_1(u,v) \quad G_2^*(u,v)}{M_1(u,v) \quad M_2^*(u,v)}$$

Es ist offensichtlich, daß für $g_1 = m_1$ und $g_2 = m_2$ sich für $k(x,y)$ eine Deltafunktion mit Maximum für $x = 0$ und $y = 0$ ergibt.

Sind g_1 und g_2 genügend ähnlich zu den Mustern m_1 und m_2, so ergeben sich damit auch deltafunktionsähnliche Korrelationsfunktionen im Ortsraum, auch wenn g_1 und g_2 völlig verschieden sind.

Vorteilhaft gegenüber Paßpunktmethoden ist dabei, daß zum Aufbau der Korrelationsfunktion der gesamte Bildbereich beiträgt und nicht nur ein kleiner Teil wie bei Paßpunktmethoden.

Für praktische Anwendungen der Methode muß jeweils geprüft werden:

- Wie stark dürfen die aktuellen Bilder von den Mustern m_1 und m_2 abweichen?
- Es ist sehr bequem, nach der Fouriertransformation der Einzelbilder mit anschließender Phase-only-Filterung auch die Korrelation direkt im Frequenzraum vorzunehmen; man erhält dadurch direkt die Fouriertransformierte der Korrelationsfunktion. Die Rücktransformation in den Ortsbereich ergibt direkt die gesuchte Korrelationsfunktion.
  Bei Anwendung der FFT (Fast Fouriertransformation) ist jedoch auf Einflüsse der Begrenzung durch das Korrelationsfenster zu achten (evtl. Anwendung eines Hamming-Fensters !) /OPPENH 75/. Dieser Effekt kann jedoch vermieden werden, wenn die Korrelation im Ortsraum durchgeführt wird; dazu sollte jeweils ein kleineres Teilbild mit dem Ausschnitt des anderen Gesamtbildes korreliert werden, damit Bildrandeffekte ausscheiden.
- Weiter muß bedacht werden, daß durch die relative Anhebung hoher Ortsfrequenzen auch das Bildrauschen angehoben wird.

Beispiele

Zur Erläuterung der Methode sind in Fig. 1-8 2 Beispiele abgebildet. Beispiel 1 soll die Unempfindlichkeit gegenüber Beleuchtungsänderungen zwischen Muster und aktuellem Bild zeigen. In Fig. 1 und 3 sind die beiden Muster in der richtigen Paßlage dargestellt. Fig. 2 zeigt das dem Muster aus Fig. 1 entsprechende aktuelle Bild; es unterscheidet sich vom Muster durch eine geänderte Beleuchtung der Szene. Fig. 4 zeigt die Korrelationsfunktion (Betrag) der gefilterten Bilder von Fig. 2 und Fig. 3. Da sich lediglich die Beleuchtung gegenüber dem Muster geändert hat, tritt keine Verschiebung der Paßlage auf. Beispiel 2 zeigt den Einfluß einer Verschiebung zwischen Muster und aktuellem Bild. Fig. 5 und 7 stellen die Muster in richtiger Lage dar. Fig. 6 zeigt das aktuelle Bild. In diesem Beispiel wurde das aktuelle Bild gegenüber dem Muster um 10 Pixel in x- und y-Richtung verschoben; die Korrelationsfunktion zeigt das Maximum daher ebenfalls bei der entsprechenden Verschiebung.

Literatur

/TOM83/ V.T. Tom, G. K. Wallace, G. J. Wolfe (1983) Image Registration by Statistical Method, in Proc. SPIE, Vol. 432, pp 240-243 San Diego 1983

/OPPENH75/ A. V. Oppenheim and R. W. Schafer, Digital Signal Processing, Prentice-Hall Englewood Cliffs N.Y. 1975

Fig. 1

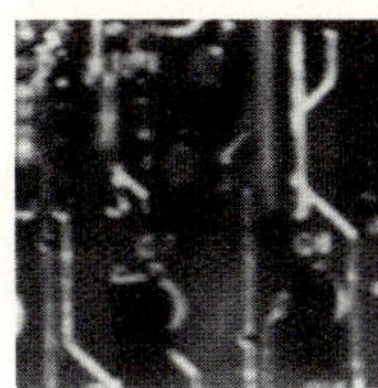

Fig. 2

Fig. 3

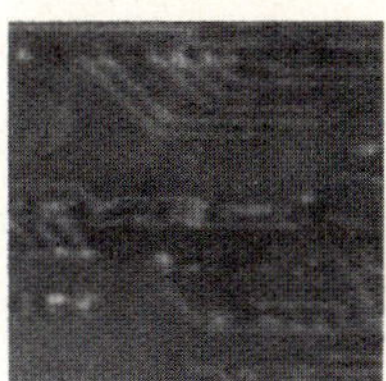

AUSSCHNITTE AUS TV-AUFNAHMEN EINER BESTUECKTEN LEITERPLATINE
zur Berechnung der Korrelation zwischen unterschiedlichen Bildern

Fig. 1 : Muster 1 (Vorderseite Original)
Fig. 2 : aktuelles Bild 1 (Vorderseite, mit veraenderter Beleuchtungsrichtung aufgenommen
Fig. 3 : Muster 2 (Rueckseite Original)
Fig. 4 : Ergebnis der Korrelationsberechnung (die Lage des Maximums zum Mittelpunkt gibt die Verschiebung zwischen Muster 1 und aktuellem Bild 1 an; da hier nur die Beleuchtungsrichtung geaendert wurde, betraegt die Verschiebung 0)

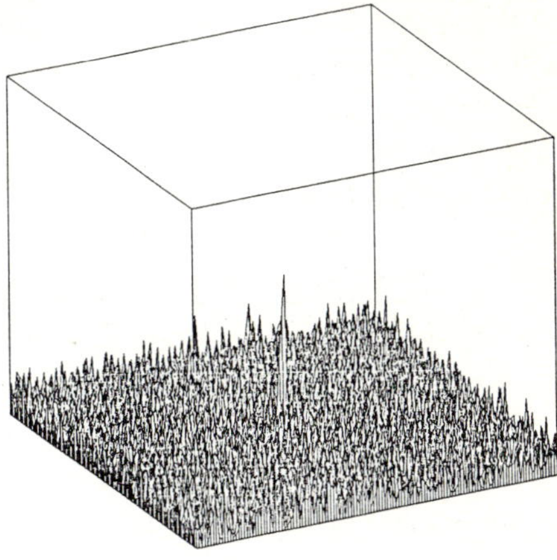

Fig. 4

Fig. 5

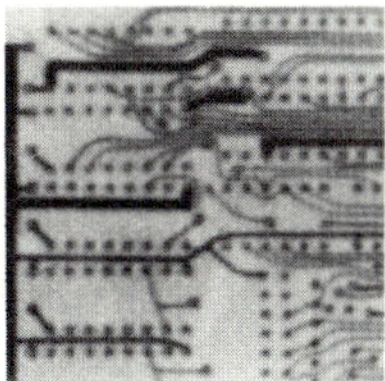

Fig. 6

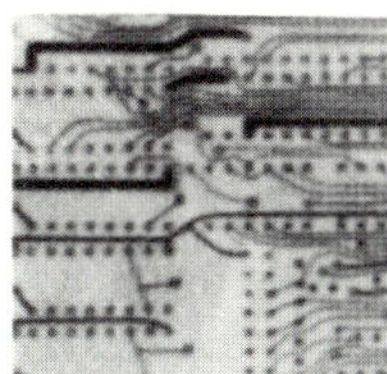

Fig. 7

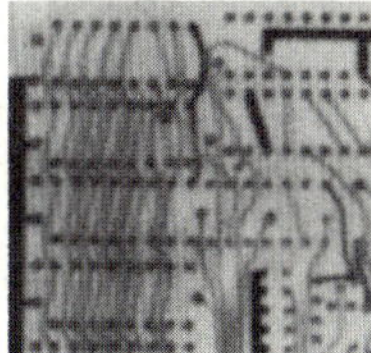

AUSSCHNITTE AUS DURCHLICHTAUFNAHMEN VON LEITERBAHNENVORLAGEN
zur Berechnung der Korrelation zwischen unterschiedlichen Bildern

Fig. 5 : Muster 1 (Vorderseite unverschoben)
Fig. 6 : aktuelles Bild 1 (Vorderseite um +10/+10 Pixel verschoben)
Fig. 7 : Muster 2 (Rueckseite unverschoben)
Fig. 8 : Ergebnis der Korrelationsberechnung (die Lage des Maximums zum Mittelpunkt zeigt die Verschiebung zwischen Muster 1 und aktuellem Bild 2)

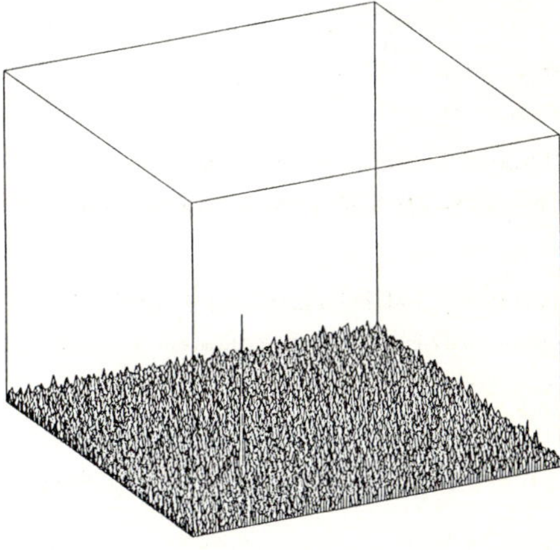

Fig. 8

Chrominanzsignalquantisierung unter Berücksichtigung
des Beitrags der Chrominanzsignals zur Wiedergabeluminanz

Guido Bruck
Fachbereich Elektrotechnik, Universität Duisburg
Bismarckstraße 81, D-4100 Duisburg 1

1. Einführung und Problemstellung

In den standardisierten Fernsehsystemen geschieht die gesamte Signalverarbeitung in einem nichtlinear vorverzerrten Bereich. Die nichtlineare Verzerrung der Rot-, Grün- und Blau-Signale, die sog. Gradations- oder Gammavorverzerrung, geschieht im Hinblick auf Verzerrungen, die in den standardisierten Wiedergabeeinheiten prinzipbedingt auftreten. Diese allgemein in der Bildverarbeitungstechnik eingeführte Methode der Gradations-Vorverzerrung hat zur Folge, daß die sog. Luminanz- und Chrominanzsignale nicht mehr nur von der Bildluminanz bzw. der Bildchrominanz abhängen. So läßt sich allgemein zeigen, daß bei den standardmäßig verwendeten Gradationsvorverzerrungen mit Gamma-Werten zwischen 2,2 und 2,8 das Chrominanzsignal nicht nur Chrominanz, sondern in erheblichem Umfang auch Luminanz transportiert. Eingriffe in das Chrominanzsignal haben somit auch immer einen Einfluß auf die Wiedergabeluminanz.

Solche Veränderungen können sein:

a) Tiefpaßfilterung der Chrominanzsignale, wie sie bei den standardisierten Farbfernsehsystemen vorgenommen wird. Darüber wurde bereits auf dem Aachener Kolloqium 1984 berichtet (/2/).

b) Quantsierung der Chrominanzsignale, wie sie für zukünftige digitale Fernsehstandards denkbar und aus Gründen der Datenreduktion wünschenswert wäre.

Die Wiedergabeluminanz ist proportional dem Farbwert Y . Dieser wird daher auch als relative Luminanz bezeichnet. Y bestimmt sich zu

$$Y = a_B(E_Y + E_U{}')^{\gamma} + a_R(E_Y + E_V{}')^{\gamma} + a_G(E_Y + E_W{}')^{\gamma}$$

mit

$$E_u' = E_B - E_Y \; ; \; E_v' = E_R - E_Y \; ; \; E_w' = E_G - E_Y$$

Dabei sind a_i die Farbbeiwerte, E_u', E_v' und E_w' die Farbdifferenzsignale und E_y das Luminanzsignal. Ist $\gamma>1$, dann gilt: Die relative Luminanz Y wird durch die Anwesenheit von Null verschiedener Chrominanzsignale gegenüber dem Unbuntfall erhöht. Dies entspricht der heutigen Realität. Daher muß davon ausgegangen werden, daß bei einem Buntbild eine relativ höhere Leuchtdichte gegenüber dem Unbuntbild vorliegt.

Auch bei einer Quantisierung der Chrominanzsignale kann es nun zu einer Beeinflussung der Wiedergabeluminanz kommen. Dies zeigt Abb. 1. Dort ist für das Beispiel einer rampenartigen Veränderung des Chrominanzsignals E_u (links oben) bei konstantem Luminanzsignal E_y=0.5 und konstantem Chrominanzsignal E_v=0 der Verlauf der Wiedergabeluminanz angegeben (links unten). Man sieht sehr deutlich, daß die wiedergegebene Luminanz von E_u abhängig ist, da E_y und E_v konstant gehalten werden. Wird das E_u nun quantisiert zu E_{uq} (rechts oben), so verändert sich auch die Wiedergabeluminanz, obwohl am Luminanzsignal selbst keine Änderung erfolgt ist. Bei einer solchen Quantisierung fallen die so entstehenden Luminanzsprünge störend auf, nicht die viel weniger sichtbaren Chrominanzsprünge.

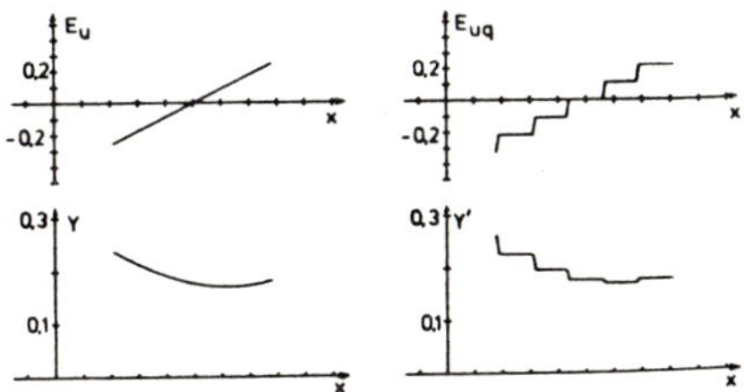

Abb. 1: Verhalten der Wiedergabeluminanz bei Quantisierung des Chrominanzsignals

2. Kompensation der Luminanzdefekte

Die durch Veränderung der Chrominanzsignale entstehenden Luminanzdefekte lassen sich durch ein dem Luminanzsignal additiv hinzuzufügendes Korrektursignal ΔE_y kompensieren. Dieses ergibt sich aus

$$Y_{soll} = a_B(E_Y + E_u' + \Delta E_Y)^\gamma + a_R(E_Y + E_v' + \Delta E_Y)^\gamma + a_G(E_Y + E_w' + \Delta E_Y)^\gamma$$

Aus dieser Beziehung läßt sich ΔE_y durch ein Iterationsverfahren mit beliebiger Genauigkeit ermitteln. Dieses zeitaufwendige Verfahren läßt sich durch geeignete Näherungslösungen umgehen.

Der Beitrag des Chrominanzsignals zur Wiedergabeluminanz ist umso größer, je größer die Farbsättigung ist. Dementsprechend macht sich die Kompensationswirkung auch in hochgesättigten Farben am deutlichsten bemerkbar.

Dies geht anschaulich aus Abb. 2 hervor. Es zeigt die Unterscheidbarkeitsschwellen für Chrominanzsignaländerungen mit und ohne Luminanzkompensation für die Grundfarbarten Rot, Grün und Blau, sowie für die einfachen Mischfarbarten Gelb, Cyan und Magenta. Auf der Abszisse ist die relative Farbsättigung aufgetragen, auf der Ordinate die Unterscheidbarkeitsschwelle, ausgedrückt durch die entsprechenden Anzahlen von Quantisierungsstufen (bei Unterstellung einer 8-bit Quantisierung der Chrominanzsignalkomponenten). So ist z.B. zu erkennen, daß für die Farbart Magenta die Unterscheidbarkeitsschwelle mit Luminanzkompenstaion um bis zu den Faktor 10 höher liegen kann als ohne Kompensation.

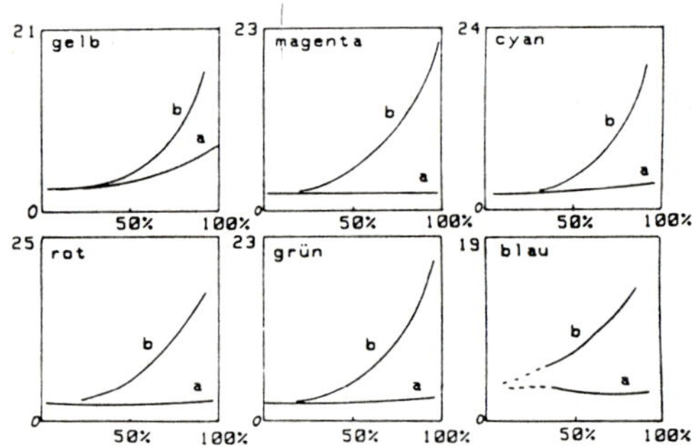

Abb. 2: Unterscheidbarkeitsschwelle mit/ohne Luminanzkompensation
a: ohne, b: mit Luminanzkompensation

3. Anwendungen der Kompensation bei unterschiedlichen Quantisierungen der Chrominanzsignale

Es wurden nun Quantisierer entworfen, die die gefundenen Zusammenhänge ausnutzen. Dabei wurden Bildvorlagen verwendet, die bei konstantem Leuchtdichtesignal einen rampenartigen Verlauf der Chrominanzsignale aufweisen, E_u in horizontaler und E_v in vertikaler Richtung. Bei solchen Bildern läßt sich eine zu grobe Quantisierung der Chrominanzsignale besonders empfindlich nachweisen.

Als Kriterium für die Akzeptanz wurde für die unterschiedlichen Quantisierer die Sichtbarkeitsgrenze bereits geringster Störstrukturen gewählt. Es wurden untersucht:

1) lineare Quantisierung von E_u und E_v in kartesischen Koordinaten ohne Luminanzkompensation (herkömmliches Verfahren)
2) lineare Quantisierung von E_u und E_v in kartesischen Koordinaten
3) nichtlineare Quantisierung in kartesischen Koordinaten mit subjektiver Optimierung der Stufenverteilung des Quantsisierers.

Die erzielbaren Datenraten in bit/Bildpunkt sind in Abb. 3 angegeben. Ebenfalls sind dort Grenzdatenraten angegeben, für die die Qualität des quantisierten Bildes gerade noch tolerabel ist.

Abb. 3: Tabelle der erzielbaren Datenraten

Verarbeitung	Datenrate mit Luminanzkompensation in bit/Bildpunkt	
	hohe Qualität	akzeptable Qualität
(1)	8 *	6 *
(2)	7	5
(3)	6	4

*-keine Luminanzkompensation

4. Zusammenfassung

Durch Luminanzkompensation lassen sich Luminanzdefekte beseitigen, die bei Farbfernsehsystemen mit Signalverarbeitung in gammavorverzerrten Bereichen auftreten. Dabei ergeben sich Bildqualitätsverbesserungen bei Tiefpaßfilterung und bei Quantisierung von Chrominanzsignalen.

Literaturangaben:

/1/ G. Dickopp: Fernsehtechnik, Vorlesung an der Universität Duisburg

/2/ G. Dickopp: Kompensation von Luminanzdefekten, die durch Tiefpaßfilterung der Chrominanzsignale entstehen, Aachener Kolloquium 1984

HQ-MAC; Ein Konzept zur schmalbandigen, kompatiblen HQTV-Übertragung

M. Silverberg, W. Boie
Lehrstuhl für Nachrichtentechnik, Universität Dortmund
Postfach 50 05 00, D-4600 Dortmund 50

Die geplante und vorbereitete Einführung von MAC-Fernsehsystemen (Multiplexed Analog Components) bringt eine Bildqualitätsverbeserung gegenüber den konventionellen Systemen PAL und Secam mit sich. Aufgrund der zeitsequentiellen Übertragung von Luminanz und Chrominanz treten bei MAC keinerlei Übersprechstörungen (Crosseffekte) zwischen Leuchtdichte - und Farbdifferenz-Signalen mehr auf. Ferner bietet das MAC-Konzept die Möglichkeit mehr Ton- und Dateninformationen zu übertragen.

Wegen der beibehaltenen Zeilensprungübertragung haften aber der MAC-Codierung nach wie vor Störeffekte, wie Kantenflackern, Zeilenwandern und Großflächenflimmern an. Ferner ist die Horizontalauflösung nur mäßig.

Für zukünftige Hochqualitäts-Fernsehsysteme ist jedoch eine Verbesserung gerade in diesen Punkten essentiell. Mit kompatibel verbesserten Systemen läßt sich in bereits bestehenden Kanälen eine beachtliche Qualitätssteigerung erzielen. Aufgrund der Kompatibilität solcher Verfahren können dann Standard- und verbesserte MAC-Empfänger nebeneinander benutzt werden.

Auf diese Weise ist ein schrittweiser Übergang zu einem Hochqualitätssystem (HQ-MAC) möglich. Durch diese Vorgehensweise wird schon sehr frühzeitig ein Potential an HDTV-fähigen Endgeräten (Kameras, Bildwiedergabesysteme) geschaffen, so daß ein späterer Übergang auf eine echte, breitbandige HDTV-Übertragung vorbereitet ist.

In diesem Beitrag wird ein HQ-MAC-Konzept vorgestellt, das zum genormten MAC-Verfahren kompatibel ist und eine erhöhte Detailauflösung bei gleichzeitiger Ausschaltung aller Zeilensprungstörungen gestattet.

Dieses wird erreicht durch eine bewegungsadaptive digitale Signalverarbeitung an Sender und Empfänger. Dabei wird, je nach Bildinhalt, entweder eine hohe örtliche oder eine hohe zeitliche Auflösung übertragen.

Die Übertragung der hohen örtlichen Auflösung gelingt durch eine Vollbildverarbeitung mit Diagonalfilterung, Offsetmodulation und synthetischem Zeilensprung. Dadurch wird eine Horizontalauflösung erreicht, die etwa der doppelten Kanalbandbreite entspricht.

Bei hoher zeitlicher Auflösung gelingt die Überwindung der Zeilensprungeffekte durch vertikale Vor- und Nachfilterung. Zwei Bewegungsdetektoren an Sender und Empfänger steuern zwischen den beiden Moden weich um. Die Detektoren arbeiten unabhängig voneinander, so daß keine Zusatzinformation für die Umsteuerung übertragen werden muß.

Das Verfahren der HQ-MAC-Codierung ist anwendbar bei terrestrischer Übertragung, wie auch in Satelliten- und Kabelkanälen.

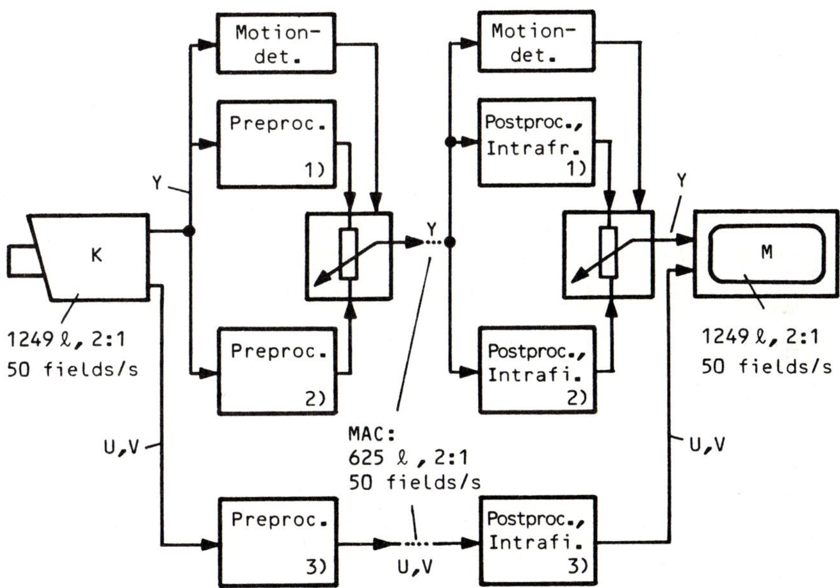

Bild 1

Literatur

/1/ Wendland, B.: High Quality Television by Signal Processing. Conference Proceedings "2nd International Conference on New Systems and Services in Telecommunication", Liège, November 1983.

/2/ Wendland, B.: Entwicklungsalternativen für zukünftige Fernsehsysteme. Fernseh- und Kino-Technik, Bd. 34 (1980) No. 2. pp. 41-48 (Paper presented at the 7th annual meeting of the FKTG in Dortmund, 1979)

/3/ Schröder, H.; Silverberg, M.; Wendland, B.; Huerkamp, G.: Scanning Modes for Flicker-Free Colour TV-Reproduction. IEEE Trans. on ICCE, Vol. CE-31, No. 4,(1985), S.627-641

/4/ Wendland, B.: High Definition Television Studies on compatible Basis with Present Standards. "Television Technology in the 80's", S. 124-131, published by the SMPTE, Scarsdale, New York, 1981, schriftliche Fassung eines eingeladenen Vortrages zur SMPTE-Konferenz 1981 in San Francisco

/5/ Schröder, H.; Elsler, H.: Planare Vor- und Nachfilterung für Fernsehsignale ntz-Archiv, Bd. 4 (1982), Nr. 10

/6/ Güttner, E.: Realtime Picture Processing System with Two-Dimensional Filtering and Offset-Modulation. European Signal Processing Conference 1986, Den Haag, September 2-5, 1986, Conference Proceedings

PROJECTION OF THE HOUGH TRANSFORM

Ulrich Eckhardt
Institut für Angewandte Mathematik
Bundesstraße 55, D-2000 Hamburg 13

Gerd Maderlechner
ZT ZTI INF 12, SIEMENS AG
Otto-Hahn-Ring 6, D-8000 München 83

The *Hough* (or *Radon) transform* in its classical form is used for identifying lines in binary images (Hough |7|, Deans |4|; Radon |8|). It can be interpreted as an evidence accumulation method. Its main advantage is that it is relatively insensitive to noise, even interrupted lines can be detected by means of it. Furthermore, it can be easily implemented on parallel computers.

The main drawback for applying the method is that it requires a huge amount of storage and computer time. This is specifically true if a generalized form of the Hough transform is considered for finding more general features needing more than two parameters for description, for example circles, ellipses etc. The computational complexity of the generalized Hough transform is the sum of the time for calculating the transform and the time needed for analyzing the transform. Let us assume that there are d parameters for describing the features and that each parameter is sampled with N discrete values. Then the time needed for performing the transform is proportional to the number of pixels in the image multiplied by N^{d-1}. The amount of storage needed for the transformed image is N^d. The time complexity for finding e.g. a maximum in the transformed image is proportional to the latter number. Hence even with moderate values of d the complexity of the generalized Hough transform becomes prohibitive.

There are numerous attempts reported in the literature to reduce the complexity of the Hough transform by clever storage techniques, by using parallel computer architectures or by projection of the transformed image onto a subspace of fewer dimension (see e.g. |1|, |6|). Obviously, such a projection will reduce the computational complexity caused by the dimension of the parameter space drastically. In this paper the latter approach is investigated more closely. For mathematical details the reader is referred to the report |5|. The discussion here is completely confined to the case of the classical transform for lines. There is no difficulty in presenting a more general theory, which will, however, be much more technical. As a practical application we assume that we want to detect within a binary picture (e.g. a document) prominent directions in order to adjust it automatically.

The classical Hough transform is an integral transform which assigns to a function $\beta(x,y)$ of two real variables the set of all integrals of this function along lines. β is understood to be the grey value function of the *image* under consideration. Hough |7| made the observation that in binary images the maxima of this transformation correspond to lines in the image. In a *binary image* β takes only the values zero and one. Within this paper we are exclusively concerned with binary images.

For investigating the question whether a line segment is contained in the image, we parametrize all lines in the plane by number pairs (a,b). The set of all parameter pairs belonging to lines is the so-called *accumulator* A. Let $\ell(a,b)$ be the line corresponding to $(a,b) \in A$. Consider the line integral (whenever it exists)

$$H_o(\beta)(a,b) = \int_{\ell(a,b)} \beta(P)\, dP$$

where P denotes a point in the plane and dP is the line element along $\ell(a,b)$. We introduce a (generally nonlinear) *filter* $F(\beta;a,b)$ which is allowed to depend on the parameters and acts on the image function β. Using this filter we define the *generalized Hough transform*

$$H(\beta)(a,b) = H_o(F(\beta;a,b))(a,b) =$$

$$= \int_{\ell(a,b)} F(\beta;a,b)(P)\, dP.$$

For projecting the Hough transform we keep one parameter, say a, fixed and project the transform with respect to the other. Therefore, the *projected Hough transform* is defined by

$$H_b(\beta)(a) = (\Pi H(\beta))(a),$$

where Π is a (linear or nonlinear) *projector* assigning to each function of two variables $f(a,b)$ a function of one variable $(\Pi f)(a)$.

The projected Hough transform should exhibit some obvious *invariance* properties in order to yield meaningful results. It is certainly necessary that the projector is invariant with respect to *translations* of the parameter b. This means that two functions $f(a,b)$ and $f(a,b+b_o)$ with fixed b_o should have the same projection. Moreover, *motions* of the picture along a line $\ell(a',b)$ should also leave the projection invariant. It is possible, to prove the following Theorem (see |5| for details):

Theorem: Assume that

1. F does not depend on b and $\Pi H_o(\beta)(a)$ is continuous and linear for each fixed $a \in A$.
2. Π is translation invariant with respect to b.
3. ΠH is line translation invariant with respect to a fixed line $\ell(a,b)$.
4. F is separable with respect to a, i.e.

$$F(u;a) = \sum_{i=1}^{n} \gamma_i(a) \cdot F_i(u)$$

with real numbers γ_i and motion invariant operators F_i.

Then

$$H(\beta)(a) = C(a) \cdot \sum_{i=1}^{n} \gamma_i(a) \cdot \int_{\mathbb{R}^2} F_i(\beta)(P)\, dP$$

is motion invariant with respect to all motions of the plane.

When the operator ΠH is invariant with respect to all motions, it does not exhibit useful information about the contents of the image. Therefore, in order to get useful results, we are required to drop either the linearity condition of the projection or its translation invariance (in the sense of conditions 2 and 3 of the Theorem) or else to use a filter which depends severely nonlinearly on a.

It is certainly not advisable to use a non translation invariant projection since the results of such an operation are hard to understand. The contribution of a specific feature in the image under a non translation invariant operator depends on its location in the picture. It is therefore impossible to distinguish faint details at a good position from strong details at a position of poor visibility. Computational experiments with non translation invariant projections performed by the authors consequently were very dicouraging |5|.

There exists a multitude of approaches in the literature using nonlinear projections. From the Theorem it becomes clear wherefore all authors preferred nonlinear approaches in spite of the fact that linear projectors can be handled easier.

The *maximum projector* which assigns to a function of two variables f(a,b) the projection $(\Pi f)(a) = \max_b f(a,b)$ is obviously translation invariant. It was successfully applied by different authors (|1| and |6|).

A different possibility is the L^p*-projector*

$$(\Pi f)(a) = \int |f(a,b)|^p \, db.$$

Biland and Wahl |3| reported good results with p between 1.5 and 4.0.

Rubart |9| used a projection of the derivative of f with respect to parameter b:

$$(\Pi f)(a) = \int \left|\frac{\partial f}{\partial b}(a,b)\right| db.$$

This approach turned out to be very useful for finding prominent directions within pictures.

Bernhardt |2| proposed a remarkable projection of the Hough transform which was reinvented quite recently |10|. It is based on a nonlinear filter with linear projector and was successfully applied in a commercially available system for reading hand-written letters on cheques.

References

1. Ballard DH, Sabbah D (1983) Viewer independent shape recognition. IEEE Trans. PAMI-5:653-660
2. Bernhardt L (1984) Three classical character recognition problems, three new solutions. Siemens Forsch.- u. Entwickl. Ber. 13:114-117
3. Biland HP, Wahl FM (1986) Cluster estimation in Hough space for image analysis. IBM Technical Disclosure Bulletin 28:3667-3668
4. Deans SR (1983) The Radon Transform and Some of Its Applications. New York etc.: John Wiley and Sons
5. Eckhardt U, Maderlechner G (1987) Projections of the Hough transform. Preprint University Hamburg
6. Gerig G, Klein F (1986) Fast contour identification through efficient Hough transform and simplified interpretation strategy. IAPR - afcet: Eighth International Conference on Pattern Recognition. Paris, France, October 27-31,1986
7. Hough PVC (1962) Method and means for recognizing complex patterns. US Patent 3,069,654. Washington: United States Patent Office, December 18, 1962
8. Radon J (1917) Über die Bestimmung von Funktionen durch ihre Integralwerte längs gewisser Mannigfaltigkeiten. Ber. Verh. Sächs. Akad. Wiss. Leipzig, Math.-Nat. Kl. 69:262-277
9. Rubart L (1986) Ermittlung von Geraden in Binärbildern durch die Hough-Transformation. Diplomarbeit Universität Hamburg
10. Sinden FW (1985) Shape information from rotated scans. IEEE Trans. PAMI-7:726-730

Ein regionenorientiertes Segmentierungsverfahren für texturierte Bildvorlagen

Rudolf Mester und **Uwe Franke**
Institut für Elektrische Nachrichtentechnik
RWTH Aachen, 5100 Aachen, Melatener Str. 23

Zusammenfassung

Das hier vorgestellte Segmentierungsverfahren für Grautonbilder entstand im Rahmen einer Untersuchung zur Textur/Kontur-Bildanalyse und -Codierung [1]. Es basiert auf einem einfachen Texturmodell und weist eine spezielle Kontrollstruktur auf, die als erweiterte quasi-parallele Variante der verbreiteten Region Growing Verfahren angesehen werden kann. Dem Problem der in natürlichen Bildern nahezu stets anzutreffenden örtlich instationären Texturstatistik kann durch eine dem 'minimal spanning tree' [2] verwandte Strategie begegnet werden.

Texturmodell

Die einfachsten und zugleich wesentlichsten Parameter eines stationären zweidimensionalen stochastischen Prozesses sind der mittlere Grauwert m und die Varianz σ^2. Im hier beschriebenen Verfahren sind diese die einzigen Texturmerkmale, die einer Region zugeordnet werden. Obwohl die Annahme eines weißen gaußverteilten Rauschens die tatsächlichen Gegebenheiten in realen Texturen eindeutig verletzt, sind diese Modellparameter zur Segmentierung der meisten natürlichen Bildvorlagen völlig hinreichend. Zur Beschreibung und Rekonstruktion der Textursignale in bereits segmentierten Bildern sind hingegen wesentlich aufwendigere Verfahren notwendig und seit kurzem auch verfügbar [1].

Regionenbildender Prozeß

Ein fundamentales Problem bei der Bildsegmentierung besteht in der Entscheidung, ob zwei Gruppen von Bildpunkten zur selben Region gehören oder nicht. Die Zuverlässigkeit, mit der diese Frage korrekt gelöst werden kann, hängt entscheidend davon ab, daß jede der beiden miteinander verglichenen Punktmengen jeweils tatsächlich eine echte Teilmenge einer "wahren" Region ist. Nur dann kann eine verläßliche Schätzung der Parameter dieser Regionen erfolgen, von deren Ergebnis die Entscheidung letzlich abhängt. Aus diesem Grunde erscheint eine 'Bottom Up' Analyse unter ausschließlicher Verwendung von Regionenverschmelzungen als besonders sinnvoll. Innerhalb solcher Verfahren werden nur Regionen betrachtet, die entweder atomar (einzelne Bildpunkte) oder aufgrund nachgewiesener Homogenität "gewachsen" sind.

Im Initialzustand bildet jeder Bildpunkt zunächst eine eigene Region. Die Verschmelzung von Regionenpaaren mit ähnlichen statistischen Eigenschaften stellt die Grundlage eines quasi-parallelen Wachstumsprozesses dar.

Verschmelzungshypothesen und Ähnlichkeitsmaß

Voraussetzung für einen Verschmelzungsvorgang ist die (Null-) Hypothese, daß die Grauwertensembles $\vec{x}_1$, $\vec{x}_2$ zweier benachbarter Regionen R_1, R_2 demselben stochastischen Prozeß entstammen. Aus dem Blickwinkel der statistischen Entscheidungstheorie führt diese Problemstellung auf die Betrachtung des verallgemeinerten Likelihood-Ratios (vgl. *Yakimovski* [3]) :

$$\lambda = \frac{max\{p(\vec{x}_1|\omega_1 \in \Omega)\} \cdot max\{p(\vec{x}_2|\omega_2 \in \Omega)\}}{max\{p(\vec{x}_1, \vec{x}_2|\omega_3 \in \Omega)\}},$$

wobei $\omega_1, \omega_2, \omega_3$ Parametervektoren innerhalb des Modellparameterraumes Ω sind. Wenn m_1, σ_1 bzw. m_2, σ_2 die Maximum-Likelihood-Schätzungen der Parameter der Einzelregionen sind, sowie m_3, σ_3 diejenigen der Zusammenfassung von Region R_1 und R_2 (Bildpunktanzahl N_1 bzw. N_2) , so ergibt sich:

$$\lambda = \frac{\sigma_3^{N_1+N_2}}{\sigma_1^{N_1} \cdot \sigma_2^{N_2}}$$

Wenn die Nullhypothese wahr ist, nähert sich die Verteilung von $2ln\lambda$ mit wachsendem Stichprobenumfang asymptotisch einer χ^2- Verteilung mit 2 Freiheitsgraden. Das Testverfahren besteht üblicherweise darin, die Nullhypothese dann anzunehmen, wenn die Teststatistik unterhalb eines kritischen Wertes λ^* liegt, der z.B. durch das *Neyman-Pearson-Kriterium* gegeben ist. Unser Ansatz besteht hingegen darin, die Größe λ primär als relatives Unähnlichkeitsmaß zu verwenden, d.h. alle anstehenden Verschmelzungshypothesen in der Reihenfolge steigender Unähnlichkeit abzuarbeiten. Hierdurch wird gewährleistet, daß unsichere Entscheidungen erst sehr spät (wenn überhaupt) getroffen werden. Zwar muß auch hier ein Grenzwert λ^* definiert sein, dessen Wahl aber relativ unkritisch ist ($ln\lambda^* \simeq 16$).

Datenstrukturen

Ein Labelfeld dient einerseits zur Repräsentierung der topologischen Struktur der Regioneneinteilung, andererseits als Indexfeld auf Speicherstrukturen, die die Bildpunktanzahl N, die Summe der Grauwerte M und die Summe der quadrierten Grauwerte Q zu jeder Region enthalten. Aus diesen Größen können m und σ^2 jederzeit berechnet werden. Jede Verschmelzungsoperation verändert den Zustand des Labelfeldes und addiert die Parameter N, M und Q der "verschwindenden" Region R_i zu denen der überlebenden Region R_j.

Als weitere wesentliche Datenstruktur wird zwischen den Knoten des Bildpunktgitters ein "Kantengitter" eingebettet (vgl. auch [4]). Dieses Kantengitter ist Träger der binären Kanteninformation sowie zusätzlich mit den Grauwertdifferenzen benachbarter Punkte attributiert.

Kontrollstruktur

Da während des Segmentierungsvorganges zu jedem Zeitpunkt eine Vielzahl von Verschmelzungshypothesen möglich sind, sollten stets die aussichtsreichsten Hypothesen zuerst generiert und überprüft werden. Man bedient sich des attributierten Kantengitters, indem alle Kantenelemente zwischen benachbarten Regionen betrachtet werden. Falls die Grauwertdifferenz über einem Kantenelement kleiner als eine Schwelle t ist, wird der Likelihood-Test für die beiden Regionen durchgeführt. Einzelnen Bildpunkten und extrem schwach strukturierten Regionen wird dabei eine Standardabweichung von einer Grauwertstufe zugeordnet. Falls die Teststatistik kleiner als λ^* ist, wird das Labelpaar und der Wert von λ in einer Liste eingetragen. Nachdem alle Kantenelemente betrachtet worden sind, wird die Liste in der Reihenfolge steigender Werte von λ geordnet. In einer zweiten Phase wird die Liste ausgelesen, und alle zulässigen Verschmelzungen sequentiell ausgeführt. Ein nochmaliger λ-Test vor der Verschmelzung trägt eventuellen zwischenzeitlichen Regionenveränderungen Rechnung. Nach dem Leerlaufen der Liste wird die Schwelle t erhöht (typischerweise um 30%), so daß weitere Kantenelemente an der Hypothesengenerierung teilnehmen können und erneut ein Durchlauf durch die beschriebenen Phasen vorgenommen. Der Segmentierungsvorgang ist beendet, wenn auch bei beliebig hohem t kein Regionenpaar dem λ-Kriterium mehr genügt.

Berücksichtigung von nicht-stationärer Texturstatistik

In natürlichen Bildvorlagen treten vielfach Regionen mit örtlich nicht stationärer Texturstatistik auf, z.B. durch die Krümmung von Objektoberflächen und variierende Beleuchtungsverhältnisse. In solchen Regionen bleiben nach der Beendigung der vorangehend beschriebenen Schritte einzelne Parzellen mit im Sinne des λ-Kriteriums unterschiedlicher Texturierung bestehen, ohne daß dort wahrnehmbare Texturgrenzen vorhanden sind. Die multiplikative Verknüpfung des senkrecht zum Kantenverlauf gemessenen Grauwertkontrastes mit der Teststatistik $ln\lambda$ ist als Maß für die tatsächlich vorhandene Kantenintensität recht geeignet. Die als Artefakte erkannten Kanten werden zunächst in Richtung ansteigender Kantenintensität geordnet und erst danach in dieser Reihenfolge eliminiert. Dabei muß beachtet werden, daß manche Regionen mehrere Kanten zueinander aufweisen können; die effektive Kantenstärke ist dann gleich der größten beteiligten Kantenstärke.

Ergebnisse

Die Abbildung 1 zeigt das Segmentierungsergebnis für zwei Bildvorlagen der Größe 256·256 Bildpunkte. Bei allen uns vorliegenden natürlichen Bildvorlagen entstehen ähnlich gute Resultate, solange keine extreme Makro-Texturierung vorliegt.

Bild 1: Bildvorlagen und zugehörige Segmentierungsergebnisse

Erweiterungen

Die Verwendung aufwendigerer struktursensitiver Texturmodelle sowie eine "Multi-Resolution" Variante des beschriebenen Verfahrens sind Gegenstand laufender Untersuchungen. Es bestehen Indizien dafür, daß durch eine grobe Vorsegmentierung eines unterabgetasteten Bildes der Rechenaufwand insgesamt entscheidend verringert werden kann.

Literatur:

[1] U.Franke, R.Mester: *Ein regionenorientiertes Bildkodierungskonzept mit sehr hoher Datenkompression.* Beitrag zum ASST'87 (gleicher Band)

[2] C.T.Zahn: *Graph-Theoretical Methods for Detecting and Describing Gestalt Clusters.* IEEE Trans. Computers, vol. C-20, no. 1, Jan. 1971, pp.68-86

[3] Y.Yakimovski: *Boundary and Object Detection in Real World Images.* Journal A.C.M., vol 23, no.4, Oct. 1976, pp.599-618

[4] C.R.Brice, C.L.Fennema: *Scene Analysis Using Regions.* Artificial Intelligence, vol. 1, 1970, pp.205-224.

Contour Extraction Of Objects Corrupted With Shadows And Reflections By Object-Adapted Filtering

Reinhard Schmidt
Institut für Nachrichtensysteme, Universitaet Karlsruhe
Kaiserstrasse 12, D-7500 Karlsruhe, W-Germany

A new image preprocessing system is presented for contour extraction of objects corrupted with shadows and reflections. The system is adapted to the objects to be identified and to the shadows and reflections on the surface of the objects. The adaption process is performed in a training phase. The structure of the system is completely parallel. A simplified classification of the objects is also enabled. Some simulation results of numbers embossed in sheet metal are presented.

1. INDRODUCTION

Industrial systems for optical recognition degrade substantially if the input images are corrupted by shadows and reflections on the surface of the objects to be identified. Industrial systems often use the contours of the objects for the classification process. The contours are conventionally extracted by edge enhancement methods, e.g. by the Laplace- or Sobel-operator [1], followed by a threshold operation. The contours extracted by these methods are acceptable as long as there are no degrading effects. The edge enhancement methods just approximate the derivation of the image and this is equal to a high-pass filtering operation. Thus the degrading effects are enhanced, too. In the following a new approach based on the signal estimation [2] is discussed. This approach takes into account the degrading effects of reflections and shadows.

2. FORMULATION OF THE CONTOUR-OBJECT MODEL

The degrading effects - shadows and reflections - are caused by the illumination. They depend on the surface form of the objects and the angle of illumination relative to the camera. Industrial applications may be characterized by the following facts: i) the number of objects to be identified is fixed, ii) the form of the objects is known to generate their correct contours, iii) the position of the camera and the illumination relative to the object plane is fixed. Furthermore the applications may often be restricted to single, non-overlapped and non-rotated objects. Under these conditions all objects corresponding to the same object class have nearly the same degradations. More than that the shadows and reflections are typical for each object class. Thus the most obvious idea is to use one filter for each object class which suppresses the typical degradations. The first aim could be now to restore the objects to their free of shadows and reflections versions in order to apply an egde enhancement method for contour extractions. However, it seems more reasonable to include already the contour extraction in the restoration problem.

Suppose that the ideal contours of the objects are given or can be generated because of the known form of the objects, e.g., by a proper graphic input system like a light-pen or a mouse by tracking the objects bounderies. Trying a linear approach a transfer function $T_i(\Omega_1,\Omega_2)$ can be defined for each object class which describes the transformation of the ideal contour into the object. Let $K_i(\Omega_1,\Omega_2)$ be the spectrum of the ideal contour and $R_i(\Omega_1,\Omega_2)$ be the spectrum of the average of all realizations of the object class i. Then the transfer function is given by definition

$$R_i(\Omega_1,\Omega_2) = T_i(\Omega_1,\Omega_2)\cdot K_i(\Omega_1,\Omega_2) \qquad (1)$$

with Ω_1,Ω_2 denoting the spatial frequencies. The average $R_i(\Omega_1,\Omega_2)$ is selected in eq.(1) because the objects are assumed to have variations in their spectrum. The variations are due to inaccuracies in the manufacturing process, the inaccuracies of the optical pick-up system, etc. Including the influence of the variations inherent the object class i in the model it is assumed that the variations can be modelled linearly. Thus the measured spectrum of an actual realization of an object $Z_i(\Omega_1,\Omega_2)$ is given by

$$Z_i(\Omega_1,\Omega_2) = R_i(\Omega_1,\Omega_2) + S_i(\Omega_1,\Omega_2) \qquad (2)$$

with $S_i(\Omega_1,\Omega_2)$ describing the actual deviation of the object $Z_i(\Omega_1,\Omega_2)$ from the representant $R_i(\Omega_1,\Omega_2)$ of the object class i. $S_i(\Omega_1,\Omega_2)$ is a random variable which is estimated from the measured data and is assumed to be a zero mean process. The power density spectrum of $S_i(\Omega_1,\Omega_2)$ can be estimated by eq.(2)

$$|S_i(\Omega_1,\Omega_2)|^2 = \frac{1}{M}\sum_{r=1}^{M} |Z_{ir}(\Omega_1,\Omega_2) - T_i(\Omega_1,\Omega_2)\cdot K_i(\Omega_1,\Omega_2)|^2 \qquad (3)$$

with $Z_{ir}(\Omega_1,\Omega_2)$ being the r-th realization of the object class i. Fig.1 shows the complete contour-object model for the object class i.

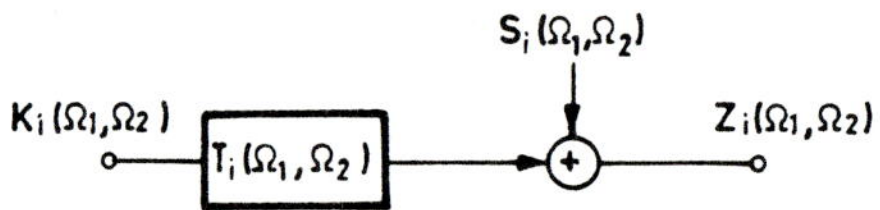

Figure: 1

3. CONTOUR EXTRACTION BY CONSTRAINED LEAST SQUARES FILTERING

In the previous section a model was derived which describes the transformation of the ideal contour into the corrupted object. The problem to solve now is to invert this model in order to obtain an estimate $\hat{K}_i(\Omega_1,\Omega_2)$ for the ideal contour $K_i(\Omega_1,\Omega_2)$. The following Fig.2 illustrates this task.

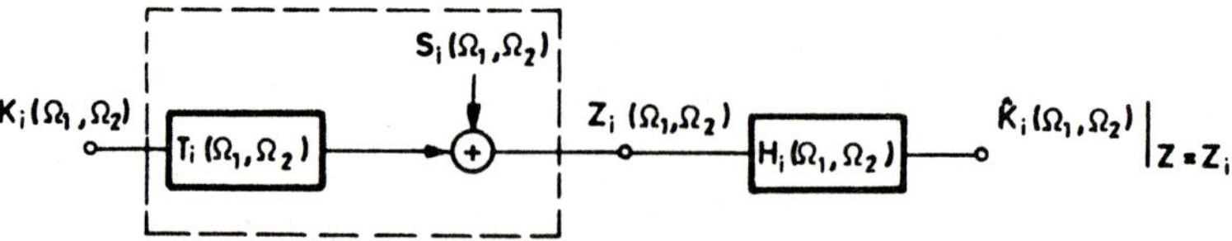

Figure: 2

A lineare approach for the estimate

$$\hat{K}_i(\Omega_1,\Omega_2) = H_i(\Omega_1,\Omega_2)\cdot Z_i(\Omega_1,\Omega_2) \qquad (4)$$

is selected. The filter $H_i(\Omega_1,\Omega_2)$ performs the contour extraction on the object class i, and its transfer function will be derived in the following. For the optimization of the transfer function the mean square error criterion is used. The criterion consists of two parts [3]. The first part minimizes the energy of the estimate.

$$\sum_{\Omega_1}\sum_{\Omega_2} |\hat{K}_i(\Omega_1,\Omega_2)|^2 = \min \qquad (5)$$

The idea behind condition (5) is that the extracted estimate should contain only the contour which results in much less energy than the unfiltered image. The second constraint adapts the filter to the object class i and represents the inverse part.

$$\sum_{\Omega_1}\sum_{\Omega_2} |Z_i(\Omega_1,\Omega_2) - T_i(\Omega_1,\Omega_2)\cdot\hat{K}_i(\Omega_1,\Omega_2)|^2 - \sum_{\Omega_1}\sum_{\Omega_2} |S_i(\Omega_1,\Omega_2)|^2 = 0 \qquad (6)$$

Equation (6) is merely a restatement of eq.(2). Conditions (5) and (6) are combined using a Lagrange multiplier λ_i:

$$\begin{aligned} U_i = & \sum_{\Omega_1}\sum_{\Omega_2} |\hat{K}_i(\Omega_1,\Omega_2)|^2 + \\ & \lambda_i\cdot\{\sum_{\Omega_1}\sum_{\Omega_2} |Z_i(\Omega_1,\Omega_2) - T_i(\Omega_1,\Omega_2)\cdot\hat{K}_i(\Omega_1,\Omega_2)|^2 - \sum_{\Omega_1}\sum_{\Omega_2} |S_i(\Omega_1,\Omega_2)|^2\} = \min . \end{aligned} \qquad (7)$$

Inserting eq.(4) into eq.(7) and minimizing U_i by differentiation with respect to $H_i(\Omega_1,\Omega_2)$ yields the optimum. The resulting transfer function is

$$H_i(\Omega_1,\Omega_2) = \frac{\lambda_i\cdot T^*(\Omega_1,\Omega_2)}{1+\lambda_i|T_i(\Omega_1,\Omega_2)|^2} . \qquad (8)$$

Finally the Lagrange multiplier has to be determined. This is achieved by differentiation of eq.(7) with respect to λ_i and setting the result to zero. This results in an implicit equation [3] which has to be solved in an iterative procedure.

4. STRUCTURE OF THE SYSTEM

The structure of the system consists of a bank of parallel branches containing the adapted filters $H_i(\Omega_1,\Omega_2)$ as given in Fig.3. The number of branches is equal to the number M of objects to be identified. Because of the parallel structure of the system the object is filtered by all filters. Each filter $H_i(\Omega_1,\Omega_2)$ yields an estimate of the contour of the average $R_i(\Omega_1,\Omega_2)$ being the representant of the object class i. But only that filter which is actually adapted to the object just being filtered, produces a good contour. Whereas all other filters result in images which after thresholding show more of an irregular pattern than thin contour lines.

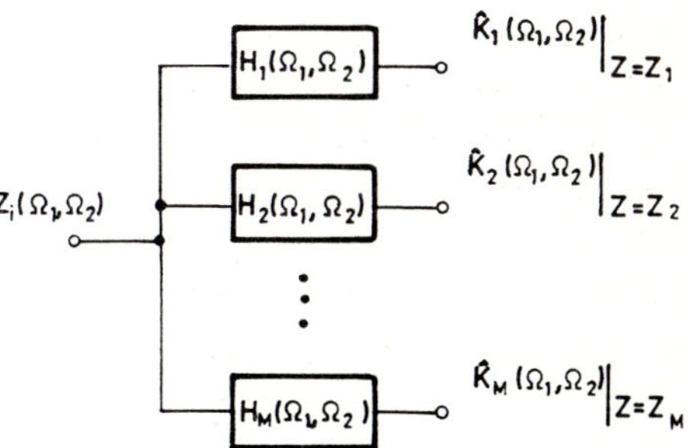

Figure: 3

5. PRACTICAL RESULTS

Some results of numbers embossed in sheet metal are presented in the following photos. The left-hand side of Fig.4 shows the objects numeral 1 and 3 corrupted with shadows and reflections. Fig.5 shows the ideal contours of the object represented in Fig.4. The contours have been generated with a graphics input system by tracking the object bounderies. In Fig.6 the object numeral 1 was processed with the filter adapted to the object class i=1. The result is shown on the left-hand side before thresholding and on the right-hand side after optimum threshold operation. In contrast to the case of adaption the case of non-adaption is shown now. In Fig.7 the result of processing the object numeral 3 with the filter adapted to the object class i=1 is shown. The filtering result on the left-hand side shows the result before thresholding and on the right-hand side after thresholding. The filtering result is rather close to an irregular pattern than to thin contour lines. In all non-adapted cases, each filtering result is similar to the example shown in Fig.7. Therefore a simple method for object classification is enabled. This can be done by discriminating the contour from the irregular patterns. One simple possibility to achieve this discrimination between adaption and non-adpations is to compare the variances of the grey-level histograms of all output images. The adapted case results in the least variance. By this way a classification rate of about 99 percent is achieved.

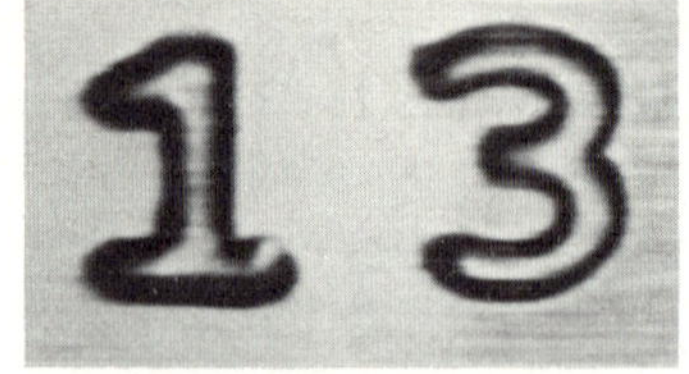

Figure: 4

Figure: 5

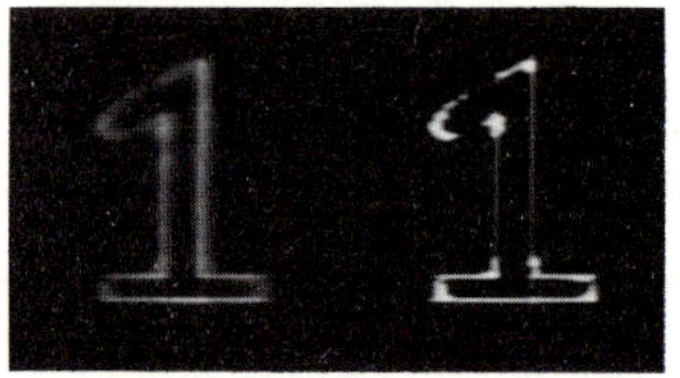

Figure: 6

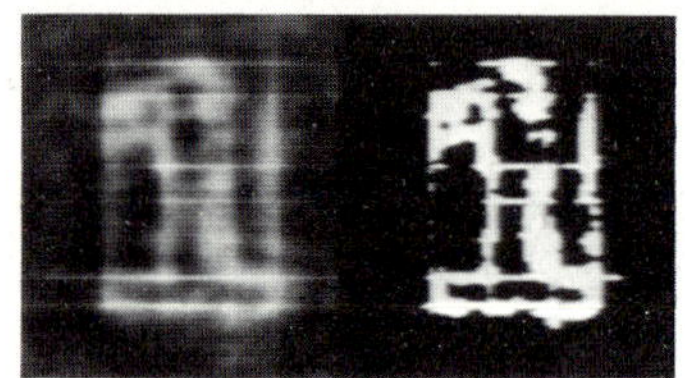

Figure: 7

6. CONCLUSION

In this paper a new approach for contour extraction has been presented. The approach takes into account the shadows and reflections on the surface of the objects. The contour extraction system is adapted to the objects to be identified and to the shadows and reflections. The system is trained by a set-up procedure. The set-up procedure consists of the following steps: i) generate the ideal contours of the objects to be identified, ii) apply a couple of realizations of the objects to the system. The system automatically designs the adapted filters which are arranged in parallel branches. A simple classification of the object is possible.

REFERENCES

[1] Pratt, W., Digital Image Processing (Wiley Interscience, New York, 1978)
[2] Kroschel, K., Statistische Nachrichtentheorie, Band 2, (Springer, Berlin, 1974)
[3] Dines, K.A.; Kak, A.C., Constrained Least Square Filtering, IEEE Trans. on ASSP-25. pp. 346-350, 1977.

KNOWLEDGE BASED PICTORIAL PATTERN RECOGNITION

H. Bunke

Institut für Informatik und angewandte Mathematik
Universität Bern, Länggass-Strasse 51, CH-3012 Bern, Switzerland

Abstract

An overview of the field of knowledge based pictorial pattern recognition is given. Methodology and implementation tools are discussed and a review of current systems and applications is given. Finally, the pros and cons of the knowledge based approach to pattern recognition as well as open problems are discussed.

1. Introduction

In this survey article, the discipline of pattern recognition is understood in a broad sense. By a pattern we mean any signal which can be measured from environment by means of a sensor or a collection of sensors. The recognition of patterns is the process of automatic inference of any desired pattern description from the raw sensory data. We will limit the discussion to pictorial patterns in this paper. This includes grey level and multi-spectral images, data from range scanners, images from medical devices, and others. Examples of recognition tasks include

- the classification of a pattern as a whole as it may occur in character recognition

- the localization of one or more predefined objects in a scene as it may occur in robotics vision and automatic visual inspection

- the inference of a symbolic description of a whole sensed image consisting of primitive image components and their mutual relations; this task may occur in automatic document analysis

- the inference of domain-specific conclusions from an input pattern; an example is the derivation of a diagnostic interpretation from medical images

In the past, statistical and decision theoretic approaches to pattern recognition have been dominating. During the last decade, however, there has been a growing interest in structural and knowledge based methods from artificial intelligence. Without doubt, the rapid development and widespread use of expert systems had a significant influence on the interest in knowledge based pattern recognition. While the main ideas in statistical pattern recognition are the representation of patterns as points in a feature space and the division of this space by means of decision functions, structural and artificial intelligence based pattern recognition relies on the explicit representation and use of knowledge about patterns. This knowledge is concerned with local and global pattern features, including their relations and their meaning. It is intented to be used for the solution of any of the above mentioned recognition tasks.

From a general point of view, there is a close relationship between the disciplines of pattern recognition and artificial intelligence. The recognition of sensory data, for example images, has always been considered as an important subfield of artificial intelligence. On the other hand, the topics of knowledge representation and inference, which belong to the center of artificial intelligence, have constantly attracted the attention of researchers working in pattern recognition. There are hints that the trend to knowledge based pattern recognition will become even stronger in the next few years.

2. Methodology

2.1 Knowledge required for pictorial pattern recognition

In a pictorial pattern recognition system, we usually find three levels of analysis processes. First, there are low-level processes like image restoration, edge detection, region extraction, texture analysis, depth from stereo, surface normal computation, etc. Typically, these processes operate on numerical quantities. Next, there are middle-level processes operating on both numerical and symbolic data. They are, like the low level processes, to a large degree domain-independent. Examples are split-and-merge algorithms on regions or the recovery of shape from shading,texture, contours, or motion. Finally, the desired pattern description is inferred by means of high-level processes, which operate on symbolic data structures and are tailored to the specific task and problem domain.

According to the different levels of analysis processes, we can distinguish between different sources of knowledge required for pictorial pattern recognition [MATSUYAMA 1984]. Low level processes mainly rely on physical knowledge. It is concerned, for example, with geometry among sensor, light sources and objects, spectral properties of the light source, reflectance properties of surfaces of objects, characteristics of the sensors, etc. Perceptual knowledge is used by middle level processes in order to group primitive pattern constituents into more meaningful global configurations. To a large degree, this knowledge is independent from particular applications. It aids in bridging the gap between the low level and high-level processes. In order to solve any of the tasks mentioned in the introduction, high-level knowledge, also called semantic knowledge, about properties of and relations between objects, which may occur in an image, is essential. It describes, for example, the appearance of houses in aerial images, or the salient features of objects to be detected in an industrial scene.

Knowledge based pattern recognition is the process of establishing a mapping between data acquired by a sensor from environment and a model, i.e. knowledge about a particular problem domain. As we have seen from the previous discussion this mapping is decomposed into different levels where at each level particular processes and sources of knowledge are utilized. In the rest of this paper we will concentrate on those aspects of knowledge based pattern recognition which are mainly concerned with symbolic data structures. The corresponding processes are primarily on the middle and high level.

2.2 Artificial intelligence based approaches to knowledge representation

Knowledge representation has been a major topic of research in artificial intelligence for many years and is far from being solved. It is an important subtask in the construction of any knowledge based system. Two important questions are: 1) what knowledge is necessary and sufficient to solve a given task, and 2) how can this knowledge be represented in a machine such that an effective system results. According to [STEELS 1984], a formalism for knowledge representation should support a number of requirements like modularity, abstraction, and open-endedness with respect to the representational formalism and the control strategy. The choice of the right representation formalism is crucial since it can have a drastic influence on the computational complexity of the problem solving procedure.

One of the classical approaches to knowledge representation in artificial intelligence is <u>formal logic</u> with first order predicate calculus, or PC1 for short, as most important subfield. In PC1, formulas constructed according to a particular syntax are used in order to represent knowledge about a problem domain. So one may write, for example,

AVERAGE-BRIGHTNESS(REGION1,46) (1)

in order to represent the fact that the average brightness of REGION1 is 46. Similarly, the formula

$$\forall x(\text{AVERAGE-BRIGHTNESS}(x,y) \wedge \text{SMALLER}(y,50) \rightarrow \text{DARK}(x)) \quad (2)$$

may be used for representation of the rule that any region x with an average brightness smaller than 50 is considered as dark. Logical conclusions can be computed from PC1 formulas by means of inference rules. One such conclusion, which follows from the two formulas given above is DARK(REGION1), i.e. REGION1 is dark. For an introduction to PC1 see [NIEMANN/BUNKE 1987, NILSSON 1982].

PC1, as an approach to knowledge representation, has a long tradition and is based on a well developed theory. Particularly it strongly supports the requirement for modularity. On the other hand there is a number of deficiencies and limitations, including semidecidability, monotony, and the frame problem [WINSTON 1984]. To overcome some of these limitations, extensions to PC1 have been proposed, for example truth-maintenance systems [DOYLE 1979], or other approaches to non-monotonic logic [BOBROW 1980, TURNER 1984].

<u>Production systems</u> is another well known approach to knowledge representation. The knowledge base of a production system consists of facts and rules where a rule has the form

<u>if</u> PREMISE <u>then</u> CONCLUSION (3)

An example from multi-spectral image analysis [CARLOTTO et al. 1984] is

<u>if</u> ((band-4>band-5) <u>and</u> (band-4>band-3)) <u>then</u> (assert vegetation) (4)

There are two major approaches to rule application, namely forward-chaining (which is also called data driven, or bottom-up, inference) and backward-chaining (goal driven, or top-down inference). For an introduction to production systems see [NIEMANN/BUNKE 1987, BUCHANAN/SHORTLIFFE 1985].

Production systems is the formalism on which most expert systems are based today. There are many similarities between PC1 and production systems. As a matter of fact, formula (2) can be understood as a rule according to (3). Conversely, the rule (4) can equivalently be represented by a PC1 formula. The advantage of modularity particularly holds for production systems. Also it is often claimed that the if-then formalism is very natural for many applications. The strict syntactic format of rules supports automatic checking for completeness and consistency. Shortcomings and limitations of production systems include the lack of powerful control structures and computational overhead.

A third class of methods for knowledge representation is <u>semantic nets</u> and <u>frames</u>. Basically, a semantic net or a collection of frames is a graph with nodes and edges representing objects and relations from a problem domain. As a generalization of graphs, the nodes and edges in a semantic net are complex data structures consisting of a number of subunits, each, thus defining all the necessary details of the objects and relations under consideration. Besides problem dependent relations of geometric, temporal, causal etc. nature, there are the standard relations "instance-of" and "is-a" for linking instances of objects to their classes, and classes to their superclasses, respectively. Along these standard relations, properties of classes are inherited. Inference and reasoning procedures in semantic nets are mainly based on matching and search (see also sections 2.3 and 2.4). An introduction to semantic nets can be found in [NIEMANN/BUNKE 1987, FINDLER 1979].

Semantic nets are particularly useful for modelling of object hierarchies and the explicit representation of relations between objects. Shortcomings include the lack of a standardized set of inference procedures.

Predicate calculus, production systems and semantic nets are not exclusive to each other. By contrast, they can be combined in a number of ways. Such a combination usually results in a more flexible and powerful approach. For a detailed discussion of this topic the reader is referred to [Bunke 1987].

2.3 Structural pattern recognition methods

The methods described in the previous section have their origin in artificial intelligence. Recently, they have been applied to a wide variety of tasks in pattern recognition. Besides, there is a number of approaches to knowledge representation and inference which have been developed in the context of "pure" pattern recognition tasks. We give a brief overview in this section.

In syntactic pattern recognition the available knowledge about patterns is represented by a formal grammar. For the recognition of unknown patterns, syntactic parsing according to this grammar is performed. Depending on the particular type of grammar and parsing algorithm, a huge number of different approaches has been proposed. For an introduction see [FU 1982]. Recent work in this field is reported in [FERRATE et al. 1987].

Another important class of methods is given by structural matching. In this approach a number of pattern prototypes is explicitly stored. In order to recognize an unknown input pattern, it is compared or matched, with all the prototypes. The most important data structures for pattern representation are strings, trees, and graphs. For more details see [SANKOF/KRUSKAL 1983, BUNKE/ALLERMANN 1983, TSAI/FU 1983, SHAPIRO/HARALICK 1981].

Relaxation is a third class of methods for structural pattern recognition. It aims at deriving an unique interpretation of all primitive parts of a pattern, which is globally consistent under a given set of constraints. There are two major lines of relaxation algorithms, known as discrete and continuous relaxation, respectively. For an introduction see [WALTZ 1975, ROSENFELD et al. 1976]. The relations which exist between the structural pattern recognition methods considered in this section and the artificial intelligence based approaches mentioned in section 2.2 are discussed in [BUNKE 1987].

Syntactic pattern recognition, structural matching and relaxation are very general and can be adapted to a wide variety of tasks. An important subfield of pictorial pattern recognition is the recognition of 3-D objects in scenes. This problem area additionally requires specialized techniques for the representation and use of 3-D information. Particular formalisms known from the literature include representations based on wire frames, constructive solid geometry, spatial occupancy, surfaces, generalized cones, multiple projections, or extended Gaussian images. A comprehensive survey is given in [BESL/JAIN 1985].

2.4 Control and search

A very important problem in the design of a knowledge based pattern recognition system is the communication among the different processes, including their execution order. There are two well known control paradigms, namely top-down (also called hypothesis-and-test or prediction-verification) and bottom-up. Both have limitations and deficiencies. In a pure bottom-up control regime all features and primitive pattern constituents must be detected in the beginning phase. This may cause many "false alarms" and lead to many redundant computations in middle-and high-level processes.

By contrast, in a pure top-down control regime, a number of wrong hypotheses may cause a tremendous waste of computational resources. So it is widely accepted that a computer vision system should incorporate both bottom-up and top-down control. For example, a first quick and approximate bottom-up processing phase can aid in reducing the number of potential high-level goals, i.e. hypotheses. Following, in a top-down phase only this reduced number of hypotheses have to be taken into account. A recursive refinement and verification of goals can be achieved by means of successive alternations between bottom-up und top-down control process.

There is number of causes which make computer vision difficult: 1) noisy and distorted input data, 2) erroneous results of low and middle level processes, i.e. errors in segmentation and primitive extraction, 3) uncertain and incomplete high-level knowledge which may be due to a wide spectrum of variations in the considered objects, for example. What results is a great degree of ambiguity and indeterminism at various stages in the recognition process. Since enormous amounts of data are usually involved in computer vision, it is necessary to avoid a combinatorial explosion, i.e. a problem solving procedure with exponential complexity. So the use of intelligent <u>search procedures</u> which are able to solve a combinatorial problem without exhaustively examining all existing possibilities is inevitable. Notice that knowledge reduces uncertainty and, therefore, search. Conversely, search can compensate for lack of knowledge.

There is a large number of search techniques known from the literature. Two standard procedures are depth-first and breadth-first search. An improvement is best-first search, which is based on a heuristic evaluation function. Other search procedures are beam search, branch-and-bound, bidirectional search, and dynamic programming. For a more detailed treatment see [NIEMANN/BUNKE 1987, NILSSON 1982, PEARL 1984].

2.5 Further issues

The emphasis in control and search is on the order in which the different processes and knowledge sources in a system are activated. Another important question concerns the static organization and the ways of communication between the different modules. There is one extreme solution to the problem where all processes are executed in a predefined order such that process i may receive input data only from process i-1 and may produce output data only for process i+1. Such a system organization is easy to implement and may be sufficient for certain applications. However, there are other tasks where a more flexible scheme is required. The other extreme solution is a completely heterarchical system where the different processes are completely independent of each other but where each process can communicate with each other. If there are many processes, control in such a system may become extremely difficult. The <u>blackboard model</u> seems to be a good compromise. A blackboard system consists of a number of independent knowledge sources with a common short-term memory, the blackboard. The only possibility of communication between the knowledge sources is via the blackboard where intermediate results and messages are stored. The blackboard model was originally developed in the context of the HEARSAY II speech understanding system [ERMAN et al. 1980]. For a general introduction see [NII 1986].

It was mentioned in section 2.4 that there is a number of causes which potentially lead to a great degree of ambiguity and indeterminism in a computer vision system. In order to cope with the computational complexity arising from this problem, the ranking of input data, (intermediate) results, and hypotheses by means of a suitable <u>measure of confidence</u>, or reliability, is inevitable. A mathematical model which has been developed and widely accepted in the statistical approach is Bayes decision theory [DUDA/HART 1973]. It is well suitable to cope with noisy and uncertain data for the purpose of classification. However there are problems with that model when it is applied in conjunction with the artificial intelligence based techniques discussed in the previous sections. From a practical point of view the correct estimation of the required statistical parameters is often impossible due to an unsufficient number of available training data. From a theoretical point of view, the Bayes model cannot adequately handle the discrimination between "disbelief" and "lack-of-belief". If a hypothesis A is given a probability p(A) it is automatically assumed that the negation B of A is given a probability p(B)= 1-p(A). So belief p(A) in A automatically implies disbelief 1-p(A) in A. However, given p(A) we cannot model lack of belief which is different from 1-p(A). Recently a number of theories have been developed where the work by Dempster and Shafer seems to be one of the important approaches [GORDON/SHORTLIFFE 1985]. Further material on this topic can be found in [NIEMANN/ BUNKE 1987, KANAL/LEMMER 1986].

Complex knowledge based systems make use of planning. Generally speaking, planning is the process of finding a sequence of actions which lead from one given situation to another. More specifically, in plan generation a sequence of processes and (sub)-goals is established, which are suitable to solve a given problem. Complex tasks require dynamic planning. That is, after execution of each process it is checked whether the corresponding (sub)goal is satisfied. If not, replanning is carried out. An early example of planning is [KELLY 1971]. Another fundamental paper on this subject is [SACEDOTI 1974].

Another very important area of research which is closely related with knowledge based systems is knowledge acquisition and learning. Generally, knowledge acquisition is the process of transforming problem dependent knowledge into a given representation formalism and bringing it on a machine. Principally, this problem can be attacked by means of different procedures, for example manual, interactive or fully automated. The latter case is also referred to as learning. In the context of expert systems the term knowledge engineering has been coined for the task of collecting and transforming knowledge from human experts in a problem domain into a complete and consistent formal framework on a machine. Knowledge acquisition can be looked at from several orthogonal views. First, it is dependent on the representation formalism. Next, there are several different fundamental approaches to the problem, like learning by analogy, by example and generalization, by trial and error, etc. Third, the particular approaches may widely vary dependent whether the problem domain is industrial scenes, natural scenes, medical images, remote sensing etc. Generally, the subject of knowledge acquisition and learning is very broad and cannot be discussed in full detail in this article. Until now, no procedures which have been applied in practice are known. For an introduction to knowledge acquisition and learning in image and speech analysis see [NIEMANN/BUNKE 1987]. A broader treatment of the subject can be found in [MICHALSKI et al. 1983, 1986].

3. Implementation

Procedural programming languages, like FORTRAN or PASCAL, have been dominating in the implementation of pattern recognition systems in the past. Under severe constraints with respect to execution speed or storage requirements, implementations were built on assembly or machine language, or directly on hardware. A main reason for using languages like FORTRAN or PASCAL is that they are usually suitable for algorithms used in low-level image processing. With respect to knowledge based processes operating on the middle- or high-level of a system, however, procedural languages have many deficiencies and limitations. For example, they don't adequately support symbolic processing on list oriented data structures. Also it is cumbersome to incorporate nondeterministic flow of control.

Up to date several programming styles besides procedural (or imperative) programming emerged. For an overview see [STOYAN 1984], for example. One of the earliest is functional programming. A functional programming language is based on the concept of mathematical function. Program execution corresponds to computing the value of a function where there may be other functions as arguments. The most prominent functional language is LISP [WINSTON/HORN 1981]. Another programming style, which has attracted much attention recently, is logic, or declarative, programming with PROLOG as the best known language [CLOCKSIN/MELLISH 1984]. Logic programming languages can be considered as implementation tools of first order predicate calculus (see section 2.2). The main emphasis in a logic program is on describing what the problem is rather on how a solution can algorithmically be obtained. Rule-based programming is another paradigm which is closely related with logic programming. There is a great number of software packages commercially available today supporting rule-based programming. They are also known as rule-based <u>expert system shells</u> and can be considered as tools for the implementation of production systems as described in section 2.2. The spectrum of machines on which these tools run ranges from small personal computers to powerful mainframe systems. Besides providing features for programming, rule-based expert system shells meet a number of further requirements resulting from expert system applications, like certainties (see section 2.5), explanation facilities, comfortable user interfaces, etc. An overview of commercially available expert system shells is contained in [HARMON/KING 1985]. In a recent article it has been reported that logic programming and rule-based expert system shells are suitable for the implementation of a number of algorithms used in pattern recognition, like syntactic parsing, string and graph matching, relaxation, and search [DVORAK/BUNKE 1987].

Another style is <u>object-oriented programming</u>. It is centered around the notion of objects and communication between objects. An object is an active data structure, arranged in a hierarchy of classes. Communication is achieved by message-passing. Object oriented programming presently receives a great deal of attention in artificial intelligence and is a very active field of research. The best known language is SMALLTALK [GOLDBERG/ROBSON 1983]. Object oriented programming has similarities with languages which strongly support the implementation of semantic nets and frames. Examples of such languages are KRL [BOBROW/WINOGRAD 1977] and KL-ONE [BRACHMANN/SCHMOLZE 1985].

From a theoretical point of view, all representation mechanisms and all programming languages principally have the same computational power. However, with respect to the effort required in system design and implementation, on the one side, and the computational complexity of the problem solving procedure, on the other side, there may be drastical differences. So the selection of the right representation formalism and implementation tool is a very important problem. Often, a pattern recognition task has many different aspects, such that a combination of different knowledge representation techniques seems to fit best. In such a case the use of a <u>hybrid expert system shell</u>, or knowledge representation language can be advantageous. A hybrid expert system shell allows, for example, to use together logic , rules, and frames for knowledge representation in one consistent framework. Examples of hybrid expert system shells are KEE and ART. For an overview see [HARMON/KING 1985].

The foregoing discussion was on software implementation tools only. Besides, a huge body of work has been devoted to architecture and hardware oriented implementations. For low level image processing, many different architectures have been proposed. Recent surveys with further references are [FOUNTAIN 1986, YALAMANCHILI et al.1985]. For middle- and high-level symbolic processing there seem to be also many useful architectures. Examples are LISP-machines [MOON 1985], PROLOG-machines [DOBRY et al. 1985], production system machines [LEHR/WEDIG 1987] and other architectures [HILLIS 1985, FAHLMAN 1979]. Further references are given in [HWANG et al. 1987].

4. Examples and Applications

In this section an overview of knowledge based pictorial pattern recognition is given. The main emphasis is on systems which make explicit use of the knowledge representation methods and inference techniques discussed in sections 2.2, 2.4, and 2.5. Applications relying on structural pattern recognition according to section 2.3 are not mentioned in the following. References to these systems can be found in [BALLARD/BROWN 1982, FU 1982, BUNKE 1985, FERRATE et al. 1987]. There is a huge number of papers describing rule-based systems for pictorial pattern recognition. Application areas are analysis of aerial images [NAGAO/MATSUYAMA 1980, SELFRIDGE/SLOAN 1981,GOLDBERG et al. 1983, CARLOTTO et al. 1984, CHESTEK et al. 1985, McKEOWN et al. 1985, PERKINS et al. 1985, LÜTJEN 1986, ZHU 1986], natural scenes [BAJCSY 1978, OHTA 1980, LEVINE/SHAHEEN 1981, KIM et al. 1984, MARQUINA 1984, DUANE et al.1985, GILMORE et al. 1985, BELKNAP et al. 1986], recognition and quality control of industrial parts [MASSONE 1983, ENDER/LIEDTKE 1986, BARLETT et al. 1987, PETKOVIC/HINKLE 1987], medical image analysis [TSUJI/NAKANO 1981, STANSFIELD 1984, BUNKE

1985, SAGERER 1985], low and middle level image segmentation [NAZIF/LEVINE 1984, VENABLE et al. 1985], document analysis [BLEY 1984, TOU et al. 1984, KUBOTA et al. 1984, NIYOGI/SRIHARI 1986], 3-D object recognition [MAGREE/NATHAN 1985], seismic image analysis [ZHANG/SIMAAN 1987], and galaxy classification [THONNAT et al. 1985].

Frame based approaches have been applied to similar domains, i.e. analysis of aerial images [BALLARD et al. 1978, BROOKS 1981, PINZ 1986, MATSUYAMA/HWANG 1985], natural scenes [HANSON/RISEMAN 1978], medical and biomedical images [BALLARD et al. 1978, TSOTSOS et al. 1980, TUCKER 1984, BUNKE 1985, SAGERER 1985], industrial parts [BOYER et al. 1984] 3-D objects [BRUNNSTRÖM/ISHIZUKA 1987] and galaxy classification [THONAT et al. 1985].

Logic programming in the context of image processing is discussed in [MOTT 1985, NIYOGI/SHRIHARI 1986] while [PHAM/DUBUISSON 1987] is on object oriented programming for 2-D part recognition. A system for natural scene recognition which is based on search through a network of compiled constraints is described in [RUBIN 1980].

5. Discussion and conclusions

Many techniques for knowledge representation, inference, control, search, system organization, etc. which have been originally developed in a "pure" artificial intelligence context have proven useful for applications in pictorial pattern recognition, too. These techniques seem to be particularly convenient for the organization of processes on the middle and high level of a computer vision system.

The advantage of using artificial intelligence techniques in pattern recognition is improved flexibility and modularity due to the explicit representation of the problem dependent knowledge and the separation of the knowledge base from the inference procedure. So a system can be easily adjusted to a new problem domain by modifying only the knowledge base while the inference procedure remains unchanged. This is in contrast with many of the earlier approaches where knowledge and control were mixed in an indistinctive fashion, and only implicitly represented by the program code. Representation of knowledge about the patterns to be recognized in an explicit way furthermore increases transparency and highly supports formal verification techniques. If a very complex system is being developed the system requirements and the domain specific knowledge cannot be completely specified beforhand. So the system must be developed in an explorative and experimental style where many refinements and modifications of the original specifications are necessary. In such a case, knowledge based techniques can help in reducing the development costs. From a certain degree of complexity on, artificial intelligence based approaches are the only feasible way of system development. On the other hand, it must be clearly pointed out that most of the knowledge representation and inference methods discussed in section 2 are somewhat unefficient with respect to execution speed and storage requirements. This is the price to be paid for increased flexibility and transparency.

In spite of the advantages discussed above, there are a number of severe problems, which exist in knowledge based pattern recognition today. A pattern recognition system has to cope with huge amounts of input data. These data are noisy and unreliable. Often, multiple sensors are involved which make the problems more difficult. Besides the unreliable nature of data, also knowledge is uncertain, imprecise, and incomplete. There are conceptually quite different levels in a pattern recognition system. Each of these level requires its own knowledge sources and inference strategies. For example, low level processes usually are extremely different from high level reasoning. Notice, however, that the different levels are highly dependent on each other. In many complex tasks a feedback from high to middle and low level, and from middle to low level is indispensable. The knowledge representation and inference methods, and also the software-tools which are available today were primarily developed with applications from "pure" artificial intelligence in mind. Consequently, they don't provide solution to the above mentioned problems. This means that current artificial intelligence techniques are useful for only a limited number of subproblems in pattern recognition. The majority of problems in knowledge based patter recognition is still unsolved, with respect to both methodology and implemention tools.

On the one hand, the people working in pattern recognition are anxiously looking at the latest developments in artificial intelligence in order to find hints to the solution of open problems. On the other hand, pictorial pattern recognition can be considered from the artificial intelligence point of view as an important and interesting paradigm for many of the above mentioned fundamental problems. Because, or in spite of, the huge number of open problems it can be expected that the trend to knowledge based pattern recognition will become stronger in the next years.

LITERATURE

BAJCSY, R. / JOSHI, A.K.: A partially ordered world model and natural outdoor scenes, in [HANSON/RISEMAN] 1978, 263-270

BALLARD, D.H. / BROWN, C.M. / FELDMAN, J.A.: An approach to knowledge directed image analysis, in [HANSON / RISEMAN 1978a], 664-670

BALLARD, D.H./BROWN, C.M.: Computer vision, Prentice Hall, Englewood Cliffs, N.J., 1982

BARLETT, S.L. /COLE, C.L. / JAIN, R.: Expert system for visual solder joint inspection, Proc. IEEE 3rd Conf. on Art. Intell. Applications, 1987, 58-63

BELKNAP, R. / RISEMAN, E. / HANSON, A.: The information fusion problem and rule-based hypotheses applied to complex aggregations of image events, Proc. IEEE Conf. on Computer Vision and Pattern Recognition 1986, 227-234

BESL, P.J. / JAIN, R.C.: Three-dimensional object recognition, Computing Surveys 17, 1985, 75-145

BLEY, H.: Segmentation and preprocessing of electrical schematics using picture graphs, Comp. Vision, Graphics, and Image Processing 28, 1984, 271-288

BOBROW, D.G. (ed.): Special issue on non-monotonic logic, Art. Intell. 13, 1980

BOBROW, D.G. / WINOGRAD, T.: An overview of KRL, a knowledge representation language, Cognitive Science 1, 1977, 3-46

BOYER, K.L. / SAFRANEK, R.J. / KAK, A.C.: A knowledge based robotic vision system, IEEE 1st Conf. on Art. Intelligence Application, 1984, 45-50

BRACHMAN, R.J. / SCHMOLZE, J.G.: An overview of the KL-ONE knowledge representation system, Cognitive Science 9, 1985, 171-216

BROOKS, R.A. / Symbolic reasoning among 3-dimensional models and 2-dimensional images, Art. Intell. 17,1981, 285-349

BRUNNSTRÖM, K. / ISHIZUKA, M.: Target-directed understanding of 3D objects in a knowledge-based vision system, Proc. 5th Scandinavian Conf. on Image Analysis, Stockholm, 1987, 201-208

BUCHANAN, B.G. / SHORTLIFFE, E.: Rule-based expert systems, Addison-Wesley, Reading, MA., 1985

BUNKE, H.: Modellgesteuerte Bildanalyse. Teubner Verlag, Stuttgart, 1985

BUNKE, H.: Hybrid approaches, in [FERRATE et al. 1987]

BUNKE, H. / ALLERMANN, G.: Inexact graph matching for structural pattern recognition, Pattern Recognition Letters 1, 1983, 245-253

CARLOTTO, M.J. / TOM, V.T. / BAIM, P.W. / UPTON, R.A.: Knowledge-based multi-spectral image classification, in TESCHER, A.G. (ed.): Applications of digital image processing, Proc. SPIE Vol. 504, 1984, 47-53

CHESTEK, R. / MULLER, H. / CHELBERG, D.: Knowledge based terrain analysis, in [GILMORE, 1985] 46-56

CLOCKSIN, W.F. / MELLISH, C.S.: Programming in Prolog, Springer-Verlag, 1984

DOBRY, T.P. / DESPAIN, A.M. / PATT, Y.N.: Performance Studies of a Prolog machine architecture, Proc. 12th Annual Int. Symp. Comp. Architecture, IEEE/ACM, 1985,

DOYLE, J.: A truth maintenance system, Art. Intell. 12, 1979, 231-272

DUANE, G.S. / VENABLE, S.F. / RICHTER, D.J. / WIEDEMANN, A.M.: A production system for scene analysis and semantically guided segmentation, SPIE Vol. 548 Applications of Art. Intell. II, 1985, 35-45

DUDA,R.O. / HART, P.E.: Pattern classification and scene analysis, J. Wiley, New York, 1973

DVORAK, J. / BUNKE, H.: Expert system shells and logic programming for the implementation of image analysis algorithms, Proc. 5th Scandinavian Conf. on Image Analysis, Stockholm, 1987, 93-100

ENDER, M. / LIEDTKE, C.-E.: Repräsentation der relevanten Wissensinhalte in einem selbstadaptierenden regelbasierten Bilddeutungssystem, in [HARTMANN 1986], 219-223

ERMAN, L.D. / HAYES-ROTH, F. / LESSER, V.R. / REDDY, R.: The HEARSAY-II speech-understanding system, Comp. Surveys 12, 1980, 213-253

FAHLMAN, S.E.:NETL - a system for representing and using real-world knowledge, MIT Press, Cambridge, MA., 1979

FERRATE, G. / PAVLIDIS, T. / SANFELIU, A. / BUNKE, H. (eds.): Syntactic and structural pattern recognition, Springer-Verlag, 1987

FINDLER, N.V. (ed.): Associate networks, Academic Press, New York, 1979

FOUNTAIN, T.J.: Array architectures for iconic and symbolic image processing, Proc. 8th ICPR, Paris, 1986, 24-33

FU, K.S.: Syntactic pattern recognition and applications, Prentice Hall, 1982

GILMORE, J.F. (ed.): Applications of Art. Intelligence II, Proc. SPIE Vol. 548, 1985,

GILMORE, J.F. /SEMECO, A.C. / EAMSHERANGKOON, P.: Knowledge-based route planning through natural terrain, in [GILMORE 1985], 46-56

GOLDBERG, A. / ROBSON, D.: SMALLTALK-80 The language and its implementation, Addison-Wesley, Reading, MA., 1983

GOLDBERG, A. / KARAM, G. / ALVO, M.: A production rule-based expert system for interpreting multi-temporal landsat imagery, Proc. IEEE Conf. on Comp. Vision and Pattern Recognition, Washington, DC., 77-82

GORDON, J. / SHORTLIFFE, E.H.: The Dempster-Shafer theory of evidence, in [BUCHANAN/SHORTLIFFE 1985], 272-292

HANSON, A.R. / RISEMAN, E.M.: Visions; a computer system for interpreting scenes, in [HANSON / RISEMAN 1978a] 303-333

HANSON, A.R. / RISEMAN, E.M. (Eds.): Computer vision systems, Academic Press, New York, 1978(a).

HARMON, P. / KING, D.: Expert-systems - artificial intelligence in business, John Wiley, New York, 1985

HARTMANN, G. (ed.): Mustererkennung 1986, Proc. 8. DAGM Symposium, Informatik-Fachberichte 125, Springer-Verlag, 1986

HILLIS, D.: The connection machine, MIT Press, Cambridge, MA., 1985

HWANG, K. / GHOSH, J. / CHOWKWANYUN, R.: Computer architectures for artificial intelligence processing, Computer 20, 1987, 19-27

KANAL, L.N. / LEMMER, J.F. (eds.): Uncertainty in artificial intelligence, North-Holland, 1986

KELLY, M.D.: Edge detection in pictures by computers using planning. In MELTZER, B. / MICHIE, D. (eds.):Machine Intelligence 6, Edinburgh, 1971, 397-409

KIM, J.H. / PAYTON, D.W. / OLIN, K.E.: An expert system for object recognition in natural scenes, Proc. 1st Conf. on Art. Intell. Applications, 1984, 170-175

KUBOTA, K. / IWAKI, O. / ARAKAWA, H.: Document understanding system, Proc. 7th ICPR, Montreal, 1984, 612-614

LAUBSCH, J. (ed.): GWAI-84, Proc. 8th German Workshop on Art. Intell., Informatik-Fachberichte 103, Springer-Verlag, 1984

LEHR, T.F. / WEDIG, R.G.: Toward a GaAs realization of a production system machine, Computer 21, 1987, 36-48

LEVINE, M.D. / SHAHEEN, S.I.: A modular computer vision system for picture segmentation and interpretation, IEEE Trans. PAMI-3, 1981, 540-556

LÜTJEN, K.: BPI - ein blackboard-basiertes Produktionssystem für die automatische Bildauswertung, in [HARTMANN 1986], 164-168

MAGEE, M. /NATHAN, M.: A rule based system for pattern recognition that exploits topological constraints, Proc. IEEE Conf. on Computer Vision and Pattern Recognition, 1985, 62-67

MARQUINA, N.: Rule-based evidence accrual system for image understanding, in CASASENT, D.P. / HALL, E.L. (eds.): Intelligent robots and computer vision, Proc. SPIE Vol. 521, 1984, 170-175

MASSONE, L.: SYRIO: a knowledge based approach to 2-D robotic vision, in: NEUMANN, B. (ed.): GWAI-83, 7th German Workshop on Art. Intell., Informatik Fachberichte 76, Springer-Verlag, 1983, 60-68

MATSUYAMA, T.: Knowledge organization and control structure in image understanding, Proc. 7th ICPR, Montreal, 1984, 1118-1127

MATSUYAMA, T. / HWANG, V.: SIGMA: a framework for image understanding - integration of bottom-up and top-down analysis, Proc. 9th IJCAI, Los Angeles, 1985, 908-915

McKEOWN, D.M. / HARVEY, W.A. / McDERMOTT, J.: Rule-based interpretation of aerial imagery, IEEE Trans. PAMI-7, 1985, 570-58

MICHALSKI, R.S. / CARBONELL, J.G. / MITCHELL, T.M.: Machine learning: an artificial intelligence approach, Tioga Publ., Palo Alto, CA., Vol. 1, 1983; Vol. 2, 1986

MOON, D.A.: Architecture of the Symbolics 3600, Proc. 12th Annual Int. Symp. Comp. Architecture, IEEE/ACM, 1985, 76-83

MOTT, D.H.(ed.): Prolog-based image processing using Viking XA. in: PUGH, A. (ed.): Proc. 2nd Int. Conf. on Machine Intell., London, North-Holland, 1985, 37-52

NAGAO, M. / MATSUYAMA, T.: A structural analysis of complex aerial photographs, Plenum Press, New York, 1980

NAZIF, A.M. / LEVINE, M.D.: Low level image segmentation; an expert system, IEEE Trans. PAMI-6, 1984, 555-577

NIEMANN, H. / BUNKE, H.: Künstliche Intelligenz in Bild- und Sprachanalyse, Teubner Verlag, Stuttgart, 1987

NII, P.: The blackboard model of problem solving, AI Magazine 7, No. 2, 1986, 38-53

NILSSON, N.J.: Principles of artificial intelligence, Springer Verlag, 1982

NIYOGI, D. / SRIHARI, S.N.: A rule-based system for document understanding, Proc. AAAI-86, Philadelphia, 1986, 789-793

OHTA, Y.: A region oriented image-analysis system by computer, Ph. D. diss., Dept. of Inform. Sciences, Kyoto Univ., Japan, 1980

PEARL, J.: Heuristics: Intelligent search strategies for computer problem solving, Addison Wesley, Reading, MA., 1984

PERKINS, W.A. / LAFFEY, T.J. /NGUYEN, T.A.: Rule-based interpreting of aerial photographs using LES, in [GILMORE 1985],138-146

PETKOVIC; D. / HINKLE, E.B.: A rule-based system for verifying engineering specifications in industrial vision inspection applications, IEEE Trans. PAMI-9, 1987, 306-311

PHAM, H.N. / DUBUISSON, B.: A 2D object oriented vision system, Proc. 5th Scand. Conf. on Image Analysis, Stockholm, 1987, 79-84

PINZ, A.: Architektur und Anwendung des bildverstehenden Expertensystems VES, in ROLLINGER, C.-R. / HORN, W. (eds.): GWAI-86 und 2. ÖAI-Tagung, Informatik-Fachberichte 124, Springer-Verlag, 1986, 212-217

ROSENFELD, A. / HUMMEL, R.A. / ZUCKER, S.W.: Scene labelling by relaxation operations, IEEE Trans. SMC-6, 1976, 420-443

RUBIN, S.M.: Natural scene recognition using locus search, Comp. Graphics and Im. Proc. 13, 1980, 298-333

SACERDOTI, E.D.: Planning in a hierarchy of abstraction spaces, Art. Intell. 5, 1974, 115-135

SAGERER, G.: Darstellung und Nutzung von Expertenwissen für ein Bildanalysesystem, Informatik-Fachberichte 104, Springer-Verlag 1985

SANKOFF, D. / KRUSKAL, B.: Time warps, string edits, and macromolecules: The theory and practice of sequence comparison, Addison-Wesley, Reading, MA., 1983

SELFRIDGE, P.G. / SLOAN, K.R.: Locating objects under different conditions: an example in aerial image understanding, Proc. IEEE Conf. PRIP, Dallas, 1981, 470-472

SHAPIRO, L.G. / HARALICK, R.M.: Structural descriptions and inexact matching, IEEE Trans. PAMI-3, 1981, 501-519

STANSFIELD, S.A.: ANGY - a rule-based expert system for identifying and isolating coronary vessels in digital angiograms, Proc. IEEE 1st Conf. on Art. Intelligence Applications, 1984, 606-609

STEELS, L.: Design requirements for knowledge representation systems, in [LAUBSCH 1984], 1-19

STOYAN, H.: Programming styles in artificial intelligence, in [LAUBSCH, J. 1984], 154-180

THONNAT, M. / GRANGER, C. / BERTHOD, M.: Design of an expert system for object classification through an application to the classification of galaxies, Proc. IEEE Conf. on Computer Vision and Pattern Recognition, 1985, 206-208

TOU,J.T. / HUANG, C.L. / LI, W.H.: Design of a knowledge-based system for understanding electronic circuit diagrams, IEEE 1st Conf. on Art. Intelligence Application, 1984, 652-661

TSAI, W.H. / FU, K.S.: Subgraph error-correcting isomorphisms for syntactic pattern recognition, IEEE Trans. SMC-13, 1983, 48-62

TSOTSOS, J.K. / MYLOPOULOS, J. / COVVEY, H.D. / ZUCKER, S.W.: A framework for visual motion understanding, IEEE Trans. PAMI-2, 1980, 563-573

TSUJI, S. / NAKANO, H.: Knowledge-based identification of artery branches in cine-angiograms, Proc. 7th IJCAI, Vancouver, 1981, 710-715

TUCKER, L.W.: Control strategy for an expert vision system using quadtree refinement, Proc. IEEE Workshop on Computer Vision: Representation and Control,Annapolis, 1984, 214-218

TURNER, R.: Logics for artificial intelligence, Ellis Horwood Ltd., Chichester, 1984

WALTZ, D.: Understanding line drawings of scenes with shadows, in WINSTON, P.H. (Ed.): The psychology of computer vision, Mc Graw Hill, 1975, 19-91

WINSTON, P.H.: Artificial intelligence (2nd ed.), Addison-Wesley, Reading, MA., 1984

WINSTON, P.H. / HORN, B.K.P.: LISP, Addison-Wesley, Reading, MA., 1981

VENABLE, S.F. / RICHTER, D. / WIEDEMANN, M.: A rule-based system for improving on image segmentation, in WEISBIN, C.R. (ed.): Proc. 2nd Conf. on Artificial intelligence applications, 1985, 94-99

YALAMANCHILI, S. et al.: Image processing architectures: a taxonomy and survey, in KANAL, L. / ROSENFELD, A. (eds.): Progress in pattern recognition 2, Elsevier Sci. Publ., 1985, 1-37

ZHANG, Z. / SIMAAN, M.: A rule-based interpretation system for segmentation of seismic images, Pattern Recognition 20, 1987, 45-53

ZHU, M.-L. / YEH, P.-S.: Automatic road network detection on aerial photographs, Proc. IEEE Conf. on Computer Vision and Pattern Recognition, 1986, 34-40

Animated 3D-Model of the Human Heart Based on Echocardiograms

Liu Jilin, Klaus Affeld*, Michael W. Engelhorn, Michael Schartl*

Technische Universität Berlin, Institut für Technische Informatik - Computer Graphics
*: Freie Universität Berlin, Abt. für Innere Medizin

1. Introduction

Two-dimensional echocardiography has become an indispensable tool of the modern cardiologist. It is noninvasive and delivers cross sections of the heart throughout the cardiac cycle, displaying the movement of the valves and cardiac walls. This technique however also has a considerable disadvantage:
only one crossection of the heart is taken and monitored at a time. If another section of the heart is to be investigated, the position of the scanner has to be changed and a new part of the heart is scanned. How does it relate to the former section? Only the experience of many previous scans permits the cardiologist to reconstruct a mental image of the complex and ever moving geometry of the heart and to detect anomalities. The question now arises how a computer can aid this mental process. The paper therefore deals with the attempt of the animated, threedimensional reconstruction of the moving heart on the basis of several two-dimensional echocardiographic scans. Unlike other authors **[1,2,3]** here it is not attempted to model the outer surface of the heart but to join several sections in the proper geometrical relation and leaving the outer contour invisible.

2. Method

2.1. Echocardiograms

Echocardiograms unfortunately contain a considerable amount of noise, which results in blurry and patchy images with the boundaries of the heart sometimes being difficult to detect even for the human eye. The automatic detection of the boundaries is a problem of the computervision and is not yet solved, this means that for the reconstruction we cannot rely upon it. Instead we rely upon the eye of the cardiologist and use manually traced cross sections from a printout of the ultrasonic scanner. Only selected images were chosen from the whole cycle, there are:

- one longitudinal section showing all four compartments of the heart in the contracted state - end of systole, shown as manual tracings in **Fig. 1**
- the same longitudinal section in the dilated state - end of diastole, **Fig.2**
- a cross section through the two ventricles in the contracted state - end of systole
- the same cross section in the dilated state - end of diastole, **Fig.3**

These sections were manually traced on a digitizer connected to an IBM PC. The curved image of the traced heart contour in this process is split up into a number of connected straight lines, so the whole image takes the shape of a polygon. Depending on the size of the contour the number of points of the polygon varied from 20 to 188.

2.2. Animation

The process to obtain a digitized image of the heart involves two manual interactions on each image -first the manual tracing of the scans and second the manual digitizing of the tracings. This introduces errors which would result in jerky movement of the outline, if this process would be applied on each scan of a whole sequence of scans. In order to obtain a smooth animation sequence the frames between the two endpositions were mathematically interpolated rather than of tracing and digitizing each individual frame of the whole sequence of one heart cycle. The position of the points of the polygon have to be selected carefully, they have to be characteristic points on both of the frames used for the interpolation. First the position points have to be determined before digitizing. Then the new contour has to be repartitioned

according to the manual tracing.
Till now we have not considered the movement of the valves. They open and close in a very short time, about 20 ms and their opening and closing movement does not show because the time interval between the scans is longer. So between the two extreme positions the valves apear to jump from the open to the closed state and vice versa, much as we would see it on the screen of the echocardiograph. Once the valve is closed it is considered as a part of the section and is deformed by the interpolation as the other structure of the heart, just as it is observed in the display of real scans.

2.3. 3D-Computer Model

For the reconstruction of the heart from sections a software package named RAVO (**R**econstruction **A**nd **V**isualization of **O**bjects) has been used. It was developed at the Computer Graphics Department of the Technische Universität Berlin. After feeding in input data this program generates a threedimensional structure and displays it on the screen. It permits hidden line removal, three axis rotation, lighting, colouring of the objects, smoothing and and even structuring of the surface. The last two features of the program were not used, since we consider the cross sections as plane objects with a certain thickness, which comes close to the physical quality - a polygon in the plane, but with rectangular edges. The program accepts the input data only as contours of parallel planes, which causes a problem here, because the two sections derived from the scanner are nearly perpendicular to each other - we have a longitudinal section and a cross section. This problem was solved by giving the cross section a certain thickness - which in reality it has - thus transforming it into a layered object built up out of rectangles. The positions of the rectangles and their sizes are interpolated as longitudinal contours. A perspective view of the two combined sections is shown in **Fig. 4** .

3. Results

With the help of a special program - RAVO - which is run on a DEC VAX 11/750 - a three-dimensional model of the human heart has been generated. The model is animated and shows the hearts movement in realtime. The model can be displayed on a vector display, where realtime transforms such as rotation, scaling and translation can be applied . On a raster display a surface shaded image can be generated with an arbitrary number of light sources, different reflection parameters and different colors applied to the model and to the light sources. The generation of such images requires about 1o min per frame, which exludes realtime operation. The images can however be generated, stored and later diplayed in a fast sequence to produce an animation. From both displays videotapes have been made for further discussion with the cardiologists. The results of this dicussion will determine the next step - the automatic tracing of the boundaries of the heart.

4. References

[1] M. Kuwahara, S.Eiho, N. Asada, G. Osakada, D. Kawai:
3D images of left ventricular myocardium reconstructed from 2D echocardiograms. Computers in Cardiology, 1984.

[2] M. H. Lewis, D. S. Schlusselberg, W. K. Smith, H. K. Hagler, D. J. Woodward, L. M. Buja:
Three-dimensional cardiac morphometry with computer graphics. Computers in Cardiology, 1982.

[3] W. E. Moritz, D. H. McCabe, A. S. Pearlman, D. K. Medema:
Computer generated three dimensional ventricular imaging from a series of two dimensional ultrasonic scans. Computers in Cardiology, 1982.

5. Keywords

Echocardiography, Medical Imaging, 3D-Model, Animation, Interactive 3D-Computer Graphics

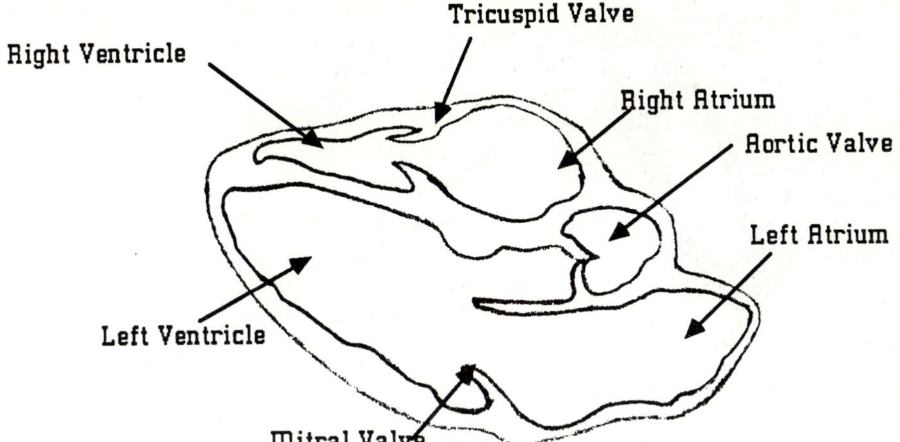

Fig.1 The end of the diasytolic phase - longitudinal section

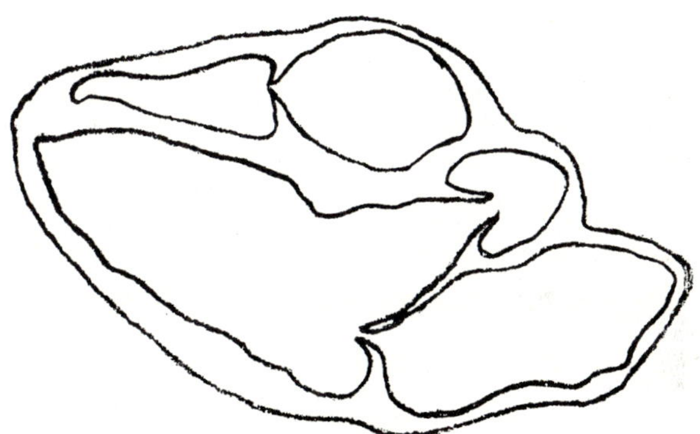

Fig.2 The end of the systolic phase

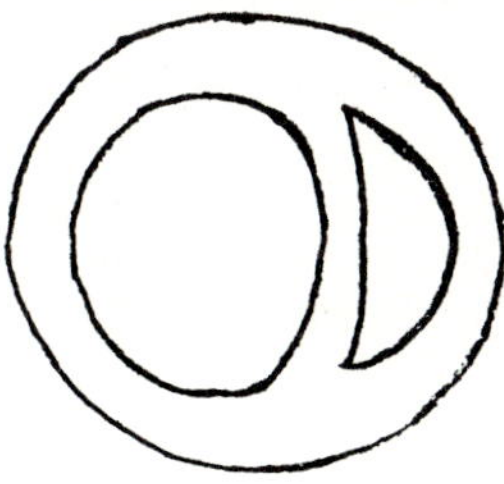

Fig.3 The cross section

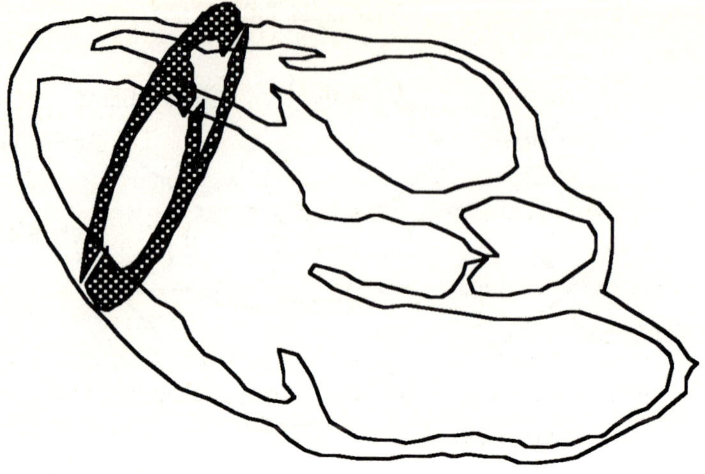

Fig.4 Perspective view of the combined sections

Mitteln bei 3-D-Rekonstruktionen biologischer Objekte

Klaus Niemann und D. Graf v. Keyserlingk
Abteilung Anatomie, RWTH Aachen
Melatener Straße 211, D-5100 Aachen

Verfahren zur dreidimensionalen Rekonstruktion von Serienschnitten im Bereich Biowissenschaften haben sich zu einer solchen Vielfalt entwikkelt, daß ihre kritische Würdigung eigene Reviews rechtfertigt [1]. Während immer aufwendigere Algorithmen immer mehr Prozessorleistung erfordern, um Einzelrekonstruktionen beliebiger Komplexität in vertretbaren Zeiten zu erhalten, ist ein biologiespezifischer Aspekt bisher kaum berücksichtigt worden: die Variation als individuell verschiedene Lage und Ausprägung von Strukturen. So werden z.B. für die Beurteilung computertomographischer Schnittserien des Gehirns als Referenz Atlanten mit zweidimensionalen Schnittserien individueller Gehirne, meist mit fester Winkelvorgabe, verwandt. Auf dem Hintergrund der zunehmend besseren Qualität digitaler Röntgenbilder ist der Wunsch des Klinikers zu verstehen, diese für eine genauere lokalisatorische Diagnostik verfügbar zu machen - bei gleichzeitig vertretbarer Strahlenbelastung für den Patienten.

Erstrebenswert wäre im Bereich kranialer Computertomographie (CCT) ein möglichst digital verfügbares (Experten-)System, das die individuelle Befunderhebung am Patienten insofern erleichtert, als es in möglichst genauer Anpassung an die spezielle Anatomie des Patienten allgemeine Aussagen über die wahrscheinliche Strukturzugehörigkeit eines in Frage stehenden Hirnareals bzw. den Ort einer Läsion macht. Hier sind nicht nur Strukturen gemeint, die im digitalen Bild "sichtbar" sind oder über Transformationen visualisiert werden können, sondern auch solche Strukturen, die sich anatomisch unterscheiden lassen, aber in den gängigen neuroradiologischen Untersuchungsverfahren nicht abgrenzbar sind, da sie sich in physikalisch-chemischen Parametern nicht von der Umgebung abheben. So sind z.B. für die Aphasiologie folgende Fragestellungen typisch: Welche Faserbahnen des Gehirns sind bei einem ischämischen cerebralen Insult von der Läsion betroffen, wie ist die genaue Lokalisation der Läsion bezogen auf das Relief der Hirnwindungen und -furchen, wie läßt sich die Angabe des Ortes einer Läsion standardisieren, um in Zusammenhang mit psychologischen Testbatterien zu einer validen Aussage über die Korrelation von pathologisch-morphologischem Befund und klinisch-psychologischer Symptomatik zu kommen?

Diese Thematik veranlaßte uns, auf relativ einfachem Niveau zunächst die Variationsproblematik zu behandeln. Die von uns vorgeschlagene Methode beruht auf der standardisierten dreidimensionalen Registrierung von Hirnstrukturen, wobei bisher hauptsächlich Furchen- und Arterienverläufe untersucht wurden [2)]. Die Summe der digitalen Strukturdaten eines Gehirns, die jeweils als Punktwolken unter mnemonischen Codes zusammengefaßt sind, bezeichnen wir als Hirnmodell. Vermessen wurden mit Hilfe eines Gerätes zur Registrierung von 3-D-Koordinaten bisher 30, der üblichen Lage der Sprachzentren entsprechend linke, autoptische Hemisphären.
Referenzsystem ist das sogenannte CACP-System der stereotaktischen Neurochirurgie, das über die Verbindungslinie zwischen der Commissura anterior und posterior in unmittelbarer Nachbarschaft des III. Ventrikels definiert ist. Dieses System ist weit verbreitet und akzeptiert, nicht zuletzt, da es ein intracerebrales Referenzsystem darstellt, das jedem knöchernen Referenzsystem überlegen ist und zudem sehr nah am Zentrum der während der Embryonal- und Fetalentwicklung stattfindenden Hemisphärenrotation liegt. Über eine - allerdings belastende und daher Spezialfällen vorbehaltene - Ventrikulographie kann dieses Referenzsystem auch beim Lebenden demonstriert werden. Der Winkel zwischen dem knöchernen Referenzsystem Orbitomeatallinie, auf das sich die Schnittebenen in CCT und NMR beziehen, und der CACP-Linie kann interindividuell um einige Grad variieren.
Die mitregistrierte Oberfläche der Hemisphären wird benutzt, um die Hirnmodelle an ein quaderförmiges Referenzvolumen, das in seinen Dimensionen (168x112x72 mm) den Ausmaßen einer durchschnittlichen Hemisphäre entspricht, anzupassen. Ausgangspunkt der Skalierung bildet der Mittelpunkt der CACP-Linie. Innerhalb dieses 3-D-Raumes können nun beliebige Hemisphärenmodelle und damit beliebige Strukturen verglichen und gemittelte Strukturen zwei- oder dreidimensional sichtbar gemacht werden. Mittelung erfolgt über die Unterteilung des Standardvolumens in kubische Voxel von 4 mm Kantenlänge und Punktzählung im jeweiligen Voxel unter Berücksichtigung der Hemisphären- und Strukturzugehörigkeit der erhobenen Daten. Das Voxel bildet damit die Beobachtungseinheit des Systems und definiert die maximale Auflösung. Die Kantenlänge wurde als Kompromiß zwischen den Anforderungen der Lokalisation in CCT und NMR und dem Punktabstand bei der Registrierung, der etwa 1-2 mm beträgt, gewählt.
Über die beschriebene Methode können mittlere Strukturverläufe und

-grenzen angegeben werden, die in beliebigen Schnittebenen aber auch direkt dreidimensional anschaulich wiedergegeben werden können. Für diese Art der Applikation hat sich die Punktgrafik in Zusammenhang mit einer Stereoprojektion nicht nur während der Registrierung sondern gerade auch bei den anschließenden Auswertungen aufgrund ihrer direkt plastischen Anschauung sehr bewährt. Gleichzeitig bedeutet sie hohe Rechenökonomie, z.B. im Rahmen der Transformation homogener Koordinaten. Direkte räumliche Vergleiche von erhobenen Befunden sind praktisch sofort, bei Verwendung standardisierter Projektionen sogar häufig über einfache binäre Addition von ein Bit tiefen Pixelbildern möglich.

Abbildung 1 demonstriert den typischen gemittelten Verlauf von 7 Primärfurchen des Gehirns, die in ihrer gesamten räumlichen Ausdehnung bis in die Tiefe des Marklagers dargestellt sind. Die Punkte markieren jeweils den Mittelpunkt eines Voxels. Es gibt zahlreiche Hinweise dafür, daß in den Tiefen der Hirnsulci funktionell wichtige Arealgrenzen verlaufen.

Die erhobenen Befunde können dann in einem zweiten Schritt auf Schnittserien der Computertomographie übertragen werden. Das dreidimensionale Atlassystem läßt sich durch Skalierung an beliebig orientierte Patientenschnittserien adaptieren und den Patientendaten direkt überlagern. Für Anpassungszwecke werden dann auch gemittelte Bestimmungen von Hemisphärenoberflächen interessant, wie sie in Abbildung 2 über Interpolation mit Cubic Splines erhalten wurden. Diese und weiterreichende Verfahren wie Surface-Patch und Polygon-Tile sollten im Rahmen der Variationserfassung allerdings eher den letzten Stufen der Analyse und Wiedergabe von Resultaten vorbehalten bleiben, da sie in den ersten Phasen der Auswertung keinen zusätzlichen Nutzen zeigen und nur die Rechenzeiten verlängern. Über eine Methode zur Bestimmung des CACP-Mittelpunktes aus Serien der kraniellen Computertomographie ohne vorliegendes Ventrikulogramm berichten wir an anderer Stelle [3].

Das hier skizzierte Voxel-Orientierte Cerebrale Informationssystem (VOCIS) ist prinzipiell offen für jegliche Art von anatomisch-histologischem Wissen, das unter Berücksichtigung des genannten Referenzsystems integriert werden kann. Umgekehrt lassen sich dreidimensionale Rekonstruktionen von CT-Herdbefunden durch Transformation im Referenzsystem standardisiert darstellen. Probleme ergeben sich momentan dort, wo - wie bei der Analyse von Faserverläufen - die Daten zwar im Sinne des Modells operationalisierbar sind, die erforderliche Datenerfassung und -reduktion jedoch an Grenzen der Hardwarekonfiguration des vorhandenen Systems stößt.

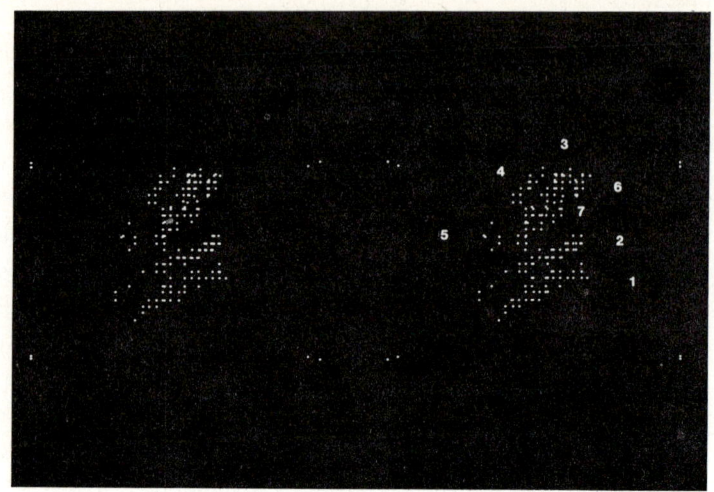

Abbildung 1: Stereobild des gemittelten räumlichen Verlaufs von Sulcus lateralis (2), S. temporalis superior (1), S. centralis (3), S. praecentralis superior (4) und inferior (5) sowie des S. postcentralis superior (6) und inferior (7). Die äußeren Punkte markieren das Referenzvolumen.

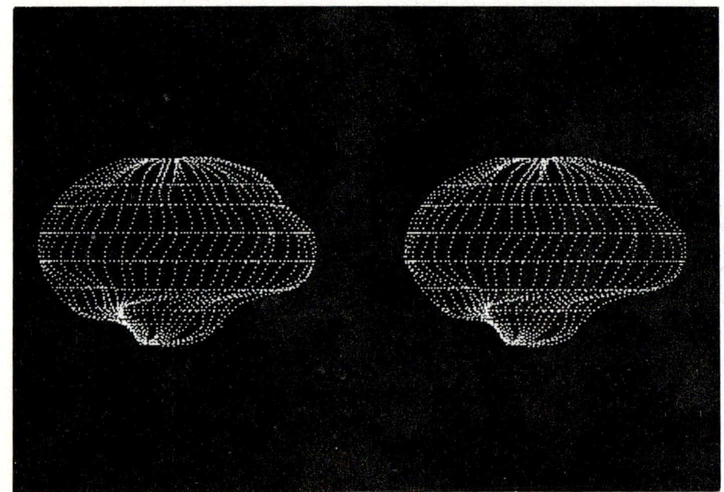

Abbildung 2: Mittlere Oberfläche von 30 linken Hemisphären nach Interpolation mit Cubic Splines.

Literatur:

1) HUIJSMANS, D.P. et al. Anat. Record 216: 449-470 (1986)
2) v. KEYSERLINGK, D.G. et al. Acta anat. 123: 240-246 (1985)
3) NIEMANN, K. et. al. Acta Neurochirurgica (eingereicht)

Animation of Medical Objects Using a Transformation Approach Between Two Data Models

L.E. Peters P. Jensch W. Ameling
Rogowski Institut fuer Elektrotechnik
RWTH Aachen

Introduction

During the last decade there was an increasing trend in using 3D reconstruction for medical applications. Until now it has been used in medical diagnosis [1], surgical planning [2] and even in special prothesis reconstruction [3]. The reconstruction methods are very different: from encoding the 3D relationship of data value ranges in different colors [1] to perhaps the most well known one based on octree encoding [4].

The rapid development of computer hardware made possible the creation of 3D high resolution images even on personal computers. The enhanced graphic cards offer hardware implementation to some very often used graphical algorithms: contour filling, geometrical figure drawing, clipping etc. These leads to quasi-real time animation on small general-purpose computers. Until now these abilities have been used in simulation and handling of solid objects. In medicine, animation of solid objects was generally implemented on special-purpose computers [4]. Due to the huge amount of data to be processed, one of the main objection in using animation for medical diagnosis is the long computing time needed to display sequences of views of the medical object. This is perhaps one of the main direction in which improvement is expected.

This paper proposes a two-model approach used to represent computer tomograms (CTs) in three dimensional space. The approach makes possible the rotation of 3D objects in quasi-real time. The speed-up is obtained without a loss of display accuracy.

Method description

The basic imaging approach is based on locating the surface of a structure in 3D and then projecting the surface onto a 2D image plane. Unless a cut is made, the structure inside the object has to be neglected by using a hidden surface removal approach [5]. The interior of the object can only be seen if a cut through it is made. After a section through the object is made the displayed cross-section becomes the new external surface for the observer.

The concept of the two model data structure has stemed from the necessity to have maximum of information and minimum of computational time. The first model contains all available information extracted from the CTs. The second one, extracted from the first, contains only the information needed to display the object on the screen. If the first model is calculated only once at the beginning of the process, the second one has to be recalculated each time the exterior surface of the displayed object is changed (for example by cutting out a piece of it). This reduction of information brings an increase in the rotation speed without influencing the accuracy of the object representation.

By choosing an appropriate encoding data structure only the changed surface of the display model has to be recalculated every time.

Preprocessing

The input data is taken from a sequence of CTs. It is assumed that the number of cross-sections through the organ is large enough to reconstruct the object from it, or to create the missing cross-section through interpolation. It is assumed that the CT cross-sections are all perpendicular to an axis and are sharp enough, that suplementary noise filtering is not neccesary.

A standard method of locating the surface of an object from given CT cross-sections is based on the transformation of gray-level images in binary ones [6], by this means extracting the object of interest, and a surface edge detection algorithm. The input data used by the two-model approach is preprocessed by using this technique [7].

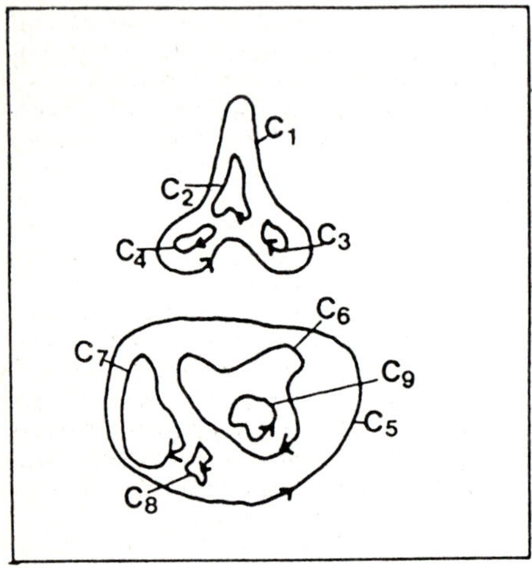

Fig.1a Contours extracted from a cross-section

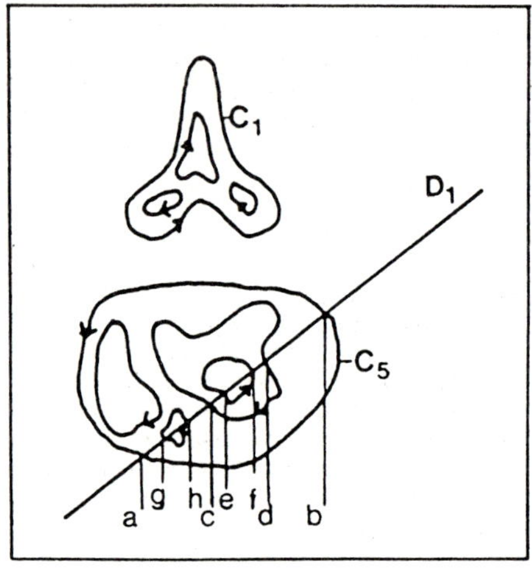

Fig.1b Cut through a cross-section; D1 is the cut-line

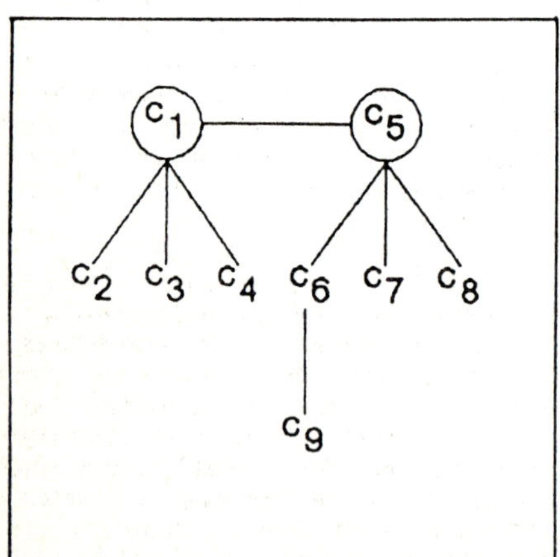

Fig.2a Hierarchical tree (model 1) for the cross-section in Fig.1a and Fig.1b

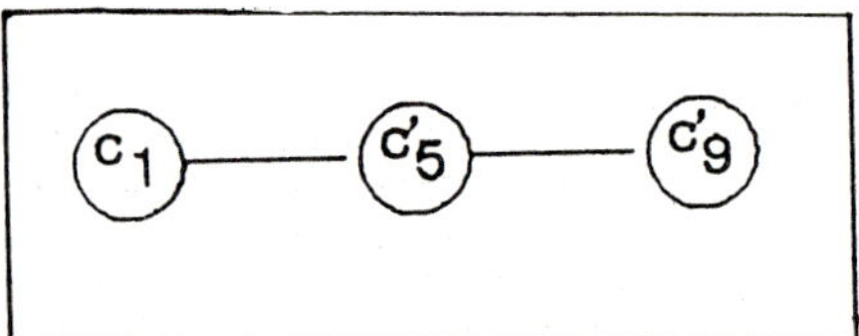

Fig.2b Hierarchical tree (model 2) for the cross-section in Fig.1b

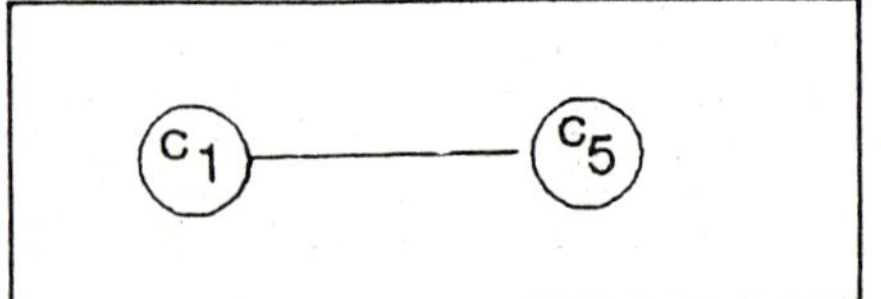

Fig.2c Hierarchical tree (model 2) for the cross-section in Fig.1a

Data structure

The extracted contours have to be stored such as quick access is possible to the contours describing the exterior of the object. This leads to an hierarchical storage of the contours. The external contours must be stored on the top and the internal ones at the bottom of the hierarchical tree. For the cross-section in Fig.1a the corresponding hierarchy is shown in Fig.2a.

The implemented data structure has to contain information about the contour type (exterior or interior) and the link beetwen the top bottom levels. One way of defining a contour as exterior or interior is through "directed contours" [2,8]. The definition of the used data structure "contour" is shown in Fig.3

```
struct contour {
                int dir;
                struct cut  *address;
                struct  contour *next;
                struct contour  *inside;
                float seed[3];
            }c1,c2,c3,c4,c5,c6,c7,c8,c9;
```

Fig.3 The implementation in C of the hierarchical tree structure

The element "dir" contains the information about the type of contour (external, internal). It has the value -1 if it is the contour of a filled element and 1 if it is the contour of a hole. The structure "cut" contains the data describing each contour. *address is the pointer to the data structure "cut". *next and *inside are pointers to the next contour on the same level, respectively the next level inside the hierarchical tree structure.

The last element of the structure "contour" is the element "seed". This element is a 4 value array (x,y,z,color). It represents the "seed pixel" of each contour [3]. This pixel is the starting point for the filling surface algorithm [5]. With the help of this "seed pixel" filled cross-section can be easily reconstructed. Different parts of the object can be outlined by changing the value of "color" in the "seed pixel". When the color information is set on background the contour is interpreted as surrounding a hole.

Display method

The two-model approach is used to represent the CT scan data in three-dimensional space and to make possible the rotation of 3D objects in quasi-real time. After the data has been preprocessed it is stored in two models using the data structure "contour" which was explained above.

The first model contains all the contours which describe the organ to be displayed. For the cross-section in Fig.1a and Fig.1b the hierarchical tree is represented in Fig.2a.

The second model is constructed using only the needed information for the display. If only the exterior of the object has to be shown, the external boundary of each cross-section is taken. For the cross-section in Fig.1a the reduced hierarchical tree is shown in Fig.2c.

When only a part of the whole has to be displayed (by showing a section through the object) the second model has to be reconstructed. Fig.1b shows the new external boundary after a cut was made to the initial cross-section along a perpendicular plane (D1). The intersection points of the plane with the contours are calculated using the hierarchical structure "contour". If a boundary on a superior level is not crossed by the cut-plane (as C1) the contours on inferior levels (C2,C3,C4) do not have to be checked for intersection. With the help of the element "dir" the new boundary can be reconstructed. If "dir" is -1 the new boundary beetween the intersection points is a line. If "dir" is 1 then the new boundary will be a part of an inner contour. For the given example (Fig.1b), the intersection points are: a, b, c, d, e, f, g, h. The new contours to be displayed for this cross section are C1, a modified C5 and a part of C9. The new boundaries have to be recalculated for all the cross sections. The reduced hierarchical tree for the Fig.1b is shown in Fig.2b.

As so far no restrictions have been imposed on the way of saving the contours, any surface reconstruction method could be applied. One of the criteria in choosing the reconstruction method was the computational time. The chosen method is triangularisation. Tiled surfaces are formed by using triangular elements [8,9].

The rotation of the displayed object is now dependent only on the number of computations needed to describe the hull of the object. A supplementary reduction of the computational time is reached by using pseudo shading [10]. The hue value is encoded in the z coordinate. The objects which are near to the viewer are lighter and those in the background darker. The hidden line removal is done by a back to front approach, the z-buffer algorithm [11].

Conclusion

The described method is currently being implemented using two workstations. While on the first is done the data preprocessing, on the second is implemented the described reconstruction method using the advantages of an enhanced color graphic card. The 3D image has a resolution of 512x512 pixel.

This approach has the advantage to give the possibility to rotate the medical object without a huge amount of computation. The physician has the opportunity to have a quick look at the whole object first and then to decide which viewpoint or cross-section he wants to see in detail.

References

[1] E.J.Farell, "Color 3-D Imaging of Normal and Pathologic Intracranial Structures ", IEEE Computer Graphics and Applications, Vol.4 Sept.1984, pp.5-16.

[2] L.J.Brewster, S.S.Trivedi, H.K.Tuy, J.K.Udupa "Interactive Surgical Planning" IEEE Computer Graphics and Applications, Vol.4 Mar. 1984, pp.31-40.

[3] M.L.Rhodes, Y.Kuo and S.L.G.Rothman "An Application of Computer Graphics and Networks to Anatomic Model and Prothesis Manufacturing", IEEE Computer Graphics and Applications, Vol. 7 Feb. 1987, pp.12-25.

[4] D.J.Meagher,"Interactive Solids Processing for Medical Analysis and Planning", Proc. Fifth Ann. Conf. Nat'l Computer Graphic Assoc., Vol. II, 1984, pp.96-106.

[5] T.Pavlidis "Algorithms for Graphics and Image Processing", Springer-Verlag Berlin-Heidelberg, 1982.

[6] G.T.Herman "Image Reconstruction from Projections", Academic Press, New York London, 1980.

[7] P.Harst, "Untersuchung von Interpolationsverfahren fuer 3D-Objekte auf der Basis von 2D-Schnittbildern, Diplomarbeit.Rogowski Institut fuer Elektrotechnik, RWTH Aachen, 1985 (unpublished report).

[8] J.K.Udupa, "Interactive Segmentation and Boundary Surface Formation for 3D Digital Images", Computer Graphics and Image Processing, Vol 12,1982,pp.213-235.

[9] L.T.Cook, S.J.Dwyer III, S. Batnitzky, and K.R.Lee, " A Three-Dimensional Display System for Diagnostic Imaging Applications", IEEE Computer Graphics and Applications, Vol.3, Aug.1983, pp.13-19.

[10] N.I.Badler, J.O'Rourke and H. Toltzis, "A Spherical Representation of a Human Body for Visualizing Movement," Proc. IEEE, Vol.67, No.10, 1979,pp. 1397-1403.

[11] D.F.Rogers "Procedural Elements for Computer Graphics", McGraw-Hill Book Company, New-York, 1985

COMPENSATION OF THE DISPERSION OF OPTICAL SYSTEMS WITH CRUDE A-PRIORI KNOWLEDGE OF THE INPUT SIGNAL

Kristian Kroschel
Institut für Automation und Robotik, Universität Karlsruhe
Kaiserstraße 12, D-7500 Karlsruhe 1

Introduction

All real optical systems are dispersive systems which can be seen from the fact that abrupt changes from white to black in gray level images are smoothed by the optical system so that the processed image shows unsharpness at its edges. In principle this disadvantage can be overcome by a more precise fabrication of the optical lens system. Since this involves a significant increase in manufacturing cost means of digital signal processing are looked for to compensate the dispersive effect of the optical system. The bandwidth of the input signal is normally unknown for the optical system. Thus the dispersion problem for optical systems has to be solved with low-level a-priori knowledge of the input signal.

In this paper the dispersion problem is solved using Wiener filter theory. For the design of this filter the power spectrum of the input signal has to be known. Because the spectrum of the image is not available the filter is derived from an estimate or an assumption about this spectrum.

Model of the Optical System

The original image to be processed is denoted by $a(x,y)$ with x,y being the coordinates of the pixels. This image is transformed by the optical system which is assumed to be linear with impulse response $h(x,y)$. The transformed and linearly distorted image is $s(x,y)$. The uncertainty in modelling the impulse response $h(x,y)$ is represented by the noise $h_n(x,y)$ which is added to $h(x,y)$ as shown in Fig. 1.

It is intended to use digital techniques to compensate the dispersion of the image $s(x,y)$. Thus the optical signal is transformed into an electrical signal whereby noise $n(x,y)$ is added. The input $r(x,y)$ of the digital processing system with impulse response $g(x,y)$ is used to generate an estimate $\hat{a}(x,y)$ of the original image $a(x,y)$ which is optimal in the minimum mean square sense.

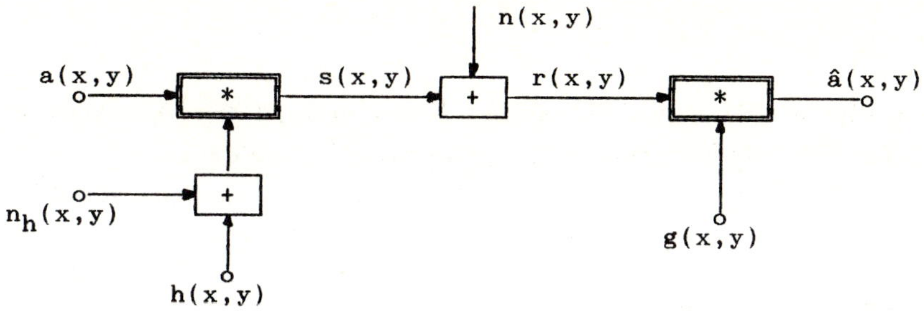

Fig. 1 Model of the optical system and the estimator

For simplicity it will be assumed in the sequel that the image is processed line by line to yield one-dimensional signals. The techniques presented in this paper therefore have to be expanded to two dimensions for practical applications.

The Wiener-Filter Method

The original image a(x,y) or one line a(x) of it is estimated with minimum mean square error by a Wiener filter. The non-causal Wiener filter is described by its transfer function G(u) [1]

$$G(u) = \frac{W(u)\cdot S_{ss}(u)}{S_{rr}(u)} = \frac{W(u)\cdot S_{ss}(u)}{S_{ss}(u)+S_{nn}(u)} \tag{1}$$

with u representing the frequency variable and $S_{ss}(u)$, $S_{rr}(u)$, and $S_{nn}(u)$ the power density spectra of the uncorrupted and corrupted images and the noise, respectively. W(u) denotes the ideal transfer function with no restrictions to realizability and no noise influences. It is equal to

$$W(u) = \frac{H(u)}{|H(u)|^2} \quad , \tag{2}$$

i.e. the inverse transfer function of the optical system. The power spectrum $S_{ss}(u)$ has to be known for the implementation of the Wiener filter G(u). For unspecified images this spectrum is either unknown or only roughly known. Therefore suboptimal solutions of the Wiener filter problem will be given in the sequel.

One approach is to assume a specific shape of the spectrum $S_{ss}(u)$:

Gaussian $$S_{ss}(u) = c\cdot\exp(-(0.2825\cdot\alpha\cdot u)^2) \tag{3}$$

$\cos^2$ $$S_{ss}(u) = \begin{cases} c\cdot\cos^2(0.25\cdot\alpha\cdot u) & |0.25\cdot\alpha\cdot u|\le\pi/2 \\ 0 & \text{elsewhere} \end{cases} \tag{4}$$

$$\text{rectangular} \quad S_{ss}(u) = \begin{cases} c & |0.5\cdot\alpha\cdot u| \le \pi/2 \\ 0 & \text{elsewhere} \end{cases} \tag{5}$$

with α a free parameter controlling the bandwidth of the spectra which is identical for all these shapes for the same value of α.
In a second approach the so called spectral subtraction method [2] is applied, i.e. the true spectrum $S_{ss}(u)$ is replaced by

$$\hat{S}_{ss}(u) = \hat{S}_{rr}(u) - \beta\cdot S_{nn}(u) \tag{6}$$

where the spectrum $\hat{S}_{rr}(u)$ is calculated as a short time estimate from the input $r(x)$ over a finite window and $n(x)$ is assumed to be a white noise process with spectral density $S_{nn}(u)=N_w$. The free parameter β controls the influence of the noise.
A further method uses a recursive algorithm for the estimation of $S_{ss}(u)$. Here the actual value of $S_{ss}(u)$ is replaced by the spectrum $S_{\hat{s}\hat{s}}(u)$. It is calculated from the power density of the estimate $\hat{a}(x)$ and the transfer function of the optical system:

$$S_{\hat{s}\hat{s}}(u) = S_{aa}(u)\cdot|H(u)|^2 \quad . \tag{7}$$

To initiate the recursive algorithm $S_{ss}(u)$ is assumed to have a Gaussian shape or is proportional to $|H(u)|^2$ which means that the input process is white.

Results

The criterion of optimality which is used to compare the methods discussed in this paper is the mean square error given by

$$F = \frac{1}{N}\sum_{x=1}^{N}(\hat{a}(x)-a(x))^2 \quad . \tag{8}$$

One line $a(x)$ of the image consits of N pixels which are used to calculate the error given above.
First the influence of the signal-to-noise ratio on the error F as a function of the different methods for reconstruction is investigated. As a reference the ideal Wiener filter with known spectrum $S_{ss}(u)$ is used which obviously yields the minimum error. From the other methods the most efficient ones are selected. Those are the recursive estimation and the one assuming a Gaussian shape of the input spectrum. Fig. 2 shows the error F as a function of the signal-to-noise ratio S/N. The recursive method is very close to the optimum for high values of S/N whereas for low values of S/N the assumption of a Gaussian shape of $S_{ss}(u)$ is better.
The approach to estimate the Spectrum $S_{ss}(u)$ from the short time

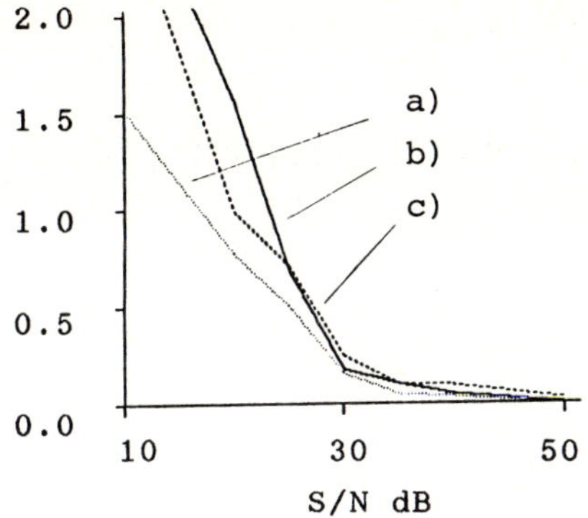

<u>Fig. 2</u> Error F as a function of the signal-to-noise ratio S/N. a) Wiener filter with known $S_{ss}(u)$, b) recursive estimation of $S_{ss}(u)$, c) $S_{ss}(u)$ with Gaussian shape

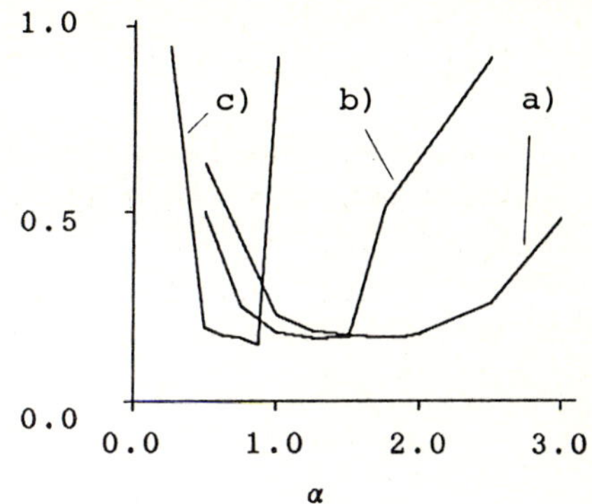

<u>Fig. 3</u> Approximation of the input spectrum $S_{ss}(u)$ by three shapes a) Gaussian, b) $\cos^2$, c) rectangular. Bandwidth parameter α

spectrum of r(x) and the white noise spectrum proved to be very sensitive with respect to the parameter ß so that it was not included in this comparison.

Sensitivity of the parameter α plays a role for the selection of the shape of the input spectrum $S_{ss}(u)$. Fig. 3 shows the mismatch of the the bandwith parameter α with respect to the bandwidth of the true input spectrum. The input signal a(x) is a line with trapezoidal shape of gray levels covering roughly one third of the line consisting of 256 pixels which corresponds with the bandwidth parameter $\alpha=2.5$. As can be seen from Fig. 3 all the methods yield more or less the same mimiumum error F. The assumption of a Gaussian shape has the advantage of being more insensitive with respect to the parameter α than the others, i.e. α can be chosen in a wide range without drifing too much away from the mimimum error. Furthermore it is interesting to notice that the optimal parameter for α in all the three cases does not coincide with the bandwidth parameter of the input spectrum. This is due to the noise component n(x) which is assumed to introduce a signal to noise ratio of S/N=30dB.

References

[1] Kroschel, K.: Statistische Nachrichtentheorie, 2. Teil Signalschätzung, Springer, Berlin 1974

[2] Reich, W.: Adaptive Systeme zur Reduktion von Umgebungsgeräuschen bei Sprachübertragung, PhD dissertation, Karlsruhe, 1985

2-D VECTOR REPRESENTATION OF MULTI-DIMENSIONAL SYMPTOM SPACE FOR COMPUTER AIDED MEDICAL DIAGNOSIS

*Tomio SEKIYA, **Akira WATANABE, **Masao SAITO and *Makoto KIKUCHI

*National Defence Medical College
3-2 Namiki, Tokorozawa, Saitama 359, JAPAN
**Faculty of Medicine, University of Tokyo
7-3-1 Hongo, Bunkyo-ku, Tokyo 113, JAPAN

1. INTRODUCTION

One of the major objectives of the computer-aided diagnosis is to construct an algorithm for medical diagnosis which is hopefully help physicians in accuracy as well as insight. Many theoretical approaches to this aim were proposed by researchers, but most of them utilize the discriminant analysis.

The disease space is multi-dimensional and composed of symptom points. The region of a disease covers points corresponding to the patients with the particular disease. When the disease region has a complicated structure, it will more efficiently be understood by representing the situation in a graphic form.

Watanabe (1965) proposed an algorithm for congenital heart diseases , where the each disease region is represented as a sum of sub-regions. However, the shape of resultant multi-dimensional disease regions are so complicated that it is difficult to represent them in a graphic form. Any efficient interactive method of 2-D or 3-D representation of the multi-dimensional disease space has not been developed yet.

In this paper, we represent the disease space by a constellation graph. Each disease region is reduced to a sector in a semicircular space by properly decreasing the spacial dimension by statistical techniques. The advantage of this method is that the physician can intuitively understand the global structure around the patient point in the multi-sector disease space with simple boundary lines. In addition, comparison of the constellation graph method has been made with other ones including the potential method(Sekiya,1983) and well known factor analysis to discuss the capability in mapping the data from the multi-dimensional space to the 2-D plane with less overlapping and convenience in intuitive classification,which are considered to be measures of merit.

2. INPUT DATA

A set of symptoms of each case is represented by a 51-dimenssional vector in a multi-dimensional symptom space. Every component of 51-dimensional vector consists of three values {-1,0,1} corresponding to normal, abnormal and unsettled situations, respectively. 190 cases four congenital heart diseases without any complication, who had been operated on at the Tokyo University Hospital; 47 cases of auricular septal defect (ASD), 36 cases of ventricular septal defect (VSD), 47 cases of patent ductus arteriosus (PDA) and 60 cases of tetralogy of Fallot (Fallot). The 51 symptoms chosen for analysis were 10 subjective symptoms (No.1-10), 6 differential diagnosis (No.11-16), 29 phonocardiograms (No.17-45) and 6 electrocardiograms (No.46-51).

3. PROCEDURES AND RESULTS

In the constellation graph method, the multi-dimensional data is represented as the connected vectors in the semicircle. Given n samples of data with p valuables($x_{1i}, x_{2i}, \ldots, x_{pi}$; $i=1,2,\ldots,p$), a valuable is represented by a vector by the following transformation.

$$g_{ja}=f_j(x_{ja}) \;;\; j=1,2,\ldots,p,\; a=1,2,\ldots,n \qquad (1)$$

where $0 \le g_{ja} \le \pi$

The transformation of f_j is defined as follows.

$$f_j(x_{ja})=(x_{ja}-x_{jv})\pi/(x_{ju}-x_{jv}) \qquad (2)$$

where $x_{ju}=\max x_{ja} \quad (1 \le a \le n)$

$x_{jv}=\min x_{ja} \quad (1 \le a \le n)$

x_{ja}: a value based on every symptom

$$\sum_{i=1}^{p} w_j = 1 \qquad w_j \ge 0 \;,\; j=1,2,\ldots,p \qquad (3)$$

w_j : length of vector

The value of w_j is suitably determined to separate as much as possible the four congenital heart diseases, by asking the views of specialists in a conversational mode. The value of x_{ja} is determined statistically on the bases of the occuring frequency of abnormality of every symptom. The resultant final vector composed of p elementary vectors is plotted in the semicircle for the constellational representation. In practice the categories to which each vector belongs correspond to the specific diseases and the vectors correspond to the results of laboratory tests. The vectors are drawn as follows. The first vector starts from the center of the semicircle, then the second vector follows starting from the end of the first vector which is in turn followed by the third one. The procedure continues to the last vector. A mark is plotted at the top of the final vector to indicate the symptom situation. Fig.1 shows the result of processing suitably the sample data to separate as much as possible the four congenital heart diseases. The final result of discrimination is shown in Table 1.

For comparison analysis have been performed using the potential method and factor analysis. Fig.2 and Table 2 show the definition and the result of analysis by the potential method, respectively. The cumulative percentage of factor by using factor analysis is shown in Table 3. Fig.3 shows each boundary line of four categories by the use of the factor score.

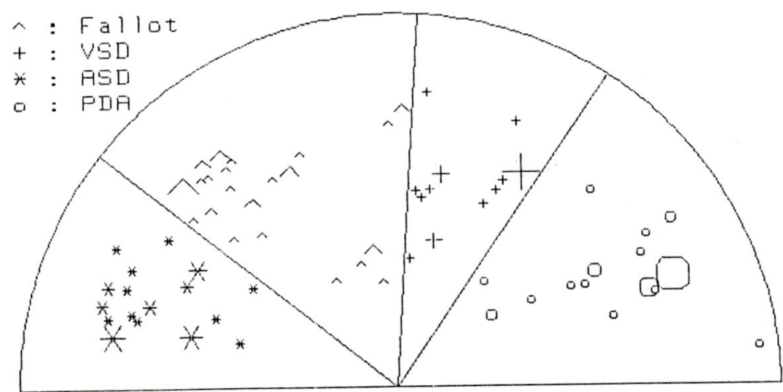

Fig.1 CONSTELLATION GRAPH METHOD

Table 1 RESULT FOR DISCRIMINATION

	FALLOT	VSD	ASD	PDA
FALLOT	93.3	6.3	0.0	0.0
VSD	5.6	94.4	0.0	0.0
ASD	10.6	4.3	85.1	0.0
PDA	0.0	0.0	4.3	95.7

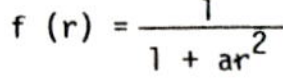

f (r) : potential function
a : constant
r : distance

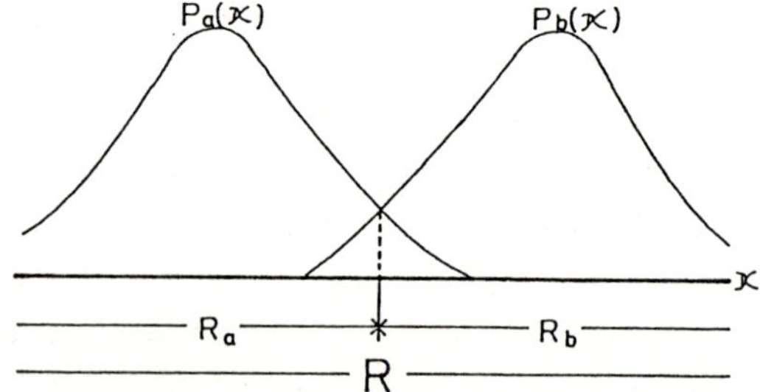

Fig.2 DIVISION INTO PARTIAL SPACES BY USING POTENTIAL METHOD

Table 2 RESULT FOR DISCRIMINATION

Disease	M	COR.	ERR.	REJ.
FALLOT	21	93.3	0.0	6.7
VSD	12	86.1	0.0	13.9
ASD	49	75.4	0.0	24.6
PDA	8	89.4	0.0	10.6
Ave.	22.5	86.5	0.0	13.5

M : Number of Sub-space
COR.: Correct
ERR.: Error
REJ.: Reject

Table 3 Cumulative percentage of factor

FACTOR	PERCENTAGE	CUMULATIVE PER.
1	22.45	22.45
2	7.27	29.72
3	6.80	36.52
4	5.59	42.11
5	4.11	46.22
6	3.74	49.96
7	3.55	53.52
8	3.02	56.53
9	2.89	59.43
10	2.73	62.16

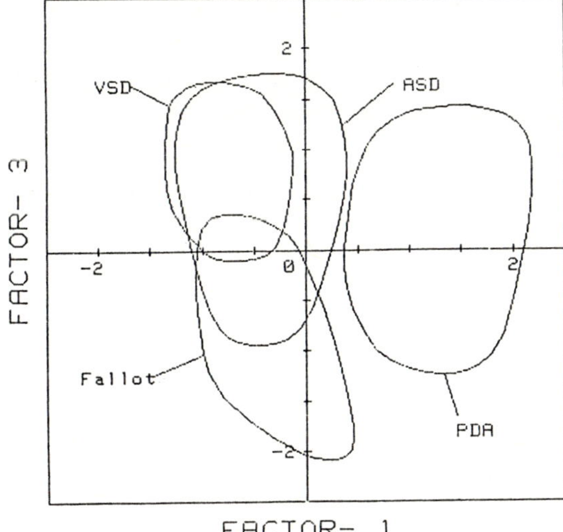

Fig.3 Factor score

4. DISCUSSION

Through these experiments, it is found that the constellational representation provides suitable means to process the sample data by asking the views of specialists in a conversational mode and to represent the given multi-dimensional data with less overlapping in the 2-D plane with the satisfactory result for discrimination. In the constellational representation, the decision-making process is indicated by drawing the vector path from the center of the semicircle to the final vector.

In case of time series data, we may be able to treat the constellational configulation by representing it in 3-D semisphere.

The main points of the results are following;

(1)Each disease space in the multi-dimensional symptom space is satisfactorily separated in the form of the sector in the constellational space with the reduced dimension.

(2)The vector path seems to visualize the decision-making process for each disease.

Automatische Detektion von Bereichen schwacher Rückstreuung in Radar-Abbildungen

Behrens K., Mauer E.

Forschungsinstitut für Informationsverarbeitung und Mustererkennung

Eisenstockstr. 12 7505 Ettlingen 6

Motivation

Das abbildende Radar hat im Bereich der Fernerkundung als neue Informationsquelle, welche wetterunabhängig und beleuchtungsfrei zur Verfügung steht, bereits einen hohen Stellenwert erreicht. In Anbetracht der großen Datenmenge ist für eine vollständige und schnelle Auswertung der Bildinhalte eine automatisierte digitale Bildverarbeitung erforderlich. Diese Arbeit beschreibt einen Baustein eines Verarbeitungsprozesses (s.a. [EBE 87,MAU 87]), der in Abb. 1-1 grob schematisiert dargestellt ist.

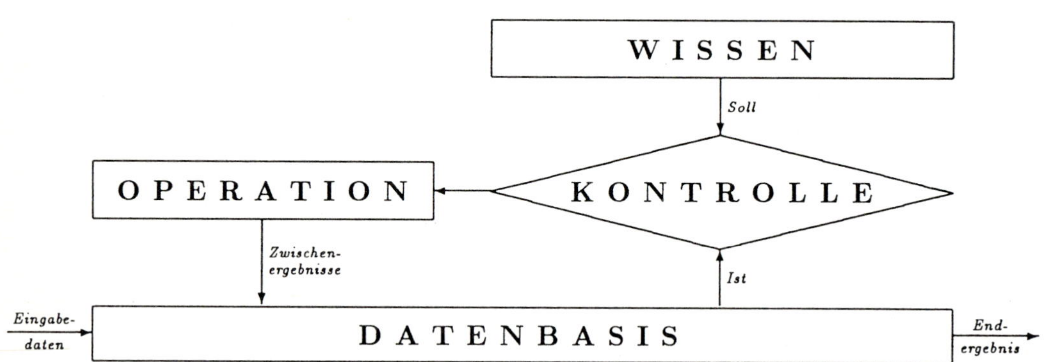

Abb. 1-1: Schema der Verarbeitung

Die Wissensbasis enthält zu Beginn der Verarbeitung sowohl allgemeines Wissen in abstrakter und modellhafter Form als auch Wissen über das System selbst. In die Datenbasis werden die Eingabedaten und die im Laufe der Verarbeitung anfallenden (Zwischen-)Ergebnisse eingetragen. Der Kontrollmodul steuert den Verfahrensablauf durch die Auswahl des jeweils für den aktuellen Zeitpunkt günstigsten Verfahrens mit den dazu benötigten Daten und Parametereinstellungen. Dazu werden die Informationen aus der Wissens- und der Datenbasis verwendet. Es entsteht ein Verarbeitungszyklus, der abbricht, wenn keine Ergebnisverbesserung mehr zu erreichen ist.

Da die Signalwerte, die durch das Eingabebild vorliegen, noch keinen Vergleich mit den abstrakten Beschreibungen in der Wissensbasis erlauben, müssen zunächst Verfahren zur initialen Signalverarbeitung eingesetzt werden, um aus den Eingabedaten durch Zerlegung des Bildes in bedeutungsrelevante Teile Informationen höheren Abstraktionsgrades zu gewinnen.

Kontrast versus Energie

Aus der Art der zu untersuchenden Daten lassen sich in der Regel einige globale Bildeigenschaften ableiten, die eine Orientierung im Bild ermöglichen. So sind z.B. in realen Multispektraldaten sowohl Gebiete mit homogener als auch Gebiete mit inhomogener Grauwertverteilung vorhanden. Zur Unterteilung der Daten in disjunkte Bereiche mit diesen komplementären Eigenschaften werden die im Bild enthaltenen Grauwertkontraste untersucht [BEH 84]. Für die in den Radardaten durchzuführende Zerlegung hat sich jedoch früh gezeigt, daß eine identische Vorgehensweise wie bei den Multispektraldaten nicht angebracht ist, da hier die Kontrastwerte wegen radartypischer Eigenschaften (wie z.B. *Systemrauschen* und *Kohärenzrauschen*) derart gestört sind, daß hier abgegrenzte Grauwertverteilungen a priori nicht zu erwarten sind. Bei den Radardaten stützt man sich deshalb auf die Eigenschaft, daß in Bereichen hoher Energierückstreuung eher relevante Informationen enthalten ist als in Bereichen schwacher Rückstreuung. Daher ermöglicht eine Abtrennung der Gebiete schwacher Rückstreuung die Konzentration auf bedeutungsrelevante Bildteile.

Die dazu notwendigen Analysen sind nicht mehr mit Hilfe von Grauwertdifferenzen (wie dies bei der Kontrastanalyse notwendig war) durchzuführen, sondern man kann sich auf die Untersuchung von Energiesummen stützen, da hierdurch der Einfluß der verschiedenen Rauscharten weitgehend reduziert wird. In beiden Fällen liegt jedoch die gleiche Zielsetzung vor, nämlich die Detektion von Gebieten mit ähnlichen Eigenschaften. Benutzt man also Intensitäts-Summen statt - Differenzen, so läßt sich der gleiche Mechanismus einsetzen, der auch bei den Multispektraldaten zur Anwendung kommt.

Vorgehensweise

Dieser Mechanismus läßt sich folgendermaßen kurz zusammenfassen: Ausgehend von dem Eingabebild (Abb. 1-2 a) wird eine automatische Unterteilung des Gesamtbildes in lokal schwach und lokal stark rückstreuende Bereiche durchgeführt. Für die dadurch entstandenen zusammenhängenden Gebiete findet ein Konsistenztest statt. Alle akzeptierten Gebiete werden einer Klasseneinteilung unterzogen. Abschließend wird bezüglich dieser Trainingsklassen eine Kreuzklassifikation durchgeführt.

Schwellwertbestimmung

Um die Trennung zwischen schwach und stark rückstreuenden Bereichen rückstreuenden Bereichen zu ermöglichen, ist es zunächst notwendig, einen globalen Schwellwert für diese Unterscheidung zu bestimmen. Dies soll nicht interaktiv, sondern an die jeweils vorliegenden Daten adaptiert geschehen. Dazu bedarf es einer statistischen Analyse der lokalen Energiesummen. Man kann a priori nicht angeben, wie groß der zu analysierende lokale Bereich ist. Daher wird, um mit möglichst hoher Wahrscheinlichkeit innerhalb dieses Bereiches zu bleiben, die Kontrastanalyse in der direkten Nachbarschaft jedes Bildpunktes durchgeführt. Diese Beschränkung ist unkritisch, da zu einem etwas späteren Zeitpunkt eine Überprüfung der Ergebnisse mit veränderlicher Nachbarschaft und einer eventuell notwendigen Korrektur stattfindet.

Um eine gesicherte statistische Entscheidung treffen zu können, werden eine Vielzahl unterschiedlicher Stichproben aus dem Gesamtbild entnommen und getrennt untersucht. Damit der Geometrieeinfluß auch in den Stichproben erhalten bleibt, werden zusammenhängende Bildausschnitte als Stichproben verwendet. Zu diesen Ausschnitten werden gemäß der Kostenfunktion $K_G(i)$

$$K_G(i) = f_G(P_i, IH_G(T_i, T_i), IH_G(T_i, R_i)) \qquad \text{mit} \qquad K_G(e) = \min_{0 \leq i \leq N} K_G(i)$$

G = Grundmenge ($= T_i + R_i$)
i = Trennstelle
T_i = abgetrennte Teilmenge von G
P_i = Anteil von T_i an G
IH = Inhomogenitätsmaß

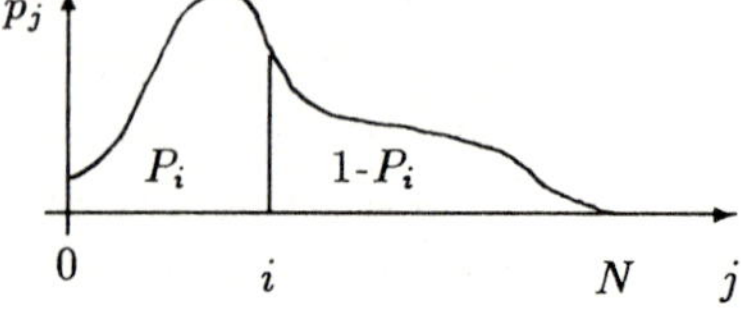

jeweils ein Schwellwert e mit einem den Verhältnissen in diesem Ausschnitt entsprechenden Vertrauenswert bestimmt. Diese Schwellwerte werden zur Kumulation, gewichtet mit ihren Vertrauenswerten, in ein Histogramm 2. Ordnung übertragen [MAU 84]. Dieses Histogramm 2. Ordnung wird dann mit einer a priori Schätzung verglichen, wobei es sich im Zustand vollständiger Unkenntnis um die Annahme einer Gleichverteilung handeln wird. Die Stelle, an der die a priori Schätzung am stärksten übertroffen wird, gibt den günstigsten Schwellwert an. Bei der Annahme einer Gleichverteilung ist dies die Position des Maximums im Histogramm 2. Ordnung.

Es ist derzeit noch nicht geklärt, ob es in Anbetracht der Kohärenzeffekte eine "beste" Stichprobengröße für diese Untersuchung gibt, wodurch der Aufwand bei der Stichprobenbildung reduziert werden kann. Denkbar ist, daß diese von den spezifischen Eigenschaften des Radarsensors abhängig ist.

Schwellenanwendung mit lokaler Kontrolle

Abhängig von diesem Schwellwert werden die Bildpunkte des Gesamtbildes in zwei Klassen eingeteilt, nämlich die *lokal schwach rückstreuenden Bildpunkte* und die *lokal stark rückstreuenden Bildpunkte* (Abb. 1-2 b). Da nicht zu erwarten ist, daß die globale Schwelle in allen Bildbereichen eine sinnvolle Unterteilung bewirkt, muß nun eine lokale Überprüfung dieser Unterteilung vorgenommen werden.

Dazu wird jede einzelne Menge zusammenhängender, schwach rückstreuender Bildpunkte daraufhin untersucht, ob die Verteilung paarweiser Energiesummen der gesamten Fläche F bezüglich des Schwellwertes auch unter Aufhebung der Forderung direkter Nachbarschaft den Anforderungen genügt:

$$K_F(e) = f_F(P_e, IH_F(T_e, T_e), IH_F(T_e, R_e)) \leq K_{F_{max}}(e) = g(K_G(e))$$

Ist dies nicht der Fall, so wird so lange eine Flächenreduktion durchgeführt, bis die Fläche die Anforderungen erfüllt oder aber bis die Fläche zu klein geworden ist, um als statistisch relevant gelten zu können. Diejenigen Flächen, die diesen Test erfolgreich bestehen, werden nun als Kandidaten für die Bildung von Ähnlichkeitsklassen benutzt (Abb. 1-3 a).

Klassenbildung

Die Bildung von Ähnlichkeitsklassen zerfällt in zwei Teilschritte. Zunächst wird jeder der Kandidaten als Repräsentant einer eigenen Ähnlichkeitsklasse betrachtet. Davon ausgehend wird untersucht, ob die gemeinsamen Eigenschaften zweier Klassen F_1, F_2 so ähnlich sind, daß sie zu einer Klasse vereinigt werden können. Dieser Sachverhalt wird durch folgende Formel wiedergegeben:

$$K_{F_1+F_2}(e) = f_{F_1+F_2}(P_e, IH_{F_1+F_2}(T_e, T_e), IH_{F_1+F_2}(T_e, R_e)) \leq K_{F_{max}}(e)$$

Ergeben sich mehrere Möglichkeiten der Zusammenfassung, so wird die günstigste Paarung bevorzugt. Dieser Vorgang wiederholt sich, bis es keine zwei Klassen mehr gibt, die eine genügend hohe Eigenschaftsähnlichkeit mehr aufweisen.

Da in dem ersten Schritt wegen der geforderten Zuverlässigkeit bei der Klassenbildung nur genügend große Flächenstücke betrachtet wurden, ist es das Ziel des zweiten Schrittes zu untersuchen, ob es noch weitere Bildpunkte gibt, die einer dieser Klassen zuzurechnen sind. Die Durchführung dieses Schrittes erfolgt mittels Kreuzklassifikation, bei der die Eigenschaften jedes noch nicht zugewiesenen Bildpunktes mit denen der definierten Ähnlichkeitsklassen verglichen werden. Die Entscheidungsfunktion entspricht der in der zuletzt gezeigten Formel, wobei hier die Fläche F_2 nur aus einem einzigen Bildpunkt besteht. Ergeben sich mehrere mögliche Zuweisungen, so wird der Bildpunkt derjenigen Klasse zugefügt, zu der seine Eigenschaften am besten passen. Diejenigen Bildpunkte, die keiner dieser Klassen zuzuordnen sind, werden in einer Rückweisungsklasse zusammengefaßt.

Das Endergebnis dieser Klassenbildung ist in Abb. 1-3 b dargestellt. Diejenige Klasse, die das niedrigste Energieniveau aufweist, repräsentiert die Bereiche ohne Rückstreuung, sofern im Bild überhaupt solche Bereiche vorhanden sind. Anhand dieser speziellen Klasse lassen sich die Systemeigenschaften des verwendeten Sensors genauer untersuchen.

Zusammenfassung

Dieses Verfahren stellt einen adaptiv und automatisch arbeitenden Baustein in einem Verarbeitungsprozeß dar, mit dem drei wesentliche Ergebnisse erzielt werden: Die Erkennung der Gebiete ohne Rückstreuung erlaubt eine detailliertere Untersuchung der Systemeigenschaften, während durch die Abtrennung aller schwach rückstreuenden Gebiete eine Reduktion des Aufwands sowie eine Reduktion der Störungen bei der Analyse der stark rückstreuenden Gebiete bewirkt wird. Auf diese Ergebnisse kann man nun zurückgreifen, um das Ziel der vollständigen, adaptiven und automatischen Auswertung der Bildinhalte zu erreichen.

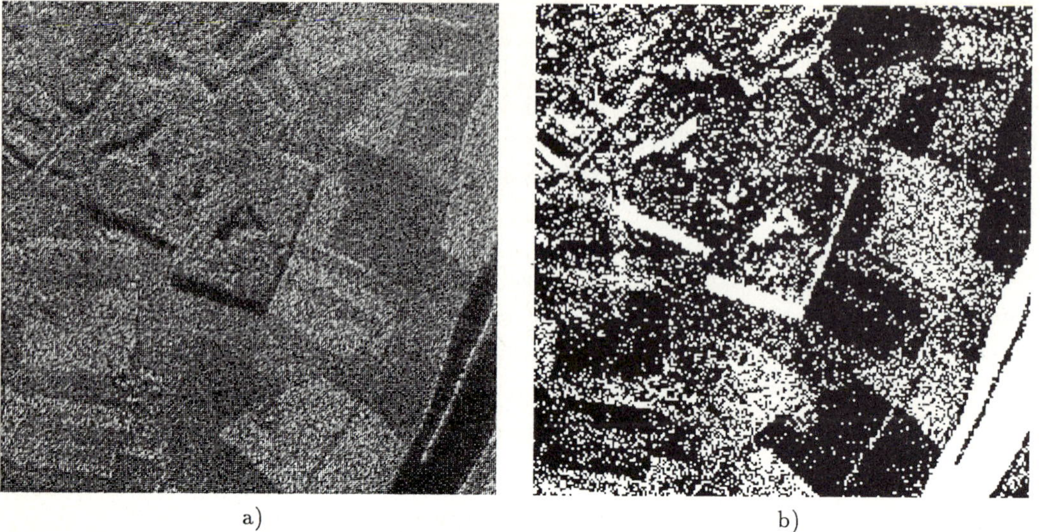

a) b)

Abb. 1-2: Eingabebild und Binarisierungsergebnis

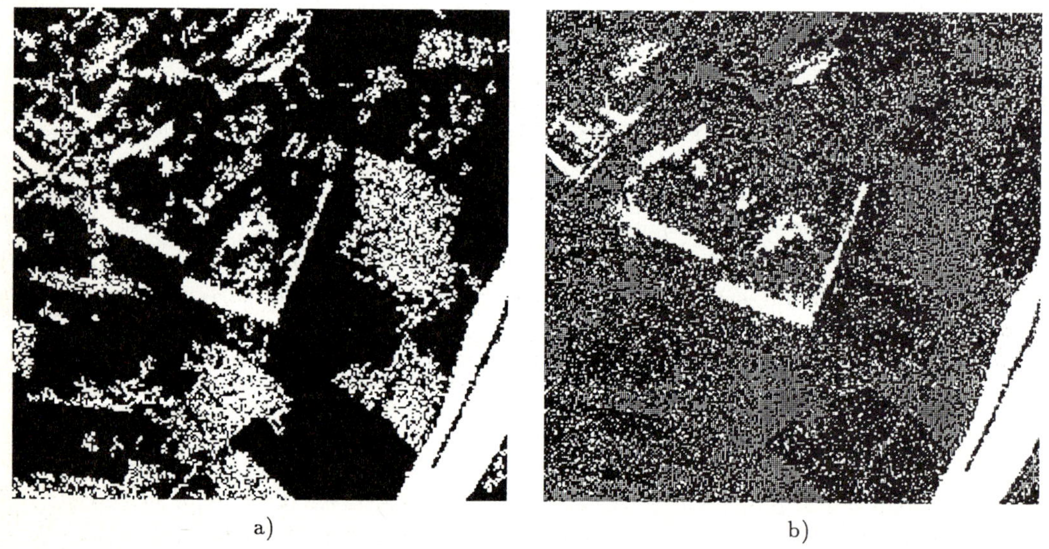

a) b)

Abb. 1-3: Klassenkandidaten und Klassen

Literatur

[BEH 84] Behrens K. *Adaptive und automatische Parametereinstellung zur Segmentation von Fernerkundungsdaten*, FIM-Bericht 131, August 1984

[MAU 84] Mauer E. *Aspects of automation in a system for remote sensing data analysis by feature combination*, ERIM-Proceedings, Oktober 1984

[EBE 87] Ebert A. *Automatische Detektion von Bereichen einheitlicher Interferenzerscheinungen in Radar-Abbildungen durch Speckle Modellierung*, ASST-Tagungsband, Sept. 1987

[MAU 87] Mauer E. *Adaptive kontextbezogene Signalanalyse zur Ermittlung von auffälligen Bildbereichen für die initiale Bildanalyse*, DAGM-Tagungsband, Oktober 1987

Automatische Detektion von Bereichen einheitlicher Interferenzerscheinungen in Radar - Abbildungen durch Speckle Modellierung

Ebert A., Mauer E.

Forschungsinstitut für Informationsverarbeitung und Mustererkennung
Eisenstockstr. 12 7505 Ettlingen 6

Motivation

Ziel einer initialen Bildanalyse ist die Aufgliederung eines Bildes in relevante Bildsegmente, welche in nachfolgenden Verarbeitungsebenen höherer Abstraktion interpretierbar sind. A priori relevant sind alle auffälligen Erscheinungen im Bilde, welchen unter Berücksichtigung der Aufgabenstellung voraussichtlich eine Bedeutung zugeordnet werden kann. Bei unseren Untersuchungen konzentrieren wir uns, unter Berücksichtigung des lokalen Kontextes, auf die Ermittlung statistischer Auffälligkeiten. Das Einbringen von Kenntnissen über spezielle Sensoreigenschaften erlaubt eine vereinfachte Bestimmung dieser ausgezeichneten Bereiche. So erwarten wir, aufgrund der Radar-Abbildung, in Bereichen ursächlich konstanter Rückstreuung einheitliche Interferenzerscheinungen. Eine Detektion derartiger Gebiete ohne Berücksichtigung der Radar-spezifischen Bildeigenschaften ist ohnehin wegen des niedrigen Signal-/Rauschverhältnisses nur eingeschränkt möglich.

a) Originalbild

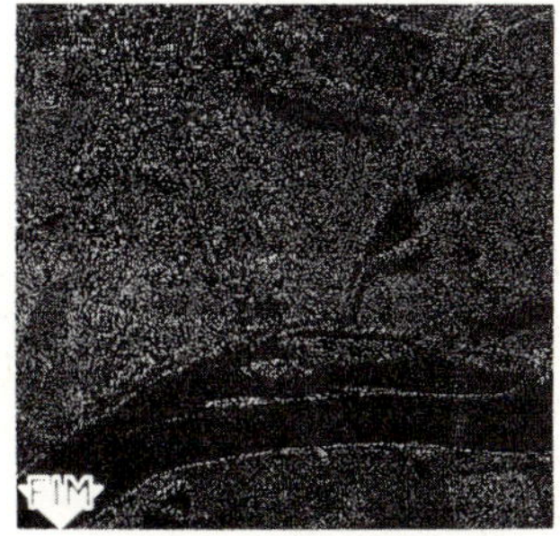

b) Regionale Maxima

c) Erw. regionale Maxima

Abb. 1: Demonstration zur Informationsverteilung im Radarbild

Abb. 1b zeigt die räumlich dichte Verteilung der regionalen Intensitätsmaxima ('Peaks') des Originalbildes aus Abb. 1a. Die Erweiterung dieser erheblich reduzierten Bildpunktemenge (hier in Abb. 1c erweitert um 20%) ergibt eine gute Wiedergabe des Originalbildes. Auf diesem Effekt aufbauend, betrachten wir im folgenden die Auflösung eines Radarbildes nicht nur aus der

Sicht der Signaltheorie, die die lokale Schärfe der Impulsantwort bewertet, sondern den durch ein regionales Maximum repräsentierten Interferenzbereich, und fassen ihn in einem neuen Meßwert, dem 'Superpixel', zusammen. Dabei wird die Impulsantwort des Radars zur Modellierung dieses Speckle-Bereiches herangezogen.

Modellierung des Superpixels

Die Definition der Superpixel basiert auf folgender Modellvorstellung: Ein Bereich konstanter Rückstreuung wird als ein Feld von 'Punktstrahlern' gleicher Intensität aufgefaßt. Da ein Punktstrahler eine Approximation eines Dirac-Impulses darstellt, erhält man nach /1/ als Abbild dessen Fouriertransformierte mit dem Funktionskern einer sin(x)/x-Funktion. Diese zeichnet sich durch einen stark ausgeprägten Haupt- und rasch abfallende Nebengipfel aus. Bei der Überlagerung der Abbilder dicht benachbarter Punktstrahler gleicher Intensität bleiben die Hauptgipfel, wenn auch leicht modifiziert, erhalten.

In Abb. 2b ist der Bereich konstanter Rückstreuung aus Abb. 2a in lokal variante Punktstrahler gemäß der Modellvorstellung zerlegt worden. Die isolierte Betrachtung eines Punktstrahlers ergäbe das in Abb. 2d skizzierte Bild der gemessenen Energierückstreuung. Bei realem Bildmaterial ist aufgrund unterschiedlicher Breiten und Dichten der Punktstrahler ein Verlauf gemäß Abb. 2c zu erwarten.

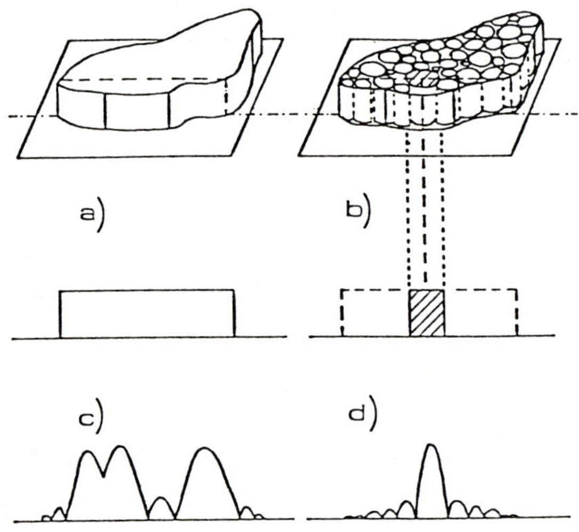

Abb. 2: Modell des Superpixels

Ermittlung des Bereiches eines Superpixels

Entsprechend obiger Modellvorstellung umfaßt ein Superpixel gerade den Hauptgipfel der Funktion sin(x)/x, sodaß das erste Minimum ausgehend vom zugehörigen Maximalwert das Superpixel abgrenzt. Bedingt durch Bildstörungen ist eine genaue Positionierung der Minima durch eine Funktionsverlaufsanalyse in digitalem Datenmaterial nur schwer möglich.

Deshalb erfolgt die Ermittlung dieser Abgrenzung mit Hilfe eines Wachstumsprozesses. Dieser generiert parallel, ausgehend vom jeweiligen regionalen Maximum, Niveaulinien, welche der jeweiligen Signalform angepaßt sind. Der Wachstumsprozess wird so lange fortgesetzt, bis ein Gleichgewicht gleichzeitig startender Niveaulinien benachbarter Maxima erreicht ist.

In /2/ werden mit Methoden der Abstandstransformation in jedem Bildpunkt der geometrische Abstand zum nächstgelegenen ausgezeichneten Bildpunkt (Objekt) ermittelt und ausgewertet. Wählt man als ausgezeichnete Punkte die regionalen Maxima und ersetzt den geometrischen durch den radiometrischen Abstand, so löst diese sehr effektive Methode gerade obige Aufgabenstellung.

Eindimensionale Schätzung der Intensität eines Superpixels

Zur Bestimmung der Intensität der Rückstreuung eines Superpixels bestünde ein erster Ansatz darin, für jeden Bildpunkt die relative Position zum jeweiligen Maximum zu bestimmen, daraus den Zahlenwert der Intensität abzuleiten, und zur Fehlerminimierung diese Resultate zu kumulieren. Da jedoch die Abbildungsfunktion nicht linear ist, kann ein korrekter Schätzwert wegen der unsicheren Funktionsstützpunkte nicht erwartet werden.

Der gewählte Ansatz basiert stattdessen auf der Integration der Bildfunktion. Dabei betrachten wir die eindimensionale Verteilung, welche sich bei Schnittbildung ab Hauptgipfelmaximum in einer beliebigen Ausdehnungsrichtung ergibt. Nach der Modellvorstellung stellt diese das Abbild eines Impulses der Intensität h und der Dauer T dar, welche als Fouriertransformierte gemäß /3/ zu folgendem Abbildungsresultat führt:

```
(1)          i(f) = h * T * sinc ( pi*T*f)
```

Die Impulsdauer T ergibt sich im Bild als reziproker Wert des Abstandes des betrachteten Randpunktes zum Maximalwert und f kennzeichnet die Ausdehnung in Schnittrichtung. Das zugehörige bestimmte Integral int(i(f)) vom Maximalwert bis zum ersten Nulldurchgang ergibt sich zu:

```
(2)          int(i(f)) = 0.587 * h
```

Dieses Integral entspricht einer normierten Summation der ursächlichen Intensitätswerte:

```
(3)          int(i(f)) = sum / n
```

Die benötigte Summe und die Anzahl n der Summationen wurden bereits bei der Bereichsermittlung bestimmt. Somit kann die Intensität h der Rückstrahlung aus folgender Beziehung ermittelt werden:

```
(4)          h = sum / (0.587 * n)
```

Da hierbei lediglich von der Position des regionalen Maximums und des Randes sowie den Intensitätswerten Gebrauch gemacht wird, ergibt sich hierdurch eine wesentlich sicherere Schätzung als beim zuerst vorgestellten Ansatz.

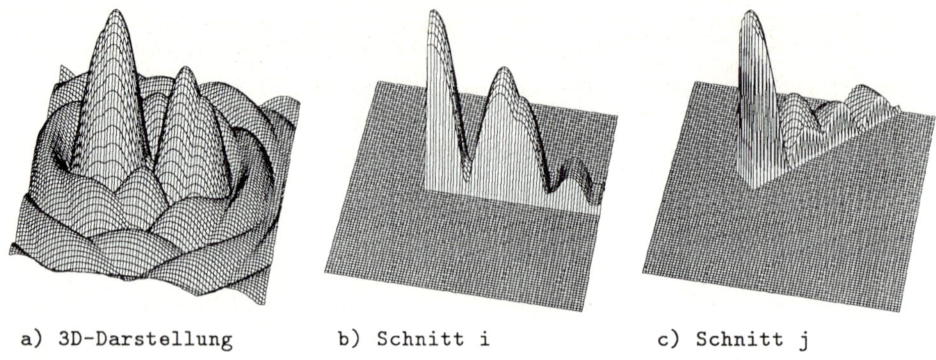

Abb. 3: Schnittbildung bei der zweidimensionalen Schätzung

Zweidimensionale Schätzung des Superpixels

Obige Berechnung der Intensität der Rückstrahlung beruht auf einem beliebigen Schnitt durch die in Abb. 3a dargestellte Intensitätsverteilung in der Umgebung eines Superpixels. Dabei wird die Integration ausgehend vom regionalen Maximum des Superpixels bis zum ersten Minimum durchgeführt, z. B. in Schnittrichtung i wie in Abb 3b. Um die vollständige zweidimensionale Bildinformation zu verwerten, werden, wie in Abb. 3c angedeutet, durch Rotation alle möglichen Schnitte berücksichtigt. Da gemäß unserer Modellvorstellung ein Superpixel von einem Punktstrahler erzeugt wird, legt ihr Mittelwert die endgültige Schätzung der Intensität der Rückstrahlung fest.

Diskussion der Ergebnisse

In Abb. 4a ist der laut Modell geschätzte ursächliche Remissionswert als konstante Grauwertintensität über den Einzugsbereich des jeweiligen Superpixels wiedergegeben. Man erkennt, daß gegenüber dem Ausgangsbild in Abb. 1a in Bereichen konstanter Rückstreuung die Kohärenzeffekte deutlich abgeschwächt sind. Dies geht nicht zu Lasten des Kontrastes zwischen diesen Bereichen. Die vorgestellte Schätzung der Rückstreueigenschaften kann somit auch als modellgesteuerter lokal adaptiver Filter mit wiederholter Meßwertstapelung aufgefaßt werden.

Das Ergebnis einer Kontrastanalyse der Superpixel, welche nicht mehr in der ursprünglichen Bildmatrix, sondern in den Nachbarschaften sich berührender Superpixel stattfindet, ist in Abb. 4b dargestellt. Eine Homogenitätsprüfung des Ensembles zusammenhängender Superpixel hat zwar hier noch nicht stattgefunden, aber man erkennt die sinnvollen Randabgrenzungen zu den Bereichen variabler Rückstreuung.

Es ist nicht zu erwarten, daß die Modellannahme in allen Bildbereichen zutrifft. Eine Überprüfung des Modells in jedem Superpixelbereich erlaubt die Detektion inkonsistenter Superpi-

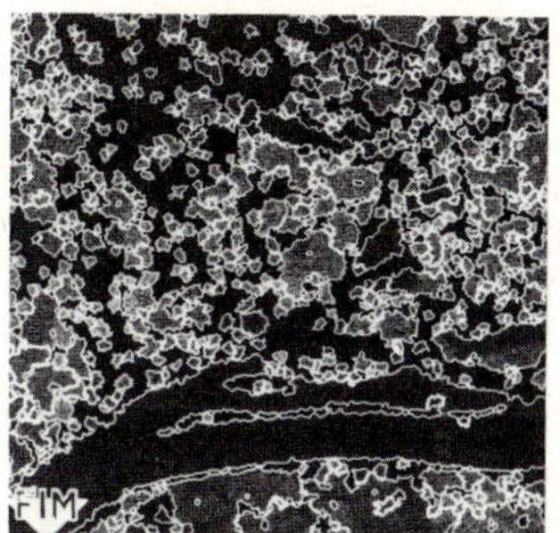

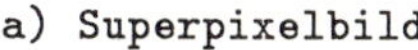
a) Superpixelbild

b) 1. Segmentationsergebnisse

Abb. 4: Ergebnisse

xel. Dazu wird die Verteilung anhand der geschätzten Modellparameter errechnet und mit der tatsächlichen Intensitätsverteilung verglichen. Inkonsistenzen können anschliessend durch erweiterte Modellannahmen näher untersucht werden oder durch Berücksichtigung weiterer Radar-Eigenschaften (s. /4/) vermieden werden.

Zusammenfassung

Ein effektives und schnelles Verfahren zur Unterdrückung des Kohärenzeffektes wurde vorgestellt und erste Anwendungen gezeigt. Weitere Einsatzmöglichkeiten dieses Integrationsfilters werden untersucht.

Literatur

/1/ H. Kazmierczak
Erfassung u. maschinelle Verarbeitung von Bilddaten
Springer-Verlag 1980

/2/ E. Mauer , R. Schärf
New Applications of Distance Transformation Methods for
Effictive Structural Image Analysis
8. ICPR 86 Paris

/3/ H. Wolf
Nachrichtenübertragung
Springer Verlag 1974

/4/ K. Behrens, E. Mauer
Automatische Detektion von Bereichen schwacher
Energierückstreuung in Radar - Abbildungen
6. ASST-Symposium Aachen 1987

Bildcodierung

IRRELEVANZREDUKTION VON BILDINFORMATION DURCH UNSCHARFFILTERUNG

Ferdinand Arp
Fachbereich Elektrotechnik, Universität GH Wuppertal
Postfach 100127, D-5600 Wuppertal 1

Es ist wohlbekannt, daß das visuelle System des Menschen deutliche Mängel hat. Außerdem wird die vom Auge aufgenommene und zum Sehzentrum im Gehirn übertragene Informationsmenge nochmals erheblich reduziert. Trotzdem hat der menschliche Beobachter den subjektiven Eindruck der fehlerlosen Wahrnehmung einer von ihm beobachteten Szene. Eine der Ursachen dieser Informationsreduktion ist die unscharf verschmierte Abbildung der beobachteten Szene auf der Netzhaut des Auges. Sie wird verursacht von mehreren Unzulänglichkeiten des Augenapparates. Die subjektiv trotzdem als perfekt empfundene Wahrnehmung infolge nachträglicher Detailakzentuierung wird durch die sogenannte laterale Inhibition im neuronalen System bewirkt, die sowohl örtliche als auch zeitlich wirksame Korrekturen der Fehler vollzieht.

Die mit der Unscharfabbildung im menschlichen Auge verbundene Irrelevanzreduktion wird auf ihre entsprechende Eignung zur Datenreduktion von nachrichtentechnisch übertragenen oder gespeicherten Bildern in Verbindung mit der wohlbekannten DPCM (Differenz-Puls-Code-Modulation) untersucht. Das Prinzip der DPCM-Kodierung ist in Bild 1 dargestellt. Vom Eingangssignal s_i wird sein Schätzsignal $\hat{s}_i$ subtrahiert, diese Differenz wird anschließend quantisiert und zum Ausgang Δs_i geführt. Durch Addition des Differenzsignals Δs_i zum Schätzsignal $\hat{s}_i$ wird die mit dem Quantisierungsfehler behaftete Rekonstruktion s_i erzeugt. Der Prädiktor wiederum erzeugt das Schätzsignal $\hat{s}_i$ aus dem Vergangenheitsverlauf des rekonstruierten Signals. Dabei wird dieses Schätzsignal infolge Quantisierung des Differenzsignals verfälscht und allgemein sowohl im DPCM-Koder Bild 1 als auch im DPCM-Dekoder Bild 2 mit Rückführung des Quantisierungsfehlers identisch reproduziert.

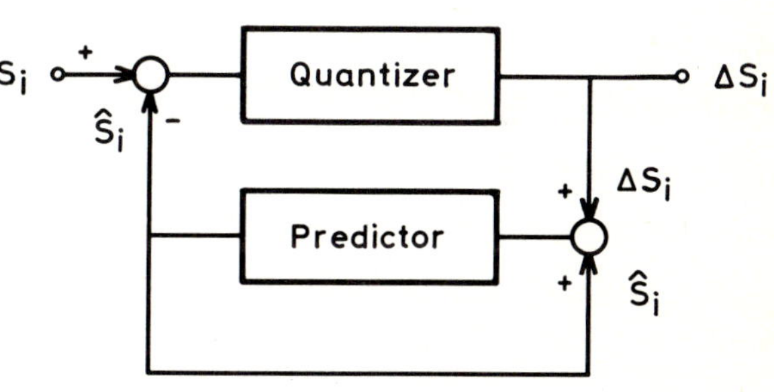

Bild 1. DPCM-Koder

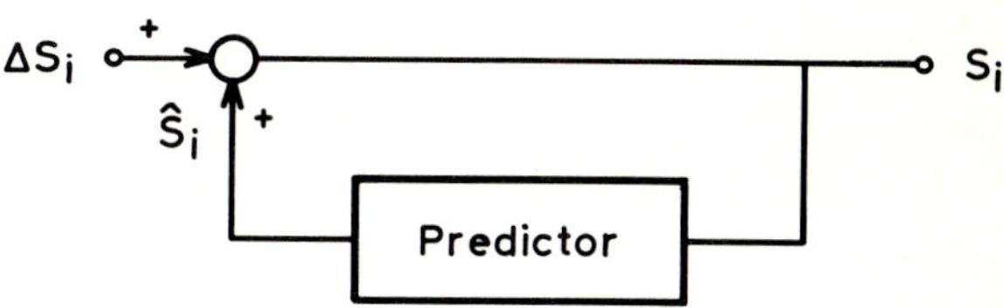

Bild 2. DPCM-Dekoder

In der neuen Strategie zur Irrelevanzreduktion wird dem DPCM-Koder das zuvor unscharfgefilterte Bildsignal statt des

Originalsignals zugeführt. Diese Filterung erfolgt in der Äquivalenz einer örtlich zweidimensionalen symmetrischen Unscharfblende. Das dem elektrischen Bildsignal äquivalent zugeordnete zeitdiskrete Unscharffilter ist nichtkausal und hat eine zeitsymmetrische Impulsantwort $H(n\tau) = H(-n\tau)$. Sein Übertragungsfaktor hat daher die Phase $\varphi(\omega) = 0$. Das Gesamtkonzept realisiert einerseits das zu diesem nichtkausalen Filterprozeß gehörende Inversfilter und berücksichtigt andererseits den Einfluß des Quantisierungsfehlers auf das aus dem DPCM-Signal rekonstruierte Unscharfsignal, indem es auf den erzeugenden Unscharffilterprozeß auf der Koderseite wieder zugeführt wird.

Zur Realisation der Unscharffilterung wird zunächst der Übertragungsfaktor $h(z)$ der Unscharfblende in der Äquivalenz des elektrischen Bildsignals unter der Bedingung

$$h(z) > 0 \quad , \quad |z| = 1 \ , \qquad (1)$$

gewählt und durch spektrale Faktorisierung

$$u(z)\ u(z^{-1}) = h(z) \qquad (2)$$

in den Übertragungsfaktor $u(z)$ der kausalen Impulsantwort $\{U(n\tau)\}$ und den Übertragungsfaktor $u(z^{-1})$ der zeitlich gegenläufig kausalen Impulsantwort $\{U(-n\tau)\}$ zerlegt. Das kausale Teilfilter wird als rekursives Filter Bild 3 entworfen, das den Übertragungsfaktor

$$u(z) = K\left(1+a(z)\right)^{-1} \qquad (3)$$

hat. Der Rückführungszweig im Teilfilter Bild 3 hat den Übertragungsfaktor

$$a(z) = \sum_{n=1}^{N} A_n\ z^{-n} \qquad (4)$$

und ist als Transversalfilter Bild 4 realisiert. Der Faktor K wird zu

$$K = u(1) = 1 + \sum_{n=1}^{N} A_n \qquad (5)$$

gewählt, so daß das Teilfilter mit $u(1) = 1$ ein gleichspannungsinvariantes und stabiles Allpolfilter ist. Das zu dem Teilfilter Bild 3 inverse Filter mit dem Übertragungsfaktor

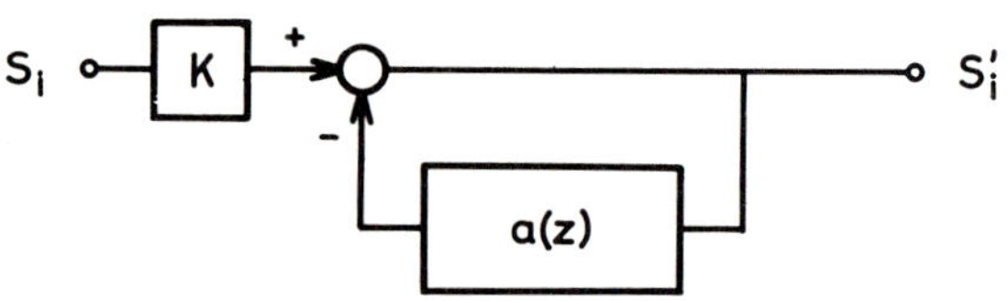

Bild 3. Kausales rekursives Teilfilter mit dem Übertragungsfaktor $u(z)$ Gl. (3).

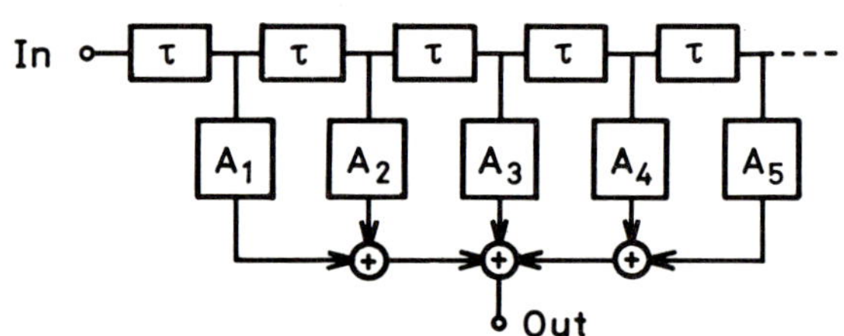

Bild 4. Blockschaltbild für den Rückführungszweig mit dem Übertragungsfaktor $a(z)$ Gl. (4).

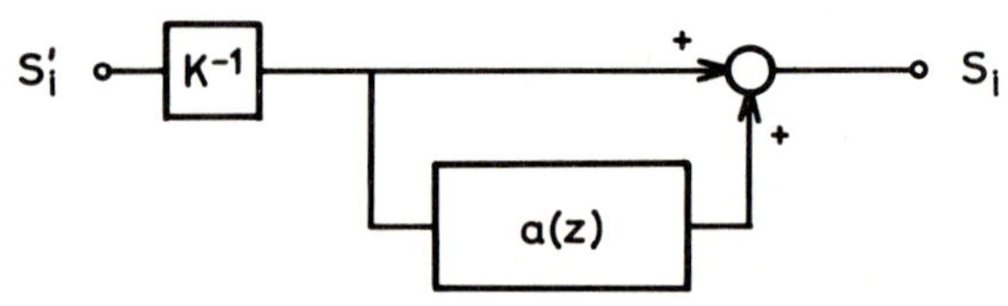

Bild 5. Kausales nichtrekursives Teilfilter mit dem inversen Übertragungsfaktor $1/u(z)$ Gl. (6).

$$u^{-1}(z) = K^{-1}\left[1+a(z)\right] \qquad (6)$$

ist in Bild 5 dargestellt. Es ist nichtrekursiv. Der Nebenzweig in dem Inversfilter hat den gleichen Übertragungsfaktor $a(z)$ Gl. (4) und die identischen Koeffizientenbewertungen A_i wie das rekursive Teilfilter Bild 3.

Die Unscharffilterung mit dem Übertragungsfaktor $h(z)$ wird nun entsprechend in zwei Schritten vollzogen. Sie sind in Bild 6 dargestellt. Zunächst wird der blockweise gespeicherte Bildinhalt s_i mit der zeitlich gegenläufigen kausalen Teilblende, die die Impulsantwort $\{U(-n\tau)\}$ mit dem Übertragungsfaktor $u(z^{-1})$ hat, in zeitlich rücklaufender Richtung zum Zwischensignal s_i' gefiltert und wieder gespeichert. Dieses Zwischensignal s_i' wird anschließend mit der zeitlich normalläufig kausalen Teilblende, die die Impulsantwort $\{U(n\tau)\}$ und den Übertragungsfaktor $u(z)$ hat, zum Unscharfsignal s_i'' gefiltert. Unter der Voraussetzung Gl. (1) der nichtnegativen Definitheit des Übertragungsfaktors $h(z)$ ist dieser Filterprozeß stabil. In der in Bild 6 dargestellten rekursiven Anordnung mit Quantisierer Q wird der Filterprozeß aus dem *quantisierten Ausgangssignal* s_{iq}'' gespeist, so daß die Quantisierungsfehler vergangener Signalwerte s_{iq}'' zurückgeführt und bei der Konstruktion jedes neuen Ausgangssignalwerts s_{iq}'' berücksichtigt werden.

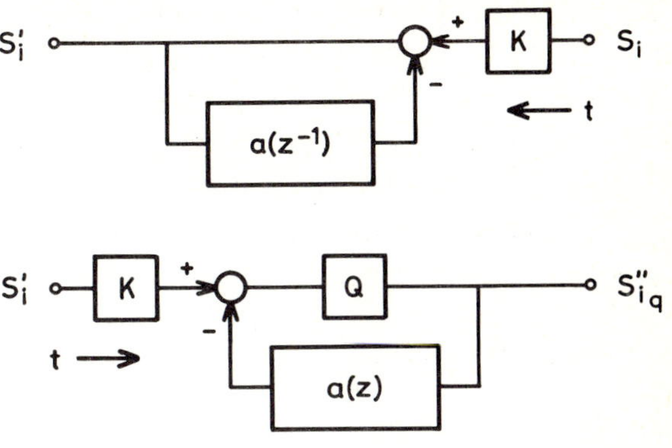

Bild 6. Schematische Darstellung der Unscharffilterung des Signals s zunächst zum Zwischensignal s' in zeitlich rücklaufender Richtung und anschließend zum Unscharfsignal s'' in zeitlich normallaufender Richtung.

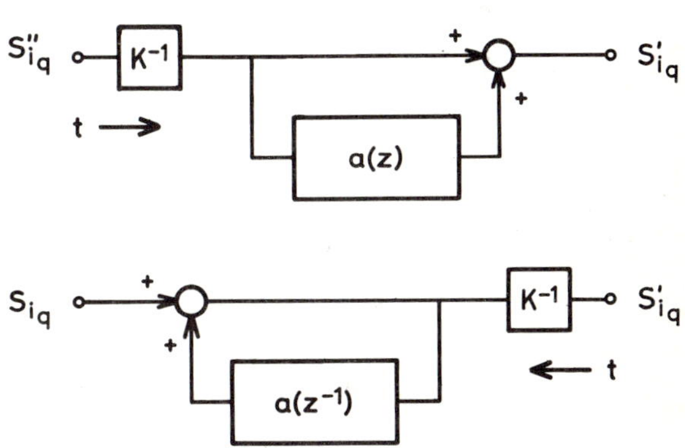

Bild 7. Schematische Darstellung der Inversfilterung des Unscharfsignals s'' zunächst zum Zwischensignal s' in zeitlich normallaufender Richtung und anschließend zum Rekonstruktionssignal s, das den Quantisierungsfehler enthält, in zeitlich rücklaufender Richtung.

Die Inversfilterung mit dem effektiven Übertragungsfaktor $h^{-1}(z)$ vollzieht sich gemäß Bild 7 ebenfalls in zwei Schritten. Das Unscharfsignal s_i'' wird zeitlich normallaufend zum Zwischensignal s_{iq}' und anschließend zeitlich gegenläufig zum Signal s_{iq} zurückgefiltert, das die Rekonstruktion auf der Empfangsseite darstellt.

Die beschriebene Anordnung wurde in einem ersten Versuch mit linear gestuftem Quantisierer Q im Rückführungszweig des zweiten Teilfilters der Unscharffilterung in

Bild 6 erfolgreich getestet. Die beschriebene Anordnung stellt ohne Quantisierung des Unscharfsignals eine lineare Filterung und anschließend die lineare und numerisch streng korrekte Inversfilterung dar. In der Praxis treten naturgemäß Rundungsfehler auf, deren Einfluß durch die Genauigkeit der gewählten Arithmetik für die internen Filteroperationen unter jede gewünschte Schranke gebracht werden kann. Die Fehler des rekonstruierten Signals s_{iq} hängen wesentlich nur von dem Quantisierer Q in Bild 6 ab, der unter dem Einfluß der Unscharffilterung den Detailgehalt des Originalbilds effektiv gröber als die visuell wichtigen großflächigen Strukturen quantisiert.

Die Kombination der Unscharffilterung mit dem DPCM-Koder ist in Bild 8 dargestellt. Diese Anordnung ersetzt das zweite Teilfilter mit Quantisierer in Bild 6. Der DPCM-Koder ist im oberen rechten Teil des Teilfilters mit seinem linearen Schätzfilter eingefügt, das den Übertragungsfaktor $p(z)$ hat. Außerdem wird der lokale Mittelwert $\overline{s''}$ des Unscharfsignals vor der DPCM-Kodierung vom Unscharfsignal s''_i subtrahiert und nach der DPCM-Dekodierung wieder addiert. In der dargestellten Weise werden die Quantisierungsfehler des rekonstruierten Unscharfsignals s''_{iq} aus der Vergangenheit zurückgeführt und im Filterprozeß berücksichtigt. Die Inversion der Differenzwerte $\Delta s''_i$ am Rekonstruktionsort vollzieht sich in drei Schritten. Zunächst wird das Unscharfsignal s''_{iq} unter Berücksichtigung des lokalen Mittelwerts $\overline{s''}$ mit dem DPCM-Dekoder nach Bild 2 rekonstruiert. Anschließend folgt die Inversfilterung in zwei Schritten nach Bild 7 zum rekonstruierten Signal s_{iq}.

Die dargestellte Anordnung zur äquivalent zweidimensionalen örtlichen Kodierung und Dekodierung wird in Rechnersimulation getestet. Die zur Zeit erreichte Entropie des Differenzsignals $\Delta s''_{iq}$ liegt in der Größenordnung von einem Bit je Bildpunkt, ohne daß visuell störende Fehler im rekonstruierten Bild auftreten.

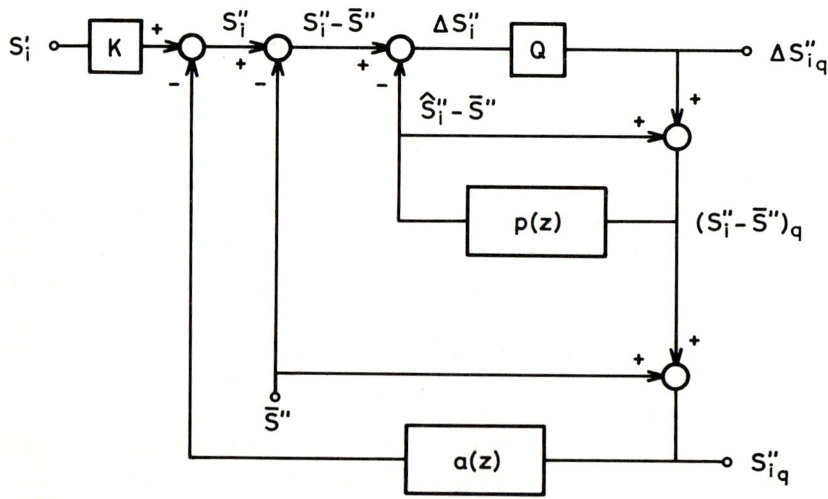

Bild 8. Kombination der zweiten Teilfilterung in Bild 6 mit DPCM-Koder Bild 1.

Auf Schätzung basierende Bufferüberwachung für einen echtzeitfähigen Transformationskoder

Yonggang Du

Institut für elektrische Nachrichtentechnik
RWTH Aachen, 5100 Aachen, Melatener Str. 23

Der gesamte Vektorraum der N×N-dimensionalen Bildblöcke wird bei einem adaptiven zonalen Transformationskoder durch eine Klassifikation in M Teilräume zerlegt. Um eine homogene Bildqualität zu erzielen, müssen die kritischeren Teilräume mit höherer Bitzahl quantisiert werden. Die Datenrate eines solchen Kodierungssystems ist deshalb i. a. von Klasse zu Klasse verschieden. Um die kodierten Videosignale trotzdem über einen synchronen Kanal in Echtzeit übertragen zu können, ist der Einsatz eines Buffers zum Datenratenausgleich unentbehrlich.

Seien R die Kanalkapazität, mit der der Buffer ausgelesen wird, und G die Buffergröße. Eine Kodierung verursacht einen Over- bzw. Underflow des Buffers, wenn der Bufferpegel P mindestens einmal die Ungleichung $P > G$ (Overflow) bzw. $P < 0$ (Underflow) erfüllt. Um Over- bzw. Underflow zu vermeiden, wird oft die Kodierungsauflösung der Klasse k durch den Bufferzustand s gesteuert. Werden die Ersatzwerte mit fester Codewortlänge b(k,s) kodiert, ergibt sich die mittlere Datenrate R(s) in einem stationären Bufferzustand s aus

$$R(s) = \frac{1}{N^2} \sum_{k=1}^{M} h(k) \cdot b(k,s) \ . \tag{1}$$

Dabei stellen h(k) die relativen Klassenhäufigkeiten dar.

Die Art und Weise, wie die Kodierung in Abhängigkeit vom Bufferzustand gesteuert wird, hat eine große Auswirkung auf die von Augen wahrgenommene Bildqualität. Weil der Gesamteindruck eines kodierten Bildes schon durch eine einzige zu schlecht rekonstruierte Bildstelle beeinträchtigt werden könnte, bewirkt eine gleichmäßig verteilte Bildverfälschung oft eine geringere Sinneswahrnehmung als eine impulsartige Störung bei gleichgebliebenem MQF (mittleren quadratischen Fehler). Unter diesem Aspekt sollte eine gute Bufferüberwachung in der Lage sein, die Koderparameter beim Bildwechsel so an die Bildstatistik anzupassen, daß die Kodierung jedes kompletten Bildes immer in einem fixierten optimalen Bufferzustand erfolgen könnte. Solche Überlegung geht aus den bisher aus der Literatur bekannten Verfahren [1,2,3] leider nicht genügend deutlich hervor.

Ziel dieser Arbeit ist es, ein neues Verfahren zur Bufferüberwachung vorzuschlagen, das die subjektive Auswirkung der Bufferzustandsumschaltung mit berücksichtigt. Es soll die Anzahl der Umschaltung innerhalb der Kodierung eines kompletten Bildes minimiert werden, jedoch ohne auf die maximale Ausnutzung der Kanalkapazität zu verzichten. Um dies zu ermöglichen, werden die ***kumulativen*** Klassenhäufigkeiten H(k,i) für L äquidistante Prüfstellen i gemessen und abgespeichert, i=1,...,L. Sei J die Anzahl der Pixel zwischen zwei benachbarten Prüfstellen. Wenn ein Bild in einem festen Bufferzustand s kodiert wird, läßt sich der Bufferpegel in Prüfstelle i aus den kumulativen Klassenhäufigkeiten $H_v(k,i)$ des diesem Bild vorausgegangenen Bildes schätzen, und zwar mit

$$P_e(i,s) = \sum_{k=1}^{M} H_v(k,i) \cdot b(k,s) - R \cdot J \cdot i + \frac{G}{2} . \tag{2}$$

In Gl.(2) bedeutet G/2 den Pegelbezugspunkt. Der '*optimale*' Bufferzustand ist als derjenige Zustand s_0 definiert, bei dem der geschätzte Pegel $P_e(i,s_0)$ maximiert wird, jedoch in keiner einzigen Stelle i die reduzierte Buffergröße G-η überschreitet. Dabei ist η ein genügend groß gewählter Sicherheitsabstand. Ferner muß der geschätzte Pegel nach der Kodierung des letzten Blocks eines Bildes mindestens um η unter dem Bezugspegel G/2 liegen, da in diesem Verfahren der Datenratenausgleich immer innerhalb eines jeden Bildes erfolgen soll. Seien Q Bufferzustände definiert, s=1,...,Q. Mit der Vereinbarung R(s) > R(s+1) heißen die obigen zwei Bedingungen auch, daß ein *minimales* s_0 gesucht wird, das die folgenden zwei Ungleichungen

$$\max_{1 \le i \le L} P_e(i,s_0) < G - \eta , \tag{3}$$

$$P_e(L,s_0) < \frac{G}{2} - \eta \tag{4}$$

erfüllt.

Weil bei einem schnellen Bildszenenwechsel die Bildstatistik zwischen zwei aufeinanderfolgenden Bildern doch einen starken Unterschied aufweisen kann, muß die Statistikänderung während der Kodierung detektiert werden. Die Kodierung jedes Bildes beginnt deshalb mit s_0 als dem Anfangszustand des Buffers. Es wird also $s(1) = s_0$ gesetzt. Weicht der tatsächliche Bufferpegel $P_a(i)$ in i-ter Prüfstelle zu stark von $P_e(i,s_0)$ ab, wird der aktuelle Zustand s(i) umgeschaltet. Inwieweit s(i) umgeschaltet werden soll, hängt sowohl vom geschätzten Pegelverlauf $P_e(i,s_0)$ als auch von s(i) selbst ab. Die maximal erlaubte Abweichung A von der Kurve $P_e(i,s_0)$ mit i=1,...,L wird begrenzt durch den Abstand

$$D_1 = G - \max_{1 \le i \le L} P_e(i,s_0) \tag{5}$$

vom größten Schätzpegel zum Bufferanschlag und den Abstand

$$D_2 = \frac{G}{2} - P_e(L, s_0) \tag{6}$$

vom geschätzen Endpegel zum Pegelbezugspunkt. Es muß daher gelten

$$A = \min\{D_1,\ D_2\}\ . \tag{7}$$

Um die Gefahr von Over- bzw. Underflow frühzeitig zu detektieren, wird die maximal erlaubte Abweichung A in Abhängigkeit vom Bufferzustand s(i) nach

$$W_o = \frac{A}{\max\{1,\ Q-s\}} \quad bzw. \quad W_u = \frac{A}{\max\{1,\ s-1\}} \tag{8}$$

verkleinert. Es tritt in Prüfstelle i eine Overflow-Gefahr ein, falls die Pegelabweichung

$$\delta(i) = P_a(i) - P_e(i, s_0) \tag{9}$$

die Schwelle W_o überschreitet. Der Bufferzustand s(i) wird dann um t erhöht, falls W_o von δ t-fach überschritten wird:

$$s(i+1) = \min\{s(i)+t,\ Q\} \quad mit \quad t = [\frac{\delta(i)}{W_o} + 0,5] \tag{10}$$

In Gl.(10) bedeutet [x] den ganzzahligen Teil von x. Für den Fall einer Underflow-Gefahr gilt

$$s(i+1) = \max\{s(i)-t,\ 1\} \quad mit \quad t = [-\frac{\delta(i)}{W_u} + 0.5]\ . \tag{11}$$

Für einen 8×8-DCT-Koder mit 15 Klassen und 4 Bufferzuständen wurde der oben beschriebene Umschaltungsalgorithmus getestet. Die Prüfstellen i wurden am Ende jeder 8-ten Bildzeile gesetzt. Die Kanalkapazität R betrug 1,1 bit/pel. Die Bitanzahlen b(k,s) mit k=1,...,15 in jedem Bufferzustand s wurden so festgelegt, daß einerseits die Bildqualität aller Klassen in jedem festen s fast gleich ist, andererseits aber die mit Gl.(1) über 23 reale Bilder gemittelte Datenrate R(s) genau einen vorgegebenen Wert beträgt. Die gewählten Datenraten waren 1,0, 0,9, 0,8 bzw. 0,7 bit/pel für s=1,...,4. Zwei ganz extreme Bildkombinationen wurden untersucht. Im ersten Versuch wurde das sehr unkritische Porträtbild 'Jeanin', charakterisiert durch einen großflächigen, ruhigen Hintergrund, als Referenzbild zur Statistikschätzung herangezogen, während das sehr kritische Affenbild 'Baboon', charakterisiert durch viele hochfrequenten Haare, kodiert wurde. Im zweiten Versuch wurden die Rollen vertauscht. 'Baboon' wurde als das Referenzbild und 'Jeanin' als das zu kodierende Bild benutzt. In Abb. 1 ist die Umschaltung des Bufferzustands s und ihre Auswirkung auf die Datenrate dargestellt. Es ist zu beobachten, daß trotz der Fehlschätzung für s_0 sich s(i) immer Schritt für Schritt einem stationären Bufferzustand annähert. Es tritt keine Oszillation oder impulsartige Änderung der Bildqualität auf.

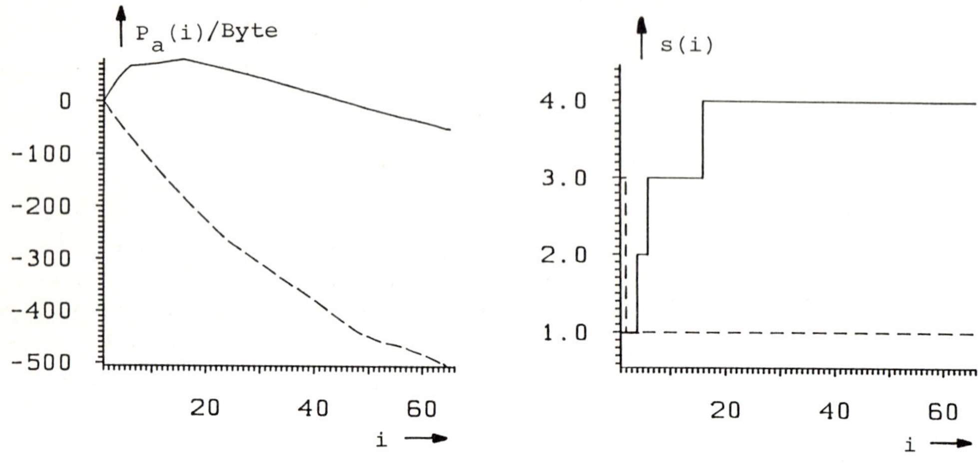

Abb. 1 Das dynamische Verhalten des Umschaltungsalgorithmus bei zwei kritischen Bildkombinationen
——— 'Jeanin' als Referenz
— — — 'Baboon' als Referenz

Bei den obigen Simulationen wurde der Sicherheitsabstand auf $\eta = 0$ und die Buffergröße auf $G = \infty$ gesetzt. In einer anderen Simulation wurde aber auch festgestellt, daß bei jeder beliebigen Bildkombination aus den 23 verschiedenen Bildvorlagen ein Sicherheitsabstand von $\eta = 24$ Byte ausreicht, um den Bufferpegel $P_a(L)$ nach der Kodierung eines Bildes wieder unter den Bezugspegel G/2 zu bringen. Es wurde ferner gezeigt, daß ab G > 92 Byte kein Overflow und ab G > 540 Byte auch kein Underflow mehr auftritt.

Literatur:

[1] Farvardin, N.; Modestino, J.W.: Adaptive Buffer-Instrumented Entropy Quantizer Performance for Memoryless Sources.
IEEE Trans. Inform. Thoery, vol. IT-32, pp. 9-22, Jan. 1986

[2] Chen, W.H.; Pratt, W.K.: Scene Adaptive Coder
IEEE Trans. Commun., vol. COM-32, pp. 225-232, March 1984

[3] Pratt, W.K.: Image Transmission Techniques.
Academic Press, New York, San Francsico, London, pp. 126-151, 1979

Ein regionenorientiertes Bildkodierungskonzept mit sehr hoher Datenreduktion

Uwe Franke und **Rudolf Mester**
Institut für Elektrische Nachrichtentechnik
RWTH Aachen, 5100 Aachen, Melatener Str. 23

Übersicht

Im vorliegenden Beitrag wird ein neues Bildkodierungskonzept vorgestellt, das sowohl den Belangen der Bildkodierung als auch einer objektorientierten Bildverarbeitung Rechnung trägt. Nach einer Zerlegung des Bildes in homogen texturierte Regionen werden Form und Inhalt dieser Segmente getrennt voneinander beschrieben, wobei die Parameter entsprechend ihrer Relevanz hierarchisch geordnet vorliegen. Eine Bildrekonstruktion auf der Basis der wichtigsten Informationen resultiert in einer Rekonstruktion des Bildes mit sehr hoher Datenreduktion, ohne daß dabei die wesentlichen Bildinhalte (die 'messages') verloren gehen.

1. Einleitung

Die etablierten Bildkodierungsalgorithmen wie Transformationskodierung und Vektorquantisierung stoßen aufgrund ihrer blockweisen Arbeitsweise, die den Inhalt des Bildes weitestgehend ignoriert, schnell an ihre Grenzen. Darüberhinaus sind sie als Vorverarbeitung für die zunehmend wichtiger werdende *objektorientierte* Bildverarbeitung nicht geeignet.

Der Wunsch nach ökonomischer Speicherplatzausnutzung wie auch schnellem Bildzugriff (man denke z.B. an das Suchen bestimmter Objekte in Bilddatenbanken) macht daher alternative datenkomprimierende Kodierungsstrategien erstrebenswert.

Ansätze mit dem Ziel extremer Kompressionsfaktoren sind bei Kunt [1] zu finden. Bei dem dort beschriebenen Verfahren wird das Bild mittels eines Region-Growing Algorithmus in einzelne Regionen mit geringen Grauwertschwankungen zerlegt, die anschließend durch Polynome nullter bis zweiter Ordnung approximiert werden. Ein neueres Verfahren segmentiert die Bildvorlage bereits so, daß die entstehenden Regionen mit minimalem Fehler durch diese Polynome beschrieben werden können [2]. Mit diesen Techniken lassen sich hohe Datenkompressionsfaktoren erzielen; sie sind jedoch nicht in der Lage, feine Details wie z.B. Haare, Gras und ähnliche Texturen effizient zu beschreiben, da solche Signale von der Approximation als 'Rauschen' interpretiert werden.

Im folgenden soll ein Bildkodierungskonzept vorgestellt werden, das diese Nachteile nicht aufweist und dabei neben den Aspekten der Bildkodierung auch die der Bildverarbeitung in Betracht zieht.

2. Das Konzept

An eine solche flexible Bildbeschreibung sind - unabhängig von den gewählten Algorithmen - die nachstehend aufgeführten Forderungen zu stellen (vgl. [3]):

- Die notwendigerweise regionenorientierte Beschreibung der Bildvorlagen sollte möglichst *kompakt* (zur Erzielung hoher Datenkompressionsfaktoren bei guter Bildqualität) und *vollständig* sein, letzteres insbesondere dann, wenn zum Zeitpunkt der Kodierung nicht bekannt ist, wie genau das Bild rekonstruiert bzw. welche Bildverarbeitungsalgorithmen auf es angewandt werden sollen.
- Die Beschreibung muß *explizit* (wichtige Informationen müssen einfach zugreifbar sein) und *hierarchisch strukturiert* sein, wobei die höchste Hierarchieebene die wichtigsten Merkmale der Bildvorlage enthalten und damit eine grobe Beschreibung des Bildes darstellen soll.
- Innerhalb dieser Ebenen sollte sie einem *Linearitätsprinzip* gehorchen, um die Möglichkeit eines progressiven Bildaufbaus zu bieten.
- Kontextunabhängige Suchoperationen verlangen nach einer *kontextfreien* Beschreibung, die robust gegenüber geringen Veränderungen der Objekte ist.
- Nicht vergessen werden darf die Tatsache, daß die Beschreibung eine effiziente *Interaktion* mit den kodierten Bildern ermöglichen sollte.

Keines der in der Literatur beschriebenen Verfahren ist in der Lage, die oben skizzierten Forderungen zu erfüllen. Das Kernproblem stellt dabei die kompakte Beschreibung der Texturen dar.

Im folgenden werden Algorithmen aufgezeigt, die eine den erhobenen Forderungen gerecht werdende Bildcodierung ermöglichen. Hierzu wird die Bildvorlage im ersten Schritt in *homogen texturierte* Regionen zerlegt. Diese Art der Bildzerlegung unterstützt nicht nur die Aufgaben aus dem Bereich des 'computer vision', sondern ist auch in besonderem Maße der menschlichen Wahrnehmung von Bildern angepaßt. Es ist offensichtlich, daß die Zerlegung des Bildes in stationäre Signale der Kodierung ebenfalls zugute kommt. In dem vorgestellten Ansatz wird das in [4] beschriebene Segmentierungsverfahren verwendet.

Entsprechend dem eingangs zusammengestellten Anforderungskatalog werden die Formen der resultierenden Regionen separat und von ihren Inhalten (Texturen) getrennt beschrieben.

3. Beschreibung der Konturen

Für die Beschreibung der Konturen haben wir eine hierarchisch organisierte polygonale Konturapproximation gewählt. Sie ermöglicht eine Beschreibung der Konturen mit wachsender Auflösung und verhält sich robust gegenüber verrauschten Konturen und kleinen Deformationen.

Nach Bestimmung der größten Ausdehnung des Segmentes erhält man in Verbindung mit einem dritten ausgezeichneten Konturpunkt ein segmentbezogenes

Koordinatensystem, was eine translations- und rotationsinvariante Beschreibung der Form ermöglicht.

Die bei einer groben Approximation aller Regionen eines Bildes auftretenden Mehrdeutigkeiten aufgrund sich gegenseitig überlappender Segmente bzw. mögliches 'Niemandsland' können unter Berücksichtigung einfacher Regeln beseitigt werden.

4. Beschreibung der Texturen

Die meisten in der Texturanalyse üblichen Verfahren lassen sich der Klasse der strukturellen bzw. der stochastischen Ansätzen zuordnen. Die Algorithmen beider Richtungen können in Bezug auf Analyseaufgaben sehr effizient sein, sind jedoch mit dem Nachteil behaftet, eine gegebene Textur nicht exakt rekonstruieren zu können.

Im vorliegenden Verfahren werden die Texturen daher im Fourier-Spektralbereich kodiert, um eine kompakte (Leistungskonzentration!) und vollständige Beschreibung zu ermöglichen. Schreibt man den gegebenen Ausschnitt g als Produkt einer fiktiven Texturfunktion f mit einem binärwertigen Fenster w:

$$g(m,n) = f(m,n) \cdot w(m,n) \quad \circ\!\!-\!\!\bullet \quad G(k,l) = F(k,l) * W(k,l)$$

wird deutlich, daß das durch DFT erhaltene Spektrum G durch die Faltung $F * W$ stark verschmiert ist. Die Verwendung dieser Spektralbeschreibung hätte gravierende Nachteile:

1) die Form des betrachteten Segmentes würde ein zweites Mal beschrieben
2) eine exakte Rekonstruktion der Textur würde aus diesem Grund mehr Spektrallinien als im Originalbereich bekannte Abtastwerte erfordern
3) eine nur grobe Approximation der Konturen könnte in Verbindung mit dieser Texturbeschreibung schwarze 'Löcher' in der Rekonstruktion nach sich ziehen.

Diese aufgrund der beliebigem Berandung der Segmente auftretenden Probleme bei der DFT können durch Anwendung des Verfahrens der 'selektiven Entfaltung' [5] gelöst werden. Dieser auf der FFT basierende iterative Algorithmus eliminiert durch Extrapolation des gegebenen Texturausschnittes den störenden Einfluß der als binäres 'Fenster' fungierenden Segmentform.

5. Ein Beispiel

Das skizzierte Kodierungskonzept generiert eine Datenstruktur, die aufgrund ihres regionenorientierten und hierarchisch organisierten Aufbaus sowohl einen progressiven Bildaufbau ermöglicht als auch als Basis für eine Vielzahl einfacher (z.B. Bildverbesserung) wie höherer (z.B. modellgestützte Analyse) Bildverarbeitungsoperationen dienen kann. Das folgende Beispiel zeigt verschiedene Stufen einer Bildrekonstruktion.

Die Segmentierung der verwendeten Vorlage (256^2 Bildpunkte) mittels des Algorithmus nach [4] ergab 117 Regionen. Bild 1a zeigt eine sehr grobe Approximation, bei der nur der Mittelwert eingeblendet wurde und Bild 1b eine verbesserte Texturbeschreibung, die auf weniger als 0.5% der Spektrallinien beruht. Bild 1c gibt die Rekonstruktion nach verfeinerter Konturapproximation und Bild 1d das Resultat nach nochmaliger Verbesserung der Texturbeschreibung wieder.

Bei der Beurteilung der angegebenen Zahlenwerte muß man berücksichtigen, daß die Spektrallinien paarweise konjugiert komplex zueinander sind und noch keine (komprimierende) Quantisierung und Kodierung durchgeführt worden ist.

Bild 1: Verschiedene Stufen einer Bildrekonstruktion
a) 615 Konturpunkte, 117 Spektrallinien, b) 615 Konturpunkte, 1241 Spektrallinien, c) 1747 Konturpunkte, 1241 Spektrallinien, d) 1747 Konturpunkte, 6309 Spektrallinien

Literatur:

[1] M. Kunt et al.: *'Second-Generation Image Coding Techniques'*, Proc. of the IEEE, Vol.73 (1985), S.549–574

[2] M. Kocher, R. Leonardi: *'Adaptive Region Growing Technique Using Polynomial Functions for Image Approximation'*, Signal Processing 1986, S.47–60

[3] P.J. Burt: *'The Pyramid as a Structure for Efficient Computation'*, in: Multiresolution Image Processing and Analysis, Springer Verlag, 1984, S.6–35

[4] R. Mester, U. Franke: *'Ein regionenbasiertes Segmentierungsverfahren für texturierte Bildvorlagen'*, Beitrag zum ASST '87 (gleicher Band)

[5] U. Franke: *'Selective Deconvolution: A New Approach to Extrapolation and Spectral Analysis of Discrete Signals'*, Proc. ICASSP '87, Dallas, S.1300–1303

MULTIPLE CODING OF IMAGES WITHOUT DISTORTION ACCUMULATION

Michael Gilge and Wolfgang Guse
Institut für elektrische Nachrichtentechnik
RWTH Aachen, Melatener Str. 23, 5100 Aachen

1) Introduction: With the advent of ISDN and other network services the use of codecs for data compression for image storage and/or image transmission becomes more widespread. The image material to be processed can be separated into two classes:

- Generic material coming from sources like a TV-camera, an analog video-tape-recorder or broadband digital sources without any digital signal processing.
- Non-generic material which has been filtered or transform-coded beforehand e.g. image sequences transmitted via narrow-bandwidth channels.

Although the two classes may visually not be separated, the difference will appear by strong degradation in image quality when applying filtering or compression algorithms to non-generic images without taking the previous processing step into account. This is important as images are no longer transmitted and successively consumed at a TV-screen only. Thinking of laying-out a magazine page electronically gives a pretty good idea of what may happen to an image between coding steps: An image, coded for efficient storage or transmission, is decoded and then translated and/or rotated and/or scaled before it is coded again for storage or transmission. Without any consideration of the block raster introduced by transform coding for example, the sequence coding/decoding-moving-coding/decoding usually will yield unacceptable picture quality. The algorithm introduced in this paper will detect the location of the grid of image-subblocks as an input information for the next processing steps. For the sake of simplified discussion, only translatory movements of the images between transform coding steps will be considered in the following.

2) Multiple Coding of Images: The reason for the image degradation under multiple coding becomes quite apparent by considering the following example: Figure 1 shows an 8x8 image block which is represented by just a single spectral coefficient, which is supposed to be above the threshold. Now in the original domain the image block is moved by one pixel to the right and one pixel down. Again the spectral domain of the now shifted block is computed and depicted in Fig. 2. The

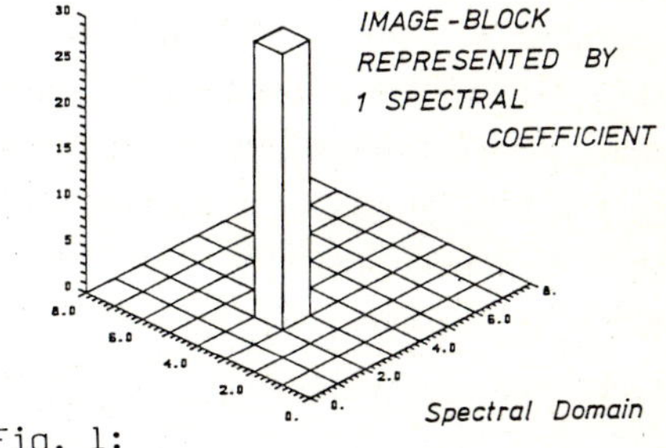

Fig. 1:

energy is now split and distributed between several spectral coefficients. Naturally the amplitudes of the coefficients will become smaller. Now a thresholding operation takes place and most of the coefficients are set to zero, while the few coefficients above the threshold are quantized. After inverse transform the image block is shifted back to it's original position, in order to compare it to the original block. Therefore the shifted back block is transformed again as shown in Fig. 3. The block was originally represented by ofe cgefficient only. Now the ampdatude of the dominant coefficieft became smaller and there are a number of additional nonzero coefficients accounting for the unaccurate image reconstruction. Naturally, without moving tè image in between processing steps and keeping-up the block-saze$ always the same coefficients are obtained for every following transform process. But if on the other hand any geometric change of the image may happen between coding processes, only two methods will keep-up the picture quality:

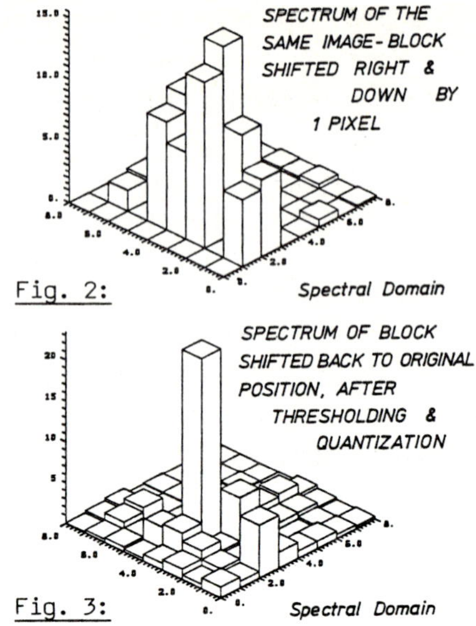

SPECTRUM OF THE
SAME IMAGE-BLOCK
SHIFTED RIGHT &
DOWN BY
1 PIXEL
Spectral Domain
SPECTRUM OF BLOCK
SHIFTED BACK TO ORIGINAL
POSITION, AFTER
THRESHOLDING &
QUANTIZATION
Spectral Domain

Fig. 2:

Fig. 3:

- Lowering the threshold to retain more spectral coefficients, which will become important in a future representation of the image. Drawback: The image data reduction is directly proportional to the value of the threshold. Therefore this method will strongly decrease compression performance. Besides, this method can only lower the slope of image degradation with regard to the number of processing steps but cannot avoid distortion accumulation.
- Block raster detection will keep up the compression performance by not changing any coder parameters. No additional distortion is introduced as the same coefficient representation of the image subblocks is always regained in the consecutive coding processes.

3) Difference between Original and Coded Images: Figure 4 shows the principle flow of operations. The difference between an original and a coded/decoded image is due to the quantization in the spectral domain. Originally the probability density function of the spectral coefficients after the discrete cosine transform is continuous. By quantizing the spectral coefficients the probability density function becomes dis-

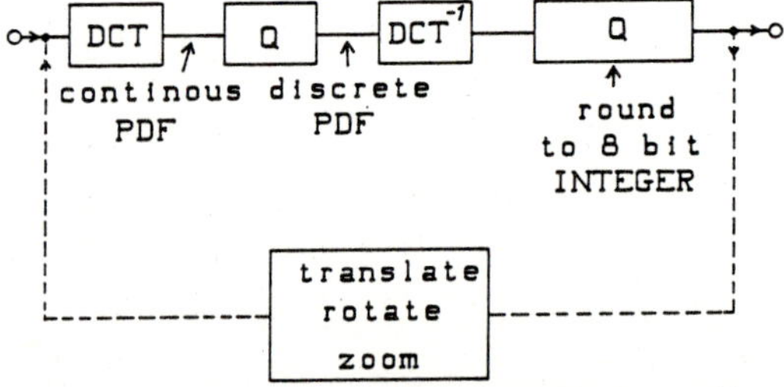

Fig. 4: Operation flow for multiple coding

crete as shown in Fig. 5. The distance between the peaks corresponds to the quantization intervals, here 30. Considering an unknown image, the DCT of an image subblock can be computed. Then the probability density function of the spectral coefficients can be tested for discreteness. If it is discrete, the picture has been coded and the block-raster has been found. If it is not, the picture may either not have been coded, or the block taking as a test block does not match the raster of a previous computing step. This ambiguity may be eliminated by trying another block, maybe 1 pixel to the right and so on.

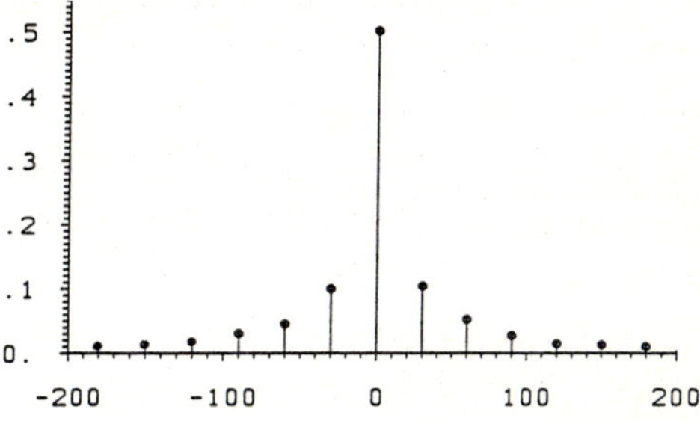

Fig. 5: PDF of quant. spectral coeffi.

4) Quantization in the Original Domain: Unfortunately there is another quantizing step involved in the chain of processing. This round-off operation happens after the inverse DCT in order to convert the picture back from floating point format to 8 bit integer, see Fig. 4. Thereby a round-off error is introduced, which has a uniform probability density function with a variance of 1/12. Assuming natural images, it can be shown that the single errors are uncorrelated. The DCT can be considered as a linear combination therefore and due to the central limit theorem the error in the spectral domain (after the next DCT) approaches a discrete gaussian distribution with the same variance /2/. The probability density function of the error due to the rounding in the original domain is shown in Fig.6 both after the round-off and after the consecutive DCT. Considering a picture with known block-raster we get the PDF as shown in Fig. 7 for one spectral coefficient of all raster-blocks. The diagram is plotted modulo the quantization interval. So if the value matches the quantization level exactly, it will appear in this diagram at the position 0. Besides the diagramm shows the modulo-PDF without rounding to 8 bit integer as indicated by the dashed line. Considering alternatively the blocks which are not aligned with the raster, we obtain a discrete uniform distribution (indicated by the dashed-dotted horizontal line in Fig. 7) if the quantization is not too coarse. The decision offers no problem if the statistic is computed on a complete picture for all possible block locations.

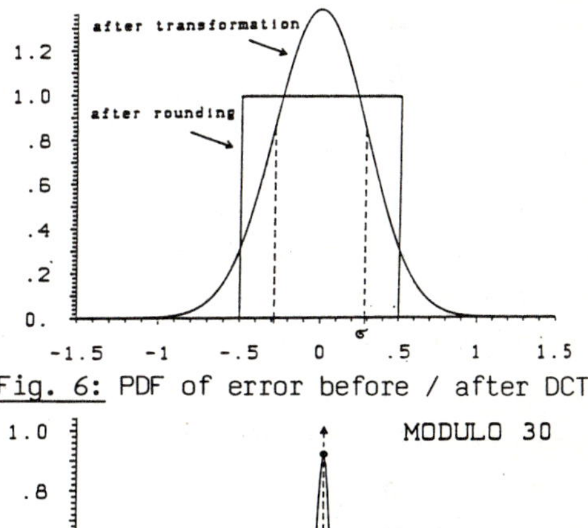

Fig. 6: PDF of error before / after DCT

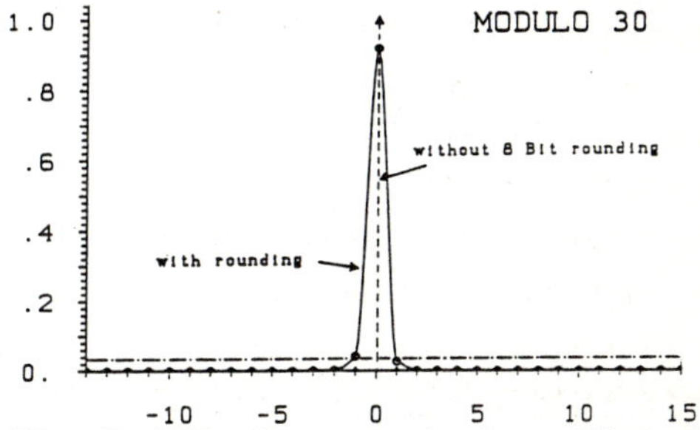

Fig. 7: PDF of one spectral coefficient

But it is an interesting question, what the smallest size of an image needed to detect the raster is.

5) Block Raster Detection Algorithm: Unfortunately there is no answer in numbers. Consider an image with no activity and the DC-coefficient happens to be on a quantization level. There is no way to detect any raster, but also there is no need to do so, as image degradation due to multiple coding only happens in regions with activity. And there the block raster can be detected:

a) Starting with four adjacent blocks, the coding-likelihood ratio is computed. Measures associated with the coding probability are for example the number of coefficients matching the quantization intervals, the number of zeros in the spectral representation due to the thresholding, or a periodic characteristic function due to a discrete probability density function.
b) The likelihood ratio is computed 64 times for all possible starting locations of the block-raster.
c) If the biggest of the 64 likelihood ratio exceeds 50 %, the block raster is said to be found for the coordinates corresponding to the location associated with that likelihood ratio.
d) If the biggest likelihood ratio is smaller than 50%, the likelihood ratio is updated by including additional adjacent blocks. If the number of blocks taken for the likelihood computation is exceeding a selected threshold, the search is stopped and the image is said to be generic. If on the other hand the likelihood ratio is exceeding 50 % the search is stopped and the block-raster has been found.

6) Results: Taking as measures the two spectral AC-coefficients (1,2) and (2,1) only, and testing whether they match the quantization intervals, the detection performance of the algorithm is shown in Fig. 8 for different thresholds. Usually a subimage of size 24x24 pixels is sufficient for the detection task. The detection speed can be increased by incorporating more measures into the likelihood computation. Current investigations include the extension of the algorithm to a) images consisting of two and more different parts with inconsistent block-raster and b) detection of the block-raster when the image is rotated or/and scaled in between coding steps.

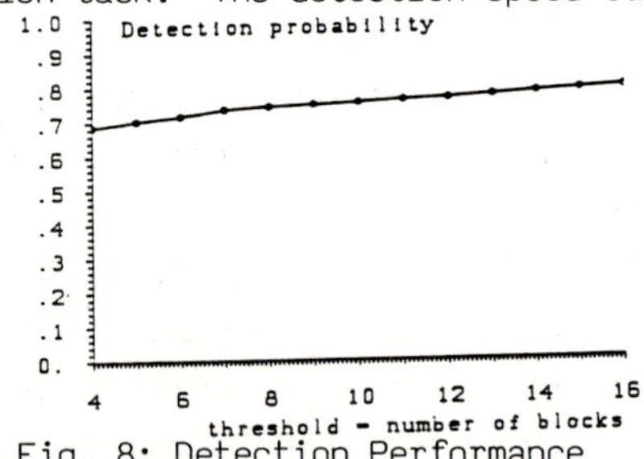

Fig. 8: Detection Performance

References:

/1/ M. Gilge : Block Boundary Detection in Transform Coded Images, Proceedings Picture Coding Symposium PCS-87, Stockholm, June 1987, pp.100-101

/2/ M. Gugliemo: An Analysis of Error Behavior in the Implementation of 2-D Orthogonal Transforms, IEEE-Trans. on Communications, Sept. 86, p.973

CODIERUNG VON VIDEOSEQUENZEN MIT NIEDRIGER DATENRATE DURCH VEKTORQUANTISIERUNG UND BEWEGUNGSKOMPENSATION

Bernard Hammer, Achim von Brandt
Siemens AG, ZT ZTI INF 121
Otto-Hahn-Ring 6, D-8000 München 83

Einleitung

Zur Datenkompression von Bildfolgen für Videokonferenz- oder Bildfernsprechanwendung mit Übertragungsraten im Bereich von 384 kbit/s und 2 Mbit/s haben sich orthogonale Transformationscodierungen (OTC) in Verbindung mit Bewegungskompensation als Standardverfahren durchgesetzt. Ihr Einsatz für 64 kbit/s Bildfernsprechen, das z.Zt. als möglicher Dienst in einem Schmalband-ISDN zur Diskussion steht, zeigt jedoch einige prinzipielle Schwächen des Verfahrens. Bedingt durch die Verarbeitung von Bildern in einem festem Blockraster von voneinander unabhängigen Blöcken der Größe 16 x 16 oder 8 x 8 Bildpunkten, reicht die zur Verfügung stehende Datenrate nicht mehr aus, um bei hohen Bewegtanteilen der Bildfolge die Stetigkeit der Bildrekonstruktion an Blockgrenzen zu gewährleisten. Diese Blockartifakte wirken, nicht zuletzt wegen der Größe der Blöcke, sehr störend. Als Alternative zur OTC wird hier ein auf einer Pyramidentransformation basierender Algorithmus vorgestellt, der eine 'glatte' Bildrekonstruktion sicherstellt und eine hierarchische Codierung mit variabler Blockgröße durch Bewegungskompensation und Vektorquantisierung vornimmt.

Coderkonzept

Die Pyramidentransformation ist eine aus der progressiven Übertragung von Standbildern her bekannte Technik [1] zur hierarchischen Darstellung von Bilddaten durch eine pyramidenförmige Datenstruktur. Die Pyramide besteht aus mehreren Stufen, die mit $l = 0$ bis L numeriert sind. Hierbei bezeichnet die Stufe L die Eingangsdaten, während alle weiteren Stufen l tiefpaßgefilterte Versionen der Stufe L mit je $2^l \times 2^l$ Abtastwerten repräsentieren. Die Elemente der Stufe l-1 stellen in der von uns verwendeten Version den Mittelwert von jeweils vier benachbarten Elementen der Stufe l dar. Zur Verarbeitung eines Bildes werden quadratische Blöcke mit jeweils $2^L \times 2^L$ Bildpunkten pyramidentransformiert. Daneben werden auch die Elemente der Pyramidenstufen in sogenannten Stufensegmenten der Größe $n \times n$, mit $n = \min(2^l, 4)$ zusammengefaßt. Damit läßt sich, wie in Bild 1 dargestellt, die Datenstruktur eines Blocks durch einen Baum beschreiben, dessen Knoten die Stufensegmente bilden.

Die Grundidee des hier vorgestellten Coderkonzept geht von der Beobachtung aus, daß sich bei der Bild-zu-Bild-Differenz bewegungskompensierter natürlicher Bildfolgen signifikante Fehlerenergien hauptsächlich im Bereich hochfrequenter Details ergeben. Der Coder

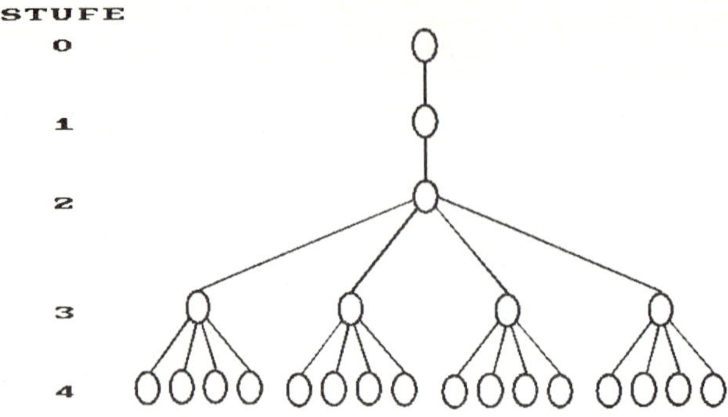

Bild 1: Datenstruktur eines Bildblocks der Größe 16 x 16 (L = 4)

nutzt diesen Effekt zur Datenkompression, indem er korrespondierende Pyramidenstufen des Eingangsbildes und des bewegungskompensierten rekonstruierten Vorgängerbildes beginnend mit Stufe 0 eines Bildblocks sukzessive miteinander vergleicht.

Überschreitet die Varianz der Interframestufendifferenzen einen festen Schwellwert, der proportional zur Sichtbarkeitsschwelle des Kompensationsfehlers in der jeweiligen Stufe ist, dann wird zusätzlich zu den Bewegungsvektoren als Korrekturterm die Information dieses Stufensegments und aller davon abhängigen Stufensegmente in den folgenden Pyramidenstufen des Eingangsbilds codiert und übertragen. Die Berechnung der Bewegungsvektoren erfolgt durch Blockmatching von Blöcken der Größe 16 x 16 Bildpunkten nach dem in [2] veröffentlichten Algorithmus.

Zur Korrektur des Kompensationsfehlers wurde ein Intraframecodierverfahren mit Vektorquantisierung gewählt. Bild 2 zeigt für L = 2 das Blockschaltbild dieses Verfahrens, das nach [3] als hierarchisch mehrstufige Vektorquantisierung (HMVQ) bezeichnet wird.

Ausgehend von der Pyramidentransformation des Bildblocks $\underline{X}$ durch rekursive Anwendung des Mitttelwertoperators REDUCE beginnt die Codierung mit der skalaren Quantisierung des Elements in Stufe 0. Die Stufensegmente aller nachfolgenden Stufen l werden nach Subtraktion des durch Interpolation des zugeordneten Segments in Stufe l-1 gewonnenen Tiefpaßanteils vektorquantisiert, falls die Spannweite des verbleibenden Strukturanteils einen Schwellwert überschreitet.

Das mehrstufige hierarchische Prinzip der Anordnung ermöglicht eine Anpassung der Blockgröße der Quantisierung an den Detailgehalt der Eingangsdaten im Bereich von 32 x 32 (Stufe 0, 1, 2), 16 x 16 (Stufe 3), 8 x 8 (Stufe 4) und 4 x 4 Elementen (Stufe 5).

Die Interpolationsopration EXPAND wurde zweistufig ausgelegt, um den Gleichanteil im Struktursignal klein zu halten. Diese Maßnahme sowie die Verwendung eines blockübergreifenden Interpolationsfilters vermeiden Unstetigkeiten des Tiefpaßanteils an den Grenzen der Stufensegmente. Die zweidimensionale Interpolation wurde als separierbares FIR-

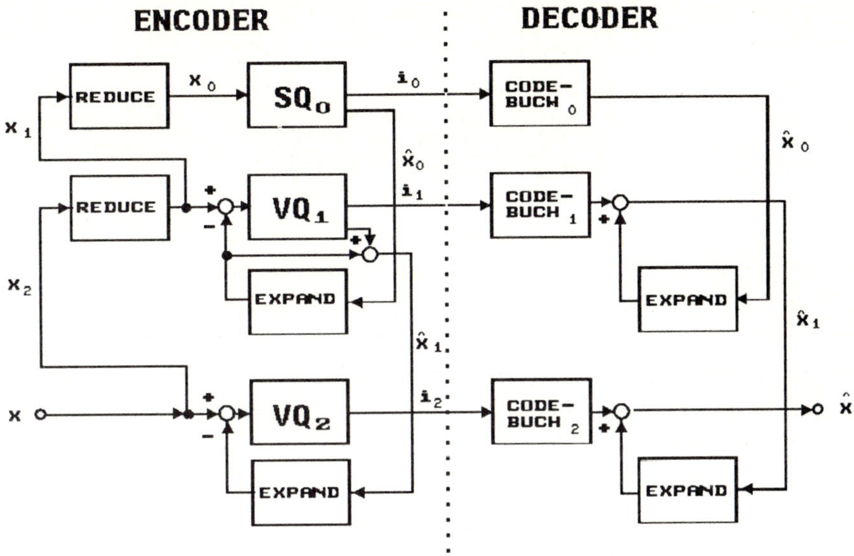

Bild 2: Blockschaltbild einer hierarchisch mehrstufigen Vektorquantisierung mit 3 Stufen.

Filter mit 6 Taps realisiert. Bild 3 zeigt das Blockschaltbild des Opeators EXPAND sowie in graphischer Form die eindimensionale Interpolationsfunktion.

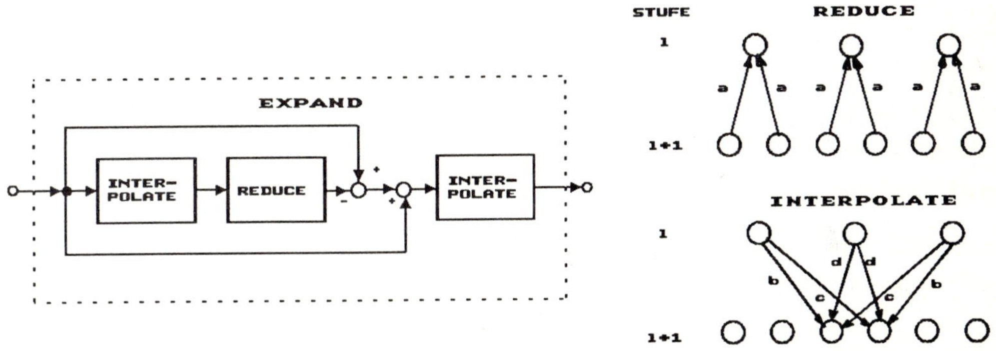

Bild 3: Blockschaltbild des Interpolationsoperators EXPAND (Filterkoeffizienten a = 1/2, b = 9/32, c = 1/32, d = 22/32)

Zur Vektorquantisierung werden 'Gain/Shape'-Quantisierer nach [4] mit baumstrukturierten Codebüchern eingesetzt. Die Codewortlängen für die Codebuchindizes liegen im Bereich von 4 bit/Element (Stufe 1) bis 0.5 bit/Element (Stufe 5). Da die Codebücher in diesem Konzept mittelwertfreie und auf Varianz 1 normierte Strukturkomponenten enthalten, ergibt sich ein sehr robustes Verhalten der Vektorquantisierung gegenüber Änderungen in der Signalstatistik.

Der Codiermodus eines Stufensegments in Stufe l wird bestimmt durch den Codiermodus seines in der Datenstruktur zugeordneten Segments der Stufe l-1. Bild 4 verdeutlicht die Regeln anhand eines Übergangsdiagramms.

Bild 4: Übergangsdiagramm der Codiermodi von Stufe l-1 zu Stufe l.

Neben dem bereits beschriebenen Übergang vom Bewegungskompensationsmodus (MC) zur Vektorquantisierung (HMVQ) des Restfehlers gibt es einen Interpolationsmodus (INT). Dieser Modus dient zur Anpassung der variablen Datenrate des Coders an die feste Kapazität des Übertragungskanals. Hierbei wird durch Schwellwertsteuerung des HMVQ-Modus die Übertragung signifikanter Strukturanteile unterdrückt, was einer Tiefpaßfilterung des betroffenen Bildbereichs entspricht. Diese Anteile werden in nachfolgenden Bildern entsprechend der durch die hierarchische Anordnung festgelegten Reihenfolge übertragen, sobald die Datenrate hierzu ausreicht.

Ergebnisse

Das Verhalten des Coderkonzepts mit L = 5 wurde für eine 64 kbit/s Bildfernsprechanwendung durch Rechnersimulation überprüft. Die Auflösung der Bildsequenzen betrug 288 Zeilen x 352 Spalten für die Luminanz und 72 x 88 für die Chrominanzanteile bei einer Bildfrequenz von $8\frac{1}{3}$ Hz.

Unter Berücksichtigung eines Fehlerkorrekturanteils von 5% verbleibt zur Übertragung der Bildinformation eine Rate von 61 kbit/s. Dies entspricht einer mittleren Codewortlänge von 0.062 bit/Bildpunkt für den Luminanzanteil.

Die Ergebnisse bestätigen, daß unabhängig vom Bewegtanteil der Sequenz keine Blockartefakte auftreten. Hochfrequente Artefakte sind ausschließlich auf die Blockgröße von 4 x 4 Bildpunkten der Pyramidenstufe 5 begrenzt. Diese Fehler treten jedoch nur im Bereich von steilen Luminanzkanten auf und werden hier weitgehend maskiert.

Literatur

[1] K.R. Sloan, S.L. Tanimoto, "Progressive Refinement of Raster Images", IEEE Transactions on Computers, Vol C-28, No. 11, Nov. 1979, pp. 871-874.

[2] A.v. Brandt, "Obtaining Smoothed Optical Flow Fields by Modified Block Matching", Proc. 5th Scand. Conf. on Image Analysis, Stockholm, June 1987, pp. 523-529.

[3] B. Hammer, A.v. Brandt, M. Schielein, "Hierarchical Encoding of Image Sequences using Multistage Vector Quantization", Proc. IEEE Int. Conf. Acoustics, Speech, and Signal Proc., April 1987, pp. 1055-058.

[4] M.J. Sabin, R.M. Gray, "Product Code Vector Quantizers for Waveform and Voice Coding", IEEE Trans. Acoustics, Speech, and Signal Processing, Vol. ASSP-32, No. 3, June 1984, pp. 474-488.

MULTI-SPECTRAL DATA COMPRESSION USING AN ADAPTIVE SPECTRAL TRANSFORM*

W.C. Huisman
National Aerospace Laboratory NLR
Anthony Fokkerweg 2
1059 CM AMSTERDAM

1 INTRODUCTION

Multi-spectral data consists of radiation measurements in a number of spectral channels. In general, the radiation measurements from the different spectral-channels will be correlated. Therefore, the performance of data compression algorithms operating on each single channel can be improved by making use of the spectral correlation. To this end a transform should be used that decorrelates each multi-spectral pixel. The obtained decorrelated images can then be encoded separately (as well the transform that has been used for the spectral de-correlation) resulting in efficiently compressed multi-spectral data. Due to the non-stationair behaviour of the correlations between the channels it is necessary to adapt the spectral transform. Application of the Karhunen-Loeve (KL) transform is then impractical for two major reasons: firstly, the large overhead that is needed to encode this transform (in order to make it possible for the decoder to apply the inverse transform), and secondly the computational difficulties that arise when obtaining the transform in (real-time) from the estimated interchannel covariance matrix.

In the sequel a very efficient unitair spectral transform is presented with the following properties (let S denote the number of spectral channels):

- easy to compute and easy to carry out;
- adaptive on the basis of N x N x S pixelcubes; N = 8, 16, ...;
- needs a very small overhead for encoding this transform;
- performance of the data compression system will be close to the one that is equipped with the (adaptive) KL transform (without encoding the transform).

2 DERIVATION OF THE SPECTRAL TRANSFORM FROM LOCAL STATISTICS

The adaptive spectral transform is computed from the interchannel covariance matrix ($\hat{C}$) as estimated from a N x N x S pixelcube. Estimating the covariance matrix is efficiently done, as will appear lateron, with the spatially transformed pixels

* This research has been performed under European Space Agency Contract no. 5724/83/NL/BI

that result when each image block of N x N pixels is first transformed using the two-dimensional discrete cosine transform (DCT):

$$\hat{C}_{rs} = \frac{1}{N^2} \sum_{\ell=2}^{m} Z_\ell^{[r]} Z_\ell^{[s]}, \quad r,s = 1, \ldots.S; \; m \in \{2, \ldots N^2\} . \tag{1}$$

Herein, $Z_\ell^{[r]}$ and $Z_\ell^{[s]}$ denote the indexed spatially transformed pixels (obtained after reordening the two-dimensional array of the transformed pixels according the zig-zag pattern, i.e. meander reordening) of channel r and channel s respectively; m is a selectable number which determines the accuracy of the estimate. When $m = N^2$ the highest accuracy is obtained but a lot of effort has to be spend on computing the estimate; however when m is taken smaller the accuracy decreases very slowly (which is due to the energy packing property of the DCT) while less effort has to be spend on computing the covariance matrix. It will be investigated lateron how sensitive the data compression system will be on the value of m. Note that the estimator given in (1) is a consistent estimator of $\lambda_m C_{rs}$ (see [1]) where λ_m is a scaling factor not depending on r and s; it will appear that the value of λ_m has no impact on the derivation of the spectral transform, hence the value of λ_m is of no importance.

Having estimated a scaled version of the interchannel covariance matrix from a N x N x S pixelcube, the adaptive spectral transform is obtained in three steps:

a. assign a sign (+/-) to each spectral channel;
b. permutate the signed channels;
c. apply a basic transform to each of the multi-spectral pixels (N x N) of the permutated spectral channels.

Ad a.

The sign to be assigned to each spectral channel is determined by the following algorithm: compute $J \in \{1, \ldots S\}$ that maximizes

$$\alpha_J \triangleq \sum_{r=1}^{S} \frac{|\hat{C}_{rJ}|}{\hat{C}_{JJ}^{\frac{1}{2}} \; \hat{C}_{rr}^{\frac{1}{2}}} . \tag{2}$$

The sign s_i for channel #i is given by s_i = sign $(\hat{C}_{iJ})$; this means that all (N x N) pixels from this channel get sign s_i.

Ad b.

The permutation π of the signed channels is found with rearranging the correlations

$$\rho_{iJ} \triangleq \frac{|\hat{C}_{iJ}|}{\hat{C}_{ii}^{\frac{1}{2}} \; \hat{C}_{JJ}^{\frac{1}{2}}} , \; i = 1, \ldots S \tag{3}$$

in decreasing order.

Ad c.

The basic transform U_S is for S = 2, 3 and 4 defined by:

$$U_2 = \begin{pmatrix} \sigma_1 & -\sigma_2 \\ \sigma_2 & \sigma_1 \end{pmatrix} \begin{pmatrix} n_2 & 0 \\ 0 & n_2 \end{pmatrix} \tag{4}$$

$$U_3 = \begin{pmatrix} \sigma_1 & \sigma_1 & -\sigma_2 \\ \sigma_2 & \sigma_2 & \sigma_1 \\ \sigma_3 & -t_2/\sigma_3 & 0 \end{pmatrix} \begin{pmatrix} n_3 & 0 & 0 \\ 0 & m_3 & 0 \\ 0 & 0 & n_2 \end{pmatrix} \tag{5}$$

$$U_4 = \begin{pmatrix} \sigma_1 & \sigma_1 & \sigma_1 & -\sigma_2 \\ \sigma_2 & \sigma_1 & \sigma_1 & +\sigma_1 \\ \sigma_3 & \sigma_1 & -t_2/\sigma_4 & 0 \\ \sigma_4 & -t_3/\sigma_4 & 0 & 0 \end{pmatrix} \begin{pmatrix} n_4 & 0 & 0 & 0 \\ 0 & m_4 & 0 & 0 \\ 0 & 0 & m_4 & 0 \\ 0 & 0 & 0 & n_2 \end{pmatrix} \tag{6}$$

The matrices given above indicate how U_S, S > 4 should be defined. The parameters given in eq. (4) - eq. (6) are defined as the standard deviation of the permuted signed channel #i, i.e. $\sigma_i = \hat{C}^{\frac{1}{2}}_{\pi_i \pi_i}$ and $t_i = \sum_{k=1}^{i} \sigma_k^2$, and n_i and m_i (i= 2, 3, 4) are normalization constants that yield unitair matrices U_s, e.g. $m_3 = (\sigma_1^2 + \sigma_2^2 + t_2^2/\sigma_3^2)^{-\frac{1}{2}}$. Encoding the spectral transform is achieved by coding the S signs s_i (1 bit/-channel), coding the permutation π ($\frac{1}{S}.{}^2\log S!$ bits/channel) and coding the basic transform U_S, i.e. encoding the values σ_i, i = 1, ...S. Encoding of σ_i is achieved with applying an (8 bit) uniform quantizer (0 - 255). If for example S = 5, encoding the spectral transform would require 0.04 bit/pixel/channel.

Note that generation of U_S should be based upon the quantized values σ_i^*. If follows that the ultimate spectral transform comprising the steps a, b and c, is unitair.

The data compression technique that will be used for encoding the spectral decorrelated pixelcube consists of applying a DCT that will be applied on N x N pixels from each spectral decorrelated channel, followed by adaptive quantizing plus coding of the coefficients then obtained [2]. It is easily seen that first applying the spectral transform and then applying the DCT gives identical results as would be obtained when the spectral transform is applied after the DCT. In view of the above mentioned determination of the adaptive spectral transform it is therefore easier to apply the DCT first, resulting in spatially decorrelated pixels followed by 1) determination of the spectral transform and 2) application of the spectral transform on the spatially decorrelated pixels. Hence in this set-up the DCT's used for determination of the spectral transform and as part of the data compression method have been joined (Fig. 1).

3 EVALUATION RESULTS AND CONCLUSIONS

Encoding a N x N x S pixelcube consists of encoding the spectral transform (see Par. 2) and encoding the decorrelated coefficients. Encoding the S sets of meander-rearranged decorrelated coefficients is archieved by encoding each set separately according the following strategy: all coefficients (except the DC coefficient) are multiplied with Q^{-1} followed by rounding and reversible coding of the obtained integers; the DC coefficient is represented in 16 bits when N = 16. The parameter Q determines the root-mean-square error (RMSE) in each decompressed spectral channel (RMSE = $Q/(2\sqrt{3})$). It follows that each spectral pixel block (N x N pixels) will have a fixed reconstruction error after decompression but the bitrate depends on the entropy of the pixels and will be variable. Note that by adjusting Q a fixed compression ratio can be obtained.

The data compression results presented are obtained with the Meander-2 data compression method in the fixed error mode for N = 16; this method has the coding strategy mentioned above [1]. Simulation results obtained with the application of the adaptive spectral transform to the compression of data from several multi-spectral sensors on-board satellites, showed a decreased bitrate compared with omitting a spectral transform.

Reproducable results can be obtained when multi-spectral Markov images are compressed. Figure 2 and figure 3 are related to Markov images with normally distributed pixels having spatial correlations 0.9 (horizontal) and 0.9 (vertical) respectively and the following interchannel covariance matrix:

$$C = \begin{pmatrix} 2000.6 & 1183.4 & 1514.0 & 919.3 & -250.0 & -69.2 & -76.8 \\ 1183.4 & 793.3 & 1018.1 & 624.5 & -146.8 & -43.1 & -53.5 \\ 1514.0 & 1018.1 & 1318.2 & 799.2 & -158.3 & -51.8 & -54.6 \\ 919.3 & 642.5 & 799.2 & 521.3 & -136.3 & -35.8 & -54.4 \\ -250.0 & -146.8 & -158.3 & -136.6 & 331.5 & 38.3 & 156.4 \\ -69.2 & -43.1 & -51.8 & -35.8 & 38.3 & 13.2 & 18.1 \\ -76.8 & -53.5 & -54.6 & -54.4 & 156.4 & 18.1 & 80.6 \end{pmatrix}$$

Figure 2 shows the rate-distortion relations obtained with the Meander-2 method with application of the adaptive- and KL spectral transform, as well no application of a spectral transform. The number of encoded channels is 7 and encoding the spectral transform is not taken into account. It follows that the adaptive spectral transform yields a rate that is 0.8 bit/pixel/channel lower compared with omitting the spectral transform. The KL-transform yields a rate that is 0.45 bit/pixel/-channel lower moreover. Also the KL-transform yields a rate that is 0.75 bit/-pixel/channel higher than the Shannon lowerbound that has been computed for the 16 x 16 x 7 pixelcubes. This latter difference is caused by the use of a uniform quantizer in the Meander-2 method (0.25 bit), the penalty for the adaptivity of this method, coding the DC coefficients always in 16 bits and the individual determination of the compressed data for each pixelcube (0.5 bit).

Figure 3 shows the rate-distortion results in case only the first 4 channels have been compressed. It follows that the adaptive spectral transform yields equal bitrates as obtained with the KL transform. The parameter m (see Par. 3) has almost no impact on the results obtained, e.g. for m = 16 the bitrates will be 0.03 - 0.08 bit/pixel/channel higher.

4 REFERENCES

[1] Börger, J.B. et al. "Final Report on Study, Design and Evaluation of Data Reduction by Encoding or information Extraction", NLR TR 85099 L, 1985

[2] Roefs, H.F.A. "CADISS: An Image (De) Compression System for Deep Space Application". Fourth Symposium on Information theory, Leuven, 1983, NLR MP 83028 U

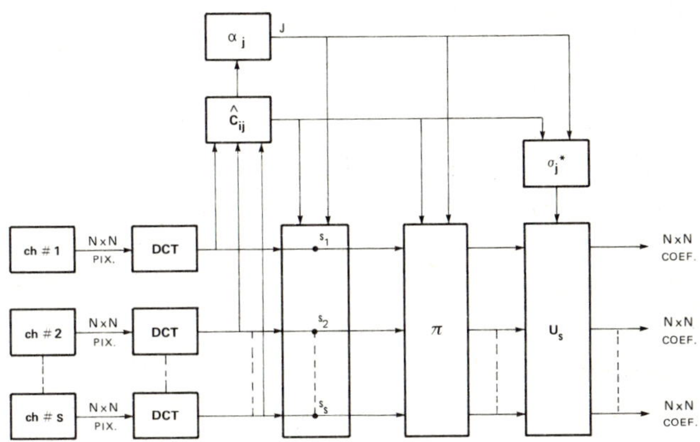

Fig. 1 Processing steps associated with the adaptive spectral transform which precedes the quantization and coding process

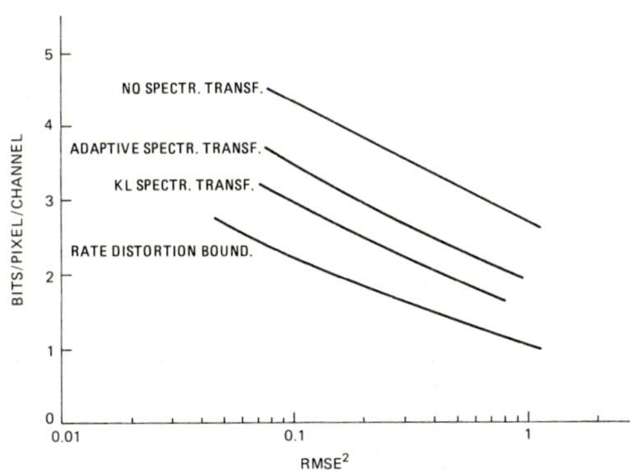

Fig. 2 Rate distortion relations when compressing 7 channels (m = 256)

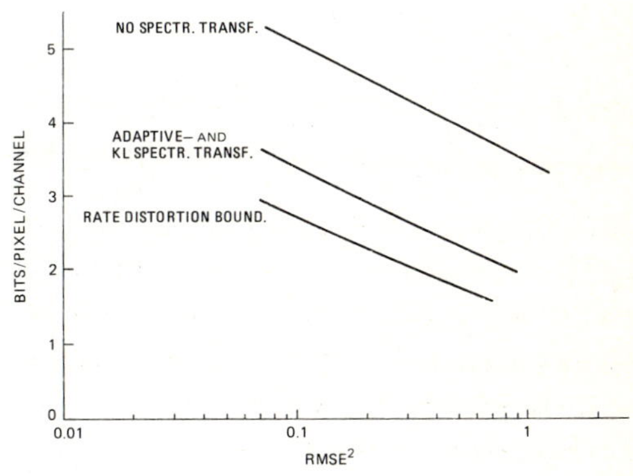

Fig. 3 Rate distortion relations when compressing 4 channels (m = 256)

EXAKTE BILDCODIERUNG UND HUFFMAN-CODES

Dietrich Manstetten
Lehrstuhl für angewandte Mathematik insb. Informatik
RWTH Aachen, Templergraben 64, D-5100 Aachen

1. Einleitung

Eine fehlerfreie Codierung von Grauwertbildern läßt sich mit den unterschiedlichsten Prädiktionsmethoden und anschließender Huffman-Codierung der Differenzwerte verhältnismäßig einfach durchführen. In dieser Arbeit werden an einem konkreten Beispiel die Möglichkeiten ausgefeilterer Techniken ohne Verzicht auf die Exaktheit der Codierung untersucht. Dabei werden im Fehlerbild, das man durch ein beliebiges Prädiktionsverfahren erhält, zusammenhängende Züge von Pixeln mit großer Codewortlänge gesucht, die sich in realistischen Bildern aufgrund von Fehleranhäufungen in Konturbereichen ergeben. Diese Züge können aus dem Fehlerbild eliminiert und unter Angabe eines Startpunktes mittels gewöhnlicher Pulse Code Modulation sowie Chain Code angegeben werden.

Bevor in Abschnitt 3 auf die Resultate praktischer Tests dieses Verfahrens eingegangen wird, werden seine Grenzen in Abschnitt 2 in einem allgemeinen Rahmen unter spezieller Berücksichtigung der Anwendung in der Bildcodierung herausgearbeitet.

2. Theoretische Untersuchungen

Wie groß kann die Einsparung sein, die mit dieser Methode maximal erreicht werden kann? Im Falle eines Bildes mit 128 Grauwerten benötigt man für jedes Pixel eines Zuges 7 Bit für den Grauwert sowie 3 Bit für den Chain Code. Somit können im Extremfall, daß alle Pixel mit einer Codewortlänge größer als 10 in einem Zug untergebracht werden können, alle Codewortlängenanteile eingespart werden, die über 10 Bit hinausgehen. Die zusätzliche Side Information beschränkt sich auf die Angabe des einen Startpunktes. Ist $(p_1,\dots,p_n)$ die dem Code zugrundeliegende Wahrscheinlichkeitsverteilung und sind $(l_1,\dots,l_n)$ die zugehörigen, durch den Huffman-Algorithmus erzeugten Codewortlängen, so ist der zu erzielende "Gewinn" beschränkt durch

$$\sum_{i=1}^{n} p_i \cdot (l_i - 10)_+ \qquad , \text{wobei } a_+ = \begin{cases} a \ , & a \geq 0 \\ 0 \ , & \text{sonst} \end{cases}$$

Eine obere Schranke für die mögliche Einsparung erhält man somit durch Maximierung dieses Ausdruckes über alle möglichen Wahrscheinlichkeitsverteilungen.

Die Verallgemeinerung dieses Ansatzes führt damit zu folgender Problemstellung:
Gegeben seien eine Anzahl n von Codewörtern sowie eine "Gewinngrenze" $g \in \mathbb{N}$. Maximiere den Ausdruck

$$\sum_{i=1}^{n} p_i \cdot (l_i - g)_+$$

über alle Wahrscheinlichkeitsverteilungen $(p_1,\ldots,p_n)$ und ihre zugehörigen Codewortlängen $(l_1,\ldots,l_n)$. Zur Erlangung eines positiven Gewinns ist dabei $g \le n-2$ notwendig.

Die exakte Bestimmung des Maximums ist in diesem allgemeinen Rahmen nur schwer durchführbar. Im folgenden wird daher nur eine obere Abschätzung entwickelt. Die dazu erforderlichen Sätze können hier wegen des beschränkten Platzes nicht bewiesen werden.

In Satz 1 wird der Wahrscheinlichkeitseintrag an einer beliebigen Stelle des Codebaumes nach oben abgeschätzt.

Satz 1: Sei s eine beliebige Stufe des Codebaumes. Der Wahrscheinlichkeitseintrag an einem in dieser Stufe auftretenden Knoten ist beschränkt durch

$$p_{max}(s) := 2 / F_{s+3}$$

wobei F_i die i-te Fibonacci-Zahl bezeichnet mit

$$F_0 = 0 \,,\; F_1 = 1 \,,\; F_{n+2} = F_{n+1} + F_n$$

Die Klasse der für die Maximierung zu betrachtenden Codebäume kann wesentlich eingeschränkt werden, denn es gilt:

Satz 2: Beim gesuchten Maximum kann keine Codewortlänge $l \le g$ mehr als einmal auftreten.

Von Stufe g-1 des Codebaumes an läßt sich der Gewinn auf einfache Weise abschätzen.

Satz 3: Existieren in der Stufe g-1 genau a Knoten des Codebaumes, aus denen noch $\bar{n}$ Codewörter entstehen, so ist der Gewinn beschränkt durch den Gewinn der Gleichverteilung der Restwahrscheinlichkeit (d.h. der Summe der Einträge an den a Knoten) auf die $\bar{n}$ Codewörter.

Die eigentliche Abschätzung erfolgt nun zunächst für einen leicht zu verallgemeinernden Spezialfall.

Satz 4: Unter der Voraussetzung, daß es kein Codewort der Länge 1 gibt, d.h. in Stufe 0 existiert eine "Gabel aus zwei Knoten", und der Beschränkung auf die Codebäume aus Satz 2 gilt:

a) Die Anzahl an Knoten in der Stufe g-1 ist $a(x) := 2^{g-1}-x$, wobei jeder x-Wert mit $0 \le x \le 2^{g-2}-1$ möglich ist.
Die zu festem x gehörige Anzahl von Codewörtern, die aus den a(x) Knoten entstehen, ist $\overline{n}(x) := n-c(x)$, wobei c(x) die Anzahl von Einsen in der Binärdarstellung von x ist.

b) Für festes x in a) mit $0 \le x \le 2^{g-2}-1$ läßt sich der Gewinn abschätzen durch

$$G(x) := \begin{cases} G_1(x) & , x < 2^{g-1}-n/2^{l+1} \\ G_2(x) & , x \ge 2^{g-1}-n/2^{l+1} \end{cases}$$

wobei $G_i(x)$ für i = 1,2 definiert ist durch

$$G_i(x) := (2^g-2x)/(2^g-x) \cdot (l+i-2^{l+i} \cdot (2^{g-1}-x)/n)$$

und mit der Gaußklammerfunktion [] gilt:

$$l := [\log(n/2^{g-1})]$$

Für i = 1,2 nimmt $G_i(x)$ sein Maximum an der Stelle

$$x_{i,max} = 2^g - \sqrt{2^{2g-2} + 2^{g-l-1-i} \cdot n \cdot (l+i)}$$

an, und $G_i(x)$ ist monoton steigend auf $(-\infty , x_{i,max}]$ und monoton fallend auf $[x_{i,max} , 2^g]$.
Das Maximum der Funktion G(x) für $0 \le x \le 2^{g-2}-1$ ist demnach leicht zu bestimmen.

c) Für $l \ge 5$, d.h. für $n \ge 2^{g+4}$ gilt:

$$G(x) \le G(0) \quad , x = 0,\ldots,2^{g-2}-1$$

Die allgemeine Abschätzung wird dann folgendermaßen ermittelt: Die von der Wurzel aus erste Stufe des Codebaumes, in der eine "Gabel aus zwei Knoten" auftritt, werde mit s bezeichnet. Wende dann Satz 4 auf die Größen $n_s = n-s$ und $g_s = g-s$ an und multipliziere den hierdurch erhaltenen Maximalwert mit der Größe $p_{max}(s)$ aus Satz 1. Schließlich werden die Ergebnisse über die möglichen s-Werte maximiert. Dabei muß man jedoch, aufgrund des Steigungsverhaltens der Fibonacci-Zahlen ($F_{s+4} \ge (3/2) \cdot F_{s+3}$), beginnend mit dem kleinstmöglichen s dieses nur solange um 1 vergrößern, bis der sich aus Satz 4 b) ergebende Wert größer als 2 wird (siehe Anwendung).

Bemerkungen:

1. In Satz 4 b) wurde zur Vereinfachung die Anzahl Codewörter $\overline{n}(x)$ durch n abgeschätzt. Verwendet man die etwas bessere Abschätzung $\overline{n}(x) \le n-1$ für $x \ge 1$ oder gar den tatsächlichen Ausdruck $\overline{n}(x)$, so läßt sich das Resultat evtl. noch verbessern.

2. Während der Maximalwert aus Satz 4 b) im allgemeinen für keine Wahrscheinlichkeitsverteilung angenommen wird, ist die Berechnung aus Satz 4 c) für $n \geq 2^{g+4}$ nicht nur besonders einfach sondern auch exakt, d.h. in diesem Falle wird das Maximum für die Gleichverteilung aller n Codewörter angenommen.

3. Die obige Theorie läßt sich ohne große Schwierigkeiten auch auf Gewinngrenzen $g \notin \mathbb{N}$ erweitern.

Anwendung auf den für die Bildcodierung relevanten Fall:

Für den Fall eines Bildes mit 128 Grauwerten (d.h. für n=128 und g=10) erhält man aus Satz 2 zunächst, daß eine "Gabel aus zwei Knoten" in den Stufen s = 0,1,2,3 nicht vorkommen kann. Für die weiteren s-Werte ergibt sich aus Satz 4:

$$s=4 : \quad G(x) \leq G(13.88) = 1.3239$$
$$s=5 : \quad G(x) \leq G(4.65) = 2.0945$$

Da der letzte Wert größer als 2 ist, erübrigt sich eine weitere Erhöhung von s. Die Multiplikation mit $p_{max}(s)$ aus Satz 1 liefert:

$$s=4 : \quad 1.3239 \cdot 2/13 = 0.2037$$
$$s=5 : \quad 2.0945 \cdot 2/21 = 0.1995$$

Damit ist in der Bildcodierung die Einsparung nach oben beschränkt durch 0.2037 Bit/Pixel. Durch Anwendung der in Bemerkung 1 aufgeführten Vorschläge läßt sich diese Abschätzung über 0.2026 (bei $\overline{n}(x) \leq n-1$ für $x \geq 1$) noch auf 0.2005 (bei $\overline{n}(x)=n-c(x)$) verbessern.
Auch der Wert 0.2005 ist dabei noch eine obere Schranke für den Maximalwert und wird nicht angenommen. Als untere Schranke erhält man durch explizite Konstruktion den Wert 0.1848, der sich bei einer ersten "Gabel aus zwei Knoten" in Stufe 5 und anschließender Gleichverteilung der restlichen 123 Codewörter ergibt.

Das erhaltene Resultat variiert nur geringfügig mit der Grauwertanzahl. Für n=256 und entsprechend g=11 erhält man beispielsweise ohne Verwendung von Bemerkung 1 eine Schranke von 0.2058.

3. Praktische Ergebnisse

Das in der Einleitung beschriebene Verfahren wurde experimentell auf vier Bilder der Größe 512 x 512 mit 128 Grauwertstufen angewendet. Jedes der Bilder wurde viermal behandelt (als Originalbild, ein- bzw. zweimal angewendeter Low Pass Filter zur Kontrastverminderung, Directional Filter zur Kontrastverstärkung). Außerdem wurden drei unterschiedliche Vorhersagemethoden verwendet (Linker Nachbar, Mittelwert aus linkem und oberem Nachbarn, adaptive Vorhersage aus linkem und oberem Nachbarn). Da sich die Ergebnisse bei den vier Bildern nur wenig unterschieden, sind in der angefügten Tabelle nur die jeweiligen Durchschnittswerte verzeichnet, wobei alle Angaben in Bit/Pixel erfolgen. Spalte 1 gibt den Codierungsaufwand des Prädiktionsverfahrens an. In Spalte 2 ist die maximal mögliche Einsparung aufgeführt, d.h. der Anteil der über 10 Bit hinausgehenden Codewortlängen. Ein Vergleich mit der in Teil 2 errechneten Schranke von 0.2005 zeigt, daß diese in der Praxis bei weitem nicht erreicht wird. Spalte 3 gibt die tatsächlich erzielte Einsparung an, wobei zur Auffindung der Züge von Pixeln großer Codewortlänge ein einfaches, jedoch hinreichend gutes Verfahren verwendet wurde, da das Resultat dem Maximalwert aus Spalte 2 unter Berücksichtigung der Side Information zum Teil recht nahe kommt. Die erzielte Einsparung sinkt dabei mit einer Verbesserung der Prädiktion. Der Wert aus Spalte 3 wird in Spalte 4 durch eine der neuen Bildstatistik angepaßte Huffman-Codierung noch erhöht.

Methode		Aufwand	Maximale Ersparnis	Erzielte Ersparnis	Verbess. Ersparnis
Linker Nachbar	2 x LPF	3.0746	0.0120	0.0081	0.0153
	1 x LPF	3.2532	0.0171	0.0121	0.0196
	Original	4.1824	0.0219	0.0144	0.0210
	Dir.Fil.	3.9080	0.0219	0.0152	0.0223
Mittelwert	2 x LPF	2.6321	0.0036	0.0019	0.0035
	1 x LPF	2.7977	0.0070	0.0039	0.0064
	Original	3.7879	0.0186	0.0065	0.0088
	Dir.Fil.	3.5269	0.0128	0.0060	0.0087
Adaptiv	2 x LPF	1.9199	0.0033	0.0010	0.0014
	1 x LPF	2.1923	0.0064	0.0019	0.0026
	Original	3.6038	0.0136	0.0023	0.0036
	Dir.Fil.	3.2056	0.0100	0.0020	0.0028

Der Effekt des untersuchten Verfahrens ist somit zwar offensichtlich, aufgrund seiner Größe für eine praktische Datenkompression jedoch im Normalfall wenig nützlich. Aus diesem Grund wurde der Schwerpunkt dieser Arbeit auch auf die allgemeinen theoretischen Untersuchungen in Teil 2 gelegt.

Schlußbemerkung

Die vorliegende Arbeit wurde im Rahmen einer vom BMFT (Bundesministerium für Forschung und Technologie) geförderten Forschungskooperation zwischen der Honeywell Bull AG, Köln und der RWTH Aachen angefertigt.

HIERARCHICAL IMAGE CODING

Th. Wendler
Philips GmbH, Forschungslaboratorium Hamburg
Vogt-Koelln-Str. 30, D-2000 Hamburg 54

ABSTRACT

Besides of the well known techniques for image source coding, the concepts of hierarchical coding have become more and more popular. The basic idea of these coding schemes is not only to compress images, but additionally to give the compressed image data a destinct structure. The pixel matrix is decomposed or transformed into a hierarchically ordered set of separated data blocks, each of which represents a meaningful part of the total image information (e.g.different parts of spatial resolution). These blocks can be stored, accessed and transmitted separately and sequentially so that a partial or progressive image reconstruction becomes possible.

One of the major application areas of hierarchical coding techniques are pictorial information systems with access to large image data bases, where the separation into meaningful data blocks turned out to be a powerful method for a user- or device-adapted way of reducing the transmission of irrelevant image information. Hierarchical coding schemes allow the userfriendly design of image workstations, with features such as very quick retrieval of survey images, browsing through pictorial data bases and progressive transmission with continuously increasing spatial resolution. Time periods for image retrieval can be objectively and subjectively shortened. A hierarchical separation of image data blocks supports the compatibility and integration of different image terminal devices in distributed pictorial information system.

Some of the introduced methods of hierarchical decomposition of still images are reviewed with emphasis on hierarchical transform coding techniques, i.e. block-transform and S-transform coding. The latter will be described as a hierarchical image coding scheme which is particularly suited for large pictorial information systems. It allows real-time operation, supports a VLSI-friendly implementation and can be used both for reversible or irreversible image data compression.

INTRODUCTION

For many years, urgent requirements of fast image transmission and efficient image storage stimulated investigations on problem oriented techniques of digital image data compression. Using the notation of entropy H, the choise H_0 generated by the stochastic process of an image source can be split into relevant (H_R), irrelevant

(H_I) and redundant (R) components:

$$H_0 = H_R + H_I + R \qquad \text{bit/pixel.}$$

The optimization strategy of traditional image source coding techniques is to remove part of the redundancy R and/or to eliminate parts of the entropy H_I which are considered to be irrelevant according to objective or subjective image quality criteria. Taking this approach, H_R appears to be constant, once an image, a compression algorithms and an image quality measure are given.

Hierarchical image coding techniques proceed one step further. They take additionally into account, that relevant parts of the entropy H_R must not necessarily be relevant in all image viewing situations or at any point of time in the image transmission and decoding process. Thus, in combination with the property of image data compression, hierarchical coding schemes attempt to give the compressed data set a distinct and useful **structure** that allows the partial or sequential reconstruction of the total relevant image entropy:

$$H_0 = H_{R_m} + H_{I_m} + H_I + R \qquad \text{bit/pixel}$$

$$H_{R_m} + H_{I_m} = H_R \qquad m = 0 \ldots M$$

Using this notation, H_{R_m} represents relevant parts of the entropy at a given level m of the decoding process. H_{I_m} denotes information which is suggested to be irrelevant at level m, but may be added to the image in subsequent (future) reconstruction steps. The optimization strategy in hierarchical coding is to transmit only H_{R_m}. $H_{R_m}(m=0)$ represents an initial stage of image reconstruction, which will in the following be referred to as the 'basic image'.

According to the requirements of a stepwise reconstruction process, hierarchical image coding schemes decompose (or transform) the image pixel matrix into an ordered set of M+1 separated data blocks which can be coded, transmitted, stored and accessed sequentially, so that partial reconstructions or progressive image transmission techniques become possible. Each data block is supposed to represent a meaningful part of the total image information with respect to the actual needs of a human observer or an image processing or display device. Most hierarchical image decomposition schemes define the set of separated data blocks according to different levels of spatial resolution or of spectral components. Problem-oriented hierarchical image retrieval procedures allow explicit access to

* survey information (the basic image) H_{R_0} $m = 0$
* partial information H_{R_m} $m = 1..i$
* additional information H_{R_m} $m = i..j$
* total information H_R $m = M$

A general distinction can be made between hierarchical coding methodes that generate multiple- or variable-resolution image data structures. Multiple resolution data structures trade the advantages of hierarchical decomposition against reduced compression (e.g. various pyramid- or tree decomposition schemes that introduce additional redundancy R). Variable resolution hierarchical image structures represent the relevant part of image entropy as

$$H_R = H_{R_0} + \sum_{m=1}^{M} H_{I_m}$$

and allows a reconstruction process of the form

$$\hat{H}_{R_i} = \hat{H}_{R_0} + \sum_{m=1}^{i} F_m(\hat{H}_{R_{m-1}}, H_{I_m}) \qquad i = 1..M,$$

where the reconstruction algorithm Fm in step m uses successively the result of the previous reconstruction level m-1. In this paper, the term 'hierarchical image coding scheme' will be only used for techniques that generate variable resolution hierarchical data structures with more than 2 level ($M \geq 2$).

BENEFITS OF HIERARCHICAL IMAGE DATA STRUCTURES

Partial image recontructions and progressive image transmission techniques appear to be an efficient way of reducing irrelevant data transfers and offer a user- and/or devive adapted way of image communication. Thus, hierachical image coding techniques proofed to be useful in application areas such as

* image transmission via very low bandwidth channels in interactive environments
* interactive access to large pictorial information systems with real-time response as a critical requirement (e.g.in medical diagnostics).

Particularly image transfers in Picture Archiving and Communication Systems (PACS) can benefit from hierarchical data structures. A number of advantages are obvious:

1. Problem-oriented image retrieval procedures that adapt to the needs of human observers in different work situations with images:
   * Instantanious presentation of retrieved image surveys which are already good approximations of the final image, earliest possible recognition of the principle image content
   * Ultimate fast retrieval of collections of survey images (e.g.folders, pictorial directories) from which full resolution image retrievals can be initialized (image data base access in the pictorial context)
   * Browsing through stacks of high resolution images in large pictorial data bases
   * Subjectively shortened image retrieval time (in addition to the objectively shortened retrieval time gained by data compression)

2. Retrieval procedures adapted to system requirements of distributed pictorial information system:

   * Reduced system traffic by earliest possible detection and interruption of unnecessary retrievals

   * Partial reconstructions contribution to the compatibility of terminal devices with different image handling and display capabilities.

All of these featues contribute to an overall optimization of the image dataflow in distributed pictorial information systems. In summery, hierarchical image data structures in combination with progressive transmission techniques offer a methodology for the efficient reduction of irrelevant image transmissions without (necessarily) effecting the image quality of the final image reconstruction.

HIERARCHICAL IMAGE CODING TECHNIQUES

A number of hierarchical image coding schemes are known from the literature. From the viewpoint of pictorial information systems, they can be assessed according to the following list of evaluation criteria :

* Image decomposition into a set of M+1 hierarchically ordered data blocks that (preferably) represent different levels of spatial resolution, where, for the ease of data management, M should be restricted to $2 \leq M \leq 10$
* Progressive transmission capability (with gross information first), a stepwise reconstruction process with increasing spatial resolution
* Meaningful image approximations in early reconstruction steps
* General applicability to grey level images
* Fast algorithm, esp.for the reconstruction process
* Simple (VLSI-friendly) implementation
* Data compression (with the option to compress lossless)

Hierarchical image coding approaches that fulfill most of these requirements can be roughly grouped into:

* Predictive coding in multi-resolution pyramids, eg.[1]
* Quadtree image representations [2] and predictive coding in quadtrees
* Binary tree encoding using the decomposition method of Knowlton [3]
* Subsampling schemes with encoding of predicted interpolation errors, e.g.[4]
* transform techniques (orthogonal block-transform [5]-[8] or S-transform [9][10] coding)

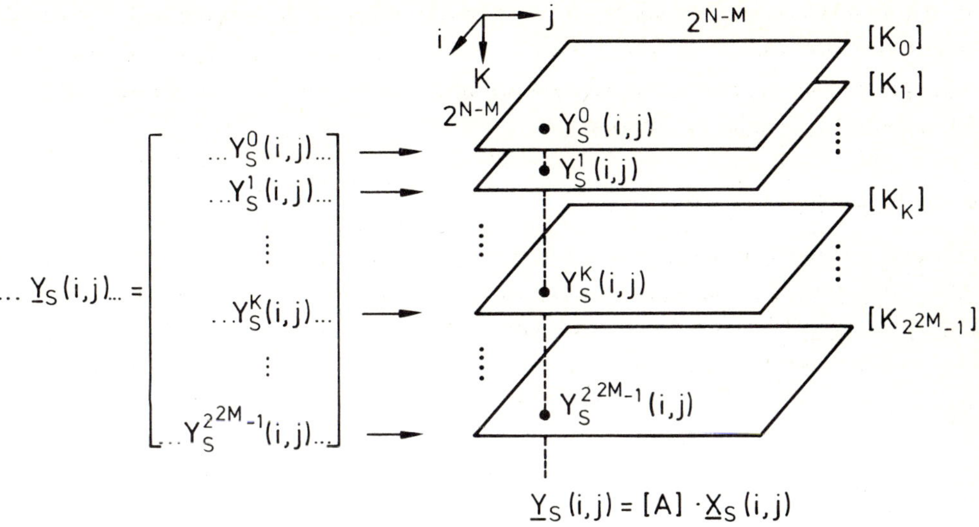

Fig.1: Hierarchical image data structure by orthogonal block transform coding, separating coefficients of the same order k into 2^{2M} matrices $[K_k]$

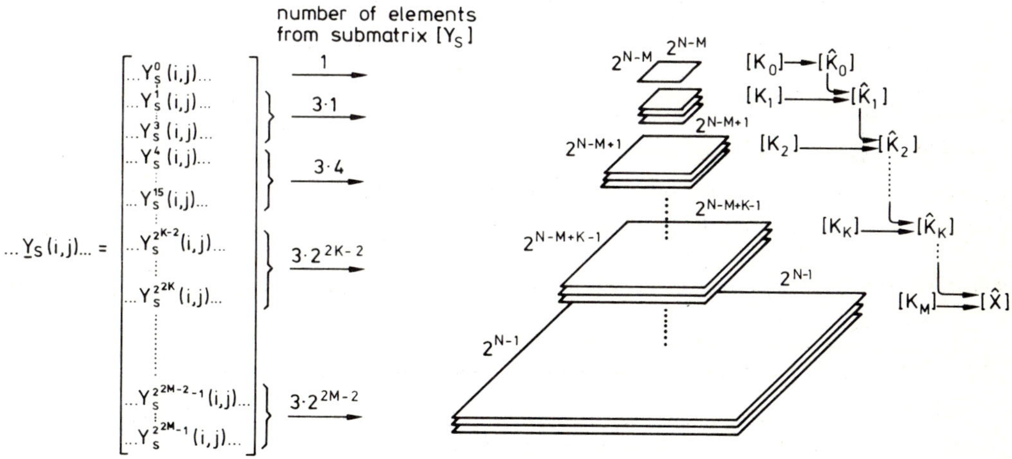

Fig.2: Hierarchical image data structure by S-transform coding after M transform steps

The use of transform techniques for hierarchical image decomposition has to be emphasized. They generally provide better compression than predictive methods and their suggested drawback of greater computational complexity can be compensated by compact and fast VLSI-implementation.

HIERARCHICAL BLOCK-TRANSFORM CODING

In orthogonal block-transform coding the original image $[X]$ (for simplicity a size of $2^N \times 2^N$ pixel is assumed) is decomposed into $(2^{N-M})^2$ submatrices $[X_s]_{ij}$ of size $2^M \times 2^M$, each of which is separately transformed into $(2^M)^2$ coefficients, in vector notation expressed by

$$\underline{Y}_s(i,j) = [A] \cdot \underline{X}_s(i,j)$$

$$\underline{X}_s(i,j) = (x_s^0(i,j),\ x_s^1(i,j), \ldots,\ x_s^k(i,j), \ldots\ x_s^{2^{2M}-1}(i,j))^T$$

$$\underline{Y}_s(i,j) = (y_s^0(i,j),\ y_s^1(i,j), \ldots,\ y_s^k(i,j), \ldots\ y_s^{2^{2M}-1}(i,j))^T$$

where $[A]$ is any transform matrix derived from orthogonal basis functions. For the inverse transform and image reconstruction process, the set of coefficients $y_s^k(i,j)$ is used as weighting factors for linear combinations of the set of orthogonal basis vectors $\underline{A}_k$ (or 2-dimensional basis pictures):

$$\hat{\underline{X}}_s(i,j) = \sum_{k=0}^{2^{2M}-1} \underline{A}_k \cdot y_s^k(i,j)$$

A hierarchical data structure for progressive image tranmsmission can be obtained, if coefficients of the same order k from all blocks (i,j) are combined to matrices $[K_k]$ as shown in fig.1. For most investigated orthogonal transforms, the level k=0 yields an approximate basic image, while higher order coefficients add detail.

Some general difficulties in the separation of hierarchical levels exist with block-transforms:

* Coefficient levels are not inherently 'meaningful' for a human observer of the reconstruction process. Several alternatives exist for the sequence of coefficients in progressive transmission systems.
* The number of 2^{2M} levels (typ.: M = 2...6) cannot be efficiently handled as separate data blocks in pictorial information systems, so that merging of several levels k becomes necessary.

Several approaches have been taken, trying to solve these problems:

* Re-ordering of coefficients and merging 2-dimensional coefficient areas to

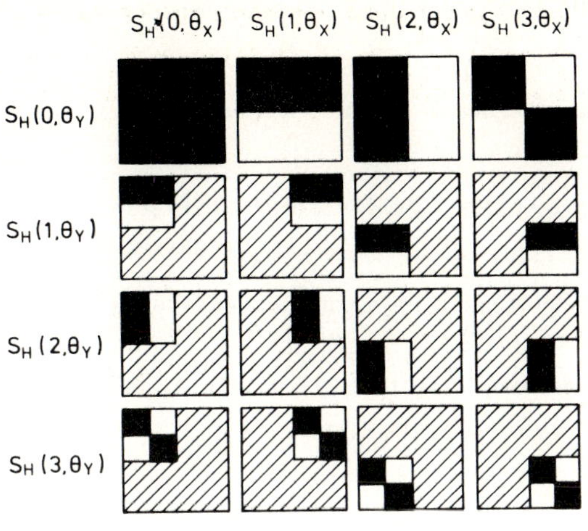

Fig.3: 2-dimensional arrangement of othonormal S-basis-functions for subpictures of 4x4 elements (black: +1, white: -1, shaded: 0)

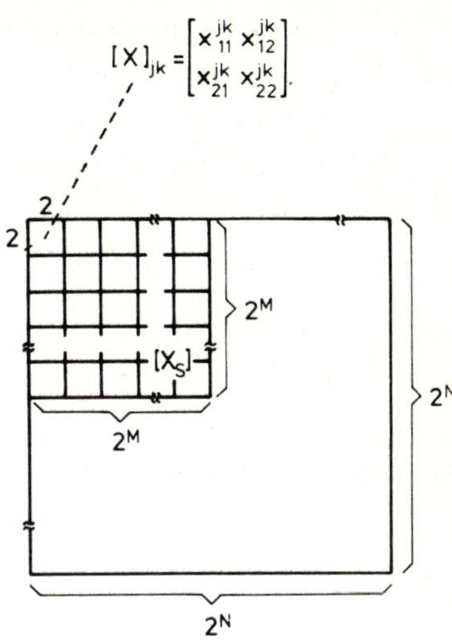

Fig. 4: Sub-matrix partitioning for S-transform coding

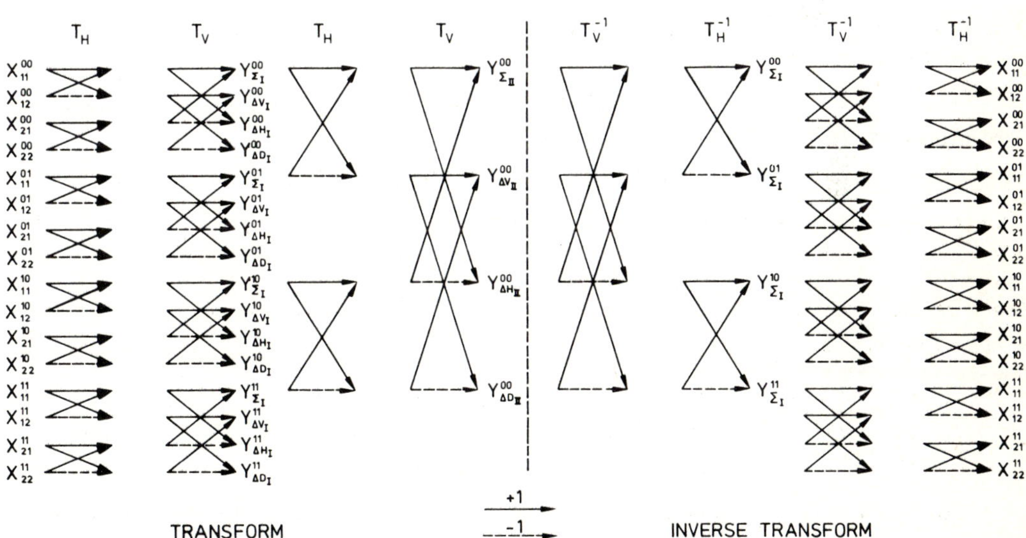

Fig.5: Signal flow diagram for S-transform coding applied to a sub-picture of 4x4 elements (M=2)

larger blocks for improved progressive transmission (in DCT and WHT systems)[5]

* Transmitting coefficients in the order of increasing k or using several other schemes for re-ordering according to different scan patterns (e.g.zig-zag ordering)[6]
* Re-ordering into larger groups according to the coefficients' statistical properties [7]
* Re-ordering according to visibility thresholds based on subjective tests [8].

None of these approaches proved to be a general solution to adapt hierarchical block-transform coding to the abovementioned evaluation criteria.

HIERARCHICAL S-TRANSFORM CODING

S-transform coding is a hierarchical image decomposition scheme which is particularly designed to be applied in pictorial information systems. The orthogonal set of 2-dimensional S-functions (in fig.3 shown for a subpicture of 4x4 elements) has been described by Lux [9] using the terminology of block-transformation. The main advantage of S-transform coding, however, is its feature to be computable following a stepwise algorithm for the transform and reconstruction process, which generates highly regular and meaningful hierarchy levels (fig.2).

Applying the forward transform, the $2^N \times 2^N$ image matrix is, according to block-transform notation, devided into 2^{2P} submatrices $[X_S]$ of size $2^M \times 2^M$ with $P = N-M$. These submatrices are further split into $(2^{M-1})^2$ matrices $[X]_{jk}$ of size 2x2 (fig.4). The first transform step now applies the Hadamard transform on all of the $[X]_{jk}$, resulting in a set of matrices $[Y]_{jk}$:

$$y_{11}^{jk} = \frac{1}{4}\left[(x_{11}^{jk} + x_{12}^{jk}) + (x_{21}^{jk} + x_{22}^{jk})\right] = y_{\Sigma}^{jk} \rightarrow [k_0^I]$$

$$y_{12}^{jk} = \frac{1}{4}\left[(x_{11}^{jk} + x_{12}^{jk}) - (x_{21}^{jk} + x_{22}^{jk})\right] = y_{\Delta V}^{jk} \rightarrow [k_1^I]$$

$$y_{21}^{jk} = \frac{1}{4}\left[(x_{11}^{jk} - x_{12}^{jk}) + (x_{21}^{jk} - x_{22}^{jk})\right] = y_{\Delta H}^{jk} \rightarrow [k_2^I] \qquad \text{for all } j,k$$

$$y_{22}^{jk} = \frac{1}{4}\left[(x_{11}^{jk} - x_{12}^{jk}) - (x_{21}^{jk} - x_{22}^{jk})\right] = y_{\Delta D}^{jk} \rightarrow [k_3^I]$$

The matrix elements y_{mn}^{jk} represent the mean (Σ) and vertical (ΔV), horizontal (ΔH) and diagonal (ΔD) differences of $[X]_{jk}$. All computed differences form the lowest level matrix $[K_M]$ of the hierarchical data structure in fig.2. Only the matrix of all sum-coefficients y_{Σ}^{jk} is processed further, using successively the identical algorithm.

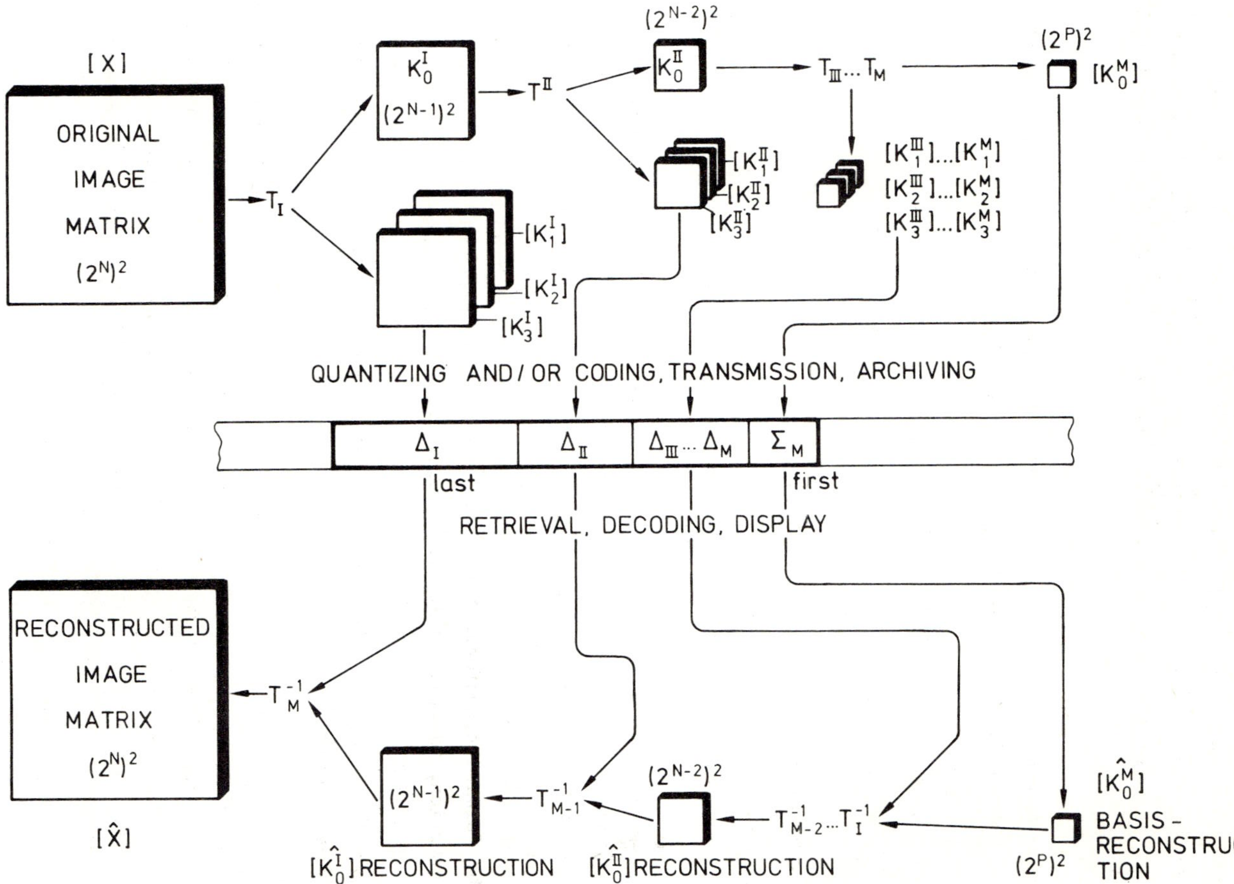

Fig.6: Hierarchical image decomposition and reconstruction process applying the S-transform in pictorial information systems

After M transform steps the process is completed, the remaining set of y_{Σ}^{jk} coefficients serves as a basic image $[K_o]$. Fig.5 shows the dataflow 'butterfly'-diagram of the algorithm for a subpicture of 4x4 elements (M=2). Only 2 different types of computation, T_H and T_V, have to be performed, which ultimately simplifies hardware implementation.

The inverse transform uses the identical stepwise algorithm in the reverse order, successively recontructing levels of x_{mn}^{jk}:

$$\hat{x}_{11}^{jk} = \frac{1}{4}\left[(y_{\Sigma}^{jk} + y_{\Delta V}^{jk}) + (y_{\Delta H}^{jk} + y_{\Delta D}^{jk})\right]$$

$$\hat{x}_{12}^{jk} = \frac{1}{4}\left[(y_{\Sigma}^{jk} + y_{\Delta V}^{jk}) - (y_{\Delta H}^{jk} + y_{\Delta D}^{jk})\right]$$

$$\hat{x}_{21}^{jk} = \frac{1}{4}\left[(y_{\Sigma}^{jk} - y_{\Delta V}^{jk}) + (y_{\Delta H}^{jk} - y_{\Delta D}^{jk})\right]$$

$$\hat{x}_{22}^{jk} = \frac{1}{4}\left[(y_{\Sigma}^{jk} - y_{\Delta V}^{jk}) - (y_{\Delta H}^{jk} - y_{\Delta D}^{jk})\right]$$

With every inverse transform step, the resolution of the partially reconstructed image is progressively increased by a factor of 2. Fig.6 shows the complete hierarchical decomposition and reconstruction process.

Fig.7 illustrates the intermediate reconstruction steps during progressive transmission of an S-transformed X-ray image. The 1024x1024 pixel original has been transformed in 6 steps down to a 16x16 basic image, the reconstruction has been stopped intendedly at a resolution on 512x512 pixel. Transmission rates (in bit/pixel) are shown respectively for all reconstruction levels, based on the lossless compression of difference coefficients by a factor of 2.5 (applying Huffman coding).

S-transform coding supports both, reversible and irreversible data compression. For the application in picture archiving and communication systems in medicine, lossless compression has been investigated for various kinds of medical images [10] showing compression ratios between 2 and 4. Fig.8 gives an example for cumulative compression ratios C(s) (=compression, if the transform is stopped at stage s) for a number of different high resolution digital radiographs (DR) generated by a large area detector system, and for digital subtraction angiograms (DSA).

Hierarchical image coding by S-transformation is proposed to be used as a standard for data compression in pictorial information systems in medicine. It has been taken as an example for a candidate standard method in an extention of the ACR/Nema image data format standard [11].

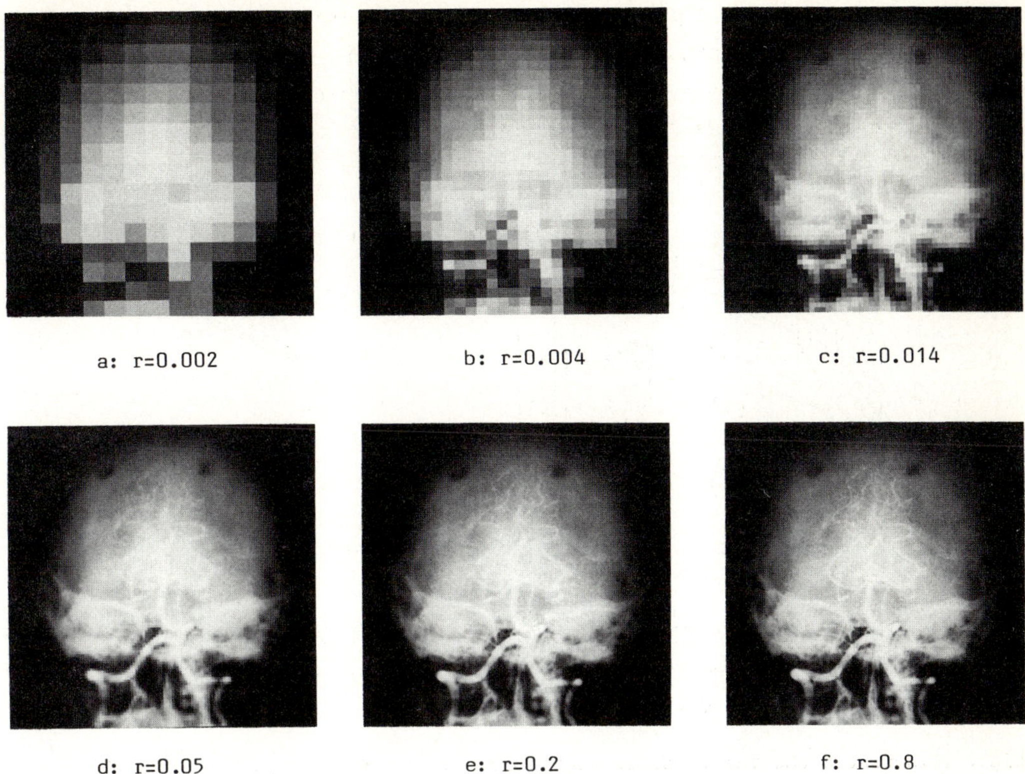

a: r=0.002　b: r=0.004　c: r=0.014

d: r=0.05　e: r=0.2　f: r=0.8

Fig.7: Intermediate reconstruction steps during progressive transmission of an S-transformed image (original data 1024x1024 pixel) with according transmission rates in bit/pixel (a: 16x16 basic image....f: 512x512 reconstruction)

REFERENCES

[1] P.J. Burt, E.H. Adelson: The Laplacian Pyramid as a Compact Image Code. IEEE Tr.COM-31, No.4, April 1983, pp.532-540

[2] H. Samet: The Quadtree and related Hierarchical Data Structures. Computing Surveys, vol.16, no.2, June 1984, pp.187-270

[3] K. Knowlton: Progressive Transmission of Grey-Scale and Binary Pictures by Simple, Efficient, and Lossless Encoding Schemes. Proceedings of the IEEE, Vol.68, No.7, July 1980, pp.885-896

[4] G. Seelmann: Progressive Farbfernseh-Standbildübertragung mit 64 kbit/s Übertragungsbitrate. NTG-Fachtagung: Wege zum integrierten Kommunikationsnetz, 25.-27. März 1984, Berlin, pp./304-315

[5] K. Takikawa: Fast Progressive Reconstruction of a Transformed Image. IEEE Tr.-IT-30, No.1, Jan.1984, pp. 111-117

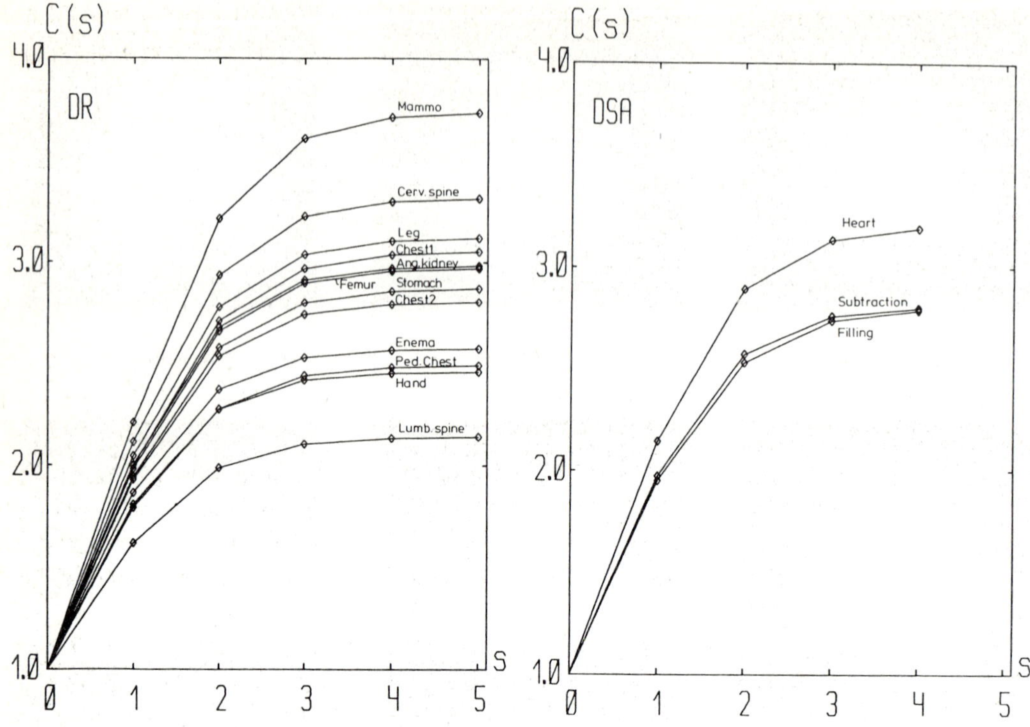

Fig.8: Cumulative compression ratios C(s) after s transform steps and uncompressed basis matrix for different digital radiographs.

[6] K.N. Ngan: Image Display Techniques Using the Cosine Transform. IEEE Tr.ASSP-32, No.1, Feb.1984, pp.173-177

[7] E. Dubois, J.L. Moncet: Transform Coding for Progressive Transmission of Still Pictures. Globecom '85, IEEE Global Telecom.Conf., New Orleans, Dec.2-6, 1985, pp. 348-352

[8] H. Lohscheller: Vision Adapted Progressive Image Transmission. Proc.Signal Processing II: Theories and Applications, EURASIP 1983, pp.191-194

[9] P. Lux: A Novel Set of Closed Orthogonal Functions for Picture Coding. AEÜ, Band 31 (1977), Heft 7/8, pp.267-274

[10] Th. Wendler: Verfahren für die hierarchische Codierung von Einzelbildern in medizinischen Bildinformationssystemen. Dissertation, Fakultät für Elektrotechnik, RWTH Aachen, 1987

[11] ACR-NEMA Standards publication, No.300-1985, Digital Imaging and Communications, Washington, 1985, Extention for coded image formats to be published.

DATA COMPRESSION AND DECORRELATION IN DIGITAL SIGNAL PROCESSING OF RANDOM DATA

Mohammad Maqusi, and Ibrahim Makhamreh
Electrical Engineering Department
University of Jordan
Amman, Jordan

Abstract

The work of this paper is concerned with certain investigations in data compression and decorrelation for application in the digital signal processing of random data utilizing linear predictive coding and discrete orthogonal transforms.

I. Generation of Pseudorandom Data

A useful method for the generation of uniform random data is called the multiplicative congruential method [1]:

$$x_{i+1} = c\, x_i \pmod{m}; \quad \text{max.cycle} = m/4 \tag{1}$$

Proper choices of parameters are $m = 2^b$ (for b 2); and $c = 8t \pm 3$, $t=1,2,3,\ldots$ It is further recommended to choose a value of c near $\sqrt{m}$. The generation of exponentially distributed data [2] may be achieved by applying the transformation

$$y_i = -\ln x_i \tag{2}$$

Generation of Gaussian data may, on the other hand, be achieved by invoking the central limit theorem, and in accordance with the transformation

$$y_i = 1/n \sum_{k=i}^{i+n-1} x_k, \quad i=1,2,..,N. \tag{3}$$

where n is the uniform data size, and N is the Gaussian data size. In considering random data, a quantity of special significance is the autocorrelation function (acf) defined by

$$R_{xx}(k) = 1/N \sum_{i=1}^{N} x(i)\, x(i+|k|), \tag{4}$$

In particular, for zero-mean data, the variance is $\sigma_x^2 = R_{xx}(0)$. Fig.1 shows the acf behavior for the various indicated distributions, including speech data (s) which seems to bear close resemblance to the Gaussian distribution.

A useful measure in assessing the performance of the processing techniques is a signal-to-noise error ratio (SNR), equivalently represented by a normalized mean-square error (MSEn),

$$\mathrm{MSEn} = \sum_{i=1}^{N} (x_i - \hat{x}_i)^2 \Big/ \sum_{i=1}^{N} x_i^2 \qquad (5)$$

where $\{\hat{x}_i\}$ is the reconstructed data sequence subsequent to data compression.

II. The LPC Technique

Linear predictive coding (LPC) relies on exploiting the correlation properties of the data in order to predict a present value form P past values [3]

$$\hat{x}(i) = \sum_{k=1}^{P} c(k)\; x(i-k) \qquad (6)$$

The optimum values of the predictor coefficients are specified by minimizing the mean-square error, and are further determined by

$$R_{xx}(k) = \sum_{j=1}^{P} c(j)\; R_{xx}(k-j);\; k=1,2,\ldots,N \qquad (7)$$

In applying LPC, we quantize the error (difference) signal, e(i) $e(i)=x(i)-\hat{x}(i)$ for data transmission. For signal reconstructiuon, quntization error, $e_q(i)$, is added to e(i), thereby resulting in a quantization SNR, $SNR_q = \sigma_e^2/\sigma_q^2$, and designates a ratio of the variances of the difference and quantization error signals, respectively. The overall SNR is given by

$$\mathrm{SNR} = \sigma_x^2 / \sigma_q^2 = (\sigma_x^2/\sigma_e^2)\;(\sigma_e^2/\sigma_q^2) \qquad (8)$$

$$= G_p + \mathrm{SNR}_q$$

where G_p is the LPC gain over straightforward PCM.

Fig.2 illustrates performance improvement achieved by LPC over PCM. It is interesting to note that, while LPC outperforms PCM for exponential and Gaussian distributions, PCM does better for uniform data

III. Transform Coding

The transform coding (TC) techniques of interest in our invetigations are the discrete Fourier transform (DFT), discrete cosine transform (DCT), and the Hadamard transform (HT), [3], [4] We consider their data compression performance with regard to variations in data block size N, and proposed compression ratios CR. In general, it is found that all transforms do better with larger N. However, as the illustrating results contained in Figures 3-5 indicate, we observe that the

HT seems to offer a good choice for uniform data, while the DCT and DFT may be more appropriate for processing exponential and Gaussian data. We should also keep in mind the complexity of imlementation of the transforms. In this regard, HT is the simplest, and the DFT is the most complex. The DCT is a compromise between the two.

IV. Data Decorrelation

The significance of data decorrelation by a transform coder is linked to its performance in data compression; the greater the decorrelation, the better the data compression. A useful, and rather simple measure for decorrealtion is suggested by the residual correlation in the tranformed data

$$r^2 = \frac{1}{N(N-1)} \sum_{\substack{i,j \\ (i \neq j)}} \frac{R^2_{yy}(i,j)}{R_{yy}(i,i)\ R_{yy}(j,j)} \qquad (9)$$

where $y(s)$, $s=1,2,\ldots,N$ denotes the output sequence of TC. In applying (9), it is found that the stated transforms attain varying degrees of decorrelation. In particular, our investigations for the Gaussian distribution, for instance, have revealed better decorrelation by the DCT. And if we refer to Fig.5, we observe better data compression performance by the DCT.

References

1. R.F. Chambers, "Random-Number Generation", IEEE Spectrum, PP. 48-56, Feb. 1967.
2. R.W. Hamming, Numerical Methods for Scientists and Engineers, McGraw-Hill Book Co., New York, Ch. 32, 1962.
3. N.S. Jayant, and P. Noll, Digital Coding of Waveforms, Prentice-Hall, Englewood Cliffs, NJ, 1984.
4. K.R. Rao, "Theory and the Applications of the Discrete Cosine Transform", Proc. JIEEEC, Amman, Jordan, 1985.
5. M. Maqusi, Applied Walsh Analysis, Heyden-Wiley (John Wiley), London, 1981.

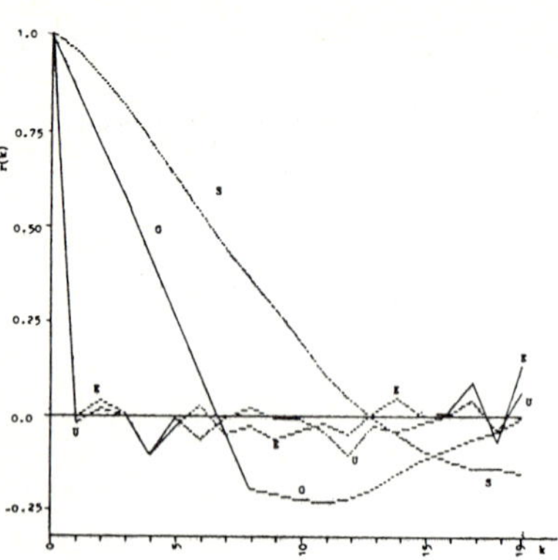

Fig.1 Random data acf's

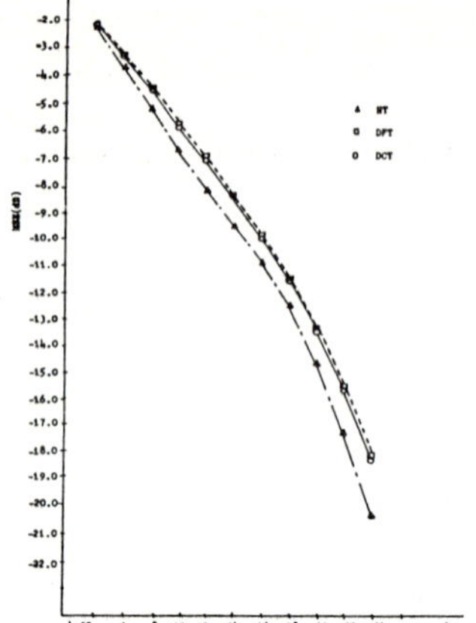

Fig.3 TC: uniform data; N=32

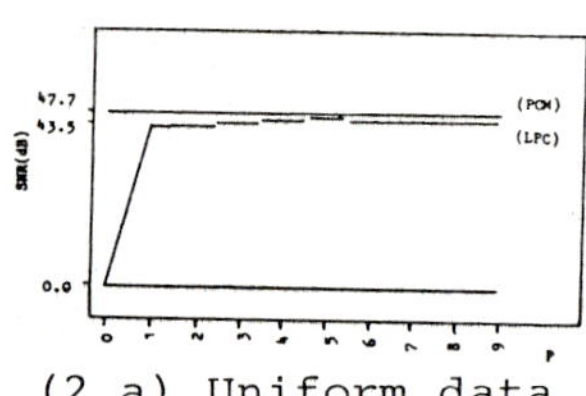

(2.a) Uniform data

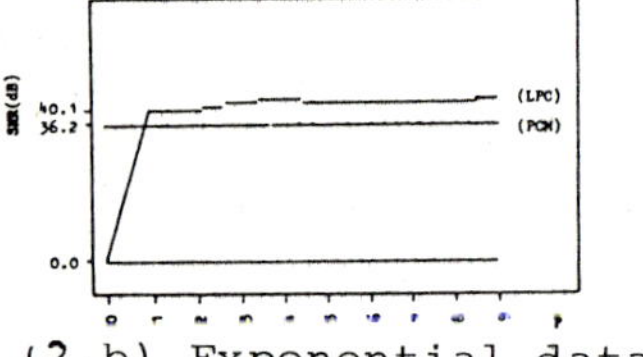

(2.b) Exponential data

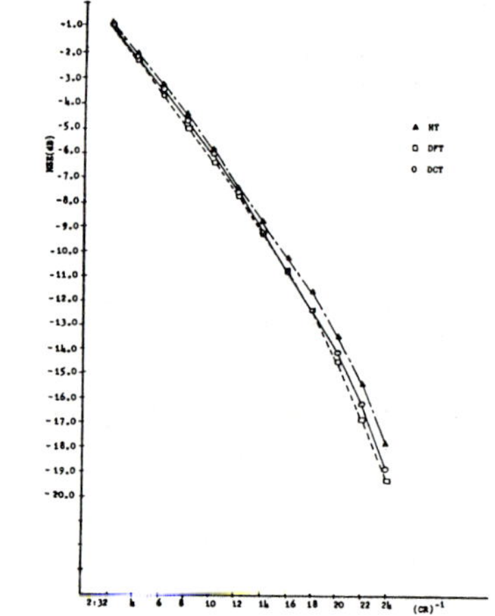

Fig.4 TC: Exp. data; N=32

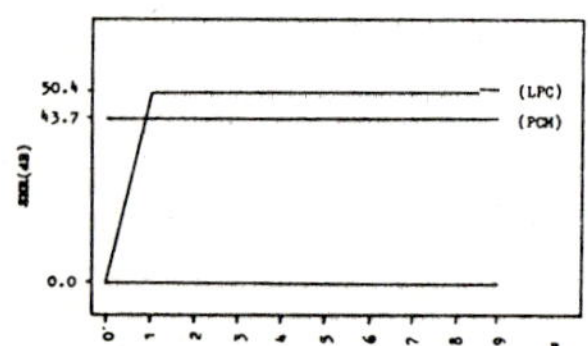

(2.c) Gaussian data

Fig.2 LPC Performance

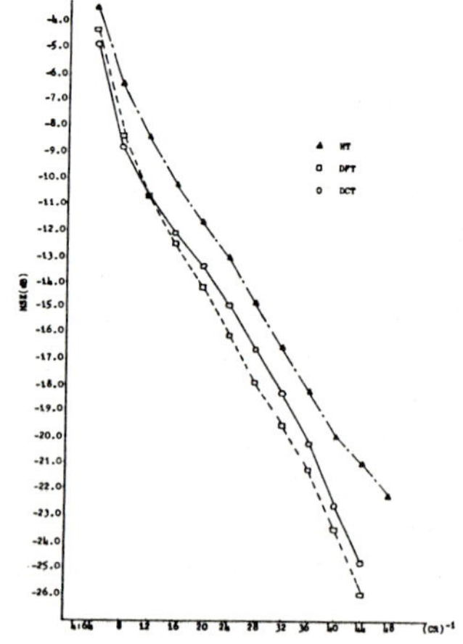

Fig.5 TC: Gaussian data; N=32 →

SEGMENTIERUNG UND VEKTORQUANTISIERUNG VON BILDSIGNALEN AUF DER GRUNDLAGE EINES COMPOSITE SOURCE MODELLS

H. Bohlmann, P. Meissner

Institut für Allgemeine Nachrichtentechnik
Universität Hannover

1. Einleitung

Die statistische Struktur von Bildsignalen legt es nahe, als Quellenmodell eine zweidimensionale Composite Source (CS) zu verwenden, die sich unter Verwendung von Zufallsfeldern definieren läßt. Das in diesem Zusammenhang entstehende Problem der Bildzerlegung wird durch ein Region Growing / Region Merging Verfahren (Segmentierung) mit anschließender Klassifikation gelöst. Auf der Grundlage des CS-Modells werden Vektorquantisierer entworfen, die gemeinsam mit einem Kantencodieralgorithmus zur Übertragung von Bildern mit stark reduzierter Rate Verwendung finden.

2. Composite Source Modell

Die zweidimensionale Abtastung eines Bildes überführt eine kontinuierliche Ortsfunktion f(x,y) in ein ortsdiskretes Signal f(m,n) mit $\{0 \le m \le M-1 \ , \ 0 \le n \le N-1\}$. Dies läßt sich auffassen als Realisierung eines Zufallsfeldes $F_{m,n}$. Eine der statistischen Struktur von Bildsignalen adäquate Modellbildung folgt aus dem Ansatz, $F_{m,n}$ aus einem Ensemble von Teilfeldern $G^1_{m,n},\ldots,G^K_{m,n}$ entstehen zu lassen, indem ein Schaltfeld $S_{m,n}$ mit dem Alphabet $\Lambda=\{1,\ldots,K\}$ in jedem Punkt (m,n) ein geeignetes Teilfeld auswählt. Ausgehend von dieser Grundannahme definiert

$$F_{m,n} = G^{S_{m,n}}_{m,n} \quad \text{mit } F_{m,n} = G^k_{m,n} \text{ für } S_{m,n} = k \in \Lambda$$

das Ausgangsfeld eines Composite Source Modells für Bildsignale (Bild 1). Um mit diesem Modell Bildvorlagen beschreiben zu können, d.h. statistische Parameter zur Definition geeigneter Zufallsfelder messen zu können, sind einschränkende Annahmen notwendig. So ist es zweckmäßig, nur stationäre Zufallsfelder zuzulassen; für erste statistische Aussagen ist sogar die Annahme sinnvoll, daß die Zufallsvariablen $G^k_{m,n}$ der k-ten Teilquelle gegenseitig statistisch unabhängig sind und die gleiche Verteilung aufweisen. Die stochastische Struktur des Quellenmodells betreffend sind z.B. die Definitionen von zerlegbaren bzw. vollständig zerlegbaren CS-Modellen aus dem eindimensionalen Fall /1/ übertragbar. Das zerlegbare CS-Modell besitzt keine statistischen Bindungen zwischen den Teilfeldern, während beim vollständig zerlegbaren Modell zusätzlich noch das Schaltfeld unabhängig vom Ensemble der Teilfelder ist. Damit steht eine Familie von Modellen zur Verfügung , die ein breites Spektrum von Anwendungen abdeckt. Wichtigste Voraussetzung ist allerdings, daß das Partitionierungsproblem für eine gegebene Bildvorlage und damit die Schätzung einer Realisierung des Schaltfeldes gelöst wird.

3. Segmentierung und Klassifizierung von Bildsignalen

Die Aufgabe, aus einer Bildvorlage die Anzahl der Teilfelder und die Realisierung des Schaltfeldes zu schätzen, zerfällt in die Schritte "Segmentierung", d.h. homogene Teilbereiche zu erzeugen und damit Objekte zu definieren, und "Klassifizierung", d.h. für Objekte, die einem gemeinsamen Teilfeld angehören, die Klassenzugehörigkeit ($S_{m,n} = k \in \Lambda$) zu bestimmen.

Die Suche nach homogenen Bereichen im Signal muß unter der Randbedingung erfolgen, daß sie dem Schaltprozeß der Quelle zugeordnet werden können. Dies ist gewährleistet, wenn eine vollständige Segmentierung /2/ vorliegt, das ist eine Unterteilung des Signals f in disjunkte, nichtleere Teilmengen $f_1,\ldots,f_T$, sodaß mit einem zu definierenden Homogenitätsprädikat $P(f_i)$ gilt:

(i) $\bigcup_{i=1}^{T} f_i = f$ (ii) f_i ist zusammenhängend

(iii) $P(f_i)$ =WAHR für i = 1,...,T (iv) $P(f_i \cup f_j)$ = FALSCH für $i \neq j$, f_i, f_j benachbart.

Je nach Art der Vorlage können zur Trennung der Bereiche grauwerthomogene oder texturorientierte Prädikate verwendet werden, wobei komplexere Szenen aufwendigere Kriterien erfordern.

Mit den Verfahren des Region Growing (RG) werden automatisch vollständige Segmentierungen erzeugt, indem man von einem Initialpunkt i ausgehend solange Bildpunkte vereinigt, wie für das wachsende Gebiet $P(f_i)$=WAHR gilt. Dabei wird das im Laufe des Verfahrens akkumulierte "Wissen" über ein Gebiet zur Entscheidung herangezogen. Aus der Vielzahl möglicher Algorithmen /3/ wurde wegen des geringen Speicherplatzbedarfs und einer gewünschten hohen Rechengeschwindigkeit ein sequentielles "Row-by-Row"-Verfahren ausgewählt.

Von großer Bedeutung ist die Wahl eines geeigneten Homogenitätsprädikates als bestimmender Parameter des Wachstumsprozesses. Verschiedene Prädikate (auf Grund des Datenmaterials grauwertorientiert) wurden getestet, wobei sich ein auf dem statistischen T-test /4/ basierendes Kriterium als besonders geeignet herausgestellt hat. Unter der Annahme, daß alle Bildpunkte in f_i und ein Testpunkt y identisch und unabhängig normalverteilt sind, gilt $P(f_i \cup y)$ = WAHR, falls:

$$T = \sqrt{\frac{(N-1)N(y-\overline{x})^2}{(N+1)s^2}} < t_{N-1}(\beta) \quad \text{mit} \quad \overline{x} = \frac{1}{N}\sum_{(m,n)\in f_i} f(m,n) \quad \text{und} \quad s^2 = \sum_{(m,n)\in f_i} (f(m,n)-\overline{x})^2$$

β - Signifikanzniveau , N - Zahl der Punkte in f_i

Die Testgröße T folgt einer "Student"-Verteilung mit N-1 Freiheitsgraden. Das Signifikanzniveau β ist ein wählbarer Parameter. Die Werte $t_{N-1}(\beta)$ sind tabelliert; $t_{N-1}(\beta)$ ist groß für kleines N. Je größer N wird, umso kleiner muß $(y-\overline{x})$ sein, um f_i weiter wachsen zu lassen.

Der Nachteil aller RG-Verfahren besteht darin, aufgrund von Rauschen und realen Kanten viele kleine, nicht interpretierbare Gebiete zu erzeugen. Außerdem bewirkt das visuelle System des Menschen, daß viele Bereichsgrenzen als künstlich empfunden werden. Durch Zusammenfassen benachbarter Gebiete (Region Merging = RM) läßt sich eine verbesserte Segmentierung mit weniger Bereichen erzeugen. Da sich die Bereichskennwerte während des RG-Prozesses verändern können, wird im RM-Prozeß geprüft, ob für zwei Nachbarregionen $P(f_i \cup f_j)$ =WAHR gilt. Es lassen sich jetzt z.B. auch abgeschwächte Forderungen an die Homogenität stellen.

Für die in Bild 3 gezeigte, mit 8 bit/Bildpunkt (8 bpp) linear quantisierte Vorlage, wurde eine grauwerthomogene Segmentierung berechnet. Das Signifikanzniveau wurde für den RG-Schritt zu β=0.5 gewählt. Dieser Wert gewährleistete eine sehr gute Homogenität der Gebiete, wobei allerdings war ihre Anzahl hoch ($\approx$ 20700). Im RM-Prozeß wurde β im Zuge mehrerer Verarbeitungsschritte verändert (β=0.5,..,β=0.0001). Diese Strategie bewirkte eine gleichmäßige Vergröberung der Segmentierung, die sich - gemessen an der Komplexität der Szene - durch wenige kleine Regionen auszeichnete. Die Gesamtzahl der Bereiche betrug am Ende 2700.

Der nächste Schritt erfordert die Klassifizierung der Bereiche, wobei die Anzahl unterschiedlicher Bereichstypen die Anzahl der Teilquellen des Modells bestimmt. Im einfachsten Fall ist eine Bereichsklassifikation mit einem Regionsmerkmal möglich. Hier wurden die mittleren Grauwerte $\bar{x}_i$ der Gebiete f_i mit einem skalaren Optimalquantisierer auf das Alphabet $\Lambda = \{1,...,4\}$ abgebildet. Die Bilder 2a - 2d zeigen die Bildbeiträge der 4 Teilquellen. Es wird deutlich, daß große Bereiche guter Homogenität gefunden wurden. Langsame Grauwertübergänge verbunden mit Hintergrundrauschen erzeugen jedoch viele zerfranste Bereichsgrenzen.

Wenn ein Merkmalvektor eine Region charakterisiert, läßt sich das Klassifikationsverfahren auf den mehrdimensionalen Fall erweitern. Der K-MEANS Algorithmus /5/, der hier auch für den eindimensionalen Sonderfall verwendet wurde, liefert dann eine Partitionierung des Merkmalraumes in K Gebiete.

4. Vektorquantisierung der Teilquellensignale

Eine mögliche Anwendung eines Quellenmodells ist es, den Entwurf von Quellencodern zu optimieren. Dies bedeutet in unserem Fall, jeder Teilquelle ein eigenes, an ihre statistischen Eigenschaften angepaßtes Coder/Decoder Paar zuzuordnen. Der geschätzte Zustand der Quelle wird als Nebeninformation zum Decoder übertragen (Bild 1). Als Codierverfahren für niedrige Übertragungsraten ist die Vektorquantisierung /6/ geeignet. Das Bild wird in Teilmatrizen der Größe a · b zerlegt, wobei das Schaltfeld die Zuordnung der Matrizen zu den Teilcodern TQ_k bestimmt. Der Teilcoder TQ_k bildet aus einer Teilmatrix einen Vektor $\mathbf{X}$ und sucht im Codebuch einen Repräsentativvektor $\hat{\mathbf{X}}_j$ ($j=1,...,J_k$, J_k = Anzahl der Codeworte des k-ten Codebuches), sodaß der quadratische Fehler $d(\mathbf{X},\hat{\mathbf{X}}_j) = \|\mathbf{X} - \hat{\mathbf{X}}_j\|^2$ minimal wird. Der Index des Vektors $\hat{\mathbf{X}}_j$ genügt dem Decoder zur Rekonstruktion des Bildes.

Besondere Berücksichtigung bei der Codierung erfordert das visuelle System des Menschen mit seiner ausgeprägten Kantenempfindlichkeit. Der Entwurf von speziellen Teilcodern für Kantenbereiche ist sinnvoll /7/, da ein Vektorquantisierer im rekonstruierten Bild eine ausgeprägte Blockstrukur erzeugt und diese sich störend in Kantenzonen auswirkt. Wenn die Teilquellen grauwerthomogene Bildanteile liefern, sind Kante und Bereichsgrenze äquivalent. Dann genügt es, für Teilmatrizen mit Bereichsgrenzen und ihren unterschiedlichen Richtungen Kantenteilcoder zu entwerfen, die das Codiersystem ergänzen.

Die Berücksichtigung der Raten zur Übertragung der Codewortindizes und der Symbole des Schaltprozesses führt auf die Gesamtübertragungsrate R . Es gilt:

$$R = \frac{1}{d} \log_2 \sum_{k=1}^{K} J_k$$ in bit/Symbol, $d = a \cdot b$ - Vektordimension , K - Anzahl der Teilcoder, J_k - Anzahl der Codeworte des k-ten Teilcoders.

Es wurden Teilcoder für vier Teilquellen entworfen (s. Bild 2 a-d), sowie Teilcoder für Teilmatrizen mit Bereichsgrenzen, die durch vier Vorzugsrichtungen (0 , 45 , 90 , 135 , s./8/) unterschieden wurden. Die Vektordimension betrug d = 4· 4 , die Zahl der insgesamt für alle Teilquellen zur Verfügung stehenden Codeworte J = 512 (≅ 0.526 bpp, Bild 4). Es zeigte sich, daß eine Umverteilung der Codeworte zwischen den Teilquellen sich dann visuell positiv bemerkbar macht, wenn der Anteil der Codeworte für Kantenteilquellen steigt. Ein Vorteil des CS-Codiersystems ist, daß die Teilquellen mit variablem Codieraufwand verarbeitet werden können. Da außerdem die Seiteninformation stark redundante Anteile enthält, bieten sich weitere Untersuchungen zur Senkung dieser Teilrate an.

/1/ R.J. Fontana
"Limit Theorems for Slowly Varying Composite Sources"
IEEE Transactions on Information Theory
Vol. IT-26, No. 6, Nov. 1980, p. 702-709

/2/ Th. Pavlidis
"Structural Pattern Recognition"
Springer Verlag 1977

/3/ R.M. Haralick
"Survey Image Segmentation Techniques"
CVGIP Vol. 29, pp100-132, 1985

/4/ L. Sachs
"Applied Statistics: A Handbook of Techniques"
2-nd edition, Springer Verlag 1982

/5/ R.O. Duda, P.E. Hart
"Pattern Classification and Scene Analysis"
J. Wiley Inc., New York, 1973
Kap. 6, S. 250

/6/ R. Gray
"Vector Quantization"
IEEE ASSP Magazine, April 1984

/7/ B. Ramamurthi, A. Gersho
"Classified Vector Quantization of Images"
IEEE Trans. on ASSP, No.6 1986

/8/ W.K. Pratt
"Digital Image Processing"
J. Wiley Inc., New York 1978
Kap. 17, S. 493

Unser Dank gilt der Deutschen Forschungsgemeinschaft und der Fulbright-Komission für die Unterstützung dieser Arbeit.

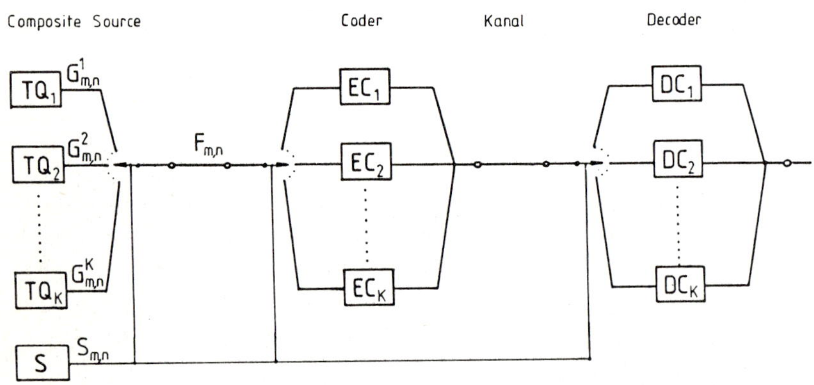

Bild 1 : Composite Source Modell, verbunden mit einem CS-Quellencodiersystem

Bild 2a : Bildbeitrag TQ_1

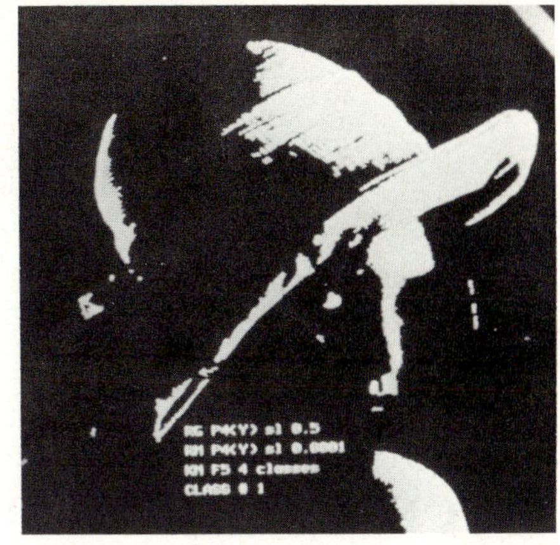

Bild 2b : Bildbeitrag TQ_2

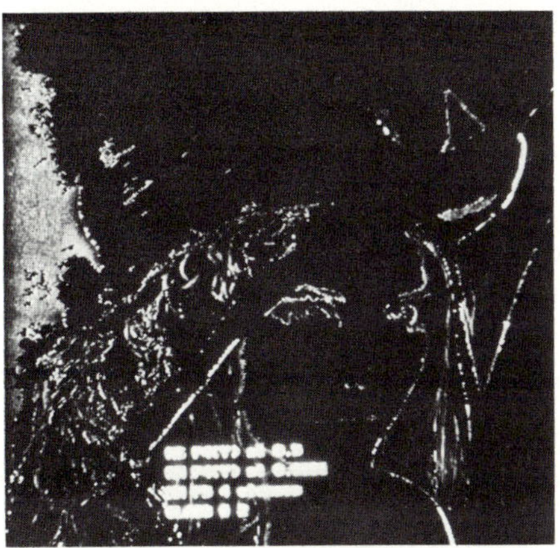

Bild 2c : Bildbeitrag TQ_3

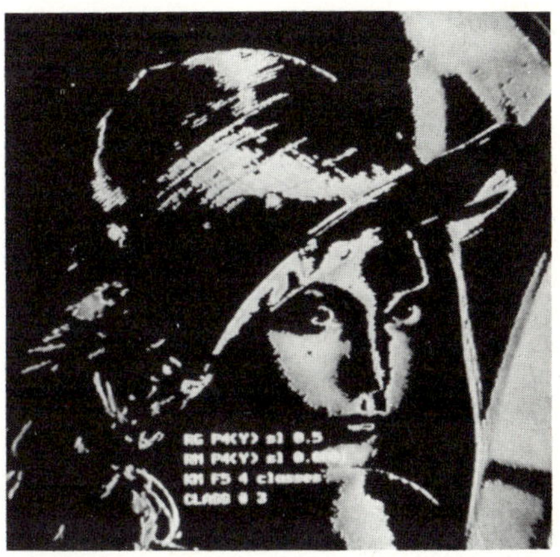

Bild 2d : Bildbeitrag TQ_4

Bild 3 : Original

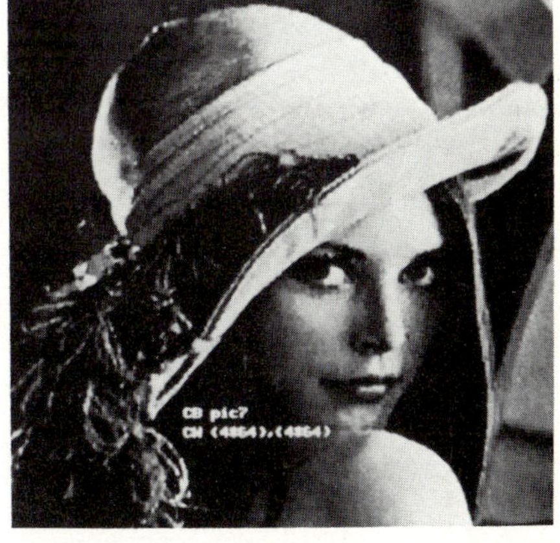

Bild 4 : Rekonstruiertes Signal

Technologie der Bildverarbeitung

NATURAL BASIS FUNCTIONS FOR IMAGE ANALYSIS

Per-Erik Danielsson
Department of Electrical Engineering
Linköping University
S-581 83 Linköping, Sweden

ABSTRACT

The search for generally applicable operators for image processing is important. The class of operators based on circular harmonic functions (spherical harmonics for 3D-signals) are strong contenders. They form the basis for optimal rotation invariant feature detectors and they lend themselves to separable kernel design in the discrete case. An interesting feature of these natural basis functions is that they are similarly shaped in signal and frequency domain. In the largely unexplored three-dimensional case, the natural basis functions might be the only road to conceptually simple tasks like edge and line detection.

1. INTRODUCTION

We hold these things to be selfevident.

i) Most, if not all tasks in computer vision start with <u>estimations of local events</u>, i.e. features or patterns of interest.

ii) An estimator $b_i(x,y)$, or operator as we will call it, should be <u>linear</u>: the output response f_i is proportional to the amplitude of the input image $f(x,y)$.

$$f_i(0,0) = \iint_{-\infty}^{\infty} f(u,v) \cdot b_i(-u,-v)\, du\, dv \tag{1.1}$$

$$f_i(x,y) = \iint_{-\infty}^{\infty} f(u,v) \cdot b_i(x-u, y-v)\, du\, dv \tag{1.2}$$

$$= f(x,y) * b_i(x,y) \tag{1.3}$$

iii) We often need <u>several</u> linear operators $b_i(x,y)$, $i = 1, 2, \ldots, N$. All of these are <u>local</u>, i.e. they are zero for $|x| < X$, $|y| < Y$

iv) Optimally, every two operators should produce fully independent measurements of the same neighborhood. This means that the operators b_i should form an <u>orthogonal set of basis functions</u>. The basis set is <u>orthonormal</u> if for all i, j

$$b_i * b_j = 0 \qquad \text{for } i \neq j$$

$$\iint_{-\infty}^{\infty} [b_i(x,y)]^2 \, dx \, dy = \iint_{-\infty}^{\infty} [b_j(x,y)]^2 \, dx \, dy = 1$$

v) Given a basis set $[b_1, b_2, \ldots, b_N]$ the function $f(x,y)$ is <u>expanded</u>, or at least approximated around the origin as

$$f(x,y) \approx \hat{f}(x,y) = \sum_{i=1}^{N} f_i(0,0) \cdot b_i(-x,-y) \tag{1.4}$$

where $f_i(0,0)$ is now a coefficient obtained from (1.1). More generally, $f(x,y)$ is expanded around any point u, v in the xy-plane according to

$$f(x,y) \approx \hat{f}(x,y) = \sum_{i=1}^{N} f_i(u,v) \cdot b_i(u-x,\ u-y) \tag{1.5}$$

where $f_i(u,v)$ is obtained from (1.2)

vi) Equation (1.1) computes the <u>projection</u> f_i of the neighborhood of (0,0) of $f(x,y)$ onto the basis function b_i. Likewise equation (1.2) computes the projection of the neighborhood around (x,y) onto b_i.

vii) If the basis functions form a <u>complete orthonormal set</u>, (to be defined), then we can define the <u>magnitude</u> response as the Euclidean norm

$$\| f_1, f_2, \ldots, f_N \| = \left[\sum_{i=1}^{N} (f_i)^2 \right]^{\frac{1}{2}} \tag{1.6}$$

for each point (x,y) where the primary responses f_i have been computed.

viii) From the primary response set $[f_1, f_2, \ldots, f_N]$ we should also be able to compute the <u>direction</u> of a local event. Magnitude and direction are complementary or rather, orthogonal features of a pattern.

iv) It follows from vii) and viii) that in order to estimate direction, the magnitude should not depend on direction: <u>For any input pattern</u> a rotation of this input around the point of interest (x,y) in equation (1.2) should not affect the outcome of (1.6). An operator set (a set of basis functions) $[b_1, b_2, \ldots, b_N]$ for which this is true is said to be <u>rotation invariant</u>.

x) In practice, the basis functions has to be implemented by discrete convolution kernels.

It has been shown [4] that one family of functions that can meet the requirement for rotation invariance is given in polar coordinates by

$$\{h_0(r)\}$$

$$\{h_1(r)\cos\varphi,\ h_1(r)\sin\varphi\}$$

$$\{h_2(r)\cos 2\varphi,\ h_2(r)\sin 2\varphi\}$$

$$\{h_3(r)\sin 3\varphi,\ h_3(r)\sin 2\varphi\}$$

etc

or more generally

$$h_n(r)\cos n\varphi + j\,h_n(r)\sin n\varphi = h_n(r)\,e^{jn\varphi} \quad n = 0,1,2,\ \ldots \tag{1.7}$$

For each order n there is one complete rotation-invariant set of basis functions with two members.

These functions have one angular variation (cosnφ sinnφ) and one radial variation $h_n(r)$. The angular variations are the <u>circular harmonics</u>. In fact, using these basis functions means that equation (1.1) computes the circular-harmonic expansion coefficients of f(x,y) and that (1.4) is the <u>circular-harmonic expansion</u> of f(x,y). It has been shown very convincingly by Arsenault et al [5], Wu and Stark [6], [7] that this technique provides a very direct path to rotation-invariant pattern recognition.

The radial variation $h_n(r)$ can be chosen quite arbitrary without violating the rotation invariant property. In reality, however, the functions are supposed to be local detectors which means that $h_n(r) = 0$ for $r > R$ (selfevidence iii) above) where R is a "radius of interest". A pictorial description of the operator set is given in Figure 1. The function values are indicated by zero-crossing lines and signs of polarity. A formal proof of the rotation invariance property of these functions is given in section 2 below. Also, the reason for grouping the 0-order and the two 2nd order operators together will be clear in Section 3.

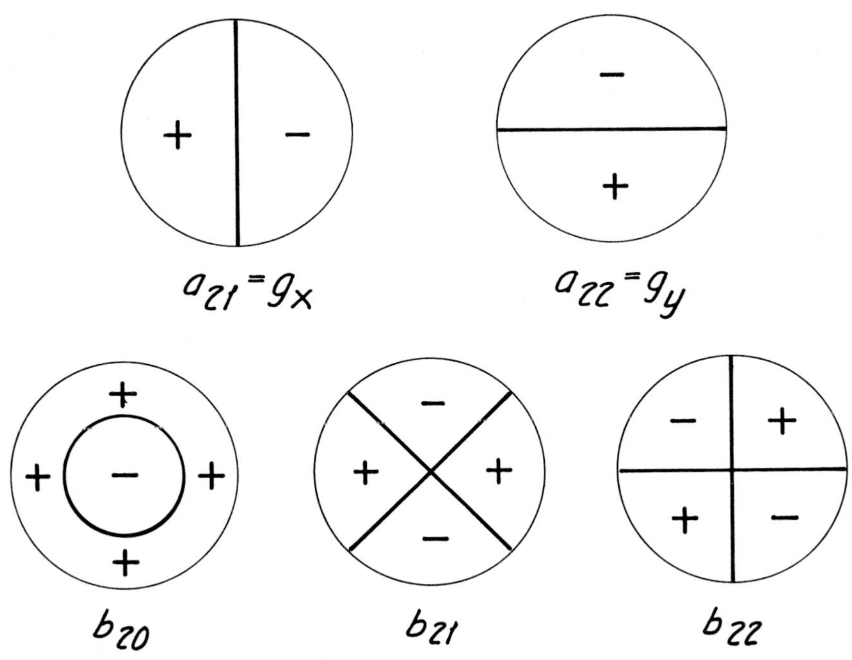

Figure 1

The generality and optimality of these operators have been discovered by several authors. Hueckel [1] used a set of eight operators that include those of order 0, 1 and 2 in Figure 1. Frei and Chen [2] suggest a basis set that consists of orthogonal 3×3 kernels, the zero-crossings of which strongly resembles those of Figure 1.

Like Hueckel, Hummel [3] searched for an optimal set of basis functions to detect ideal edges (= step edges). Given a fixed number N of basis functions, the set of functions is said to be optimal if it can approximate an arbitrarily rotated target pattern f(x,y) (in this case an edge) with a minimum error ε:

$$\varepsilon = \| f(x,y) - \hat{f}(x,y) \| = \| f(x,y) - \sum_{i=1}^{N} f_i \cdot b_i \| \qquad (1.8)$$

where f_i is obtained from (1.1).

The result is quite natural. Since no specific orientation of the target pattern is favoured over others the optimal basis functions b_i must be sets of rotation invariant operators.

Recently, Lenz [10] has shown that the spherical harmonics (generalization of circular harmonics to three dimensions and higher) are optimal in the very same sense. This proof is a considerable extension of the more limited proof for the specific case of 3D-edges by Zucker and Hummel [11].

2. CIRCULAR HARMONICS ARE ROTATION INVARIANT

The rotation invariance property of the functions in Figure 1 can be shown as follows.

From now on, we will use equation (1.1), knowing that we for most applications are applying it throughout the image like in (1.2). Furthermore, we will now switch to polar coordinates and assume that the convolution operators are already mirrored when written in the form $h_n(r)e^{jn\varphi}$. Then equation (1.1) becomes

$$f_{n1} = \int_0^{2\pi}\int_0^{\infty} h_n(r) \cos n\varphi \cdot f(r,\varphi) \cdot r \, dr \, d\varphi \qquad (2.1)$$

$$f_{n2} = \int_0^{2\pi}\int_0^{\infty} h_n(r) \sin n\varphi \cdot f(r,\varphi) \cdot r \, dr \, d\varphi \qquad (2.2)$$

$$f_{n} = \int_0^{2\pi}\int_0^{\infty} e^{jn\varphi} h_n(r) \cdot f(r,\varphi) \cdot r \, dr \, d\varphi \qquad (2.3)$$

The contribution $f_{nl}(r)$ from a thin concentric ring of the basis function is

$$f_{nl}(r) = h_2(r) \cdot r\,dr \int_0^{2\pi} \cos n\varphi \cdot f(r,\varphi)\,d\varphi \qquad (2.4)$$

See Figure 2.

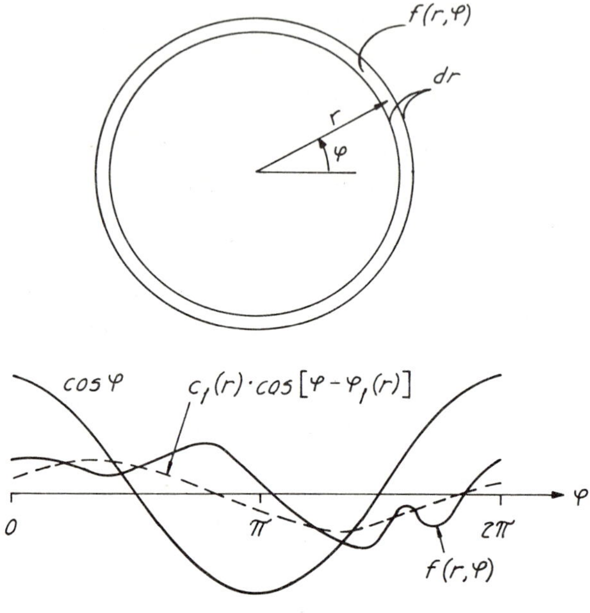

Figure 2

Clearly, by evaluating the integral in (2.4) we sift out from $f(r,\varphi)$ the harmonic component of order n. Only this component will contribute to the integral. The component has a certain phase $\varphi_n(r)$ and a magnitude $c_n(r)$ which means that we can write the integral

$$\int_0^{2\pi} \cos n\varphi \cdot f(r,\varphi)\,d\varphi = \int_0^{2\pi} \cos n\varphi \cdot c_n(r) \cos\left[n(\varphi - \varphi_n(r)\right]\,d\varphi$$

which simplifies to

$$\pi\, c_n(r) \cdot \cos\left[n\, \varphi_n(r)\right]$$

so that we get

$$f_{n1}(r) = h_n(r) \cdot r\, dr \cdot \pi\, c_n(r) \cdot \cos\left[n\, \varphi_n(r)\right] =$$

$$= \|f_n(r)\| \cdot \cos\left[n\, \varphi_n(r)\right] \tag{2.5}$$

For symmetry reasons we get immediately

$$f_{n2}(r) = h_n(r) \cdot r\, dr \int_0^{2\pi} \sin n\varphi \cdot f(r,\varphi)\, d\varphi =$$

$$= \|f_n(r)\| \sin\left[n\, \varphi_n(r)\right] \tag{2.6}$$

(2.5) and (2.6) combines into

$$f_n(r) = \|f_n(r)\|\, e^{jn[\varphi_n(r)]} \tag{2.7}$$

Clearly, if the input image is rotated an arbitrary angle Ψ we have

$$\mathrm{Rot}_\Psi\left(f(r,\varphi)\right) = f(r,\varphi-\Psi)$$

and the result (2.7) is seen to be modified to

$$f_{n,\Psi}(r) = \|f_n(r)\|\, e^{jn[\Psi+\varphi_n(r)]} \tag{2.8}$$

Now, for different radii we will get identical results except for individually different $\|f_n(r)\|$ and phase angle $\varphi_n(r)$. The total response $f_{n,\Psi}$ can be written

$$f_n(\Psi) = \|f_n\|\, e^{jn\varphi_n} \cdot e^{jn\Psi} \tag{2.9}$$

where φ_n is the phase angle of the sum of contributions (2.7). Since the magnitude of (2.9) is constant the basis set is rotation invariant.

The above shows in the first place that the projection onto the basis functions <u>commutes</u> with rotation of the input image. See Figure 3, which can be summarized in the statement.

$$\mathrm{Proj}_n\left(\mathrm{Rot}\left[f(r,\varphi)\right]\right) = \mathrm{Rot}_n\left[\mathrm{Proj}_n\left(f(r,\varphi)\right)\right] \tag{2.10}$$

The "n-projection" of a rotated pattern (left-hand side of (2.9) and (2.10)) equals the n-rotation of the n-projection of the unrotated pattern.

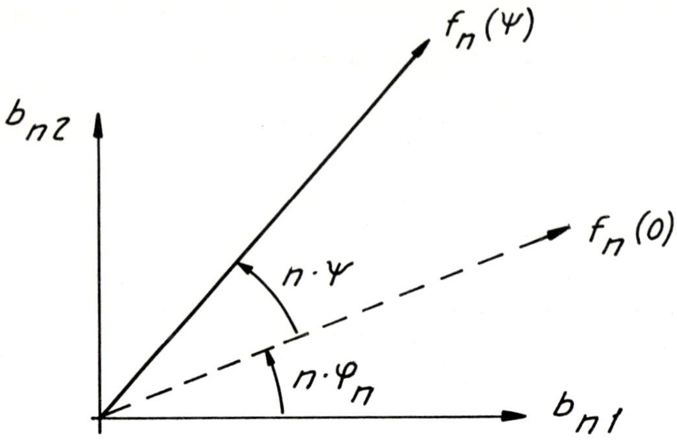

Figure 3

Another way to put it is to say that the angular variation of a basis function carries over to angular response variation with the orientation of the image. And since the angular part of basis functions are a complete set of harmonic functions

$$(\cos n\,\varphi)^2 + (\sin n\,\varphi)^2 = 1$$

we also have

$$\|f_n(r)\| = [(f_{n1}(r))^2 + (f_{n2}(r))^2]^{\frac{1}{2}}$$

independently of the rotation Ψ.

As demonstrated by Lenz [10] the generalization to higher dimensions of the result (2.7) is given by the Funk-Hecke theorem [12], [13].

3. EDGES AND LINES. FOURIER TRANSFORMS

A curve in the image plane (x,y) as seen from distance but in high resolution can be immensely complicated. However, as we limit our scope to local observations any part of the curve tends to take the form of a straight line. The phenomenon is a general law for images:

> With increasingly local observations (smaller neighborhood support) follows that all structures are 1-dimensional.

Thus, unless we attempt to detect fairly large complicated patterns in one master stroke as in [5], we are primarily interested in 1D-structures embedded in a 2D- (or 3D-) signal.

A pure one-dimensional variation in $f(x,y)$ $(\text{or } f(x,y,z))$ means that the signal has the form of a *wave*. Along the wavefront the signal is constant; across we find the profile $p(r)$ of the wave. All such profiles can be separated in one odd part $p_o(r)$ and one even part $p_e(r)$.

$$p(r) = p_o(r) + p_e(r) \tag{3.1}$$

$$p_o(r) = - p_o(-r)$$

$$p_e(r) = p_e(-r)$$

A wave in two or three dimensions with a purely odd profile is called an *edge*. A wave with a purely even profile is called a *line*. The 2D-case of an edge and a line is illustrated by Figure 4.

Let (ρ,Ψ) be the polar coordinates in the Fourier domain.

The *Fouriertransform $P(\rho,\Psi)$ of a wave* with a profile $p(r,\varphi_1)$ is zero except on a line through the origin of the Fourier plane at the angle $\Psi=\varphi_1$ as shown in Figure 4.

$$P(\rho,\varphi_1) = \quad [p(r,\varphi_1)] \tag{3.2}$$

Obviously, we can simplify (3.2) to a purely one-dimensional transform

$$P(\rho) = \quad [p(r)] \tag{3.3}$$

The odd and even components of $p(r)$ give rise to one real even and one imaginary odd transform respectively.

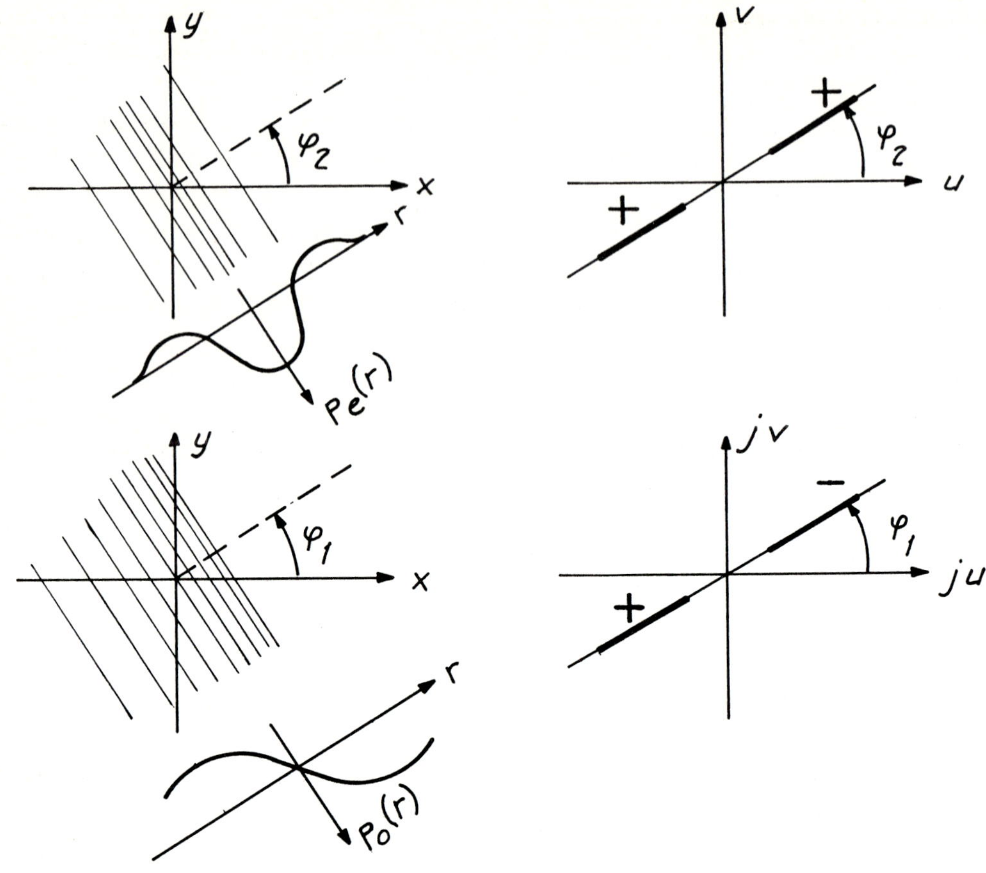

Figure 4

$$P(\rho) = P_e(\rho) + P_o(\rho) = \quad [p_e(r) + p_o(r)] \tag{3.4}$$

<u>Theorem 3.1</u>. The Fourier-transform of a circular harmonic-function has the same angular variation as the function itself, i.e.

$$\mathcal{F}\,[h_n(r)\cos n\varphi] = H_n(\rho)\cos n\Psi$$

A proof of this fairly wellknown theorem is found in [8].

The radial weight function $h_n(r)$ in Theorem 3.1 is linked to the function $H_n(\rho)$ over the <u>Hankeltransform</u> $\mathcal{H}_n$ of order n.

$$H_n(\rho) = \mathcal{H}_n\,[h_n(r),\ \rho] \tag{3.5}$$

which is shorthand for

$$H_n(\rho) = 2\pi \int_0^\infty h_n(r)\, J_n\, (2\pi r\rho)\, r\, dr \tag{3.6}$$

where J_n is the Bessel function of order n.

From the appearance of the basis functions of Figure 1 and knowing that they look in principle the same in the Fourier domain, we can now quickly convince ourselves that edges expands into odd order components lines into even orders.

Assume that we want to employ only the 1st order set of basis functions for edge detection/estimation and the 0 and 2nd order sets for line detection. Since the 1st order operators have the general appearance of derivators/gradient detectors we call them g_x and g_y respectively.

$$a_{21} = \frac{\partial}{\partial x} = g_x = h_1(r) \cos\varphi \qquad G_x = H_1(\rho) \cos\Psi \tag{3.7}$$

$$a_{22} = \frac{\partial}{\partial y} = g_y = h_1(r) \sin\varphi \qquad G_y = H_1(\rho) \sin\Psi \tag{3.8}$$

$$f_x = f*g_x \qquad F_x = F\cdot G_x \tag{3.9}$$

$$f_y = f*g_y \qquad F_y = F\cdot G_y \tag{3.10}$$

$$f_{xx} = f*g_x*g_x = f*g_{xx} \qquad F_{xx} = F\cdot G_x^2 = F\cdot G_{xx} \tag{3.11}$$

$$f_{xy} = f*g_x*g_y = f*g_{xy} \qquad F_{xy} = F\cdot G_x\cdot G_y = F\cdot G_{xy} \tag{3.12}$$

$$f_{yy} = f*g_y*g_y = f*g_{yy} \qquad F_{yy} = F\cdot G_y^2 = F\cdot G_{yy} \tag{3.13}$$

Using (3.7) - (3.13) we can now produce

$$\frac{1}{\sqrt{2}} (G_{xx} + G_{yy}) = \frac{1}{\sqrt{2}} [H_1(\rho)]^2 \tag{3.14}$$

$$G_{xx} - G_{yy} = [H_1(\rho)]^2 \cos 2\Psi \tag{3.15}$$

$$2\, G_{xy} = [H_1(\rho)]^2 \sin 2\Psi \tag{3.16}$$

which is the natural basis functions of order 0 and 2 given in the frequency domain. Using the inverse Hankel transform of order 0 for (3.14) and of order 2 for (3.15) and (3.16) we get

$$b_{20} = \frac{1}{\sqrt{2}} (g_{xx} + g_{yy}) = \frac{1}{\sqrt{2}} h_0(r) \tag{3.17}$$

$$b_{21} = (g_{xx} - g_{yy}) = h_2(r) \cos 2\varphi \tag{3.18}$$

$$b_{22} = 2\, g_{xy} = h_2(r) \sin 2\varphi \tag{3.19}$$

Thus, linear combinations of second derivative operators constitute the basis functions of second order. Furthermore, we see from (3.11) - (3.13) that these operators are separable. We are to compute the responses as follows

$$f_{20} = f{*}b_{20} = \frac{1}{\sqrt{2}} (f_{xx} + f_{yy}) = \frac{1}{\sqrt{2}} (f_x{*}g_x + f_y{*}g_y)$$

$$f_{21} = f{*}b_{21} = f_{xx} - f_{yy} = f_x{*}g_x - f_y{*}g_y$$

$$f_{22} = f{*}b_{22} = 2\, f_{xy} = 2\, f_x{*}g_y$$

To illustrate the above we now propose two kernels g_x and g_y in Figure 5.

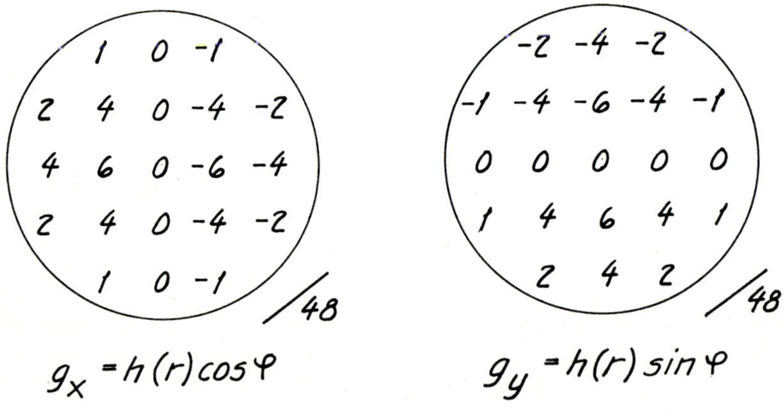

	1	0	-1	
2	4	0	-4	-2
4	6	0	-6	-4
2	4	0	-4	-2
	1	0	-1	

/48

$g_x = h(r) \cos \varphi$

	-2	-4	-2	
-1	-4	-6	-4	-1
0	0	0	0	0
1	4	6	4	1
	2	4	2	

/48

$g_y = h(r) \sin \varphi$

Figure 5

A virtue of these kernels are their separability. For instance, g_x can be separated as follows into a sequence of 7 additions.

$$g_x = \begin{bmatrix}1\\1\end{bmatrix} * \begin{bmatrix}1\\1\end{bmatrix} * \left\{ \begin{bmatrix}1 & 0 & -1\end{bmatrix} * \begin{bmatrix}1\\1\end{bmatrix} * \begin{bmatrix}1\\1\end{bmatrix} + 2 \begin{bmatrix}1 & 0 & 0 & 0 & -1\end{bmatrix} \right\}$$

At the same time, however, we require that g_x should be a sampled version of a rotation invariant operator, i.e.

$$g_x \approx h_1(r)\cos\varphi$$

$$g_y \approx h_1(r)\sin\varphi$$

The $h_1(r)$-function implicitly defined by the kernels of Figure 5 is plotted in Figure 6.

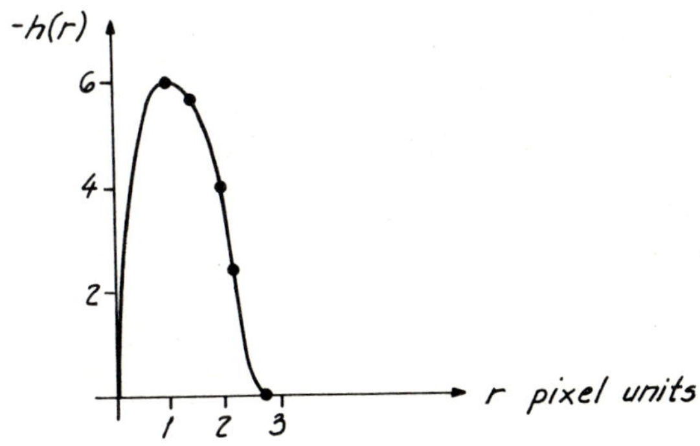

Figure 6

It seems possible to fit a fairly smooth h(r)-curve through the given points. Enough smoothness indicate that the h(r)-function is confined to the Nyquist bandwidth given by the pixel distance.

From g_x and g_y, we can now compute the g_{xx}, g_{yy} and g_{xy}-functions and from (3.17), (3.18) and (3.19) we get the basis funtions b_{20}, b_{21} and b_{22} shown in Figure 7.

By summing the squared coefficients in these kernels we find that relative difference in their norms is ±5 % which is probably due to violations of the Nyquist criterion for the $h_1(r)$-function.

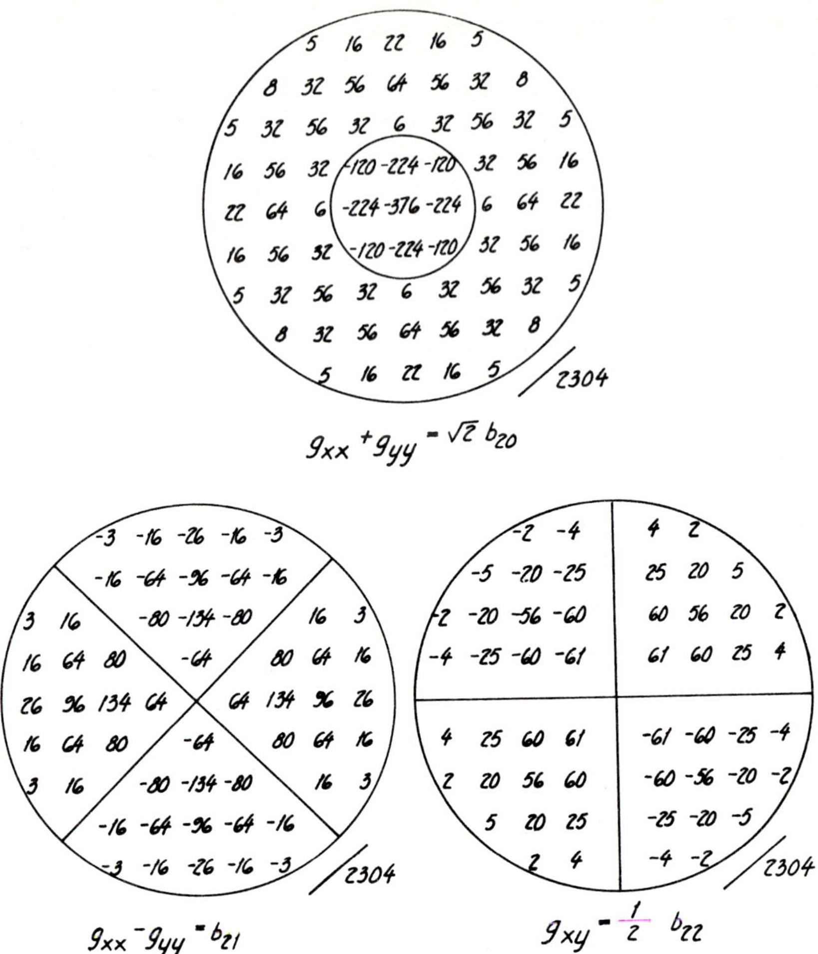

Figure 7

CONCLUSIONS

We started from certain axiomatic assumptions. Foremost among them were that our operators should form a complete orthonormal set of basis functions. Furthermore, the primary responses should lend themselves to estimate the orientation of the image feature as well as a rotation-invariant magnitude. The only operators that meet these requirements are those that have an angular variation in the form of circular harmonics. The complete functions with a choosen radial weight function

$h_n(r)$ are called natural basis functions.

Other authors [3], [10] have shown that these operators are optimal for approximation of a given pattern (image feature). Another characteristic follows from the circular harmonic property: the operators have the same appearance in signal and frequency domains except for the radial variation. We used this to obtain an attractive separability of the 0 and 2nd order function sets into the first order set.

This paper can only be a short introduction to the very rich and rewarding area of natural basis functions and rotation invariant operators. We believe that these concepts are the key to a general approach in low-level computer vision, and more generally, to multi-dimensional image processing. In a forthcoming monography will be demonstrated how the basic concepts are applied to edge and line detection, how the radial weight functions are matched to a specific pattern, how to design quadrature filters, how to apply the technique to image enhancements and how the theory relates to the GOP-concept [9]. Last but not least, we will extend the theory and its applications to 3D-signals. While a gradient/edge detector seems to be trivially extended from 2D to 3D a line detector is not. For instance, in 3D the complete set of basis functions of 2nd order consists of not three but five elements. Thus, to compute magnitude and direction for a three dimensional line feature we need five kernels producing five primary responses.

REFERENCES

[1] Hueckel, Manfred H., "An operator which locates edges in digitized pictures". Journ. ACM, Vol. 18, pp. 113-125, 1971.

[2] Frei, Werner and Chen, C-C, "Fast boundary detection: A Generalization and a new algorithm". IEEE TC, Vol. C-26, pp. 988-998, 1977.

[3] Hummel, Robert A., "Feature detection using basis functions", Computer Graphics and Image Processing, Vol. 9, pp. 40-55, 1979.

[4] Danielsson, Per-Erik, "Rotation-invariant linear operators with directional response". Proc. 5th International Conference on Pattern Recognition, pp. 1171-1176, 1980.

[5] Hsu, Y-N, Arsenault, H.H. and April G., "Rotation-invariant digital pattern recognition using circular harmonic expansion". Applied Optics, Vol. 21, pp. 4012-4015, 1982.

[6] Wu, Ronald and Stark, Henry, "Rotation-invariant pattern recognition using a vector reference". Applied Optics, Vol. 23, pp 838-840, 1984.

[7] Wu, Ronald and Stark, Henry, "Rotation-invariant pattern recognition using optimum feature extraction". Applied Optics, Vol. 24, pp. 179-184, 1985.

[8] Stein, Eias M. and Weiss, Guido, "Introduction to Fourier Analysis on Euclidean Spaces", Chapter 4. Princeton University Press 1971

[9] Knutsson, Hans and Granlund, Gösta H., "Texture Analysis using two-dimensional Quadrature Filters". Proc. IEEE Computer Architecture for Pattern Analysis and Image Database Management, pp. 206-213, 1983.

[10] Lenz, Reiner, "Reconstruction, Processing and Display of 3D-images". Linköping Studies in Science and Technology. Dissertations No. 151, Linköping 1986.

[11] Zucker, Steven V. and Hummel, Robert A., "A Three-dimensional Edge Operator", IEEE Trans., Vol. PAMI-3, pp. 324-331, 1981.

[12] Funk, P., "Beitäge zur Theorie der Kugelfunktionen", Matematische Annalen, Vol. 77, pp. 136-152, 1916.

[13] Hecke, E., "Über orthogonale-invariante Integralgleichungen", Matematische Annalen, Vol. 78, pp. 398-404, 1918.

"IPAS - Ein Pipeline-Bildverarbeitungssystem"

K. Gütschow
AEG Aktiengesellschaft, Fachgebiet Optronik
Bildverarbeitungsmethoden
Industriestraße 29, D-2000 Wedel/Holstein

Einleitung

Der Stand der Technik im Bereich der Bildverarbeitung ist im wesentlichen durch folgende Situation gekennzeichnet:

Es ist eine Vielzahl unterschiedlichster Verarbeitungsmethoden bekannt, die zufriedenstellende Ergebnisse allerdings meist nur in einem kleinen Anwendungsbereich liefern. Der Einsatz der Methoden ist im allgemeinen durch hohen Rechenzeitbedarf geprägt, der sich in der ikonischen Bildverarbeitung überwiegend aus der Menge der zu bearbeitenden Daten ergibt.

Ein Bildverarbeitungssystem, das für den industriellen Einsatz konzipiert wird, ist Anforderungen ausgesetzt, die den genannten Aspekten nahezu diametral entgegenstehen:

- Die Ergebnisse einer Bildauswertung sollen in - durch die jeweilige Anwendung definierter - "Echtzeit" für eine weitere Verarbeitung zur Verfügung stehen.

- Die kostenintensive Entwicklung von Spezialsystemen für Einzelanwendungen ist durch Verwendung "universeller" Systeme zu vermeiden.

Das im folgenden beschriebene IPAS-Bildverarbeitungssystem (IPAS: "Image Processing and Analysis System") ist mit Blick auf diese Problematik entworfen worden.

Hardware-Konzept

Die Hardware des IPAS besteht aus einem Rahmensystem, das von einem Hostrechner kontrolliert wird. Das Rahmensystem kann je nach

Erfordernis mit unterschiedlichen Bildverarbeitungsmodulen bestückt werden. Die Schnittstellen zu den einzelnen Modulen sind - gegliedert in zwei Gruppen - identisch ausgelegt, so daß die in dem Rahmensystem vorhandenen Steckplätze frei genutzt werden können.

Die Module sind als Pipeline-Prozessoren ausgelegt, die im Videotakt die Verarbeitung eines Bildpunktdatums durchführen. Sie verfügen je nach Modul über ein bzw. zwei Dateneingänge und ein bzw. zwei Ausgänge. Verarbeitungsmodule in diesem Sinn sind z. B. Kamera-/Monitor-Interface, Bildspeicher, Faltungs-Modul, Pixel-ALU.

Die Konfigurierung einer Verarbeitungs-Pipeline geschieht durch ein System von Video-Datenwegen. Jedem Datenausgang an den Steckplätzen ist ein Video-Datenweg fest zugeordnet. Die Eingänge können auf die Busse beliebig aufgeschaltet werden. Somit ist eine freie Verknüpfung aller vorhandenen Verarbeitungsmodule möglich.

Wird durch die Konfigurierung einer Verarbeitungs-Pipe das Rahmensystem nicht voll ausgenutzt, ist es möglich, mit den verbleibenden Steckplätzen weitere Pipes aufzubauen und gleichzeitig zu betreiben.

Ist Verarbeitung in den konfigurierten Pipes abgeschlossen, kann die Verschaltung der vorhandenen Module zum Aufbau neuer Pipelines durch Software-Steuerung geändert werden.

Die Synchronisation des Bilddatenstromes innerhalb einer Pipe erfolgt dezentral durch die Verarbeitungselemente selbst: Der Pixelstrom wird von einem gesonderten Signal begleitet, das den empfangenden Modulen Anfang bzw. Ende eines Bildtransfers anzeigt.

Angestoßen wird die Verarbeitung in einer Pipeline, indem die jeweilige Bildquelle ein Startsignal erhält, das sie veranlaßt, ein Bild auf ihren Ausgang zu legen. Eine Pipeline hat ihre Verarbeitung abgeschlossen, wenn auf den betroffenen Datenwegen keine Bildströme mehr erfolgen.

Das Aussenden von Startsignalen, die Überwachung von Datenwegen und die eventuell erforderliche Umkonfigurierung des Verbindungsnetzes, sobald Verarbeitungen abgeschlossen sind, geschieht durch ein zentrales Steuer-Modul. Es kontrolliert auf diese Weise die Bearbeitung in den einzelnen aktiven Pipes und liefert damit die Grundlage zur

Ablaufsynchronisation mit dem Hostrechner.

Um mehrere Pipeline-Abläufe nacheinander ohne Eingriff des Hostrechners durchführen zu können, verfügen alle Module über lokale Befehlsspeicher. Die enthaltenen Befehle legen fest, in welcher Weise das jeweils folgende Bild zu prozessieren ist, bzw. beim Steuer-Modul, in welcher Weise das Verbindungsnetz zu verschalten ist. Der Zugriff auf die Befehlsspeicher und die Synchronisation mit dem Steuer-Modul erfolgen über einen Steuer-Bus (VME-Bus), auf dem alle Module liegen.

Der Hostrechner (SUN-/68K-Systeme) hat folgende Aufgaben:
- Laden der Befehlsspeicher aller Module
- Ggf. Laden des zu verarbeitenden Bildes in einen Bildspeicher
- Anstoßen des Steuer-Moduls zwecks Beginn der Verarbeitung
- Nach Abschluß der Verarbeitung: Auslesen von Ergebnisdaten

Software

Die notwendigen Steuer- und Überwachungsfunktionen des Hostrechners sind in einer Programmbibliothek zusammengefaßt. Auf Benutzerebene sind sie als Befehle verfügbar, die im Rahmen einer Interpretersoftware abgesetzt werden können.

Integriert ist dieser "IPAS-Interpreter" in ein komplexes Bildverarbeitungssystem, das eine Vielzahl von Bildverarbeitungsmethoden zur Verfügung stellt. Den Grundstock für die Bildverarbeitungs-Software bilden Arbeiten des AEG-Forschungsinstituts in Ulm. Durch die Integration beider Systeme ist ein fließender Übergang von hardware- und softwaregestützten Bildverarbeitungsverfahren möglich.

Der Interpreter und das Bildverarbeitungssystem werden über die gleiche menueorientierte Benutzerschnittstelle bedient. Sie gestattet sowohl interaktiven als auch prozedurgesteuerten Betrieb.

Das Bildverarbeitungssystem stellt mit seinen Softwaremethoden zum einen eine Erweiterung der in der Hardware verfügbaren Verfahren dar. Es dient zum anderen als Entwicklungs- und Simulationsumgebung bei der Erarbeitung und Erprobung von Bildverarbeitungsabläufen.

Ausblick

Auf der Hardware-Seite zeichnet sich das IPAS-System durch schnelle Prozessoren aus, die in einem Rahmensystem flexibel konfiguriert werden können /Gr/. Durch den modularen Aufbau des Systems sind die Entwicklung und Integration weiterer Prozessoren in einfacher Weise möglich. Es lassen sich so leistungsfähige dedizierte Bildverarbeitungssysteme für unterschiedlichste Anwendungen zusammenstellen. Ein Beispiel für eine Anwendung ist in /Sc/ beschrieben.

Auf der Software-Seite steht ein System mit einer durchgängigen Funktionalität und einem einheitlichen Benutzerinterface zur Verfügung.

Literatur

/Gr/ Granlund, G. H.; Arvidsson, J. B.
"Computer Architectures for Image Processing"
Proceedings of the 4th Scandivavian Conf.
on Image Analysis, Trondheim, N, June 17-20, 1985

/Sc/ Schmid, R. A.
"Eine Hardware-Architektur zur ikonischen Bildverarbeitung auf der Basis der hierarchischen Formcodierung"
Proceedings des 8. DAGM-Symposiums, Paderborn, 1986,
Informatik-Fachberichte 125, Springer-Verlag

CONTEXTUAL IMAGE PROCESSING IN MRI - APPLICATIONS

Helmut Schwarz
CONTEXT VISION Systems GmbH, Subsidiary Munich
Dietlindenstr. 15, D - 8000 München

Abstract

Conventional Digital Image Processing reaches its limits in those cases, where spectral characteristics (grey levels, colours, spectral bands) cannot be properly associated with specific object - or structural properties. Examples are manifold: Typical scanners in medicine - like ultrasound, CT, MRI, or even light microscopy - frequently produce images, where the grey level distribution is not related to the global or local image characteristics. In other words, simple thresholding algorithms are in such situations not powerful enough to reveal structural properties.

The clue for solving the problem is to take the contextual information into account. This is exactly the key feature behind the GOP theory,developed by Prof. G. Granlund at the Technical University in Linkoping, Sweden: The information contents of every pixel is not only dependent of its own spectral value, but related to its embedding in neighbourhoods of varying size. The processing is hierarchical, with different levels being combined either top-down, as well as bottom-up. The resulting pixel values are represented as 16 bit in a polar coordinate system, with the low byte giving the contextual feature (e.g. orientation), and the high byte the certainty ("strength") of that feature.

Specific hardware and software has been developed in order to realize the ideas of Prof. Granlund.The result, the GOP 300, may be trained, using spectral and contextual features, in order to identify and classify virtually any type of texture or structure that characterize the objects or regions of interest in a tissue.

Besides classification, applications include 3D reconstruction, context - controlled image enhancement (an operation analogue relaxation), and postprocessing within the field of Magnetic Resonance Imaging, Computer Tomography, PET investigations, Nuclear Medicine, Ultra sound examinations, Mammography, as well as light and electron microscopical applications.

In the following, the major features of the GOP in Magnetic Resonance Imaging are presented.

GOP Technology In Magnetic Resonance Imaging

MRI is no doubt the most expanding field in medical imaging today, and new techniques continously have lead to new applications in that area. So far the method primarily has been used to detect anatomic structures in competition with other imaging techniques. The specific GOP - package for Magnetic Resonance Imaging opens up extended diagnostic possibilities with a new and unique range of analysis tools applicable in that field. The main features comprise:

* Tissue characterization using automatic or interactive context controlled classification routines
* Adaptive image enhancement using context controlled filtering
* Geometric correction and alignment to enable operations and arithmetics between different images
* Interactive image synthesizing, increasing patient throughput

* Dynamic reconstruction with variable parameter definitions
* Flow analysis using the phase information the image reconstruction routine

Tissue Classification Analysis

Very complex decision models can be defined interactively using classification procedures including training of the computer system.
Training areas are defined on reference objects like phantoms or known pathological structures. Image information from these areas then is fed into the system to be analyzed with respect to the features used in the classification procedure.
The system allows to choose from various strategies, so that the computing power can be adjusted to the complexity of the problem, allowing to differentiate between normal and pathological tissue.
A variety of different image properties, like intensity, orientation, rotational consistency, spatial frequency and variance, can be used in and between different relaxation images, in order to quantify and evaluate certain types of tissues. The automatic procedures supporting these tasks easily may be modified to suit specific needs.

Image Enhancement Operations

Adaptive context - controlled enhancement procedures provide local filtering of different image regions simultaneously Homogeneous regions undergo a smoothing (noise reduction) procedure, whereas lines and edges are sharpened by means of a contour enhancement.
Interpolation routines avoid a "digital appearance" of the result.

Geometric Correction

Geometric correction and resampling of images can be performed either by highly effective software algorithms, or, even faster, by means of the Geometric Transform Processor (GTP) The resampling can either be supervised by a mathematical model, or by a displacement image, which is resulting of the comparison of two images.
This feature enables image subtraction, division, or any other comparison method, where the two or more images have to perfectly geometrically aligned.
Images from different sources, like CT, Nuclear Medicine, PET, etc. can be aligned, in order to provide anatomical references which give new tools for specific diagnoses.
Using the GTP, a typical time for resampling or rotation is about 1 second for an image of 512*512 pixels.

Synthesis Of Images

By using the calculated values of T1, T2, and PD, an interactive synthesis of new SE and IR images is possible, with continously selectable parameters TE, TR, and I.

Instead of the time - consuming search for the correct parameters on the MR - unit, the GOP - system calculates this pulse sequence which gives optimum contrast between different areas, based on a pair of fast recorded, noisy images.
Analysing this "artificial" image the result can be verified by taking another MR image of the patient using the sequence just calculated. In this way the number of images necessary for a correct diagnosis is dramatically reduced.

MRI Statistics

Numerical local estimates of T1, T2, and PD are calculated, for interactively or predefined regular or irregular regions of the image. In combination with the geometrical correction facility, this gives extended possibilities for the comparison of different slices or regions of a reference material, or earlier studies.
Apart from that, a complete statistical and presentation package offers features like histograms, area and distance measurements, standard deviations, variances, intensities, etc, within certain regions.

Dynamic Reconstruction

The system offers the possibility to feed the raw signal of the MRI - unit directly into the GOP, to enable a FFT reconstruction of the image.
This means that there is a free choice of reconstruction parameters, including the signal phase information, and magnetic field characteristics. A typical reconstruction time for a 256*256 matrix is only 6 seconds. This opens up new facilities to interactively optimize the reconstruction routine on site.

Image Calculator

For clinical and research purposes the system contains a complete, easy to use arithmetical and trigonometrical image calculator. Functions frequently used may be defined as macros. The GOP floating point array processor leads to extremely short calculation times: a trigonometric function takes less than a second for a 512*512 image!

Image Data Compression

Images contain a great deal of information, which has consequences for storage, transmission, and processing. The redundant, insignificant part of the image information can be removed in order to achieve a more data compact version of the image without any loss of visual perception. Compression rates of up to 30 times without noticeable deterioration in image quality lead to a dramatic reduction in storage space and transmission time.

Contrast Enhancement

Display and windowing functions to enhance the contrast include gamma functions and multiple windows as well as negative images and quantification routines. The result of such a modification always is graphically presented in the menue area, so that misunderstandings are avoided, and reproducability is guaranteed.
Apart from that, the "Wallis" operation provides local contrast stretching in order to achieve adaptive histogram equalization in different areas of the image.

Pseudocolouring

For increased contrast separation, a colour table may be used. The look up table can be cyclically rotated; this makes it easy to find maximum contrast differences in the image. It can either be applied to the full dynamic range, or only to a subset of it (20 %).
Besides the three predefined tables, the user may produce his own ones, and store them for future use.

Patient ID Routines And Collection Data

This module allows to display important patient, collection and reconstruction data directly within the image.

Three Dimensional Display

Using highlighting techniques, a stack of slices can be presented, giving a visual perception of a completely reconstructed object. The 3D - image can be rotated by any angle up to 360 degrees.
It is also possible to use the 3D - image as a reference or scoutview when selecting the different slices.
Automatic cost functions are implemented as part of highly efficient contour following algorithms.

Summary

The GOP 300 specialized MRI - package offers a comprehensive set of tools for any user of that technology, either in a routine, or clinical environment. The unique textural features of the system - in combination with easy and user friendly handling procedures - make solutions possible which are both, advanced and efficient.

PixelPipe – A New Bus Concept for a High Speed Image Processor

D. Meyer–Ebrecht, Th. Schilling, W. Winkler
Lehrstuhl für Meßtechnik
RWTH Aachen

For a fast processing of large image matrices machines, are required which may contain a variety of programmable and/or special purpose hardwired processors and a large shared image memory with random pixel access. The internal data transport mechanism will be the backbone of the machine. Three principles have been applied to design a pixel bus which is efficient but still simple and transparent:

Strict separation of image data:
The exclusive purpose of the pixel bus is to transport pixels between a large image RAM and one of the image processors. Control data, progammes or any other type of support data will be transferred by means of a separated control bus.

Strict isochronous operation
Instead of the commonly applied asynchronous handshake operation every bus transfer is performed within one time slot of fixed duration. Bus drivers as well as bus receivers are synchronized by a master bus clock. Time slots may be arbitrarily dedicated to different processors.

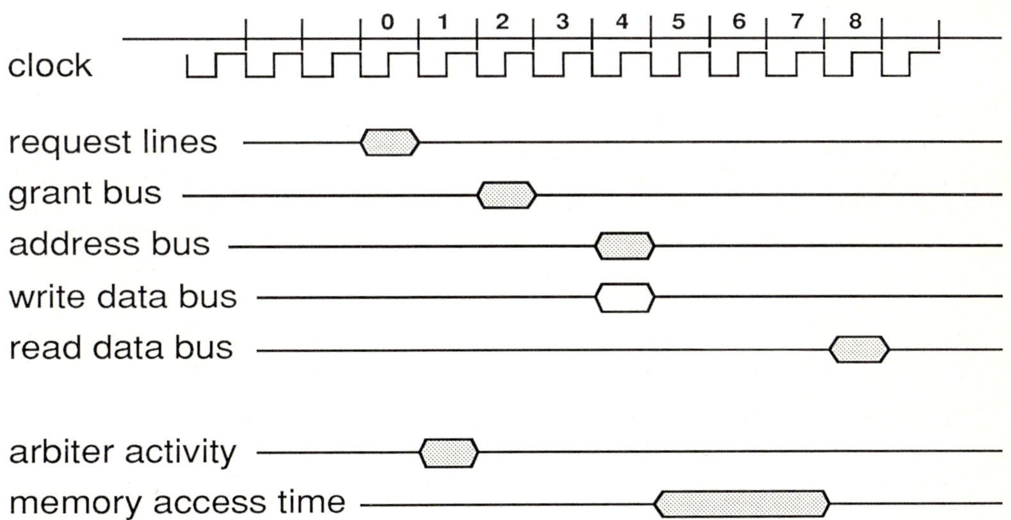

Fig. 1: Isochronous operation

Pipelining of the complete memory access sequence
For the sake of a high pixel troughput, the bus time slot shall be as short as necessary to stabilize the logical state on the buslines. Leading activities (bus request - arbitration - bus grant), memory access (address and write data transfer) and trailing activities (read data delayed by the memory response time) are packed into a fixed sequence of successive bus slots each dedicated to a separate portion of the buslines (see Fig. 2). Therefore, inspite of a memory cycle time of a multiple of the bus time slot and additional processing time of the bus arbiter, the bus behaves as a *pipe* with clock frequency throughput of random pixel sequences.

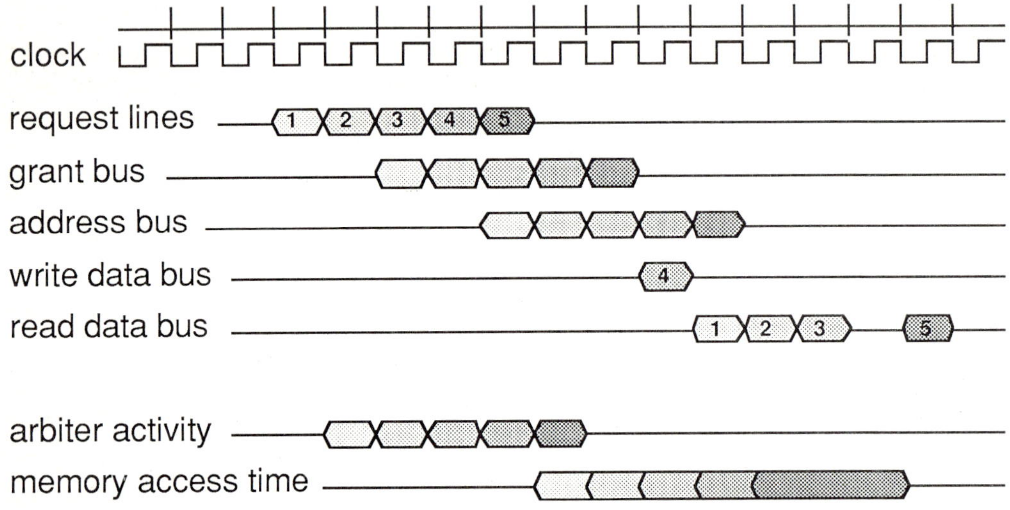

Fig. 2: Pipclining the access

With an experimental system, a continous 10 Mpixel/s (16-bit pixel) throughput rate is achieved though using only standard bus electronics and bus backplanes and dynamic RAMs for the image memory. Due to a decoupling logic at each bus port, each processor may handle the bus as its private bus with an asynchronous single pixel or pixel sequence access. Instead a processor may also adapt its internal operation to the bus by making use of its strict temporal determination and transparency of operation. The *PixelPipe* satisfies the requirements for image workstations (high-resolution image processing within a fraction of a second, handling of image sequences at video rate) and offers processing power for fast sophisticated processing such as image reconstruction or pattern recognition.

Final clause

The authors hope that the results of their work will not be misused for military purposes. To the best of our knowledge we will not participate in any activities concerning military applications.

Bildverarbeitung mit Datenflussrechnern

Dr. Johannes Baston
NEC Electronics (Europe) GmbH
Oberrather Str. 4

D - 4000 Duesseldorf 30

Abstract

A new data-flow processor uPD7281 is the base of an image processing board for standard PCs. All parts needed are implemented. An input multiplexer selects one of 16 CCIR sources (camera or VTR) with either internal or external synchronization. This signal is converted to digital with an 8 bit flash-converter and can be manipulated by an input look-up-table. Four image memory planes are available with each 512*512*8bit. Three analog composite video outputs with output look-up-tables allow false color representation of the selected memory plane. Four data-flow processors and a chip for memory and host interface provide a speed up to 20 MIPS. All functional blocks are accessed via the host PC bus. An internal bus transfers the data between memory and dataflow processors to avoid any performance degradation. The programming technique for the dataflow processors will be explained in detail.

Einleitung

Ein neuartiger Datenflussrechnerchip, der uPD7281, ist das Herz eines Bildverarbeitungssystems, das mit allen notwendigen Teilen einschliesslich der Prozessoren mit 20 MIPS Rechenleistung auf einer Standard-PC Karte aufgebaut wurde. Dieser Beitrag erklaert kurz die Architektur des Prozessors und den Aufbau der Karte. Ein Schwerpunkt ist die Software bzw. die Programmierung von Karte und Datenflussrechner.

Datenflussprozessor

Mit dem uPD7281 wurde zum ersten Mal eine Datenflussrechner-Architektur in einen kommerziell verfuegbaren Siliziumchip integriert. Dieser zu den konventionellen Mikroprozessoren voellig unterschiedliche Rechnerchip ist ein idealer Subprozessor fuer Anwendungen mit relativ vielen Daten und kleinen Programmen. Der Datenflusschip ist charakterisiert durch eine interne Ringstruktur mit mehreren internen Stationen (Abb. 1). Die Daten werden zusammen mit Steuerinformationen (Daten + Steuerzeichen = Token) durch diesen Ring gefuehrt. An allen internen Stationen werden gleichzeitig Token bearbeitet (Interne Parallelverarbeitung, Pipelining). Das Programm ist im lokalen RAM in Linktable und Funktiontable auf dem Chip enthalten, dadurch entfallen Zeitverluste und Busbandbreitenbegrenzung durch Instruktionstransfers. Bis zu 512 Daten koennen in einem integrierten Datenmemory gespeichert werden. Das Herz des Chips ist die Prozessoreinheit, welche die Token, die bei Zweioperandenbefehlen auch beide Daten enthalten, mit einem Durchsatz von 200 ns/Befehl entsprechend 5 Millionen Befehlen pro Sekunde bearbeitet. Die Befehle umfassen arithmetische (incl. Multiplikation) und logische Operationen sowie Bitmanipulationen. Besondere Instruktionen sind zur Adressierung der Speicher vorhanden, die durch Software erfolgt. Ein Supportchip uPD9305 stellt das Interface zum Hostsystem und zum Subsystemspeicher dar und dekodiert u.a. die Adresstoken.

Ein wesentlicher Vorteil dieser Struktur ist die einfache Kaskadierbarkeit mehrerer Datenflussrechner (externe Parallelverarbeitung). Die Rechenleistung wird ohne Hardware- oder Softwareoverhead fast linear erhoeht. Das Supportchip kann vier Prozessoren komplett unterstuetzen. Es sind aber auch groessere Kaskaden moeglich, um einen noch hoeheren Grad externer Parallelverarbeitung zu erreichen.

Die groesste Leistungsfaehigkeit dieses Systems wird bei der schnellen Bearbeitung von Daten erzielt, die regelmaessig im Subsystemspeicher abgelegt sind. Solche Aufgaben findet man typisch in der Bildverarbeitung, so dass der uPD7281 auch den Namen Image Pipelined Processor (IMPP) bekam. Durch die externe Parallelverarbeitung in Multiprozessorsystemen koennen sowohl verschiedene Bearbeitungsstufen, aber auch unterschiedliche Bildabschnitte gleichzeitig verarbeitet werden.

Bildverarbeitungskarte

Fuer eine typische Applikation wurde eine komplette Bildverarbeitungskarte entwickelt. Dieses Board im Standard-PC Format enthaelt alle fuer die Bildverarbeitung notwendigen Teile (Abb. 2). Als Eingang koennen Kameras, aber auch Videorecorder nach CCIR-Norm angeschlossen werden. Ein vom PC gesteuerter Multiplexer waehlt einen von 16 Eingangskanaelen aus. Das Eingangssignal kann von der Karte synchronisiert werden, aber die Karte laesst sich auch auf die Eingaenge synchronisieren. Dem 8-bit Flashwandler ist eine Eingangs-Look-up-table (LUT) nachgeschaltet, die Daten nach programmierbaren Kennlinien veraendern kann. 4 Speicherebenen mit je 512 x 512 x 8 bit nehmen die Bildinformation oder die verarbeiteten Daten auf. Zwei Speicherebenen lassen sich auch zu einem 512 x 512 x 16 bit Speicher organisieren. Nach der Wandlung durch 3 Digital-/Analogwandler mit Ausgangs-Look-uptable werden die Bilder fuer einen Monitor ausgegeben. Mit der Ausgangs-LUT lassen sich Falschfarben zur Hervorhebung geringer Kontrastunterschiede ueber den PC programmieren.

Mit einem direkten Bus ist der Bildspeicher durch den Interfacechip uPD9305 mit den vier Datenflussrechners verbunden. Die Bildverarbeitung belastet dabei weder den PC-Bus noch den Hostprozessor und wird mit hoechster Geschwidigkeit durchgefuehrt. Ein besonderer 16 bit * 64 K Speicher (ROM oder RAM) enthaelt die Anwendungsprogrammerogramme fuer die Datenflussrechner, die durch einen kurzen Befehl des PC getriggert automatisch in die IMPPs geladen werden. Dieses Selbstladen von Programmen kann aber auch durch eine einfache externe Schaltung erzeugt werden, wodurch die Bildverarbeitungskarte praktisch als "stand-alone" System in industriellen Applikationen arbeitet.

Software

Fuer das Board steht ein Softwarepaket zur Verfuegung. Eine interaktive menuegesteuerte Software mit Mausinterface erleichtert die Bedienung ueber den PC. Im Aquisition Menue werden Kanalauswahl, Bildspeicherung und Kopieren, Programmierung der Look-up-Table sowie Bildausgabe interaktiv gesteuert. Im Prozessormenue werden verschiedene Programme fuer die Datenflussrechner gestartet. Ein ausfuehrlich dokumentiertes Beispielprogramm fuer eine 3 x 3 Convolution liegt vor.

Programmierung

Die Datenflussarchitektur verlangt natuerlich eine spezielle Art der Programmierung . Ein funktioneller Assembler - diese Bezeichnung stammt aus der Datenflusstheorie - uebersetzt die Quellenprogramme in die Object-Codes. Mit dem Softwaresimulator koennen die Programme ohne die Zielhardware getestet und Fehler beseitigt werden. Dabei laesst sich auch eine Laufzeitoptimierung erreichen.

Um den Source-Code zu erstellen, wird zunaechst der Algorithmus in ein Datenflussdiagramm umgesetzt (Abb. 3). Dabei kann schon die Parallelverarbeitung beruecksichtigt werden. Die Knoten oder Funktionsbloecke definieren die Befehle, die als FUNCTION-statements im Sourcecode erfasst und im IMPP in der Functiontabelle abgelegt werden. Die Verbindungen zwischen den Knoten, die Links, werden als Netzwerksbeschreibung in der Link-Table gespeichert.

Der Datenflussrechner verfuegt ueber 32 Befehle (Prozessor) und weitere 16 fuer den Adressgenerator und Flusscontroller. Beide zusammen werden in einen Befehl eingebaut und im internen Pipelining verarbeitet. Auch die beiden Ausgangsbefehle werden in das Pipelining einbezogen. Die Adressierung des Bildspeichers erfolgt per Software. Dafuer sind drei weitere Befehle vorhanden, die aber auch in anderen Programmteilen benutzt werden.

In der binaeren Bildverarbeitung ist der Bildschirmspeicher im 16 bit Modus eingesetzt. Bei jedem Datentransfer werden 16 Pixel in die Datenflussrechner uebertragen. Alle Beispielprogramme aus der Applikationsschrift sind direkt lauffaehig.

Zusammenfassung

Die Bildverarbeitungskarte erfuellt zwei Funktionen: Sie kann zum einen als vollwertiges, durch die Datenflussrechner aeusserst schnelles Bildverarbeitungssystem fuer vielfaeltige Aufgaben in der optischen Qualitaetspruefung, zur Verbesserung von schwachen Bildsignalen (z.B. Medizintechnik, Werkstoffpruefung), bei der Bilderkennung und Auswertung z.B. in Robotersteuerungen dienen. Dazu steht eine Bibliothek von Anwendungsprogrammen zur Verfuegung. Zum anderen ist die Karte - zusammen mit dem Softwarepaket - ein ideales Trainingsmittel zum Erlernen der Datenflussarchitektur und Programmierung.

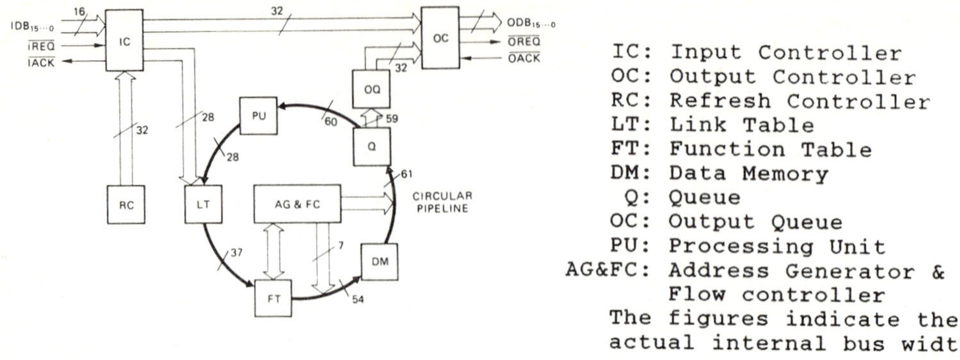

Abb. 1: Blockschaltung des Datenflussrechnerchips uPD7281

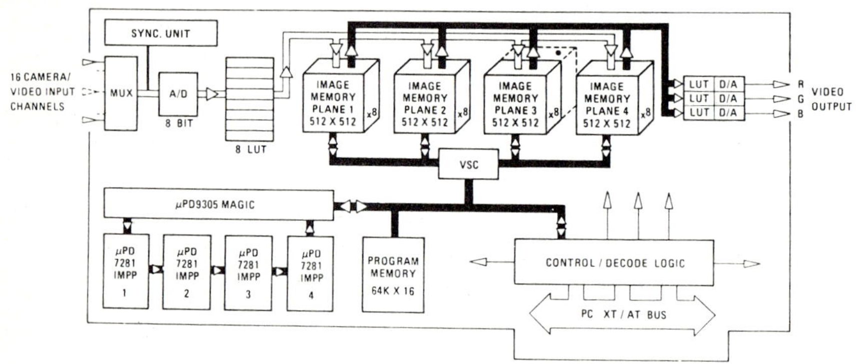

Abb. 2: Blockschaltung der Bildverarbeitungskarte

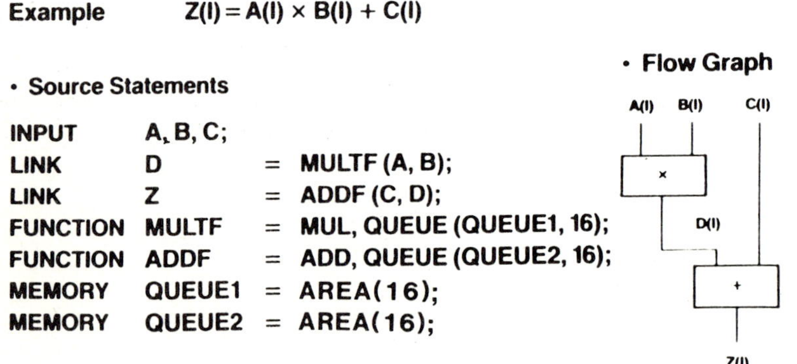

Example Z(I) = A(I) × B(I) + C(I)
• Source Statements
INPUT A, B, C;
LINK D = MULTF (A, B);
LINK Z = ADDF (C, D);
FUNCTION MULTF = MUL, QUEUE (QUEUE1, 16);
FUNCTION ADDF = ADD, QUEUE (QUEUE2, 16);
MEMORY QUEUE1 = AREA(16);
MEMORY QUEUE2 = AREA(16);
• Flow Graph
A(I)
B(I)
C(I)
×
D(I)
+
Z(I)

Abb. 3: typisches Datenflussdiagramm zur Software-Erstellung

DISTANCE TRANSFORMS WITH DATA FLOW TECHNIQUES

S.T. Dekker, P.P. Jonker, F.C.A. Groen
Department of Applied Physics
Delft University of Technology
The Netherlands

Abstract: This paper describes a dataflow processor board and its application in image processing. The board consists of 4 NEC IMPPs (IMage Pipelined Processors) and is a slave processor board of an 68010 UNIX/VME-bus system. The IMPP is an efficient single chip data flow system aimed for signal processing applications, where mostly fast and small though non trivial algorithms are used. The performance of the board was tested using distance transformation algorithms.

1. Introduction

Image processing is inherently time consuming but has a potential for parallel execution. Often image processing is SIMD processing; the same operation has to be applied to each pixel. The time constraints in many practical applications of image processing may make special image processing hardware necessary. A relatively new technique in trying to obtain a high degree of parallelism is the data flow concept. Whereas in the conventional control flow concept the sequence of operations is defined by the programmer, in the data flow concept the sequence of operations is defined by the data dependencies of the operations and need not explicitly be specified by the programmer. There are two forms of the data flow paradigm: Data driven computation, where an operation is executed as soon as the operands are available, and demand driven computation, where an operation is executed as soon as its result is demanded. The NEC IMage Pipelined Processor is based on a data driven model of computing. Data flow programs can be represented by flow graphs [2]. The nodes of the directed graph represent the operations, the branches form the conceptual medium on which the data flows. The dataflow on the branches can be seen as a token flow in which each token represents a data entity. A branch may contain more than one token, although passing each other is often not allowed. If all the necessary input tokens of a node are present, a result is calculated and an output token is fired. The most practical data flow hardware architecture is the token ring structure as used in the Manchester dataflow machine [3][6]. The ring is a circular pipeline structure; the data in the ring represent the tokens on the branches of the program graph. The NEC IMage Pipelined Processor [4] is also based on a token ring structure. However, the IMPP uses the branches of the graph as token identification instead of the node numbers as in the Manchester design. Hence the data flow program is split into a graph description table and a node description table (see figure 1).

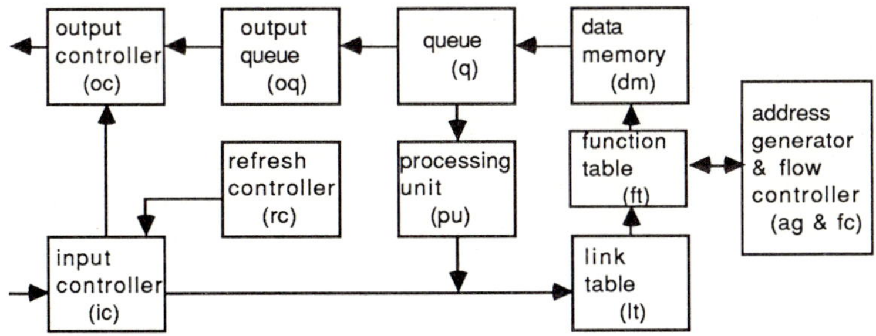

Figure 1 The NEC IMage Pipelined Processor (IMPP)

The IMPP concists of a:

Link table: The branch number is an index to the Link Table. In the Link Table the token receives the number of the node where the branch directs to, a Left/Right indicator (specifying which of the two inputs of the node has to be used) and a specification of the branch number at the output of the node. Note that the same node can thus be used on several places in the graph.

Function Table: It specifies the operation code for the Processing Unit and the AG & FC unit.

Address Generator and Flow Controller: In the AG & FC unit the addresses for the Data Memory are generated (e.g. for token matching) or token flow control is realized, such as splitting of token streams.

Data Memory: Memory used for the storage of constants, for temporary memory and for storage of tokens that wait for its partner at two input nodes.

Queue: Queue for the Processing Unit and the Output Queue. Tokens are queued here if either the Processing Unit is busy or the Output Queue is full.

Processing Unit: The arithmetic and token copy unit. It can perform one and two operand operations such as arithmetic and logic operations, comparison, adjustment, generation of numeric values and copying of tokens. If the operation of the PU cannot be completed within one pipeline cycle, it sends a signal both to the Queue and to the Input Controller inhibiting token flow into the ring or to the PU.

Input controller: This block inserts tokens from other IMPPs or the host interface into the ring, or passes tokens not meant for this IMPP to the Output Controller. It inserts refresh tokens from the Refresh Controller in the Ring. It samples the PU and inhibits insertion if the PU is busy. It restricts or inhibits token insertion if the Queue is nearly full.

Output controller: This block outputs tokens from the Output Queue, tokens from the Input Controller and Error Tokens generated by the OC internally.

Output queue: Small fifo queue for tokens leaving the ring.

Refresh controller: Sends refresh signals to the Input Controller to refresh the dynamic RAMs in the ring.

The architecture provides no facilities for subroutines (the Manchester machine does). The loading of the tables is performed by using special table write tokens. If there are no tokens in the queue the ring can be passed in 7 steps, each of 200ns (at 10Mhz). Token I/O takes 320ns. The maximum size of the Link Table is 128 branches and the maximum size of the Function Table is 64 nodes.

2. Implementation of distance transform algorithms on a data flow test system.

Figure 2 shows the architecture of a data flow test system based on a ring of 4 IMPPs and a MAGIC (Memory Access and General bus Interface Chip). An IMPP can read or write image data by sending READ or WRITE tokens to the MAGIC. Special DMA tokens let an IMPP read or write data from or to the VME memory. The System software for this board was developed in C and consists of an Assembler (a81) and a Debugger/Loader (db81). The input of a81 is formed by a textual representation of the data flow graph. The output of a81 is an object file with table contents for the IMPPs. One of the advantages of a81 over the original NEC assembler is that symbolic addressing of all branches in the entire data flow system is allowed.

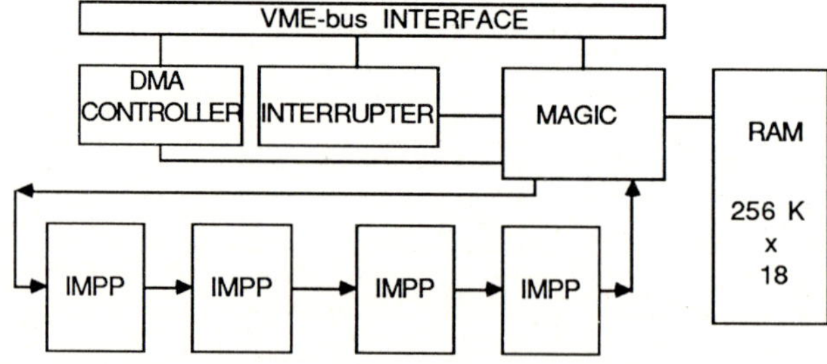

Figure 2 The VME dataflow test system

The Borgefors [1][5] algorithm produces the distance transform by means of a 2-pass recursive filter operation with a filter of size 3x2 or 5x3 depending upon the desired accuracy. It uses simple integer operations and has 3 different integer coefficients which allow a close approximation of the euclidean distance. The result is an image where each pixel contains the minimum 'chamfer' distance to a background pixel. When we indicate the Borgefors distance with BDT(d1,d2,d3), the grid constant is d1, the diagonal distance is d2 and a knights move is d3. A local correct euclidean distance is then BDT(1, $\sqrt{2}$, $\sqrt{5}$), or as a good rational approximation BDT(5,7,11).

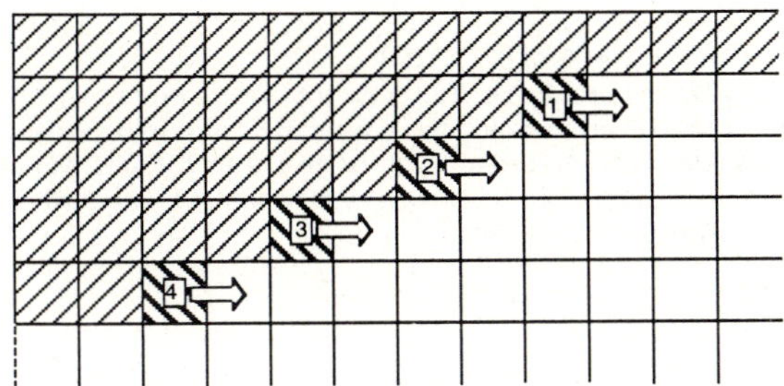

Figure 3 Downward scan with 4 processors

An algorithm for the 3x2 distance transform BDT(5,7, ∞) of an 256x256 image was implemented. The basic idea of the algorithm was that each IMPP processes a line of the image, passing information of already processed pixels to the IMPP that processes the next line in the image. Figure 3 shows the principle for the downward scan. The values of the shaded pixels have already been calculated. The processors 1 to 4 calculate a new front. The advantage of this algorithm is that the more IMPPs are added to the system, the more parallel processing will take place. As is illustrated in figure 4, the four IMPPs form a process pipeline in which the first IMPP in the pipe generates a continues pixel stream. This pixel stream is fed into the pipe and results of one process are passed to the next process in the pipe. Each process reads its own original values from memory. In the same way each process writes its own results to the memory. In order to make efficient use of the pipelining, the address generators in the calculation units generate bursts of addresses. To get the maximum speed, the first address generator in the pipe is pushed up until one of the IMPPs starts sending error tokens due to buffer overflow. Considering that IMPPs are asynchronous processors and that at random moments the refresh controller sends refresh tokens in the ring, the algorithm (1) need to be tuned to a speed that does

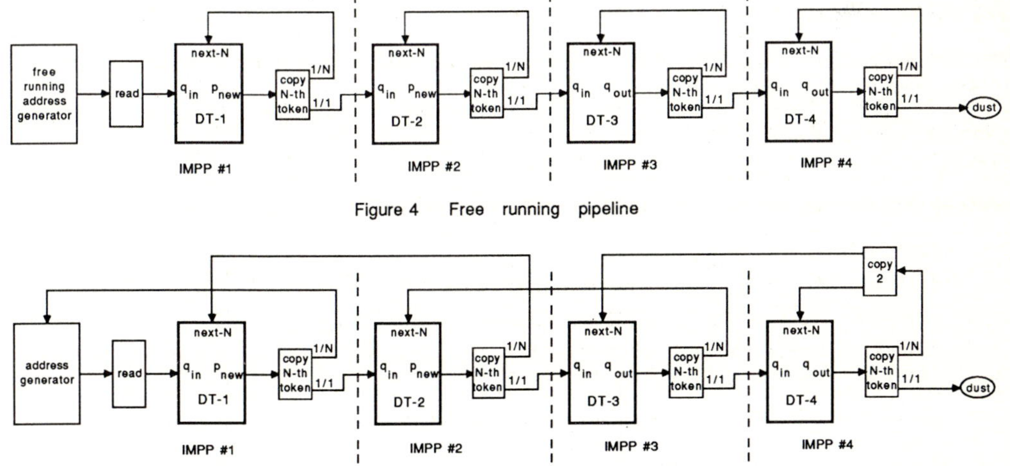

Figure 4 Free running pipeline

Figure 5 Pipeline with synchronization

not cause overflow. Figure 5 shows a version of the algorithm (2) where each process performs a handshake with its neighbours, thus making the pipeline self stabilizing. Speed improvement of the algorithm (3) was obtained by loading each IMPP with two calculation processes instead of one process. The speed figures of the algorithms are:

1	2	3	C	ASS
0.193s.	0.223s.	0.184s.	5.5s.	1.2s.

With 'C' the algorithm written in C on a 68010/8Mhz UNIX system and 'ASS' the algorithm written in assembler on a 68020/12.5Mhz UNIX system.
The optimum burst size for algorithm 1 is 2, the optimum burst size for algorithms 2 and 3 is 8. Special care should be taken by dividing the processes over the IMPPs. Tokens passing an IMPP appeared to have a higher priority than tokens in the Output Queue of the IMPP. This gave congestion in the IMPPs at Algorithm 1. Therefore the pixel stream generator was loaded in IMPP4.

3. Discussion

Programming a dataflow system is quite different from programming a conventional computer. Arithmetical operations are easily transformed into program graphs; global control and exception handling (boundary conditions) are more difficult to program. The biggest advantage of the data flow concept is that a uniform model can be used in an N-processor system. The program graph may be distributed over the N processors, not needing special process communication structures. The concepts involved in programming a single processor or an N-processor system are identical. The processing speed increases linearly with the number of processors. However, due to the scarse resources in the IMPP, the efficient distribution of processes over the processors, the elimination of data dependencies in the algorithm and the synchronization between parts of the program remain the programmers task. The performance of the system may degrade for data dependent or recursive algorithms. Before data can be used again, it must pass through at least seven pipeline delays. As shown above, in some cases the built-in hardware synchronization mechanisms are not sufficient. Then, active synchronization through demand tokens is necessary.
The NEC dataflow processor provides an efficient processing element for small computing bound applications. Although the pipeline speed is only 5 Mhz, the processing speed of one IMPP is equivalent to a 20 Mhz 68020. The IMPP instruction set proved to be very powerful. The multiplication (200ns), gives the chip its signal processing power. Notable missing is the division.

4. Acknowledgements

Without the support of Leon de Vos, Hans Schouten and Ruud van Munster this project would not have been realized.

5. References

[1] G.Borgefors, Distance Transformations in Arbitrary Dimensions, Computer Vision, Graphics and Image Processing, 27, 1984, pp. 321-345.

[2] A.L.Davis, R.M.Keller, Data Flow Program Graphs, IEEE Computer, Vol 15-2, February 1982, pp. 26-41.

[3] J.R.Gurd, C.C.Kirkham, I.Watson, The Manchester prototype Data flow Computer, Communications of the ACM, Vol 28-1, January 1985, pp. 34-52

[4] Iwashita M. et al., A data driven VLSI image processor (IMPP). In: Uhr L., Preston K., Levialdi S., Duff M.B.J., Evaluation of Multi-computers for Image Processing. Academic Press Inc., Orlando Florida USA, 1986.

[5] P.W.Verbeek, L.Dorst, B.J.H.Verwer, F.C.A.Groen, Collision Avoidance and Path-Finding through Constrained Distance Transformation in Robot State Space, Proc. Conf. Intelligent Autonomous Systems, Dec. 8-11 1986, Amsterdam, Holland

[6] I.Watson, J.R.Gurd, A Practical Data flow Computer, IEEE Computer, Vol 15-2, February 1982, pp. 51-57.

STRUCTFLOW

Ein Datenflußrechner zur Verarbeitung strukturierter und kontinuierlicher Daten

Peter Nitezki

Forschungszentrum Informatik
Haid- und Neu-Str. 10-14
7500 Karlsruhe 1

Technische Expertensysteme und Robotik
Prof. Dr.-Ing. U. Rembold, Dr. Paul Levi

ABSTRACT Mit STRUCTFLOW wird ein Ansatz zur effizienten und flexiblen Verarbeitung strukturierter und kontinuierlicher Daten vorgestellt. Es werden dabei die einfache Programmierung und die effiziente Ausnutzung von Parallelität eines Datenflußrechners mit Flußprinzipien von Hardware-Pipelines kombiniert. Daraus entsteht eine neuartige statische Datenflußarchitektur, die ohne assoziative Mechanismen und lokale Flußkontrolle auskommt. Das resultierende System kann beliebig ausgedehnte Datenströme mit prinzipiell beliebig vielen parallelen arbeitenden Prozessoren verarbeiten. Anders als bei konventionellen Datenflußrechnern kann nahezu deterministisches Antwortverhalten und Synchronität der Datenströme erreicht werden. Es wird die Architektur und der Instruktionssatz eines Rechnerknotens vorgestellt und ein Konzept zur Verschaltung solcher Knoten in ein massiv paralleles System umrissen.

Einführung

Seit einiger Zeit wird die Notwendigkeit massiv paralleler Rechner mit allgemeiner Programmierbarkeit betont, um die enormen Anforderungen an die Rechenleistung und den Softwareumfang für Computersehen und fortgeschrittene Signalverarbeitung zu erfüllen. Bisherige Systeme bieten entweder massive Parallelität (Systolische Arrays, Prozessor-Arrays) oder allgemeine Programmierbarkeit. Beide Forderungen konnten bisher noch von keinem System erfüllt werden. Deshalb werden seit einiger Zeit Datenflußarchitekturen vorgestellt /Dennis 1, Arvind, Watson/, die eine elegante und seiteneffektfreie Programmierung erlauben und eine Parallelverarbeitung bis herunter zur Ebene der einzelnen Maschinenoperation mit maximaler Parallelität bieten sollen.

Leider haben diese Entwürfe wesentliche Nachteile, was die effiziente Verarbeitung und Speicherung großer strukturierter Datenmengen angeht. Obwohl prinzipiell der Datenfluß-Mechanismus die Ausnutzung der gesamten Parallelität garantiert, haben alle bisherigen Entwürfe wenig oder keine Möglichkeit, die in den Datenstrukturen steckende Parallelität zu nutzen. Desweiteren sind Datenflußrechner zwar durch die vollständige Asynchronität der Verarbeitung dazu prädestiniert, konkurrente Prozesse ohne zusätzlichen Verwaltungsüberhang zu bearbeiten, jedoch ist es sehr problematisch, eine Synchronisierung einzelner Vorgänge mit externen Abläufen zu erzielen. Der grundlegende Nichtdeterminismus der Verarbeitung stellt die Implementierung von signalverarbeitenden Systemen auf Datenflußrechnern vor zusätzliche Schwierigkeiten.

Leitprinzipien für den STRUCTFLOW-Entwurf

Für einen Signalverarbeitungsrechner sollte die Totzeit nicht zu Gunsten des Durchsatzes erhöht werden. Die massive Parallelität erlaubt die Durchsatzsteigerung aufgrund der Replikation von Funktionseinheiten und muß nicht auf die Verschachtelung der Verarbeitung in Pipelines bauen.

Dynamische Architekturen sind zwar sehr erfolgreich bei stark laufzeitabhängigem Verhalten der Algorithmen, bei signalverarbeitenden Systemen ist die Schwierigkeit der begrenzten Problemgröße und der Notwendigkeit assoziativer Mechanismen aber eher von Nachteil. STRUCTFLOW sollte also eher auf statischen Mechanismen beruhen.

Es konnte gezeigt werden, daß sich eine große Algorithmenklasse als Pipelines darstellen läßt /Dennis 2/. Von besonderem Interesse ist dabei, daß eine wichtige Klasse von Rekursionen sich in eine Pipeline, d.h. in einen zyklenfreien Graphen abbilden läßt. Für STRUCTFLOW sind somit die Fluß- und Kontrollmechanismen der Pipeline-Rechner von besonderer Bedeutung.

STRUCTFLOW macht nun den selben Schritt wie die Pipeline-Architekturen und fügt zur Dimension der Zeit auch die Dimension des Raumes. D.h. wie in systolischen Arrays wird ein elementares Datum durch seinen Ort, den es zu einem bestimmten Zeitpunkt innehat, identifiziert. Dadurch kann ein Datenstrom in mehrere aufgespalten werden und erlaubt damit die parallele Verarbeitung eines Datenstroms in verschiedenen Verarbeitungseinheiten.

Die Flußregel der statischen Datenflußrechner, die für einen kontinuierlichen Datenstrom fordert, daß sich zwei nachfolgende Daten nie überholen dürfen, also immer einer zeitlichen Ordnung genügen, wird bei STRUCTFLOW dahingehend erweitert, daß die Daten, die auf einer Verarbeitungseinheit verarbeitet werden, nie ohne Programmkontrolle diese Verarbeitungseinheit verlassen dürfen, d.h. daß die Datenströme einer räumlichen Ordnung genügen.

Diesem zentralen Prinzip wird weiterhin angefügt, daß unter Programmkontrolle auf einfache und transparente Weise die Synchronisierung von Datenströme möglich sein soll. Zudem soll eine Typprüfung zur Laufzeit und eine einheitliche Ausnahmebehandlung vorgesehen werden, zwei Forderungen, denen bisher keine Datenflußarchitektur genügt, da eine konventionelle, d.h. zustandsabhängige Verarbeitung in einer Datenflußarchitektur nicht möglich ist. Nicht zuletzt wird in der STRUCTFLOW-Architektur ein einfacher und effizienter Mechanismus zur Darstellung von Konstanten nötig sein.

Der Instruktionssatz von STRUCTFLOW

STRUCTFLOW verwendet 16 Bit Integer-Arithmetik, was nur von der Einschränkung durch einen Entwurf mit Standardbauteilen herrührt und das Prinzip der Verarbeitung nicht berührt, den Entwurf aber extrem vereinfacht. Die vier Grundoperationen der Arithmetik werden um die Divisionrest- und die Supremum- und Infimum-Operation erweitert. Weiterhin werden alle relationalen Operationen realisiert.

Der Datentyp Boolean wird durch alle 16 zweistelligen Funktionen realisiert. Die Typen werden dabei durch ein Tagfeld, das dem Wert des Datums angefügt ist, bestimmt. Dieses Tagfeld dient zudem zur Erweiterung der Definitionsmenge der Operationen, um eine Ausnahmebehandlung für die Arithmetik zu ermöglichen. Arithmetischer Überlauf wird durch die Elemente plus und minus unendlich ($+\infty$, $-\infty$) angezeigt. Fehler durch inkompatible Operandentypen werden durch das Element ***Bottom*** dargestellt. Weitere Symbole können zur Kennzeichnung der Struktur eines Datenstroms benutzt werden. STRUCTFLOW sieht dafür ein drei Bit weites Tagfeld vor.

Weitere Operatoren dienen der Steuerung des Datenflußes und für konditionale und nichtdeterministische Konstrukte. Die Auswahlfunktion läßt ein Datum passieren, wenn der zweite Operand ***TRUE*** ist, andernfalls gibt sie ***Bottom*** aus. Die Wächterfunktion wählt ein von ***Bottom*** verschiedenes Datum aus. Falls beide Operanden den Wert ***Bottom*** haben wird ***Bottom*** ausgegeben. Die Funktionen Auswahl und Wächter dienen zur Programmierung von Alternativen (if-then-else), ohne eine dreistellige Operation zu benötigen. Dies hat sich für den Entwurf der Hardware als ganz besonders hilfreich erwiesen. Nichtdeterministische Konstrukte wie Dijkstra's-Wächter sind hiermit auch realisierbar, ohne den Flußmechanismus zu durchbrechen.

Zwei weitere Funktionen dienen der Flußsteuerung und zum Ausgleichen unterschiedlich langer Zweige in den Flußgraphen, was zur Implementierung effizienter Pipeline-Mechanismen nötig ist.

Die Identitäts-Operation gibt das Datum auf seiner linken Kante weiter. Die rechte Kante kann als Semaphor für die Synchronisation von Datenströmen dienen. Falls diese Operation als Schlangenregister dienen soll, wird die rechte Kante mit einer Konstante belegt. Diese Operation hat den Vorteil, nicht typabhängig zu sein, und kann damit ohne weiteres Betrachten des Flußgraphen für jeden Datenstrom verwendet werden.

Die Kopier-Operation dient der Replikation eines Datenstroms und wird zur Vereinfachung des Entwurfs von fast allen Datenflußarchitekturen eingesetzt. Außer ihr haben alle Operationen nur eine Ausgangskante. Diese Einschränkung erlaubt eine sehr effiziente Nutzung von Hardware und Verarbeitungszeit.

Die letzte Klasse von Operationen sind ausgabeorientiert und dienen der Verteilung von Datenströmen auf andere Verarbeitungseinheiten. Bisher ist nur eine einfache Angabeoperation implementiert, aber es sind kompliziertere Operationen mit weiterer Verarbeitung oder Flußsteuerung möglich.

Die Architektur von STRUCTFLOW

Ein Prozessorknoten besteht aus drei Hauptfunktionseinheiten. Matching-Einheit und Verarbeitungseinheit sind in dem für Datenflußrechner charakteristischen Ring angeordnet. Die Matching-Einheit übernimmt die ankommenden Daten, stellt fest, ob eine Operation ausführbar ist und speichert dann das Datum zwischen oder schnürt aus Operationscode, rechtem und linkem Datenwert und dem Ergebnisnamen einen Verarbeitungstoken. Dieser Verarbeitungstoken wird in der Verarbeitungseinheit verarbeitet und das Ergebnisdatum an die Matching-Einheit zurückgereicht.

Mit diesen Funktionseinheiten verbunden ist die Kommunikationseinheit. Sie bindet den einzelnen Prozessor an ein Netzwerk an, bildet die Flußsteuerung für eingehende und ausgehende Kanten und dient dem Laden von Flußgraphen in den Knotenspeicher.

Der Matching-Vorgang in STRUCTFLOW

STRUCTFLOW verzichtet prinzipiell auf assoziative Mechanismen und aufwändige Flußsteuerung. Im wesentlichen wurden zwei Prinzipien verwirklicht. Das erste ist eine vollständige Abbildung der Flußgraphen auf einen kombinierten Kanten/Knoten-Speicher. Jedes Datum adressiert über seinen Namen direkt eine Speicherzelle. Zweitens verzichtet STRUCTFLOW auf eine lokale Flußsteuerung. Da keine Rückwärtsverzeigerung nötig wird, halbiert sich der notwendige Speicherumfang. Da alle Flußgraphen Pipelines sind oder explizit über Semaphore synchronisiert werden, kann ein Test auf Freiheit der Ausgangskante entfallen. Er wird ersetzt durch eine quasisystolische Arbeitsweise. Die Eingangs- und Ausganskanten unterliegen einer Flußkontrolle durch die Kommunikationseinheit. Sie lädt jeden Flußgraphen in einem Polling-Modus mit Daten. Die Matching-Einheit erzeugt nur für die mit der Kommunikations-Einheit gemeinsamen Kanten eine Quittung, die in einem verschränkten Protokoll abgewickelt wird und deshalb keine Zeiger benötigt.

Der Kanten/Knotenspeicher besteht für jeden Knoten aus einem 6 Bit breiten Feld für den Operationscode, 16 Bit für den Namen der ausgehenden Kante, 19 Bit für einen Kantenpuffer und 3 Bit für die Steuerung des Matching-Vorgangs. Die 16 Bit für den Kantennamen ergeben sich aus dem derzeitigen Stand der Technologie statischer Halbleiter-Speicher und ist ein guter Kompromiss aus Aufwandsreduzierung und Bequemlichkeit der Programmierung. Der derzeitige Entwurf des STRUCTFLOW-Prozessors erlaubt deshalb 32K Knoten als maximale Größe eines einzelnen Graphen/Gallinat/

Der Kanten/Knotenspeicher braucht nur einen Kantenpuffer pro Knoten, da sichergestellt ist, daß ein Match eintritt, bevor ein weiteres Datum auf dieser Kante eintrifft. Die Pipeline-Verarbeitung und der Flußkontroll-Mechanismus sorgen hierfür. Die drei Bit der Matching-Steuerung bestehen aus dem Links/Rechts-Bit, das angibt, welche Kante gerade gepuffert wird, dem Present-Bit, das die Anwesenheit eines Datums im Puffer anzeigt und dem Konstanten-Bit, das das anwesende Datum als Konstante kennzeichnet.

Der Matching-Vorgang geht nun in zwei Stufen vor. In der Ersten wird ermittelt, wer ein Datum anbietet, Verarbeitungs- oder Kommunikations-Einheit. In der zweiten Stufe wird der Kanten/Knotenspeicher adressiert, auf Match getestet und durch Einschreiben des Matching-Bits und des Datums modifiziert. Hierbei wird bei einem Match automatisch ein Token gebildet, der gleichzeitig an die Verarbeitungs-Einheit übergeben wird.

Durch den Verzicht auf lokale Flußsteuerung fallen zwei weitere Schreibzugriffe auf den Kanten/Knotenspeicher weg. Der Matching-Vorgang läßt sich enorm beschleunigen und wird im wesentlichen durch zwei Speicherzugriffe und die Durchlaufzeit des Steuerwerkes bestimmt. Derzeitige Technologie erlaubt ca. 5 Millionen Matchings pro Sekunde mit schneller CMOS-Logik. Erkauft wird dies mit der Notwendigkeit programmgesteuerter Synchronisation oder, was häufiger ist, durch Flußgraphen, die dem Pipeline-Schema genügen.

Die Verarbeitungseinheit

Die Verarbeitungseinheit wurde als horizontal mikroprogrammiertes Bit-Slice-Rechenwerk mit einem schnellen Multiplizierer verwirklicht. Die Programmierbarkeit ist für eine Forschungsmaschine wichtiger als der Durchsatz, der bei etwa 1 MOPS liegt. Das Verarbeiten des Tagfeldes und die Boole'schen Operationen wurden in einer Funktionstafel realisiert. Hierdurch entsteht kein zusätzlicher Zeitaufwand für diese Funktion/Augenstein/.

Matching- und Verarbeitungseinheit sind über zwei FIFO-Speicher gekoppelt. Die Token-FIFO entkoppelt die Matching-Einheit von der asynchron arbeitenden Verarbeitungseinheit. Die gewonnene Freiheit kann für eine Überlappung von Verarbeitung und Programmladen genutzt werden, da während dieser Zeit das Matching ausgesetzt werden muß. Die Ergebnis-FIFO dient als Temporärspeicher für die erzeugten Daten. Da die Kopier-Operation zwei Daten erzeugt, aber nur eines konsumiert, müssen die hieraus entstehenden Daten in dieser FIFO gepuffert werden. Die halbe Kapazität dieser FIFO ist für diesen Zweck reserviert.

Vernetzung von Einzelprozessoren

Die Kommunikations-Einheit bildet die Verbindung des einzelnen Prozessors mit einem Kommunikationsnetzwerk und macht die Flußkontrolle für die eingehenden und ausgehenden Kanten. Zu diesem Zweck werden für einen festen Bereich von Kantennamen FIFO-Speicher eingerichtet, deren Füllstandssignale den Datenfluß steuern. Außerdem werden die eingehenden und ausgehenden Kanten der einzelnen Flußgraphen gruppiert und dadurch der Pipeline-Fluß gesteuert.

Für eine kleine Zahl von Prozessoren (max 16?) zeigt sich ein synchroner "Slotted Ring" als geeignetes Netzwerk, um eine Vernetzung der Prozessoren zu erreichen. Ein einfaches Protokoll läßt sich in gewöhnlicher Rückwandtechno-logie mit käuflichen Schaltkreisen zu einem Kommunikationsmedium hohen Durchsatzes kombinieren. Man kann damit maximal ca. 10 Megaworte pro Prozessor und Sekunde erreichen.

Ab etwa 16 Prozessoren wird die Totzeit eines solchen Rings jedoch viel zu hoch für die gebotene Leistung. Hier würden sich "Cube Connected Cycles" anbieten, die ein logarithmisches Anwachsen der Distanz mit der Prozessorzahl und einen festen Knotengrad vereinigen.

Wichtig für die Skalierbarkeit von STRUCTFLOW ist dabei eine Prozessor-relative Adressierung der Daten. Hiermit kann sogar bei fester Adresslänge die Anzahl der Rechnerknoten beliebig wachsen. Es lassen sich zwar nicht mehr alle Knoten direkt erreichen, aber es besteht keine prinzipielle Grenze in der Prozessorzahl.

Speicherung strukturierter Daten

Der STRUCTFLOW-Prozessor sieht keine Speicherung strukturierter Daten vor. Ähnlich wie andere Entwürfe scheint dies besser in besonderen Speichereinheiten zu erfolgen. Die bisher vorgeschlagenen I-Structures /Arvind/ bieten zwar allgemeine Strukturen und somit Flexibilität, arbeiten aber über Zugriffsbäume, die nur in den Blättern Datenwerte, in den Zweigen nur Strukturinformation speichern. Ihre Effizienz in der Speichernutzung sind daher eher klein, die Latenz des Datenzugriffs eher groß.

Für STRUCTFLOW käme daher eher der Ansatz von TIP-1 /Temma/ in Frage, wo explizite Adresströme einen linearen Speicher adressieren und aus ihm einen Datenstrom abrufen bzw. in ihn einspeichern. In einem früheren Aufsatz habe ich gezeigt, daß ein Algorithmus in Berechnung und Datenselektion zerlegt werden kann /Nitezki, Bauer/, was sich in Indizes und Laufvariablen anzeigt. Diese Zugriffsalgorithmen sind meist sehr einfach und oft nicht von Bedingungen abhängig. Man könnte damit einen Zugriffsprozessor für einfache Algorithmen mit dem Speicher vereinigen. Komplizierte Zugriffsalgorithmen könnten dann in einem gewöhnlichen Prozessor berechnet werden und auf den Speicher direkt über Adresströme zugreifen.

Danksagung

Diese Arbeit wurde im Rahmen des Forschungsschwerpunkts "Physik und Anwendung neuartiger Sensoren" ausgeführt und vom Land Baden Württemberg finanziell unterstützt.

LITERATUR

/Augenstein/ Augenstein, R.
Entwurf und Realisierung einer Verarbeitungseinheit;
Diplomarbeit, Fak. Elektrotechnik, Universität Karlsruhe, 1987

/Bauer/ Bauer, F.L.
Dualismen in der Informatik; Elektronische Rechenanlagen;
4, 1982

/Dennis 1/ Dennis, J. B.
Data Flow Ideas for Supercomputers; IEEE, 1984

/Dennis 2 Dennis, J.B.; Rong, G.G.
Maximum Pipelining of Array Operations on Static Data Flow
Machine, IEEE Parallel Processing, 1983

/Gallinat/ Gallinat, J.
Entwurf und Aufbau einer Matching-Einheit, Diplomarbeit,
Fakultät Informatik, Universität Karlsruhe, 1987

/Nitezki/ Nitezki, P.
A Dualistic Model to discribe Computer Architectures;
Architectures and Algorithms for Digital Image Processing II;
SPIE Vol. 534, 1985

/Temma/ Temma, T., Mizoguchi, M.; Hanaki, S.
Template-Controlled Image Processor TIP-1 performance
Evaluation, IJCAI,1983

/Watson/ Watson, I., Gurd, J.
A Practical Data Flow Computer, IEEE Computer, Feb. 1982

Anwendungen und Grenzen kommunizierender paralleler Prozesse in der industriellen Bildverarbeitung

Föhr, R.; Ameling, W.
Rogowski-Institut für Elektrotechnik, RWTH Aachen

Einführung

Echtzeitanwendungen von Bildverarbeitungs- und Mustererkennungsverfahren sind wegen des hohen Rechenaufwands nur in beschränktem Maße möglich. Ein Ausweg ist häufig, wegen der Regularität der verwendeten Algorithmenklassen, die Parallelisierung auf einer geeigneten Mehrrechnerarchitektur. Neben SIMD-Maschinen mit einer sehr großen Anzahl einfacher Prozessoren und einer wegen der Struktur der Rohdaten meist gitterförmigen Verbindungsstruktur werden in der letzten Zeit vermehrt Konzepte mit Netzwerken aus leistungsfähigen, mittlerweile auch verfügbaren Einzelprozessoren diskutiert.

Dieses Netzwerkkonzept unterstützt auf Prozedurniveau parallelisierte Verfahren, die über gemeinsame Variablen oder über Nachrichten kommunizieren können. Wegen des arbitrierten Zugriffs auf einen gemeinsamen Speicher wird im folgenden streng nach OCCAM-Konzept /POUN86/ eine Kommunikation mit Nachrichten innerhalb eines Transputer-Netzes durchgeführt. Beispielhaft werden drei typische Verfahren betrachtet und ihre Realisierungen mit einigen Optimierungskriterien vorgestellt.

Nachbarschaftorientierte Vorverarbeitungsalgorithmen

Das erste Beispiel, ein Template-Matching-Algorithmus mit dem Vergleich von vier gedrehten Musterkanten (Sobel-Maske), dient der Kantenextraktion und ist in einer effektiven Formulierung in /ALSF85/ beschrieben. Vorteilhaft für eine Parallelisierung ist, daß bei der Anwendung auf jeden Bildpunkt nur wenige Nachbarpunkte zur Berechnung nötig sind und daß von einem etwa gleichen Rechenaufwand je Bildpunkt ausgegangen werden kann. Das legt eine statische Aufteilung nahe, bei der jedem Rechenelement ein gleichgroßer Bereich zur eigenständigen Berechnung übergeben wird. Die Form des Bereichs hat Einfluß auf die Verbindungstruktur: quadratische Ausschnitte für eine gitterförmige, Streifen für eine eindimensionale Anordnung.

Bei solchen Filterverfahren steigt die Rechenleistung nahezu linear mit der Anzahl der eingesetzen Prozessoren, jedoch liegt das Produkt aus Rechenzeit und Prozessoranzahl noch immer im Sekundenbereich. Da die Bearbeitung von Nachbarpunkten bei begrenztem Rechenaufwand leicht durch auf dem abgestasteten Videosignal arbeitende Rechnereinheiten (diskret aufgebaut oder im VLSI-Design) realisiert werden kann, ist eine solche Konstruktion für Echtzeitanwendungen, d.h. im Bereich der Bildwiederholfrequenz, vorzuziehen.

Algorithmen mit nicht vorhersehbaren Datenaufkommen

Im zweiten Beispiel wird die klassische Hough-Transformation zum Auffinden von linienförmigen Strukturen in Bildern betrachtet. Dieses Verfahren transformiert die Punkte, die aus einer Kantenextraktion (s.o.) hervorgegangen sind, in einen Parameterraum, in dem durch Anhäufungen (relative Maxima) auf Geraden im Bild geschlossen werden kann /DUDA72/. Da die Kantenpunkte nicht mehr gleichmäßig über das Bild verteilt sind, würde eine Aufteilung wie oben zu ineffizienten Wartezeiten in Bereichen mit wenigen Kantenpunkten führen.

Hier bietet sich das Konzept des Prozessor-Farmings an, bei dem durch eine dynamische Verteilung von Elementaraufgaben mit auch unterschiedlichen Laufzeiten eine nahezu vollständige Auslastung eines Netzes erreicht wird. Die Anzahl der Prozessoren ist frei wählbar und führt zu nahezu linearem Speed-Up. Zum besseren Verständnis der Systematik werden in Bild 1 dazu fünf Prozeßtypen eingeführt.

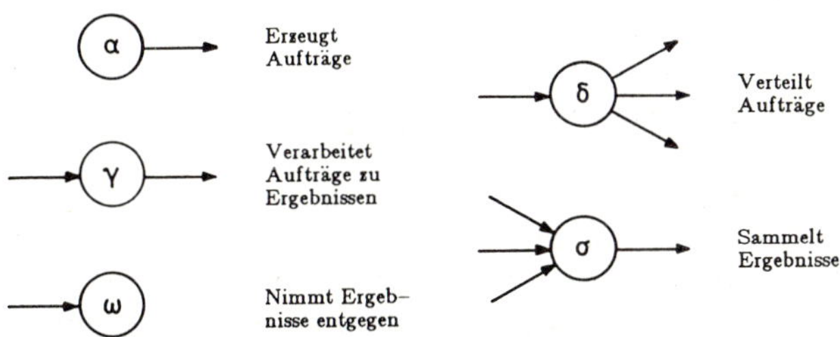

Bild 1: Prozeßtypen in einer Prozessor-Farm

α, γ und ω sind anwendungsspezifische Prozesse, die jeweils für eine bestimmte Aufgabenstellung zu formulieren sind. Die rekursiv definierte Grundkombination unter Verwendung der genannten Prozeßtypen zeigt Bild 2. Folgende Kriterien sind dabei zu beachten:

- γ-Prozesse dürfen Ergebnisse nur aus Auftragsdaten und lokalen Variablen berechnen und deshalb keine globalen Variablen benutzen.
- γ-Prozesse sollen kurz und häufig genug durchzuführen sein, um eine gleichmäßige Auslastung zu erreichen (Granularität).
- Übertragungszeiten für Auftrags- oder Ergebnisdatensätze sollten deutlich kleiner als die Rechenzeit im γ-Prozeß sein.

Die Hough-Transformation wird nun so realisiert (Bild 2), daß zuerst alle Kantenpunkte in das Akkumulatorfeld $H(r,\varphi)$ transformiert und dann in einer zweiten Phase die relativen Maxima in $H(r,\varphi)$ gesucht werden:

Transformationsphase:

- α: Auftrag: ein oder mehrere Punkte sind zu transformieren
- γ: Transformation aller aufgetragenen Punkte in lokales $H_l(r,\varphi)$
- ω: Aufaddieren der lokalen $H_l(r,\varphi)$ zum globalen $H_g(r,\varphi)$
  Verbesserung: Übertragen des Aufaddierens in die σ-Prozesse

Phase der Maximumsuche:

- α: Auftrag: Quadratisches Feld $H_q(r,\varphi)$ (⇒ minimaler Überlappungsbereich) aus $H_g(r,\varphi)$ wird zur Maximumsuche mit Parametern (Mindestwert, Mindestabstand) transferiert.
- γ: Suche nach relativen Maxima im Kernbereich durch Auswertung von quadratischen Masken über $H_q(r,\varphi)$
- ω: Entgegennahme der Maxima und Sortierung mit Hash-Verfahren.

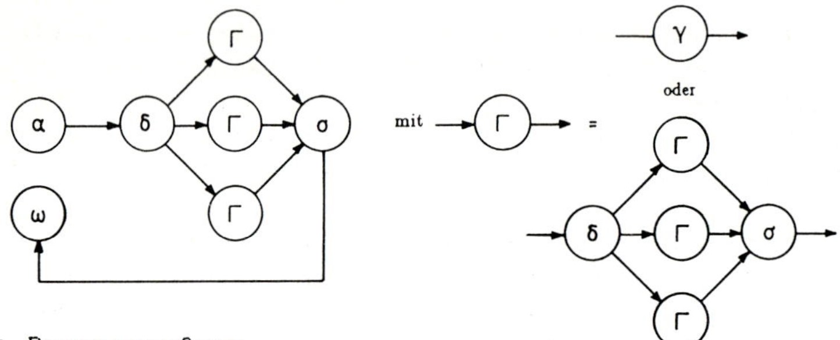

Bild 2: Prozessorfarm

Algorithmen mit stark verketteten Datenstrukturen

Zuletzt wird gezeigt, wie auch mit komplizierten Datenstrukturen (Quadtrees, Nachbarschaftgraphen) in ursprünglich rekursiv formulierten Algorithmen eine Parallelisierung durch Prozessor-Farming gelingt. Das gewählte Beispiel, Segmentierung nach dem Split-and-merge-Prinzip /HORO76/, besteht aus vier Phasen:

Initsplit-Phase: Willkürliche Anfangsaufteilung des Bildes (Größe $2^n x 2^n$ Punkte) in $2^m x 2^m$ quadratische Blöcke (Quadranten)

Initmerge-Phase: Zusammenfassung von vier ähnlichen benachbarten Blöcken zum nächstgrößeren Quadranten

Split-Phase: Wiederholtes Aufteilen von in sich inhomogenen Quadranten in vier Unterquadranten

Merge-Phase: Zusammenfassen benachbarter ähnlicher Quadranten zu in sich gleichförmigen Regionen

Das Endergebnis dieser Segmentierung ist ein markiertes Bild mit einer Liste von Merkmalvektoren der Regionen sowie ein Nachbarschaftsgraph (Region Adjacency Graph, RAG) mit Regionen als Knoten und Kanten zwischen benachtbarten Regionen. Für diese hierarchische Bearbeitung des Bildes wurde der reguläre Quadtree nach /HUNT79/ als globale Struktur während der Split-Phase und der lineare Quadtree nach /GARG82/ für die Quadrantenlisten der Regionen in der Merge-Phase verwandt. Die teilweise rekursive Definition konnte durch Extraktion unabhängiger Prozeduren und das Verwenden einer globalen Struktur zwischen α- und ω-Prozeß in eine den o.g. Farming-Kriterien entsprechende Darstellung übersetzt werden. Im einzelnen wurde implementiert:

Initsplit-Phase: Teil des α-Prozesses in der Splitphase
Initmerge-Phase: Integriert in die Mergephase
Split-Phase:

α: Alle Quadranten aus der Anfangsaufteilung werden als inhomogen angenommen. Auftrag: Quadrant wird auf Homogenität untersucht.

γ: Homogenitätsprüfung: homogene Quadranten werden mit Merkmalen, inhomogene als Unterquadranten zur weiteren Überprüfung jeweils kodiert zurückgegeben.

ω: Sammeln aller homogenen Quadranten in regulärem Quadtree, Weitergabe aller noch inhomogenen Unterquadranten an α-Prozeß.

Merge-Phase:

α: Erzeugen einer sortierten, linear quadtree-kodierten Blockliste durch depth-first-Auslesen des regulären Quadtrees (s.o); Übertragen als lokale Datenbank zu jedem γ-Prozeß; Auftrag: Suche alle rechts und unten benachbarten Quadranten eines Blocks.

γ: Suche der Nachbarn und Berechnung der Qualität der Grenzen zu den Nachbarn durch Auswerten der Merkmale.

ω: Aufstellen des RAG über alle Quadranten mit Grenzen (Kanten im RAG), Ausfiltern aller trennenden Grenzen.

Mit der Initsplit-Aufteilung kann die Granularität eingestellt werden, die der Anzahl der benutzten Prozessoren angepaßt werden sollte. Die Übertragung des gemeinsamen Datenbestandes aller γ-Prozesse zu Beginn der Merge-Phase ist der Preis für den fehlenden gemeinsamen Speicher. Sie belastet aber jeden Prozessor nur einmal und geht damit in die Zeit zum Versorgen der Farm mit Aufträgen ein.

Die genannten Beispiele wurden auf einem Netz mit 5 32-Bit-Transputern und einem VME-Bus-basierten Bildverarbeitungssystem implementiert.

Literatur

/ALSF85/ Alsford, J. et al.: "CRS Image Processing Systems with VLSI Modules" aus Kittler & Duff (Ed.), Image Processing System Architectures, RSP England 1985

/DUDA72/ Duda, R./ Hart, P.: "Use of the Hough Transformation To Detect Lines and Curves in Pictures" Comm.ACM Vol.15 p.11 1972

/GARG82/ Gargantini, I.: "An Effective Way to Represent Quadtrees" Comm. of the ACM Vol. 25, p. 905, 1982

/HORO76/ Horowitz, S.L./ Pavlidis, T.: "Picture Segmentation by a Tree Traversal Algorithm, Jour. ACM Vol. 23 p. 368, 1976

/HUNT79/ Hunter, G.M./ Steiglitz, K.: "Operations on Images using Quad Trees" IEEE Transactions Vol. PAMI-1 p.145, 1979

/POUN86/ Pountain, D.: "A Tutorial Introduction to OCCAM Programming" Preliminary Version, Inmos 1986

Diese Arbeit wurde innerhalb des SFB 208 im Rahmen des Teilprojekt G2 von der Deutschen Forschungsgemeinschaft gefördert. Den Herren Hovekamp, Große Wienker und Celler sei für ihre Mithilfe gedankt.

EIN ENTWICKLUNGSSYSTEM FÜR MULTI-PROZESSORSYSTEME MIT SIGNALPROZESSOREN FUJITSU MB 8764 UND MB 87064

Hannspeter Eulenberg
Zentrallabor für Elektronik der Kernforschungsanlage Jülich GmbH (KFA)
Postfach 19 13, D-5170 Jülich

Im Zentrallabor für Elektronik der Kernforschungsanlage Jülich wird zur Zeit ein Entwicklungssystem für Multi-Signalprozessorsysteme entwickelt und gebaut, die mit dem Fujitsu Signalprozessor MB 8764 oder seinem kompatiblen Nachfolgetyp MB 87064 arbeiten.

Die meisten käuflichen Signalprozessoren und auch der MB 8764 oder der MB 87064 [4, 5] können über einen gemeinsamen Bus miteinander kommunizieren, wenn sie zu einem Multi-Prozessorsystem zusammengeschaltet werden. Leider treten aber schon bei relativ kleinen Systemen nicht beherrschbare Bus-Verbindungsprobleme auf. Dazu kommt, daß der Geschwindigkeitsgewinn bei der Signalverarbeitung bei parallel betriebenen Signalprozessoren auch vom zu verarbeitenden Algorithmus abhängt. Nach Minsky [1] gilt für viele Algorithmen, die mit N busgekoppelten, parallelbetriebenen Prozessoren abgearbeitet werden, für den Verarbeitungsgeschwindigkeitsgewinn ein $^{2}\log N$-Gesetz. Durch mathematische Optimierungsmethoden für Parallelalgorithmen konnte eine Steigerung des Geschwindigkeitsgewinns insbesondere für größere Parallelverarbeitungssysteme erreicht werden. Für diese Systeme gilt das Amdahl-Gesetz $N \cdot {}^{2}\log N$ [1]. Eine mindestens lineare Abhängigkeit für Multi-Prozessorsysteme wäre jedoch wünschenswert.

H.T. Kung und C.E. Leierson haben 1978 [2, 3] eine Signalverarbeitungsmethode mit dieser Eigenschaft beschrieben. Sie haben ihre systolischen Systeme für die schnelle Verarbeitung digitalisierter Signale mit Rechenwerken in hochintegrierten Schaltungen der VLSI-Technik geschaffen. In einem Netzwerk gleichartiger Rechenwerke, in denen ausschließlich gleichartige Rechenvorgänge ablaufen, werden Zwischenergebnisse nach dem Pipelining-Prinzip in einem für das gesamte Netzwerk gleichen Rhythmus solange in vorgebbare, für die gesamte Architektur gültige Datenflußrichtungen in die benachbarten Prozessoren geschoben, bis das Endergebnis den Ausgangspunkt - beispielsweise einen Datenspeicher - erreicht.

Zur Steigerung der Verarbeitungsgeschwindigkeit abgetasteter und digitalisierter Signale soll dieses Signalverarbeitungsprinzip nun auch in Multi-Signalprozessorsystemen für ein- und mehrdimensionale Signale verwendet werden. Mit dem hier vorgestellten Multi-Signalprozessor-Entwicklungssystem können unter anderem sowohl semi-systolische als auch global-systolische Multi-Prozessorsysteme entwickelt werden, deren Signalprozessoren zu einer Matrix oder zu einem in modulo 2 strukturierten Baum verschaltet werden und über Speicherkopplungen (DMA) miteinander kommunizieren.

Bei **semi-systolischen Multi-Prozessorsystemen** werden die digitalisierten Eingangssignale des Systems in jeden Signalprozessor gleichzeitig eingegeben und die weitere Verarbeitung systolisch durchgeführt. In **global-systolischen Multi-Prozessorsystemen** wird das Pipelining-Prinzip auch bei der Eingabe der Eingangssignale benutzt.

Das KFA Multi-Signalprozessor-Entwicklungssystem MB 8764 unterstützt die Entwicklung von:

- global-systolisch arbeitenden Multiprozessor-Systemen in Baum-Architektur,
- semi-systolisch arbeitenden Multiprozessor-Systemen in Matrizen-Architektur,
- global-systolisch arbeitenden Multiprozessor-Systemen in Matrizen-Architektur,
- Transputer-Architekturen
- und Einprozessoranordnungen.

Transputer-Architekturen sollen solche sein, bei denen zwar, wie bei global-systolischen Architekturen, die Prozessoren mit ihren Nachbarn orthogonal gekoppelt sind, die sich von ihnen jedoch durch ihre Arbeitsweise unterscheiden. In ihren Prozessoren laufen die unterschiedlichsten Prozeduren ab. Sie tauschen ihre Daten nicht notwendigerweise in gleicher Datenflußrichtung und in einem vom Gesamtsystem vorgegebenen gleichen Rhythmus.

Der Benutzer des KFA-Entwicklungssystems MB 8764 löst seine ein- und mehrdimensionalen, vorzugsweise zweidimensionalen Signalverarbeitungsprobleme aber auch seine transputerbezogenen Aufgaben anwenderfreundlich und zeitsparend. Ermöglicht wird ihm das durch vorgefertigte Anwenderplatinen und durch ein komfortables Programmpaket zur Bedienung des Entwicklungssystems, das in FORTRAN 77 geschrieben und in einem IBM-PC unter dem Betriebssystem MS-DOS abläuft. Der Entwickler von Multiprozessorsystemen entscheidet sich zunächst in einer übergeordneten Eingabeebene dieses Programms bei Angabe eines dem zu entwickelnden Multi-Signalprozessorsystems zugeordneten Namens für

- die Programmbearbeitung und -übersetzung,
- die Emulation
- oder das Laden der Programm-ROMs.

Wählt der Entwickler die **Programmbearbeitung,** kann er in dieser untergeordneten Programmeingabeebene den Ausdruck diverser Programm- und Operations-Code-Listen auf einem Printer veranlassen oder den Editor aufrufen.

Der Editor zur Bearbeitung der Signalprozessorprogramme ist in seiner Konzeption auf die Eigenheiten der Multiprozessorsysteme abgestimmt. So wird der Entwickler zu Beginn des Programmaufbaus nach der Architektur gefragt, in der er sein Signalprozessorsystem aufbauen will. Danach wird sie auf dem Bildschirm grafisch dargestellt und bei systolischen und Transputer-Architekturen nach der Richtung der Datenflüsse gefragt. Nach Beendigung der Eingabe der jedem Signalprozessor zugeordneten maximal möglichen 100 Koeffizienten und der Vorgabe ob und wo ADCs und DACs eingesetzt werden sollen, beginnt dann die Eingabe des Programmtextes.

Eine Programmzeile beginnt mit einer Zeilennummer, danach ist Platz für eine Labelmarkierung. Es folgt der Signalprozessorbefehl und anschließend ist Raum für Kommentare. Ein Signalprozessorbefehl des MB 8764 und des MB 87064 ist ein Doppelbefehl. Dem Ladebefehl folgt alternativ ein Befehl für eine mathematische Operation. Programmzeile für Programmzeile wird während des Programmaufbaus der Prozessorbefehl auf Syntaxfehler untersucht und in ein dem Signalprozessor verständliches Standard-Format umgesetzt. Will er nachträglich Programmzeilen in das Programm einfügen, Teile des Textes einer Programmzeile ändern, einzelne Zeilen oder auch Programmbereiche löschen, den Programmtext auf dem Bildschirm des PC darstellen, mit dem Bedienungsprogramm kann er es.

Beim Verlassen des Editors wird die Logik des Programmaufbaus automatisch überprüft und bei Fehlerfreiheit dem Übersetzer zugeleitet. Es gibt bei systolischen Multi-Prozessorsystemen nur ein für alle Prozessoren gültiges Programm, für jeden einzelnen Prozessor aber wegen der ihm zugeordneten unterschiedlichen Koeffizienten einen spezifischen Operationscode.
Beim Aufbau einer Transputer-Architektur arbeitet jeder Prozessor mit einem eigenen Programm. Unter Berücksichtigung der notwendigen Synchronisation der unterschiedlichen Programmabiäufe durch automatisches Einfügen von Warteschleifen, darf die Übersetzung der Programme erst nach der Editierung aller Signalprozessorprogramme beginnen.

Zur Minimierung der Fehlermöglichkeiten beim Programmaufbau und zur Vereinfachung und Beschleunigung der Programmierung ist der vom Prozessorhersteller vorgeschriebene Befehlssatz erweitert worden. Er ist den Eigenarten der verschiedenartigen Multi-Signalprozessorarchitekturen angepaßt. So ist beispielsweise eine einfache Zuordnung von Koeffizienten, aber auch von Koeffizientenfeldern und eine automatische Label-Adressierung möglich. Unterprogramme können an das Hauptprogramm angehängt oder getrennt im Editor geschrieben und bei der Programmübersetzung an das Signalprozessorprogramm angefügt werden.

Die **Emulation** kann nun nach der Fertigstellung der Programme und ihrer Übersetzungen mit dem Laden der Programmspeicher des Emulators beginnen.

Der KFA Multi-Signalprozessoren Emulator MB 8764 bietet die Möglichkeit, Multi-Signalprozessorsysteme mit bis zu acht Signalprozessoren zu entwickeltn. Bei größeren Systemen können maximal acht Emulatoren miteinander verschaltet werden. Der Emulator selbst ist über eine V24-Schnittstelle mit dem PC verbunden. Die Organisation der Daten in seinem Inneren und ihre Aufbereitung für die serielle Übertragung übernimmt das auf dem Mikroprozessor 8085 ablaufende Betriebssystem des Emulators. In die acht Adapter des Emulators ist jeweils ein Signalprozessor MB 8764 bzw. MB 87064 eingebaut, wodurch die Prozessoren bei der Emulation mit ihrer maximalen Taktfrequenz von 10 MHz betrieben werden können. Ein sehr flexibles Bandkabel verbindet jeden Adapter mit einer gekapselten Platine, auf der die löschbaren Programmspeicher jedes Signalprozessors montiert sind. Sie ist ihrerseits mit den acht Eingängen des Emulators verbunden. Der Benutzer kann per Befehl vom PC aus den Ladevorgang für alle spezifizierten Signalprozessor-Programmspeicher einleiten und seinen automatischen Ablauf durch

Leuchtdioden im Emulator verfolgen, er kann Breakpoints setzen, er kann den Emulationsvorgang starten und anhalten, Breakpoints auslesen und löschen und auch die Taktzeit der Signalprozessoren mit 100 ns, 200 ns, 400 ns vorgeben oder auf externe Taktung umschalten.

Mit dem **Laden der Programm-ROMs** kann schließlich begonnen werden, wenn mit der fehlerfreien Emulation die Entwicklungsarbeiten abgeschlossen sind. Die Programmbefehle werden im Operationscode in drei Byte dargestellt. Jeder der drei integrierten Bausteine in der Anwenderplatine speichert ein Byte des maximal 1024 Worte langen Signalprozessorprogramms.

Für ihre engagierte Mitarbeit an diesem Projekt danke ich ganz besonders

Herrn H. Larue, Herrn H. Röder,

Herrn Dipl.-Ing. C.W. Seibel, Stipendiat der Alfried Krupp von Bohlen und Halbach-Stiftung

und Frau Dipl.-Ing. S. Seibel, beide von der Universität Florianopolis, Brasilien,

für die kollegiale Hilfe in Datenverarbeitungsfragen

Herrn Dipl.-Ing. H. Thyssen und Herrn G. Breuer.

Meinem Abteilungsleiter, Herrn Dipl.-Ing. K.F. Rittinghaus, gilt mein herzlicher Dank für die Unterstützung dieser Arbeit.

Literaturhinweise

[1] H.J. Whitehouse, J.M. Speiser, K. Bromley
Signal Processing Applications of Concurrent Array Processor Technology
VLSI and Modern Signalprocessing
Prentice-Hall INC Englewood, ISBN D-13-942699-X

[2] H.T. Kung, C.E. Leierson
Systolic Arrays for VLSI
Sparse Matrix Proceedings, 1978

[3] H.T. Kung
Why Systolic Architectures?
IEEE Computer, Vol. 15, No 1, Jan. 1982, pp 37-46

[4] Fujitsu
MB 8764 Programming Manual
Fujitsu Limited Communications and Electronics
6-1, Marunouchi 2-chome, Chiyoda-Ku, Tokyo 100

[5] Fujitsu
MB 87064 Digital Signal Processor
Advance Information, Nov. 1986
Fujitsu Mikroelektronik GmH
Lyoner Straße 44-48, Arabella-Center 9, 6000 Frankfurt-Niederrad 71

EIN MODULARES MULTI-SIGNALPROZESSORSYSTEM

Wolfgang Guse und Michael Gilge
Institut für elektrische Nachrichtentechnik
RWTH Aachen, Melatener Str. 23, 5100 Aachen

Einleitung:

Herkömmliche Computersimulationen auf Allzweckminicomputern bedingen in der Bildverarbeitung Wartezeiten bis zu einigen Stunden. Hochspezialisierte Hardware ist zwar für Echtzeitanwendungen geeignet, bietet darüber hinaus aber nicht die Flexibilität, die für die Entwicklung von Algorithmen erforderlich ist. Das hier vorgestellte Multi-Signalprozessorsystem schließt diese Lücke. Dem System liegt ein modulares Konzept zugrunde, das eine Master-Slave Konfiguration zwischen dem Steuerrechner und mehreren Signalprozessoren vorsieht. Die Signalprozessoren sind voneinander unabhängig, so daß verschiedene Verarbeitungsschritte auf die einzelnen Prozessoren verteilt werden können. Außerdem ist es ohne weiteres möglich, dedizierte Hardwarekomponenten in das System zu integrieren.

Architektur:

Die Abbildung zeigt ein einzelnes Signalprozessor-Modul am VME-Bus mit Steuerrechner und Systemperipherie:

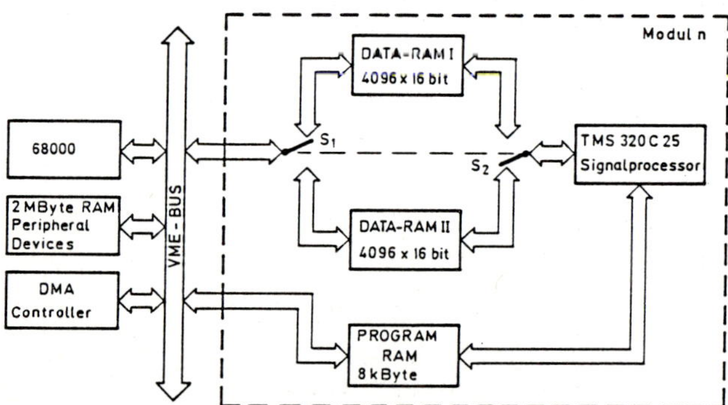

Abb.1 Blockschaltbild eines Signalprozessormodules

Der ausgewählte Signalprozessor, TMS 320C25 von Texas Instruments, besitzt eine Harvard Architektur und einen schnellen On-Chip-Multiplizierer. Der Adressraum ist unterteilt in einen Programm und Datenbereich von je 64k Worten. Die Taktzykluszeit beträgt 100 ns. Als Hauptprozessor wird der MC 68000 von Motorola eingesetzt. Die Signalprozessoren und deren Speicherbereiche werden innerhalb des Adressraums des MC

68000 direkt adressiert. Um die Rechengeschwindigkeit des Signalprozessors optimal nutzen zu können, muß dieser von jeglicher Verwaltungsarbeit, wie z.B. Datentransfer oder Kommunikation mit anderen Prozessoren, befreit werden. Aus diesem Grund besitzt jeder Signalprozessor einen eigenen Programm- und zwei umschaltbare Datenspeicher, die alle dual-ported ausgeführt sind.

Die übliche Realisation solcher Speicherbereiche mit dual-ported RAMs ist für diese Anwendung zu langsam, da für einen Betrieb des Signalprozessors ohne Wartezyklen beim Speicherzugriff die Speicherzugriffszeit 45 ns nicht überschreiten darf. Bei gleichzeitigem Zugriff beider Prozessoren auf ein dual-ported RAM wird der Zugriff durch die Zuteilungslogik langsamer. Ein Auslesen der Ergebnisse durch den MC 68000 ist so während der Signalprozessor arbeitet nicht möglich. Außerdem sind die heute verfügbaren dual-ported RAMs teuer und bieten nur einen geringen Speicherbereich (z.B. 2 kByte in einem 68(!) poligen Gehäuse). Zur Lösung dieses Problems wird hier ein double buffered memory eingesetzt. Mit einfachen bidirektionalen Bustreibern und preiswerten Standard- SRAM-Bausteinen sind auf den Signalprozessormodulen zwei Datenspeicher aufgebaut, die durch eine einfache TTL-Logik vom Hauptprozessor umgeschaltet werden. Auf einer Europakarte ist ein komplettes Modul d.h. der Signalprozessor, und drei Speicherbereiche (ein Programm- und zwei Datenspeicher jeweils bis zu 64k Worte) mit den erforderlichen Bustreibern sowie die Steuerlogik untergebracht. Dadurch kann der Signalprozessor auf einem Datenspeicher arbeiten, während der Hauptprozessor die Ergebnisse aus dem anderen Datenspeicher liest und neue Daten bereitstellt. So ist eine schnelle Verarbeitung ohne Wartezeiten für die Signalprozessoren und ohne aufwendige Arbitrations-Logik gewährleistet. Die Signalprozessoren werden für reine Rechenoperationen eingesetzt. Der Hauptprozessor übernimmt die komplette Systemsteuerung sowie den Programm- und Datentransfer, der durch den DMA-Controller beschleunigt wird.

Die Speichergröße des Programm- bzw. Datenspeichers der verschiedenen Signalprozessormodule ist abhängig von der jeweiligen Aufgabe des Moduls. Innerhalb des Gesamtsystems müssen zwei lineare Abhängigkeiten von dieser Datenspeichergröße aufeinander abgestimmt werden: Erstens die Verarbeitungsdauer der Signalprozessoren bei einer bestimmte Datenmenge und zweitens die Dauer des Datentransfers und der Systemsteuerung bei verschiedener Speichergröße. Ein Programm mit wenigen Verarbeitungsschritten würde bei einem kleinen Datenspeicher häufig Interrupts erzeugen. Andererseits ist aber auch die erforderliche Zeit für den Datentransfer kurz. Innerhalb der Zeit, die die Signalprozessoren für die Verarbeitung ihrer Daten benötigen muß der Hauptprozessor alle Module des Systems bedienen d.h. Umschalten der Datenspeicher, Starten der Signalprozessoren, Auslesen der Ergebnisse und Bereitstellen von neuen Daten. Solange die Zeit für die Steuerung und den Datentransfer des ganzen Systems kürzer ist als die Rechenzeit der Signalprozessoren, können weitere Module hinzugefügt werden, was zu einer linearen Leistungssteigerung führt. Das modulare Konzept ermöglicht auch die Erweiterung durch spezielle Hardwarekomponenten wie zum Beispiel eines DCT-Moduls zur Bildverarbeitung.

Betriebsarten:

Die Master-Slave Konfiguration ermöglicht eine Vielzahl verschiedener Arbeitsmodi, da der Hauptprozessor die einzelnen Module synchronisiert und die benötigten Daten entsprechend zuordnet. Die folgenden Beispiele beziehen sich auf Anwendungen aus dem Bereich der Bildverarbeitung auf die das System allerdings nicht beschränkt ist. Vom Standpunkt der Architektur des Systems kann man zwei Betriebsarten unterscheiden:

1. SIMD (single instruction stream multiple data stream)

Alle Module arbeiten mit demselben Programm aber auf verschiedenen Teilen des Bildes, um beispielsweise die Rechenzeit bei der DCT zu verkürzen. Modul 1 transformiert die Zeilen 1-8, Modul 2 transformiert die Zeilen 9-16 und so weiter. Wie das Bild in die Teile für die einzelnen Prozessoren unterteilt wird, hängt vom angewandten Algorithmus ab. Die Unterteilung kann zeilen-, block- oder gebietsweise sein, wobei sich die Teile auch überlappen können.

2. MIMD (multiple instruction stream multiple data stream)

Die Module arbeiten mit unterschiedlichen Programmen aber mit den gleichen Bilddaten, z.B. Transformation, Bewegungsschätzung, Quantisierung und Rücktransformation können wirklich parallel bearbeitet werden. Dies ist besonders bei Kodieraufgaben nützlich, wenn verschiedene Aufgaben unterschiedliche Teile des Bildes, Zeilen, Blöcke oder bestimmte Gebiete, benötigen.

Für die speziellen Belange der Bildsequenzverarbeitung kann das System auch je nach Einteilung des Datenflusses klassifiziert werden. Betrachtet man die Sequenz der Frames als einen globalen Datenstrom und jeden einzelnen Frame als eine lokale Datenmenge, so ergeben sich zwei andere mögliche Betriebsarten:

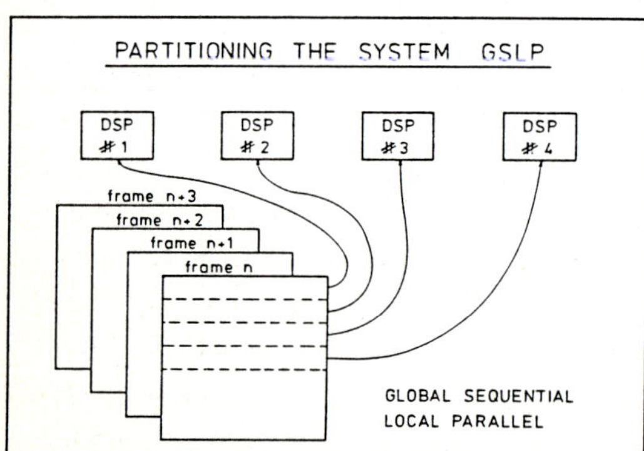

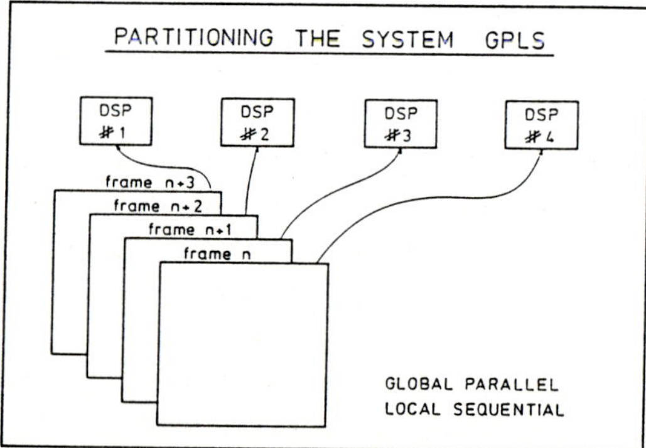

Abb.2 Betriebsart GSLP

Abb.3 Betriebsart GPLS

1. GSLP = global sequential and local parallel

In dieser Betriebsart werden die Frames sequentiell verarbeitet, indem alle Signalprozessoren parallel an einem Frame arbeiten. Denkt man dabei an Transformationskodierung, so können sich die Module die DCT eines Frames im SIMD-Modus teilen, oder auch unterschiedliche Aufgaben ausführen (MIMD-Modus).

2. GPLS = global parallel and local sequential

Werden die einzelnen Bilder der Sequenz voneinander unabhängig verarbeitet, also eine Verarbeitung ohne Bewegungsschätzung oder andere Algorithmen, die ein Vorwissen aus den vorherigen Bildern erfordern, vorgenommen, dann verarbeitet jeder Prozessor ein komplettes Bild. Mehrere Bilder einer Sequenz werden dann parallel verarbeitet, während die lokale Datenmenge sequentiell berechnet wird.

Ergebnisse:

Das modulare Konzept des Systems bietet eine große Flexibilität bezüglich der verwendeten Algorithmen und der Konfiguration der Hardware. Obwohl das Multi-Signalprozessor System ursprünglich für die Bildverarbeitung konzipiert wurde ist es überall dort einsetzbar, wo eine schnelle Verarbeitung von großen Datenmengen erforderlich ist z.B. Filteraufgaben in der Akustik, Spracherkennung und so weiter. Die Datentransferrate ist das einzige begrenzende Element in dem Signalprozessorsystem. So lange dieser Transfer schnell genug ist, um einen Betrieb des Systems ohne nennenswerte Wartezyklen zu gewährleisten, kann das System durch Hinzufügen weiterer Signalprozessormodule oder spezieller Hardwarekomponenten erweitert werden. Es wurde ein Prototypensystem aufgebaut, bestehend aus zwei Signalprozessormodulen mit dem TMS 32020 und dem MC 68000 als Hauptprozessor. Der implementierte Algorithmus ist eine schnelle DCT-Transformation /1/ mit einer nachfolgenden Klassifikation und Quantisierung im Frequenzbereich /2/ und einer abschließenden Rücktransformation (Blockgröße 8x8). Ein Bild mit einer Größe von 512x512 Pixel wird in ca. 1.47 sek. transformiert, quantisiert und zurücktransformiert. Die Verarbeitung erfolgte im SIMD-Modus.

Das System wird auf sechs Signalprozessor-Module erweitert und dann im Bereich der 64 kBit/s Codierung eingesetzt. Durch Verteilung der Aufgaben für Transformation, Codierung und Rücktransformation auf verschiedene Prozessoren wird die Verarbeitung von Bildsequenzen vorgenommen. Ist die Entwicklung eines Algorithmus abgeschlossen, so kann dasselbe System als Stand-alone-System verwendet werden, was eine kostengünstige Alternative im Vergleich zu spezialisierter Hardware darstellt.

Literatur:

/1/ W.-H. Chen et. al., "A Fast Computational Algorithm for the Discrete Cosine Transform," IEEE Transactions on Communications, Vol. COM-25, No.9 September 1977, pp.1004-1009.

/2/ M. Gilge, "Adaptive Transform Coding of Four-Color Printed Images," Picture Coding Symposium April 2-4, 1986, Tokio, Japan.

Echtzeit-Symbolextraktion aus Grauwertbildern

Massen R., Janke, P., Simnacher M., Rösch, J.
Transferzentrum Konstanz für Bilddatenverarbeitung
Reichenaustr. 81c, 7750 Konstanz

Zusammenfassung

Zur optischen Inspektion von Leiterplatten und ähnlichen strukturierten Objekten wurde ein digitaler preprocessor auf VME-Bus-Basis entwickelt, welche in Echtzeit bei einem Bildpunkttakt von 10 MHz Grauwertkanten extrahiert, diese Richtungs-gesteuert verdünnt, zu Vektoren zusammenfaßt und die Vektoren wiederum zu zusammenhängenden Konturen ordnet. Dieser pipeline-preprocessor realisiert eine Rechenleistung von mehreren Giga-OPs und läßt sich wegen der freien Programmierbarkeit für eine große Reihe linearer und nicht-linearer Echtzeit-Nachbarschaftsoperationen einsetzen.

1. Wer liefert der symbolischen Bildverarbeitung die Symbole?

Wesentliche Fortschritte im automatischen Verstehen und Auswerten von Bildern sind nur möglich, wenn es gelingt, sich von der Bildpunkt-bezogenen, ikonischen Bildbeschreibung zu lösen und Bilder auf einer höheren Beschreibungsebene durch Symbole und deren gegenseitigen Beziehungen zu beschreiben. Die zahlreichen theoretischen Arbeiten der vergangenen Jahre und die Bemühungen, die Methoden der "Künstlichen Intelligenz" in die symbolische Bildverarbeitung einzubringen, führen zu dem trügerischen Schluß, daß das Gewinnen von symbolhaften Bildelementen ein gelöstes technisches Problem sei. Tatsächlich ist es aber leider so, daß alle bildgebenden Sensoren erst einmal einen aus Bildpunkten bestehenden Datenfluß abgeben.

Bis auf wenigen Ausnahmen bei Binär-Bildern erfolgt z.Zt. die Extraktion von Symbolen aus Grauwertbildern fast ausschließlich mit Software-Methoden an den gespeicherten Grauwertbildern. Übliche Symbole sind Linien, Kreisbögen, Ecken, Verzweigungen und ähnliche einfache geometrische Elemente. Diese Software-Methoden sind nicht nur sehr langsam, sie sind auch auf als Bildpunktraster abspeicherbare Bildvorlagen begrenzt. Dies sind z.Zt. oft Bilder mit typischen Auslösungen von 512 x 512 Bildpunkten. Es gibt aber zahlreiche Anwendungen, wo mit erheblich größeren Bildpunkt-Matritzen gearbeitet werden muß:

die neuen CCD-Kameras wie die Kodak-MEGAPLUS-Kamera liefern bereits Bilder mit 1300 x 1100 Bildpunkten; Laser-Scanner zur Inspektion können bereits Bilder mit einigen 10000 Bildpunkten pro Bildzeile liefern. Hier verbietet sich daher bereits aus Gründen der Datenmenge eine direkte Abspeicherung der Bilder. Es ist im Gegenteil erforderlich, bereits unmittelbar an der Analog/Digital Schnittstelle eine massive Datenreduktion durch Extraktion der für die Bildbeschreibung ausreichenden Symbole vorzunehmen. Wir berichten im folgenden über erste Ergebnisse aus einem Projekt zur optischen Inspektion von Leiterplatten mit einer angestrebten Auflösung von 50 000 x 50 000 Bildpunkten, bei welchen die Echtzeit-Extraktion von linienhaften Symbolen mit Hilfe eines extrem schnellen preprocessors die einzige Möglichkeit ist, trotz der immensen Datenmenge die geforderten kurzen Inspektionszeiten von wenigen Minuten zu erreichen.

2. Echtzeitkanten-Extraktion -Verdünnung und -Vektorisierung

Der beschriebene Bildrechner besteht aus einem VME-Bus-System mit eigen entwickeltem universellen Video-Digitizer, pipeline-preprocessor und Symbolspeicher. Leiterplatten und ähnlich strukturierte industrielle Objekte lassen sich recht genau durch Linien und Kreisbögen beschreiben. Der preprocessor extrahiert in einem mehr-stufigen Prozess Grauwertkanten, verdünnt diese zu 1-pixel breiten Konturen, faßt diese zu Vektoren zusammen und speichert sie nach zusammenhängenden Konturen sortiert in einer Konturliste ab. Hierzu werden, wie in Fig. 1 gezeigt, folgende Schritte durchlaufen:

1. mit Hilfe von 2 parallel arbeitenden 2-D-FIR-Filter mit einem 8 x 8 Kern werden gleichzeitig horizontale und vertikale Gradienten gewonnen. Infolge dieser sehr großen Umgebung ist es möglich, sowohl die Daten als auch die Filterkoeffizienten auf 4 Bit zu begrenzen. Ausführliche Simulationen haben gezeigt, daß es in der Tat wesentlich besser ist, den Faltungskern auf Kosten der Quantisierungs-Genauigkeit zu vergrößern als umgekehrt.

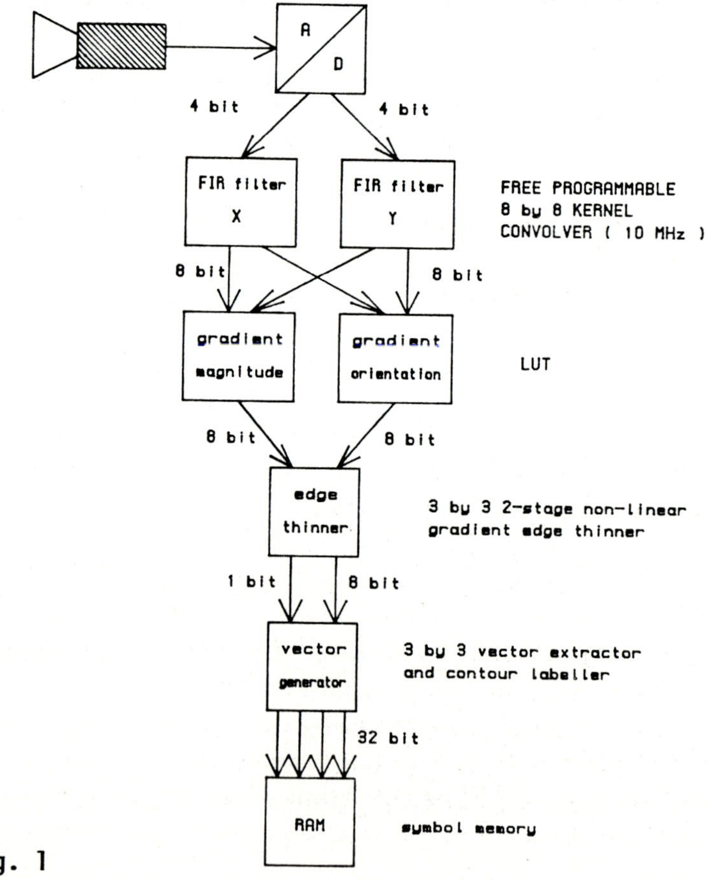

Fig. 1

2. mit Hilfe von 2 look-up-tables werden der Betrag des Gradienten und der lokale Winkel der Grauwertkante berechnet.

3. da die Gradientenbilder noch aus Kanten mit einer Breite von mehreren pixel bestehen, werden diese durch einen 2-stufigen rekursiven 3 x 3 Nachbarschaftsprozessor bis auf einen Bildpunkt verdünnt, ohne daß die Konturen aufgerissen werden. Diese Verdünnung wird durch den lokalen Winkel gesteuert und besteht aus folgenden Stufen:

   a) Der Bestimmung des absoluten Maximums in einer 3 x 3 Umgebung senkrecht zur lokalen Kantenrichtung.

   b) Der Elimination von mehrfach Maxima mit Hilfe eines rekursiven 3 x 3 Nachbarschaftsprozessors. Die resultierenden Kantenbilder sind außergewöhnlich sauber und rauscharm und zwar auch bei wenig steilen Grauwertkanten (Fig. 2). Dieser Verdünnungsprozessor führt gleichzeitig die Binarisierung dieser Kantenbilder durch und gibt einen pixel-Strom an die nachfolgende Stufe weiter welcher aus dem binären Gradienten-Maximum und einem 8 Bit-Winkel-Attribut besteht.

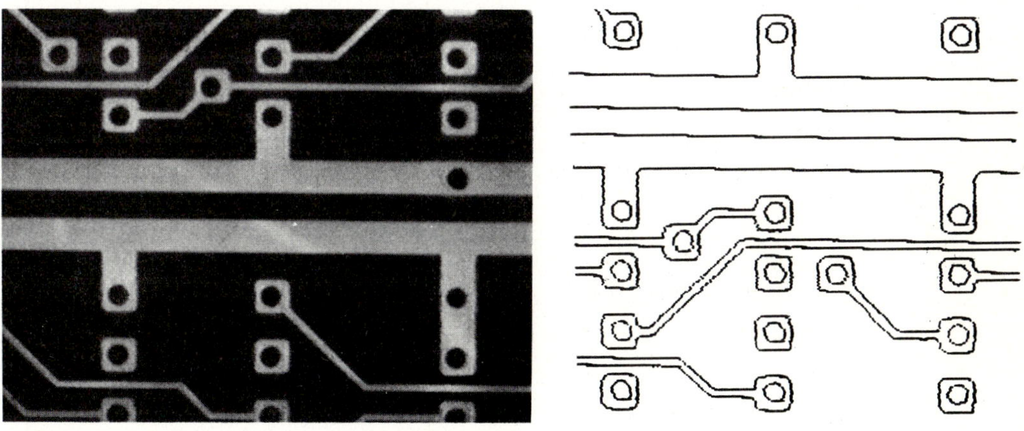

c) Eine Vektor-Extraktionstufe faßt in einer 3 x 2 Umgebung alle diejenigen Kanten zu einem Vektor zusammen, deren Richtung sich innerhalb eines vorgegebenen Toleranzwinkels befindet. Sie erkennt den Anfang und das Ende solcher Vektoren.

d) Ein Konturgenerator faßt alle Vektoren, welche auf einer geschlossenen Kontur liegen zusammen und speichert sie unter einer Liste ab, welche neben Eckpunkte enthält. Kreise werden dabei als zwei Halbkreis-Konturen abgespeichert, da wegen der Echtzeitbedingung Vektoren in der Reihenfolge erzeugt werden, wie sie in Scan-Richtung angetroffen werden.

Alle die o.g. Schritte werden in Echtzeit, d.h. im pixel-Takt von maximal 10 MHz durchgeführt. Die Schritte a und b sind bereits funktionsfähig auf 4 VME-Bus-Boards mit einer Rechenleistung von mehr als 2,5 GIGA-OPs realisiert. Die Schritte c und d sind in einer Software-Simulation, welche die Echtzeit-Bedingungen berücksichtigt, erprobt.

EIN UNIVERSELLES KAMERA-SIMULATIONSMODELL
FÜR DIE ERZEUGUNG VON BEWEGTBILDSEQUENZEN

Peter Kauff, Shing-Chi Chen, Ralf Schäfer
Heinrich-Hertz-Institut für Nachrichtentechnik GmbH,
Einsteinufer 37, D-1000 Berlin 10

Einleitung

Im Rahmen der Diskussion um einen Produktionsstandard für das hochauflösende Fernsehen (HDTV) wird im Heinrich-Hertz-Institut ein Kamerasystem untersucht, das progressiv abgetastete HDTV-Bilder liefert, ohne die Bandbreiten in den Bildaufnehmern gegenüber einer herkömmlichen Zeilensprungabtastung verdoppeln zu müssen. Dieses System arbeitet mit 4 Aufnahmeröhren, wobei ein Luminanzkanal Y_H (1250 Z, 25 Hz, 1:1) und drei RGB-Kanäle (625 Z, 50 Hz, 1:1) über einen dreidimensionalen Signalprozessor zu drei RGB-Signalen (1250 Z, 50 Hz, 1:1) kombiniert werden /1/.

Zur Untersuchung eines solchen Systems auf Simulationsbasis werden einerseits Bewegtbildsequenzen mit definierten Bewegungsabläufen und andererseits Modelle realer Bildaufnehmer wie Röhren und CCD's benötigt. Aus diesem Grunde wurde ein Simulationsmodell erarbeitet, das es erlaubt, synthetische Testbildsequenzen zu erzeugen und ihre Aufnahme mit der Vierröhrenkamera nachzubilden.

Simulationsmodell

Der Aufbau des Simulationsmodells ist schematisch in Bild 1 dargestellt. Anhand einer Gegenüberstellung mit dem physikalischen Kamerasystem ist zu ersehen, daß sich dieser Aufbau in drei Hauptbereiche aufteilen läßt.

In einer ersten Phase müssen Bilddaten erzeugt werden, die die von der Optik auf die Bildebenen projizierten optischen Intensitätsverteilungen $I(x,y,t)$ quasikontinuierlich nachbilden. Die mit $I(x,y,t)$ korrespondierenden Bildsequenzen müssen daher in einem örtlich und zeitlich überabgetasteten Raster berechnet werden. Für die Erstellung dieser Bilddateien kann man sich prinzipiell die Kombination mehrerer computergrafischer Methoden vorstellen.

Eine zweite Phase simuliert am synthetischen Bildmaterial die Aufgabe realer Aufnehmer, das analoge Signal $I(x,y,t)$ abzutasten und in die elektrischen, ortszeitdiskreten Signale $E(x_i,y_j,t_k)$ zu wandeln.

Die dritte Phase berücksichtigt den Signalprozessor, das eigentliche Herzstück der Vierröhrenkamera, in dem die benötigten Filter- und Interpolationsoperationen im dreidimensionalen Signalraum durchgeführt werden.

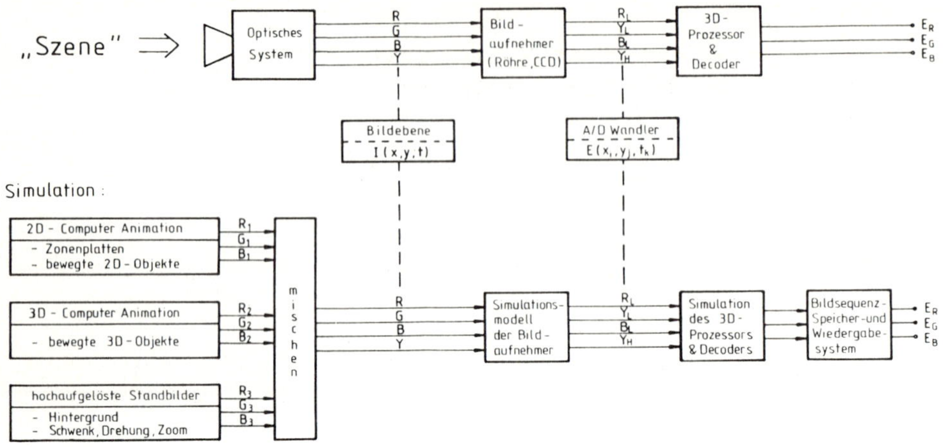

Bild 1. Blockschaltbild des 4-Kanal-Kamerasystems und des dazugehörigen Simulationsmodells

Eine besondere Schwierigkeit in diesem Konzept besteht in der Simulation der Bildaufnehmer. Während man bei der eigentlichen Bilderzeugung auf bekannte Algorithmen der Computergraphik zurückgreifen kann, ist die Nachbildung physikalischer Übertragungseigenschaften ein Thema, das im Bereich der synthetischen Bilderzeugung erst in jüngster Zeit aufgegriffen wurde /2/. Es wird daher im folgenden ein eigenes Modell für die Bildaufnahmesimulation formuliert, das auf einem Vorschlag in /3/ aufbaut. Desweiteren werden die Detailprobleme der zeitlichen und örtlichen Abtastapertur etwas genauer betrachtet.

Lineares Modell eines elektronischen Bildaufnehmers

Bild 2 veranschaulicht den Ablauf der physikalischen Vorgänge in einer Bildaufnahmeröhre.

Der mit dem optischen Bild I(x,y,t) verbundene Photostrom generiert im halbleitenden Dielektrikum der Kapazitätsbeschichtung Ladungsträgerpaare, die unter dem Einfluß des elektrischen Feldes zu den entsprechenden Elektroden wandern. Die damit verbundene Wandlercharakteristik von Lichtintensität I zu Photostrom i_p kann als linear angenommen werden, wobei der Dunkelstrom i_{po} den erzielbaren Kontrast einschränkt:

$$i_p(x,y,t) = G_p \cdot I(x,y,t) + i_{p0} \qquad (1)$$

Eine zeitliche Verzögerung des Photostroms i_p ergibt sich aus der endlichen Beweglichkeit der Ladungsträger, woraus ein erster zeitlicher Nachzieheffekt abzuleiten ist, das sogenannte "photoconductive lag":

$$\frac{d\,i_c(x,y,t)}{d_t} + \frac{i_c(x,y,t)}{\tau_c} = K_c \cdot i_p(x,y,t) \qquad (2)$$

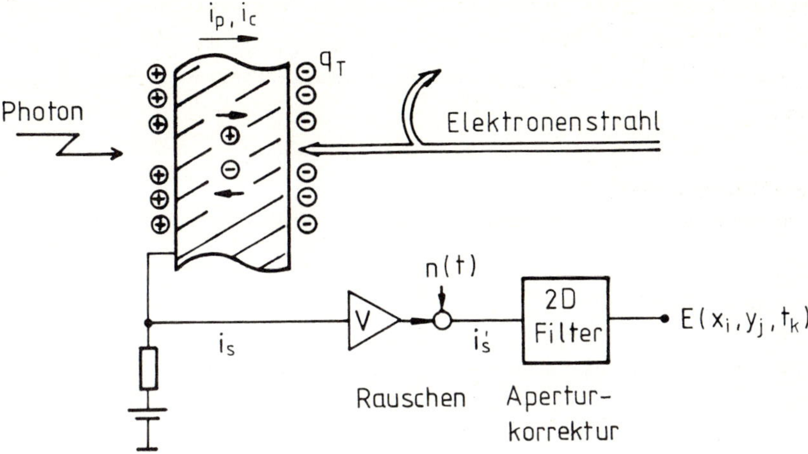

Bild 2. Prinzipschaltbild eines Röhrenbildaufnehmers mit anschließender Vorverstärkung und Aperturkorrektur

Erreichen die Löcher die dem Elektronenstrahl zugewandte Kathodenbeschichtung, so eliminieren sie die Elektronen des dort befindlichen Ladungsbelages, wodurch ein Ladungsbild q_T entsteht. Aus der Integration des verzögerten Photostroms über eine Vollbilddauer T_F mit anschließender zeitlicher Abtastung ergibt sich:

$$q_T(x,y,t_k) = K_I \cdot \int_{t-T_F}^{t} i_c(x,y,t)\,dt \Big|_{t=t_k(x,y)} + q_R(x,y,t_k) \tag{3}$$

Ein weiterer zeitlicher Nachzieheffekt, das sogenannte "beam discharge lag", ergibt sich aus dem unvollständigen Ladungsabbau durch den Elektronenstrahl:

$$q_R(x,y,t_k) = (1-\eta_R) \cdot q\ (x,y,t_{k-1}) \tag{4}$$

Das Auslesen des Ladungsbildes q_T mit dem Elektronenstrahl erzeugt letztendlich die eigentliche elektrische Ausgangsgröße des Wandlers, den Signalstrom i_s. Da die wirksamen Elektronen im Strahl allerdings nicht punktförmig verteilt sind, sondern über eine bestimmte Fläche streuen, muß q_T mit einer effektiven Stromdichteverteilung B_{eff} gefiltert werden:

$$i_s(x,y,t_k) = K_s [q_T(x,y,t_k) * B_{eff}(x,y)]_{y=y_j} \tag{5}$$

Eine wesentliche Rauschstörung n entsteht bei der Vorverstärkung des Signalstroms i_s:

$$i_s'(x(t),y_j,t_k) = V \cdot i_s(x(t),y_j,t_k) + n(t) \tag{6}$$

Das elektrische Ausgangssignal E erhält man nach A/D-Wandlung des Signalstroms i_s' und örtlicher Aperturkorrektur mit einem 2D-Hochpaß A:

$$E(x_i,y_j,t_k) = i_s'(x_i,y_j,t_k) * A(x_i,y_j) \tag{7}$$

Abstrahiert man die physikalischen Vorgänge in Gl. (1)-(7), so läßt sich daraus ein allgemeines lineares Modell erarbeiten, das neben der zeitlichen und örtlichen Apertur der Aufnehmer auch Störungen, wie Nachziehen, Rauschen und Ruhestrom berücksichtigt. Dabei beschränkt sich ein solches Modell nicht auf das einer Röhre mit progressiver Abtastung. Andere Wandler, wie z.B. CCD's bzw. anderen Abtastformate können ganz ähnlich beschrieben werden.

Nachbildung der Bewegungsunschärfe

In allgemeinster Schreibweise gilt für die zeitliche Apertur eines Bildaufnehmers [vgl. auch Gl. (3)]:

$$P'(x,y,t_k) = \frac{1}{T_s} \int_{t-T_s}^{t} P(x,y,t')dt' \Big|_{t = t_k = k \cdot T_F} \tag{8}$$

Dabei ist $T_S \leq T_F$ die Belichtungszeit, T_F die Vollbilddauer, $P(x,y,t)$ das ortszeitverständliche Bild und $P'(x,y,t_k)$ das mit Bewegungsunschärfe versehene k-te Vollbild. Bei einer Nachbildung dieser Gleichungen auf dem Rechner geht das Integral in folgende Summenform über:

$$P'(x_i,y_j,t_k) = \frac{1}{N_T} \sum_{l=0}^{N_T} P(x_i,y_j,t_{k-1} + l \cdot \Delta t) \tag{9}$$

mit $\Delta t = (t_k - t_{k-1})/N_T = T_F/N_T$

Um in Gl.(9) Abtastfehler zu vermeiden, muß eine minimale Anzahl von zusätzlichen Zwischenbildern berechnet werden, was sowohl sehr rechenzeit- als auch speicherplatzintensiv ist. Der entsprechende Überabtastfaktor N_T kann dabei aus den Bewegungsparametern abgeleitet werden.

Dieser hoher Aufwand kann im Falle der 2D-Animation allerdings erheblich reduziert werden /4/, wenn folgendes gilt:

$$P(x,y,t) = P_0(x'(x,y,t), y'(x,y,t)) \tag{10}$$

Dann nämlich läßt sich die Summe in Gl. (9) direkt im Standbild P_O berechnen, indem man unter Vorgabe des Pixels $[x_i,y_i]$ und der Bewegungsform entlang der Kurve $\underline{r} = [x'(x_i,y_i,t),y'(x_i,y_j,t)]$ eine entsprechende Anzahl von Stützstellen ermittelt und die damit korrespondierenden Intensitätswerte $P_O(x',y')$ gemäß Gl. (9) aufsummiert.

Dabei ist zu berücksichtigen, daß die Bedingung in Gl. (10) nur erfüllt werden kann, wenn entweder das gesamte Bild oder ein Objekt vor schwarzem Hintergrund bewegt wird. Im letzteren Fall kann das Objektbild P' nachträglich in ein beliebiges Hintergrundbild H' gestanzt werden, wenn man ein zusätzliches Schlüsselsignal K' zur Verfügung stellt.

$$P'_{ges}(x_i,y_j,t_k) = H'(x_i,y_j,t_k) \cdot \left\{1 - K'(x_i,y_j,t_k)\right\} + P'(x_i,y_j,t_k) \tag{11}$$

K' errechnet sich analog zu P' in Gl. (9) und Gl. (10) aus einem Kontursignal Ko(x,y), das innerhalb der Kontur des Objekts $P_O(x,y)$ den Wert 1.0 und außerhalb den Wert 0.0 annimmt.

Berechnung der effektiven Strahlapertur

Man nehme an, die Stromdichteverteilung $B_o(x,y)$ in dem auf das Target treffenden Elektronenstrahl sei gaußförmig, radialsymmetrisch und unabhängig von der eigenen Position im Bild (siehe Bild 3a):

$$B_0(x,y) = \hat{B} \cdot e^{-K \cdot \frac{x^2+y^2}{HW^2}} \tag{12}$$

Dabei ist HW die Halbwertsbreite des Strahls /5/. Da das Ladungsbild q_T durch den Elektronenstrahl progressiv abgebaut wird, muß der Auslesevorgang in GL. (5) durch eine ortsvariante Faltung beschrieben werden:

$$i_s(x_B, y_B, t_k) \sim \iint_{-\infty}^{+\infty} q_T'(x,y,x_B,y_B) \cdot B_0(x-x_B, y-y_B)\, dx\, dy \tag{13}$$

Dabei bezeichnet q_T' das veränderliche Ladungsbild in Abhängigkeit von der Position $[X_B, Y_B]$ des Strahlzentrums und von dem Ausgangszustand des Abtastvorgangs $q_T'(x,y,o,o)$ $= q_T(x,y,t_k)$.

Nach /6/ kann ein Zusammenhang zwischen q_T', q_T und B_o hergestellt werden, indem man den dynamischen Vorgang des Ladungsabbaus mit folgender Differentialgleichung beschreibt:

$$\frac{d}{dx_B} q_T'(x,y,x_B,y_B) = -\alpha \cdot q_T'(x,y,x_B,y_B) \cdot B_0(x-x_B, y-y_B) \tag{14}$$

Aufbauend auf der Lösung von Gl. (14) läßt sich der gesuchte Zusammenhang mit Hilfe einer numerisch zu lösenden Funktion F formulieren, die den progressiven Ladungsabbau in Abhängigkeit von Stromdichte B_o, Halbwertsbreite HW, Schärfungsfaktor α und Zeilenabstand Δy beschreibt:

$$q_T'(x,y,x_B,y_B) \sim q_T'(x,y,0,0) \cdot F\left\{B_0(x-x_B, y-y_B), HW, \alpha, \Delta y\right\} \tag{15}$$

Setzt man dieses Ergebnis in Gl. (13) ein, so erhält man eine ortsinvariante Faltung mit einer effektiven Strahlapertur $B_{eff}(x,y)$:

$$i_s(x,y,t_k) \sim q_T(x,y,t_k) * B_{eff}(x,y) \tag{16}$$

mit $B_{eff}(x,y) = F\left\{B_0(x,y), HW, \alpha, \Delta y\right\} \cdot B_0(x,y)$

Bild 3 b,c veranschaulicht die Bedeutung von B_{eff}. In den Überlappungsgebieten mit vorangegangener Position des Strahls wurde die Ladung schon abgebaut, so daß Elektronen in diesen Bereichen des Strahls unwirksam bleiben.

Zusammenfassung

Zur Untersuchung eines 4-Röhren-Kamerasystems, das Bildvorlagen mit mehr als 1000 Zeilen progressiv abtasten kann, wurde ein universelles Simulationsmodell entwickelt. Mit Hilfe dieses Modells ist es möglich, einerseits Bewegtbildsequenzen aus computergenerierten Objekten und hochaufgelösten natürlichen Standbildern zu generieren und andererseits Eigenschaften realer Bildaufnehmer nachzubilden. Als solche sind hauptsächlich die zeitliche Integration, das zeitliche Nachziehen, die effektive Abtaststrahlapertur, das Quellenrauschen sowie die nachgeschaltete Aperturkorrektur zu nennen.

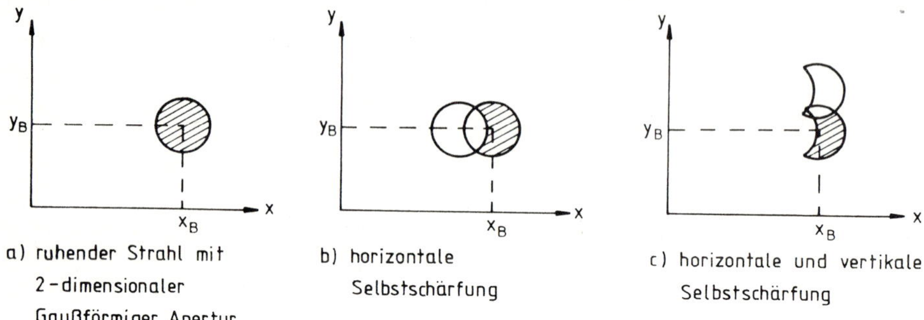

Bild 3. Schematische Darstellung der effektiven Strahlapertur bei unbewegtem Strahl (a) sowie bei Berücksichtigung des horizontalen (b) bzw. horizontalen und vertikalen (c) Selbstschärfungseffekts.

Literatur

/1/ Schäfer,R.; Kauff,P.; Chen,S.C:: "A sequentially scanned HDTV camera system exploiting the spatial-temporal resolution trade off", Picture Coding Symposium, Stockholm, June 1987

/2/ Magnenat-Thalmann,N.; Thalmann,O.: "Computer Animation - Theory and Practice", Springer-Verlag, Tokyo 1985

/3/ Isnardi,M.A.: "Modelling the Television Process", MIT, Technical Report 515, May 1986

/4/ Potmesil,M.; Chakravarty,I.:"Modelling motion blur in computer generated images", Proc.SIGGRAPH '83, Computer Graphics, Vol. 17, No. 3, pp. 389-399, July 1983

/5/ Kurashige,M.: "Effect of Self-Sharpening in Low-Velocity Electron-Beam Scanning", IEEE Trans. Electron.Devices, Vol. 29; No. 10, pp.1570-1579, Oct.1982

/6/ Miller,A.; Izatt,J.R.: "Destructive Readout in Image Tubes", Applied Optics, Vol.5, No. 12, pp. 1940-1945, Dec. 1966

Anwendungen

MEDICAL IMAGING

William J. Dallas, Ph.D.
Associate Professor
Department of Radiology and Optical Sciences Center
The University of Arizona
1501 N. Campbell
Tucson, Arizona 85724

Introduction

Medical imaging began with the discovery by Wilhelm Roentgen in the late 19th century that images of the internal structures could be formed by transmitted x-rays impinging on photographic film.[1] X-ray film has remained the main stay of medical imaging up to and including the present day. A major improvement to x-ray film imaging came with the introduction of intensification screens. These are fluorescent screens brought into contact with the photographic film during exposure.[2] The x-rays are first converted to light by the screen, this light is then recorded by a film optimized to the screen's emission wavelength. Although the resolution of such images are degraded when compared to recording directly on photographic film, the resolution is easily sufficient for almost all medical investigations. The deciding advantage is that the x-ray dose reduction is quite significant.

The second major advance in film imaging came with the introduction of anti-scatter grids. X-rays scattered from the patient are a major factor in image quality degradation. The anti-scatter grid functions like a series of very fine venetian blinds which are directed towards the x-ray source. Thus, x-rays which have been diverted from their original propogation direction are blocked and only x-rays coming directly from the x-ray source through the object impinge on the film. This reduction in scattered radiation reaching the film significantly increases the quality of the image.

For the past few years, medical imaging has begun to undergo revolutionary change; change fostered by the introduction of electronic imaging techniques and by the digital computer. The x-ray image intensifier was one of the first of these electronic imaging devices. Its output, at first, was directed to film and then to television cameras and video recording equipment. The beginnings of digital imaging came in the early 1970's with the introduction of computer tomography which required the full power of the then current digital computing technology. The introduction of digital fluoroscopy and digital subtraction angiography followed computer tomography closely in the early digital imaging. Besides acquisition of information that would not otherwise be possible, such as for computer tomography, the drive behind digital imaging comes also from the flood of information produced in the hospital. The difficulty of storing and especially of accessing this information over a long period time has become insurmountable. Digitally formatted images will be necessary for use of Picture Archiving and Communication Systems (PACS).[3]

Totally digital radiology departments are under investigation. There has been a slow pace of realization for these departments because of two limiting factors: the availability of appropriate acquisition modalities and limits on the technology, especially in the speed of available information handling systems.

In this paper we will concentrate primarily on descriptions of the digital acquisition modalities beginning with the digitized photographic film and concluding with the still somewhat speculative biomagnetic imaging.

Digitized Plain Film

The motivation for digitizing plain film is not for image processing to bring out additional information, radiologists have not accepted processed film images, but rather for storage and retrieval in digital archiving systems. The most common examination by data volume is the thorax or chest examination. In most hospitals, 80% of the data volume generated consists of chest films. There are several x-ray film-based examinations in addition to chest films, i.e., bone fracture studies. The data volume for these films is quite large. If we consider 5 lp/mm at the film, that gives for 35 x 45 cm film about 3500 x 4300 = 15 megapixels. The optical density range is between 0.1 to 3.5 with between 256 and 4096 grey levels steps to be stored.

The technology used in film digitization has been advancing steadily. Mechanical drum scanners are available which are very precise but too slow and cumbersome for routine use in a hospital. Video digitizers, on the other hand, are quite fast; they are, however, of limited spatial and density resolution. One-dimensional scanning diode arrays are also available but they have a quite limited density range. The laser scanners that are now becoming available seem to be the method of choice.[4] They will scan a maximum of 4000 x 5000 pixels at a 4096 grey level step resolution. At the present time, the processing desired by radiologists is to make the film copy of the digitized laser images as much like the original images as possible.

Psychophysical studies are in progress at several instutions to determine the necessary spatial and density (contrast) resolution. The information brought by these psychophysical studies will be crucial in designing digital systems of the future and will have a strong impact on the time scale of realization. It will also have impact on the acceptability of different types of image compression.

Computed Radiography

With computed radiography, the intermediate recording step on film is eliminated. There are two major varieties of computed radiography. One uses stimulated luminescence phosphor imaging plates.[5] There are commercial systems now available based on this principal. The second is mechanically scanned detector arrays.[6]

The stimulated luminescence phosphor (SLP) technique is widely known as the Fuji System for the first company to introduce this technique to medical imaging. The basic operation relys on use of an imaging plate. This plate is mechanically similar to film and is used in cassettes which are similar to film cassettes. However, the material differs considerably from film in that it is excitable into a metastable state. On exposure to x-rays, the molecules in this material are excited into a state whose spontaneous decay is very slow, hours or days. The decay can be stimulated by illuminating with a laser. To form an image, the plate is used as follows. The plate is exposed to x-rays. The probability at any point on the plate that a molecule will be in the excited state is proportional to the x-ray exposure. The density of stimulated molecules then stores the latent image. The plate is then introduced into a laser scanner. The stimulated emission is then recorded, point by point, by a photomultiplier as the laser scans the plate. The PMT signal is digitized and stored in the computer. The spatial resolution is typically 1700 x 2200 pixels with 1024 grey levels. A strong advantage of the imaging plate is in its wide latitude. With the common pre-reading procedures, this latitude can be four orders of magnitude on exposure. The wide latitude makes the SLP especially valuable for emergency examinations.

The simple image processing necessary is a grey-level mapping since the luminescent intensity is available and not its logarithm and a high spatial frequency boost filtering necessary to match the final image to that which would have come from a film imaging system. Optional additional high frequency boosting is becoming recognized as an effective tool, especially for finding bone fractures but processing in this area, as in most areas of medical imaging, advances only very slowly because of the dangers inherent in introducing imaging artifacts.

Another type of imaging which is being investigated but has not yet been introduced into common process is dual energy imaging with the SPL systems. This process is analagous to color photography where two different peak wave lengths in the incident x-ray distribution or in the transmitted distribution are used. For the imaging plate, two imaging plates are stacked with a separating layer being in x-ray filter, i.e., a copper plate which filters out the lower energy x-rays and so gives a spectral difference between the images produced on the top and the bottom plate.

The second type of computed radiography is done using scanning linear arrays. There are three varieties: linear, dual and slot arrays. In linear arrays, a single line of elements typically 4096 elements are scanned below the subject. The detector elements are typically charge coupled devices (CCD) or photo diode arrays with charge couple device shift registers. There is a fluorescent material covering the CCDs in order to convert the incident x-rays into light for detection by the CCDs. The movement of the linear array below the patient is coordinated with the movement of a slit above the patient in order to reduce scattered radiation and eliminate unnecessary dose to the patient. Coordination of these movements is difficult since the movements must be quite precise.

The second variety of scanning array is the dual array where two linear arrays are two linear arrays are placed parallel, one behind the other. The type is scanning array has been primarily considered for dual energy scans.

The third variety is the slot detector. This is a two-dimensional detector which is much longer than it is wide. As its detector is scanned, the image is shifted inside the detector and read-out as it over-flows the detector. The slot above the patient is scanned in coordination with the detector array as it is in the linear arrays.

Fluoroscopy

Real time medical imaging using fluoroscopy is, of course, familiar from the early days when a screen of fluorescing material was simply placed between the examining physician and the patient who was illuminated by the x-ray source. Besides giving somewhat dim images, this procedure was also quite dangerous as the physician was exposed to a considerable x-ray dose.

Fluroscopy matured with the introduction of the x-ray image intensifier.[3] This device is a large vacuum tube with an entrance window of 15 to 57 cm. This entrance window is coated with a layer of scintillating material that converts the incident x-rays into light. The next layer coated onto the extrance window is a photocathode which converts the light into electrons. The electrons are then accelerated through the tube and focussed onto a phosphor at the exit window. This phosphor converts the electron image into a light image. The light image was first captured on film. For dynamic studies, i.e., heart movement angiography, the film was a movie film. With the addition of a television camera, the image began to be recorded on video tape and finally digitzed and computer stored. The typical image resolution is 512 x 512 pixels, with recent commercial systems tending toward 1024 x 1024. 64 to 1024 grey levels are common and 25 to 30 frames per second repetition rate.

The uses of fluosocopy are manifold ranging from gastrointestinal studies to angiography. Angiography is viewing the blood vessels by means of an injected constrast medium. Digital angiography is divided into two sub categories: digital angiography and digital subtraction angiography.[7,8] In digital angiography, a bolus of contrast medium is injected arterially and the bolus is followed using the x-ray intensifier. This has the advantage of giving very high quality images but with a disadvantage of requiring a relatively high concentration of contrast medium. Digital subtraction angiography, on the other hand, makes use of two images recorded sequentially, one without the contrast medium and one with the contrast medium injected. The two images are subtracted in order to give an increased contrast for a given amount of contrast medium. Problems occur, of course, in noise sensitivity and in image registration. For perfect subtraction, the non-contrast mask must be brought not only into exact registration but also deformed so as to conform to the second picture.

Development of x-ray intensifiers has, of course, not stopped. The next stage seems to be the introduction of charge coupled devices as sensors. It may be that the development of the large field CCD's will dominate the future development, although an alternative is the use of several smaller CCD's with image dividers. The dividers can use beam splitters or fiber taper coupling. A fiber taper is a bundle of fiber-optic strands formed into a coherent image transmitting bundle. By heating and drawing the bundle, the output end of the taper can be reduced in size thus reducing the image size.

One example of such a novel image intensifier is under construction at the University of Arizona and consists of a fiber optic plate which is coated with a scintillating layer. The light coming from the scintillating layer is then directly incident on a proximity focussed image intensifier. This intensified light image is divided into six separate sections, each section is accepted by a fiber optic taper. This taper demagnifies the image so that at the exit, it is exactly the proper size for a CCD. The signal from each CCD is then digitized and the image sections are computationally recombined in a digital image memory.

Nuclear Medicine

Nuclear medical imaging is one of the earliest electronic imaging modalities to make use of digital computation. The basic operating principle is that a radioactive tracer is injected into the patient. The tracer concentrates itself preferentially in specific tissues and by this concentration indicates the functioning of that tissue. For instance, in a healthy heart muscle there are tracers which will tend to concentrate; the image formed is of the healthy tissue. The amount of damage caused by a heart attack can therefore be seen.

The gamma rays emitted from the tracer element are collimated onto a detector. The most common detector is the Anger camera.[9] The Anger camera is a solid plate of scintillator material (typically sodium iodide) with an array of photomultiplier tubes attached to its rear face. An interpolating network is attached to the output of the PMT's. This interpolating network accepts the signals from the PMT's, and using the signal strength variation between the PMT's, interpolates to give much finer position information than the tube spacing would indicate. Typical resolution is 64 x 64 pixels to 128 x 128 pixels and 64 to 256 grey levels.

The processing necessary is for Anger Camera calibration in the newer digital cameras. In older cameras, although the end result was a digital image, the interpolation network itself was entirely of passive electronics. Processing is also done for analysis of function, i.e., heart ventrical volume and wall movement or liver lesion characterization.

Computer Tomography

Computer tomography is an imaging modality that was first made practical by the digital computer.[10,11] This now common imaging modality produces an x-ray image of a transverse slice in the patient by collecting a set of slice projections. An x-ray source is collimated onto a linear (circle segment) detector. The source and detector circle the patient, collecting approximately 180 projections of one slice in the patient. The collection of projections is then converted into an image of that slice using any one of several computerized techniques. Filtered back projection is now the most common. The imnage size is commonly 256 x 256, 512 x 512 or 1024 x 1024.

The necessary processing is in calibration, especially in suppressing artifacts, i.e., ring artifacts caused by varations in detector sensitivity, and for analysis, i.e. mensuration and, as is now becoming popular, three-dimensional display.

Magnetic Resonance Imaging

Magnetic resonance imaging has only recently become the method of choice in certain examinations, i.e., brain and spinal examinations in spite of its attractiveness in using non-ionizing radiation.[12] The operating principle is, of course, well known. The subject is placed in a very uniform and strong magnetic field. This magnetic field causes the nuclear magnetic moments ot the hydrogen atoms to precess about the magnetic field axis. Very precisely controlled sequences of microwave pulses then irradiate the subject thus causing the precessing atoms to "tip" away from the magnetic field's direction thus decreasing the magnetization in the field direction. The microwaves radiating from the precessing atoms are then the detected signals. The spatial localization is obtained by imposing small variations (gradients) on the imposed magnetic field thus varying the precession frequency. Typically three images are presented: a proton density image and two magnetization decay time images, the so-called T1 weighted and T2 weighted images. Typical image resolution is 256 x 256 pixels and 256 grey levels for each of three images.

The processing necessary of the obtained signal is commonly a Fourier transform in order to obtain an image slice. However, there is great flexibility in the possible signals generated since one has control over: the field gradients, sequences of microwave pulses, and measuring techniques. Many many types of information can be extracted including spectroscopic and blood flow information. Some exciting new directions being investigated now are the interactions of magnetic resonance imaging and new contrast media, particularly fluorine base compounds for looking at glucose metabolism and brain function.

Biomagnetic Imaging

Biomagnetic imaging is a very new and as yet unproven imaging modality makes use of passive measurements of magnetic fields generated by electrical currents flowing in the body.[13] Although magnetocardiography, measurement of magnetic signals from the heart, began in the early 1960's, the progression to the much weaker fields used in magnetoencephalography and now to actual imaging using these magnetic field measurement has been slowed by difficulties in measuring the very weak fields and providing sound theoretical bases for imaging. The magnetic fields can be generated by tracers such as magnetic dust introduced into the lungs or by tracer-bound magnetite which has been used as a contrast agent in the liver for magnetic resonance imaging, but these fields can also be those generated by electrical nerve currents: nerve currents in the heart and in the brain.

A breakthrough in detection of the biologically generated magnetic fields was the invention of the Superconducting Quantum Interference Device (SQUID). This device is very sensitive and, in fact, can measure fields that are less than one-hundred millionth the strength of the earth's magnetic field. This is approximately the strength of the magnetoencephalogram fields. The SQUID does, of course, have the difficulty that it is a cryogenic device operating at liquid helium temperatures. Though, in the very recent past, the advent of high temperature superconductors (up to 90^0 Kelvin) have given hope of simplifying construction of such measuring devices.

Reconstruction processing to the present time has consisted primarily of fitting programs, that is assuming a current dipole in the nerve as a source and then fitting the measured magnetic fields to that which would be generated by a magnetic dipole leaving the direction and strength as fitting parameters. In the recent past, three techniques of imaging have been under investigations: Least Square Error Estimators, Algebraic Reconstruction Technique (ART), and Fourier Techniques.

References

1. Donizetti, P: Shadow and Substance - The Story of Medical Radiography. Pergammon Press, Long Island City, New York, 1967.

2. Buchanan, R. A: IEEE Trans. Nucl. Sci. N.S. - 19 (1972) 81.

3. Combee, J.D., Botden, P. J. M., Kuhl, W: X-ray Image Intensifiers. In Biberman, L. M., Nudelman, S. (eds). Photoelectronic Imaging Devices. New York, Plenum Press, 1971, vol. 2.

4. Lo, B. S.-C., Taira, R. K., Mankovich, N. J., Huang, H. K., Takenchi, H: Computerized Radiol. 10 (1986) 227.

5. Fajardo, L., Hillman, B. J., Hunter, T. B, Claypool, H. R., Westerman, B. R., Mockbee, B: Radiology 162 (1987) 345.

6. Sashin, D., Sternglass, E. J., Spisak, M. J., et.al: Proc. SPIE 173 (1979) 88.

7. Ovitt, T. W., Christenson, P. C., Fisher, H. D. III, et. al: A.J.N.R. 1 (1980) 387.

8. Mistretta, C. A., Crummy, A. B. and Strother, C. M: Radiology 139 (1981) 273.

9. Anger, H. O: Radioisotope Cameras. In Instrumentation in Nuclear Medicine (G. J. Hine, ed.) Academic Press, New York 1967, pg. 485.

10. Cormack, A. M: J. of Appl. Phys. 34 (1963) 2722.

11. Hounsfield, G. N., Ambrose, J., Perry, J., et. al: Brit. J. of Radiol. 46 (1973) 1016.

12. Mansfield, P. and Morris, P. G: NMR Imaging in Medicine. In Advances in Magnetic Resonance, Suppl. 2 (J. S. Waugh, ed.). Academic Press, New York, 1982.

13. S. J. Williamson and L. Kaufman: J. of Magnetism and Magnetic Materials 22 (1981) 129.

DISPLAY OF HEMISPHERIC LOCAL METABOLIC RATES FROM HUMAN BRAIN

Claude Nahmias, Martin Loken, E. Stephen Garnett

Nuclear Medicine, McMaster University Medical Centre
Hamilton, Ontario, Canada

Introduction:

Positron tomography, a technique that employs the concepts of computerized tomography in combination with specific molecules labelled with positron emitters, is now making possible the direct regional measurement of blood flow and energy requirements or the mapping of the distribution of molecules such as neurotransmitters and the drugs that affect the central nervous system. This technique allows accurate recovery of the distribution of the labelled molecules in thicknesses of grey and white matter in cross sections through the brain. Although sixteen to twenty different levels are routinely examined during a typical study, the data in each tomographic section is usually considered in isolation. A major problem with this approach is that it does not consider the major subdivisions of the brain, the various lobes and gyri, in their entirety. These structures often overlap many tomographic sections, and are not always easily traceable from section to section. We describe an approach that maps the distribution of radioactivity, and hence of the process being studied, over the whole of the cortical mantle from a knowledge of the distribution in the transaxial sections. Our approach allows a more accurate localization of defects in the brain, and a more accurate quantification of the extent of the deficit in the cortical mantle.

Method:

We analyse data obtained studying the regional distribution of glucose metabolic rates in the brain by the ^{18}F fluorodeoxyglucose method. The tomograph used has an in plane spatial resolution of 8 mm (FWHM) and an axial resolution of 11 mm (FWHM). Routinely, sixteen to twenty transaxial sections are used to map the metabolic activity of the brain. Each section overlaps its predecessor by 5 mm. Each cross sectional image is reconstructed on a 128 * 128 pixel matrix; each pixel corresponds to 2 mm in the object and is 16 bits deep.

The delineation of the cortex is achieved by pixel intensity thresholding followed by intensity contouring. The features that are extracted must display a difference in pixel intensity from the remainder of the image. In the case of the outer rim of cortex, the edge can be isolated by a thresholding technique where all pixels greater than the threshold value are set to one and all pixels less than the treshold are set to zero. The binary image is processed with a local intensity subrange filter and then an automatic contour following algorithm is applied to extract a unit width connected edge (fig. 1).

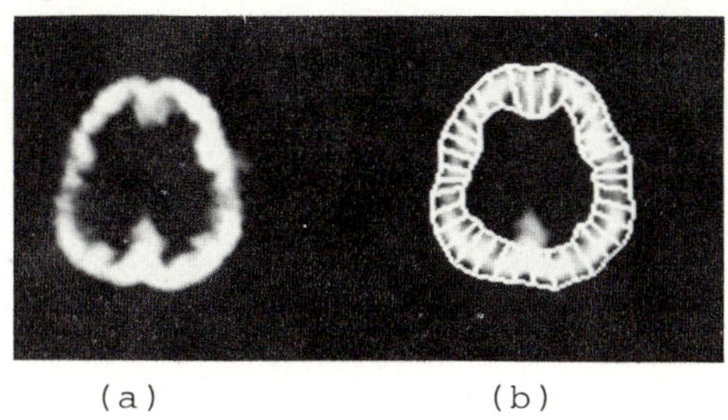

(a) (b)

Fig.1: Transaxial section at the level of the corpus callosum. Original image (a) and segmented image (b).

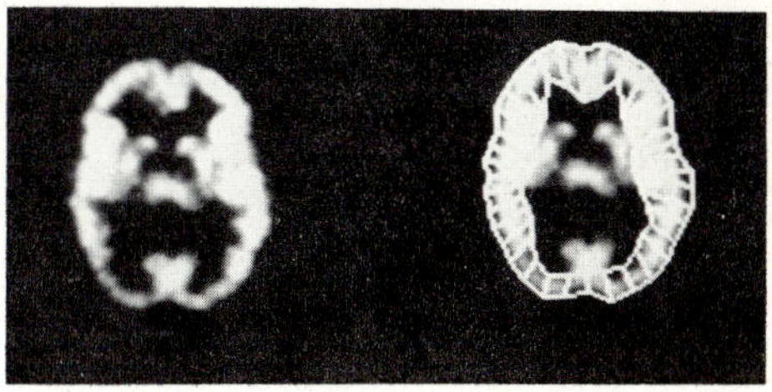

(a) (b)

Fig.2: Transaxial section at at the level of the basal ganglia. Original image (a) and segmented image (b).

Unfortunately the inner rim of cortex is rarely clearly defined. In most cases the cortex does not have a uniform intensity along its entire outline. It may contain a significant amount of noise and it is not unequivocally separated from other features of similar intensity, eg the basal ganglia and the cingulate and hippocampal gyri (fig. 2). Automatic extraction of the inner contour is difficult as these deep structures are often included with the cortex. In these cases, operator interaction is used to trace the most probable contour.

Once the inner and outer edges of the cortex have been defined, the centre of mass of the outer contour is computed. In almost all cases, this is also the centre of mass of the inner contour. The major and minor axes of the outer contour are found. The angle the major axis makes with the vertical axis provides a rotation correction angle to counter tilting of the subject's head to one side. From the major and minor axes, the perimeter of an ellipse approximating the outer extent of the cortex is computed. The distance between the posterior edge of the cortex and a fixed spatial reference point is calculated. these provide correction lengths to be used in the subsequent mapping.

Radial lines originating from the centre of mass every ten degrees of arc are traced. These divide the cortex into thirty six sectors (figs. 1, 2). The cummulative intensity of all pixels contributing to a sector and the number of pixels within that sector are computed and stored in a data file for mapping. The process is repeated for each cross sectional slice.

The new data set consists of a two dimensional array of numbers, each representing the mean glucose metabolism in a volume of cortex the dimensions of which are governed by the slice thickness, the thickness of cortex at that location and the angle between two consecutive radial lines. The axial separation of each of the cross sections is determined by the scanning protocol and is independent of the object. The perimeter of the cortex in any slice will vary from slice to slice depending on its level in the object.

In order to construct a map of hemispheric local metabolic rates, the value of each data point is assigned to a bin which corresponds to the anatomical location of the volume of cortex from which it was derived. The height of every bin is the same. The width of each bin at a given level is equal to the perimeter of the outer edge of cortex at that level divided by 36. The segmented slices are stacked taking into consideration the offset of the posterior edge of the cortex with respect to the fixed spatial reference point. The resulting map is the same that would be obtained if the cortical rim had been dissected out and flattened onto a two dimensional plane.

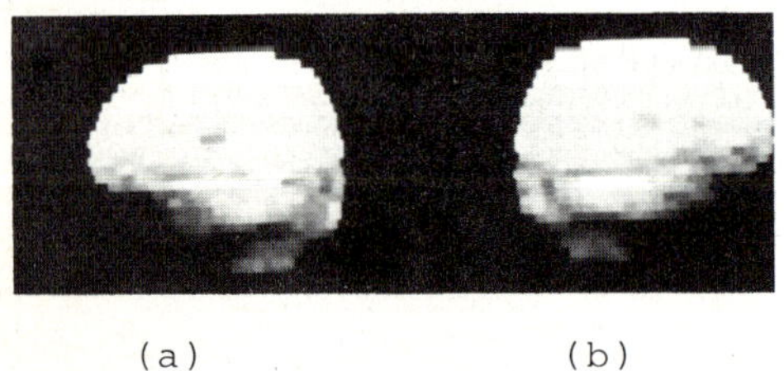

(a) (b)

Fig.3: Map of regional glucose metabolism in the left (a) and right (b) hemispheres obtained from a normal individual.
Note that the various subdivisions of the brain, such as the temporal lobes and the cerebellum, are clearly identifiable.

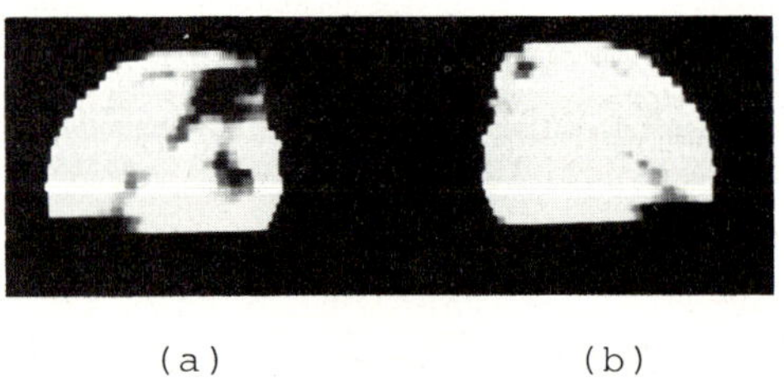

(a) (b)

Fig.4: Map of regional glucose metabolism in the left (a) and right (b) hemispheres obtained from a patient suffering from the neurological complications secondary to systemic lupus erythematosus. This patient exhibited marked deficits on neurocognitive testing involving predominantly the left tempero-parietal region.

Discussion:

Cross sectional information can be processed in order to display coronal and saggital sections through the brain. For CT and NMR data, further manipulation of the data can yield representations of the shape of any structure isolated from its surroundings. This technique relies on the detection of the surface of the structure. Relatively large differences in voxel intensity exist at structural interfaces, such as those existing at an air-tissue interface, a tissue-bone interface or a grey-white matter interface. In positron tomography, such interfaces do not exist; the technique maps variations in metabolic activity in volumes of grey or white matter. Consequently, the three dimensional reconstruction techniques developed for the other two modalities are not applicable.

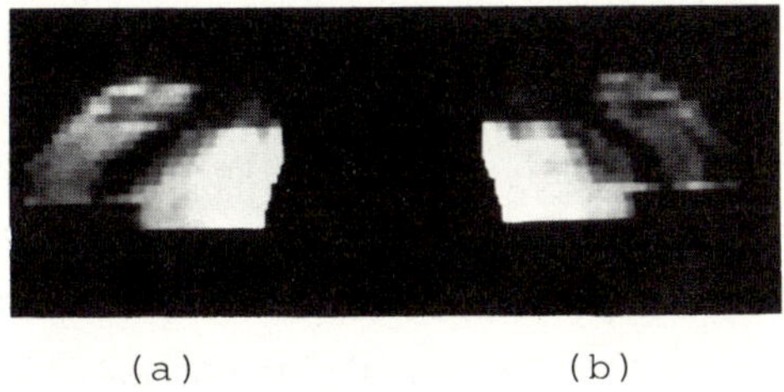

(a) (b)

Fig.5: Map of regional glucose metabolism in the left (a) and right (b) hemispheres obtained from a patient suffering from primary lateral sclerosis. Note the profound hypometabolism of all of the sensory motor areas around the central sulcus bilaterally.

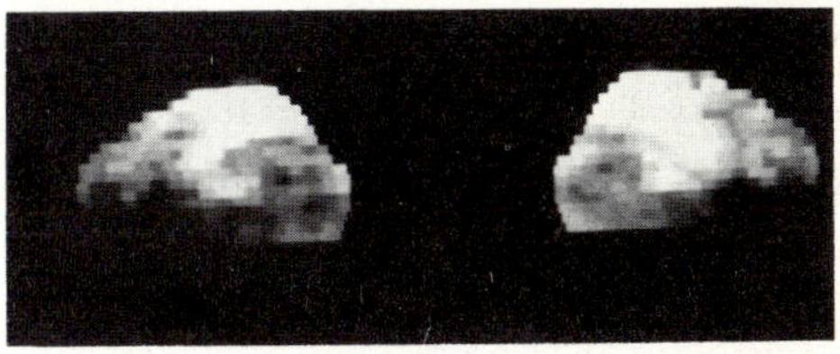

(a) (b)

Fig.6: Map of regional glucose metabolism in the left (a) and right (b) hemispheres obtained from a patient suffering from senile chorea. Note that, in contrast to the patient shown in fig. 5, this patient exhibits an area of hypermetabolism confined to the sensory motor cortex.

Using our approach, the various subdivisions of the normal brain are clearly identifiable (fig. 3). The extent and anatomical location of metabolic deficits in the diseased brain can easily be determined and correlates with results of neurocognitive assessment (fig. 4) and neuro-motor assessment (figs. 5, 6). The advantage of our technique is that it integrates data from many slices into a unique, easily interpretable image. The result is a two dimensional image similar to a lateral projection. Lateral projections usually suffer from the inclusion of information emmanating from structures beyond the cortical rim deep in the brain. Because we use cross sectional information to construct our lateral views, we essentially eliminate this problem and truly represent the process taking place in the cortical mantle.

COMPARISON OF STATISTICAL METHODS FOR THE DETECTION OF CONTRAST MATERIAL IN ECHOCARDIOGRAPHIC IMAGE SEQUENCES

E. Steinmetz, R. Brennecke, D. Jung, N. Wittlich, R. Erbel, J. Meyer
II. Medizinische Klinik, Johannes Gutenberg-Universität
Mainz, Langenbeck Str. 1

Introduction

Ultrasonic imaging of the heart is a diagnostic tool which is increasingly used in cardiology. In addition to the representation of important anatomical information two dimensional images provided by mechanical or electronically steered sector scanners can be used for the extraction of functional parameters of the heart (as e.g. enddiastolic volume or ejection fraction). A poor definition of the endocardial border especially resulting from the noisy appearance of the images and from qualitatively restricted echocardiograms leads to uncertainties in the quantitative analysis and therefore requires refined methods for the determination of functional parameters. Our investigations which are based on the consideration of statistical properties of the received ultrasonic amplitude data resulted in the development of a new method for the detection of contrast material in echocardiographic image sequences. In the following we discuss the basic ideas of this method and also present first results gained by its application to some representative samples.

Theoretical aspects

The characteristic properties of the interaction between the ultrasonic wave and the reflecting tissue are responsible for the granular appearance of ultrasonic images. In theoretical investigations on this interaction the tissue is modeled as a collection of scatterers which are so numerous that there are many within one resolution cell of the scanner. The wavelets which are returning with different phases from every scatterer within one resolution cell interfere during their propagation to the scanner, so that speckle is explained as an interference phenomenon. Assuming a large number of scatterers within one resolution cell and uniformly distributed phases of the scattered waves leads to a Rayleigh probability function (1,2) for the probability distribution of the reflected amplitude A.

$$P_{Ray}(A) = A/\sigma^2 \exp(-A^2/2\sigma^2) \qquad (A>0) \qquad (1)$$

The corresponding probability distribution for a non-uniform collection of small and large cross-section scatterers is given by the Rician probability function (2).

$$P_{Ric}(A) = A/\sigma^2 \exp\{-(A^2+S^2)/2\sigma^2\} \; I_0(AS/\sigma^2) \qquad (A>0) \qquad (2)$$

Since the signal-to-noise ratio (SNR) plays an important role in our considerations we want to point out that the Rayleigh probability function is distinguished by an amplitude-independent SNR which is defined by the ratio of the average amplitude $\bar{A}$ and the rms-deviation from the average.

$$SNR_{Ray} = \left(\frac{\pi}{4-\pi} \right)^{1/2} = 1{,}91 \qquad (3)$$

Conversely the SNR of the Rician distribution depends on the ratio of the parameters S and σ. If this ratio is assumed to be independent of the amplitude A the corresponding value of the SNR is also constant. The desired SNR can then be achieved by adjusting the ratio S/σ in a proper way.

Data Acquisition

The whole echocardiographic examination which is performed with a phased array sector scanner (Ultramark 8) at transducer-frequencies of 3.5 MHz or 5.0 MHz is recorded on a video tape. Afterwards image sequences are selected and digitized using an image processing system (VTE Picturecom). The storage capability of this system amounts to 192 images at a spatial resolution of 256*256 pixels and at a grey-level resolution of 8 bit (256 grey-levels). The image processing system is operated by a standard minicomputer (DEC PDP 11/73) via FORTRAN-programs.

Statistical Methods

A group of heart phase synchronous echocardiograms including contrast and non-contrast images is selected and a frequency distribution of the grey-levels adopted in this time sequence is attached to every pixel location. In regions of contrast material the characteristic frequency distribution of the grey-levels depends critically on the ratio of the number of contrast images (K) to the number of non-contrast images (L). Outside these regions the characteristics of the frequency distributions remain unchanged if compared with the distributions of the corresponding contrast free image sequence. Basic statistical properties such as mean value ($\bar{A}$), the rms-deviation from the average (SD) and higher moments and combinations of them as e.g. SNR are computed from the grey-level distribution of every pixel location. The behaviour of these parameters in regions of contrast-material can be roughly approximated by a model frequency distribution with L grey levels of value zero (corresponding to L non-contrast images) and K identical grey-levels with an arbitrary value (corresponding to K contrast images). The SNR obtained from this model is given by :

$$SNR_{Kon} = \sqrt{K/L} \qquad (4)$$

A comparison of this result with the SNR value of the Rayleigh distribution function (see formula (3)) shows that the choice of an unsymmetrical image sequence ($L>K$) will influence the separation capability of the SNR parameter between regions of contrast material and contrast free regions in a favourable manner.

Results

As the first four moments of the frequency distributions calculated from image sequences of non contrast images combined with contrast images do not provide definite separations in parameter areas of contrast material and tissue we will concentrate our discussion on SNR. In accordance with the theoretical models described above there exists a linear relation between the mean value ($\bar{A}$) and the rms-deviation from the mean value (SD) which is figured in the scattergram of Fig.3 for a contrast-free image sequence. Furthermore it can be shown experimentally that in general the properties of the tissue or physical parameters as e.g. amplification, wavelength or power (amplitude) have only a weak influence on the slope of the straight line fitted to the frequency distribution of the pair ($\bar{A}$,SD), whereas the zero of this line depends on these parameters. It is therefore obvious to identify the slope of this line with the most frequent SNR of the image sequence under consideration. Using this approach for the determination of the most frequent SNR from the scattergram in Fig.3 which is calculated from eight non-contrast images yields a value between 1.5 and 1.6. Mean SNR was 1.9. Corresponding to this considerations a reasonable definition for the calculation of SNR for every pixel location should be based on the shifted scattergram, which is obtained by shifting the SD parameter in such a way that the straight line adjusted to the scattergram of a contrast-free image sequence meets the X-axis at the origin. The scattergram of an image sequence including eight non-contrast images and two contrast images (Fig.4) contains a second branch belonging to the regions of contrast material. The slope of the straight line fitted to this part assumes a value of about 0.9 which differs from the value 0.5 predicted by formula 4. Since both branches of the scattergram in Fig.4 are well separated a definite detection of regions of contrast material is possible. The corresponding SNR image obtained from the shifted scattergram is shown in Fig.2. A comparison of this image with an unprocessed contrast image (Fig.1) shows that SNR enables the detection of endocardial borders although the contrast material can not be visually separated from the myocard in some regions (at the top and the bottom of the endocardial region in Fig.1).

Discussion

Independent of the settings of parameters used while obtaining the image sequences, we measured a mean SNR of 1.9 for non-contrast regions as expected from theory. The enhancement of contrast regions seems to be optimized by using unsymmetrical ratios of contrast and non-contrast images. The separation of the ventricular cross-sectional area provided by SNR-imaging simplifies the determination of functional parameters of the heart.

Acknowledgement

This work was supported by the Deutsche Forschungsgemeinschaft and the Robert Müller-Stiftung.

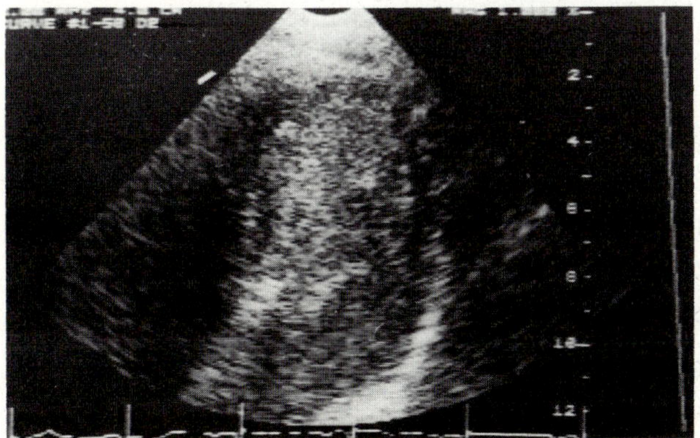

Fig.1 Unprocessed enddiastolic sector image of left heart after contrast material injection.

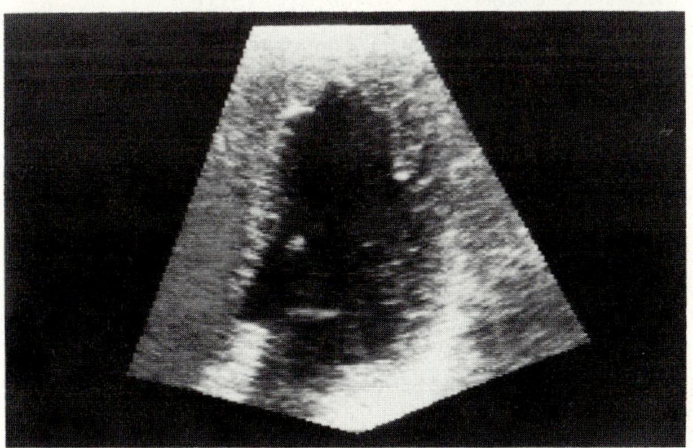

Fig.2 SNR image calculated from the shifted scattergram of Fig.4, which was obtained from an enddiastolic image sequence including eight non-contrast images and two contrast images.

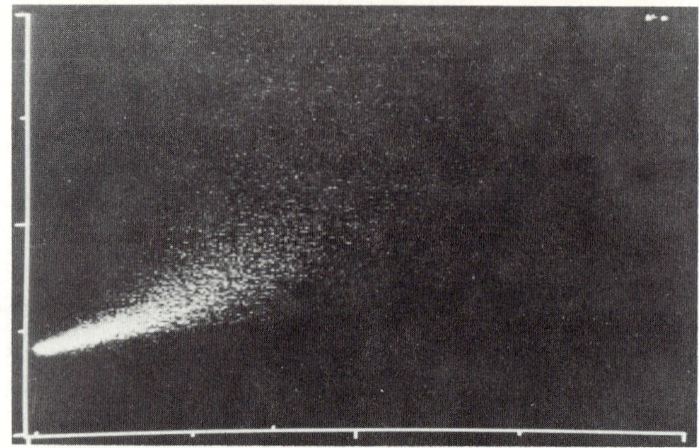

Fig.3 Scattergram (frequency distribution) of the pair ($\bar{A}$,SD) obtained from eight enddiastolic non-contrast images. X-axis: rms-deviation from the mean (0 --> 40), Y-axis: mean value (0 --> 100).

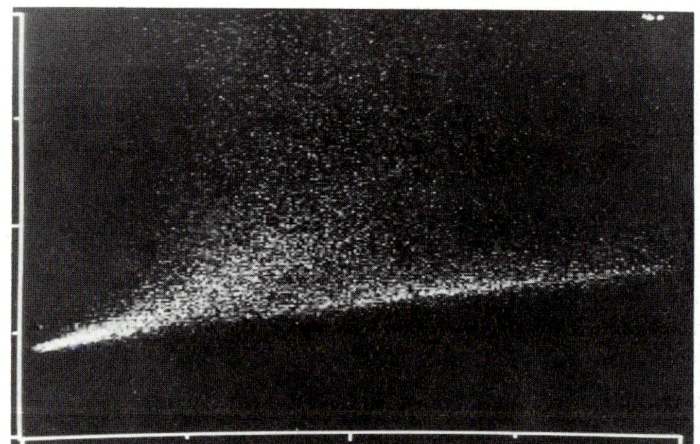

Fig.4 Scattergram of the pair ($\bar{A}$,SD) obtained from an enddiastolic image sequence of eight non-contrast images and two contrast images (same scaling as in Fig.3).

References

(1) Burckhardt CB: Speckle in ultrasound B-mode scans. IEEE Trans. on Sonics and Ultrasonics, SU 25:1, 1978

(2) Fitzgerald PJ, LF Joynt, SE Green, RL Popp: Computerized echocardiographic tissue characterization. IEEE Proc. Computers in Cardiology, p. 395, 1981

Verbesserte Strukturerkennung durch Bildverarbeitung: Anwendung beim menschlichen Gehirn

Josef Wasel und D. Graf v. Keyserlingk
Abteilung Anatomie, RWTH Aachen
Melatener Straße 211, D-5100 Aachen

Einleitung

Die in den letzten Jahren entwickelten bildgebenden Verfahren, insbesondere die Röntgen-Computer-Tomographie (CT) und die Magnet-Resonanz-Tomographie (MRT), haben der klinischen Medizin eine Vielzahl neuer Möglichkeiten erschlossen. Aber auch in der medizinischen Grundlagenforschung sind damit der Strukturerkennung neue Dimensionen gegeben worden. Als eine Forschergruppe der DFG beschäftigen wir uns mit der Verarbeitung von cranialen CT-Bildern mit dem Ziel, den untersuchenden Arzt in der Klinik bei der Beurteilung dieser Bilder durch Einsatz des Computers zu unterstützen, um so Qualität und Sicherheit der Befunde zu verbessern. Ein wesentlicher Bestandteil der dem Computer obliegenden Aufgaben ist dabei die Abgrenzung von Krankheitsherden von ihrer Umgebung sowie die Verbesserung der Strukturerkennung allgemein. Nicht die vollautomatische Verarbeitung bis hin zur Diagnosefindung steht im Vordergrund, sondern in erster Linie sollen dem Arzt Entscheidungshilfen für seine Diagnosestellung an die Hand gegeben werden. Eine große Unsicherheit bei der rein subjektiven Beurteilung des Arztes rührt von der starken individuellen Variation der Größe und Form des menschlichen Gehirns her. Daher wurden gemittelte oder standardisierte 3D-Hirnmodelle mit einer Auflösung von 2 mm entwickelt. Die Modelle wurden aus 30 anatomischen Präparaten gewonnen. Sie enthalten damit Information über den wahrscheinlichen Aufenthaltsort aller interessierender Strukturen des Gehirns. Diese werden nach 3D-Anpassung über ausgewählte Referenzstrukturen in das CT-Bild des Gehirns eines Patienten hineinprojiziert. Sie ermöglichen so eine Art von Beschriftung der Computertomogrammaufnahmen.

Strukturgrenzen in CT-Bildern

Bei der Strukturerkennung in digitalen und somit auch in CT-Bildern werden im ersten Ansatz die Relativwerte der aufgenommenen Daten, im CT Hounsfieldwerte genannt, in Betracht gezogen. Eine Gewebedifferenzierung kann verbessert werden, wenn an den Strukturgrenzen die Grau-

wertübergänge stärker betont werden. Ein relativ einfacher Weg dazu führt über Histogramme. Ein Bildausschnitt wird so gewählt, daß die zu differenzierenden Strukturen enthalten sind und sich im Histogramm ein lokales Minimum abzeichnet. Einen weiteren Vorteil dieses Verfahrens bietet die Weiterverabeitung dieses Schwellenwertes in einem Algorithmus, der sehr effizient Kontur- und Flächenberechnungen durchführt.
In Einzelfällen ist eine Abgrenzung der Strukturen vom umgebenden Gewebe auf diese Weise nicht möglich, da sich die zugehörigen Hounsfieldwerte kaum unterscheiden. Zusätzlich macht sich hier das Bildrauschen besonders bemerkbar, so daß schließlich im Histogramm kein lokales Minimum zu erkennen ist. Selbst mit dem Auge ist hier eine objektive Grenzfindung kaum möglich. Ein Beispiel zeigt das linke Bild in Abbildung 1: Ein Krankheitsherd (Läsion) hat sich in der linken Hemisphäre soweit ausgedehnt, daß er an das zentral gelegene Ventrikelsystem grenzt. Die Absorption der Röntgenstrahlung im Läsionsgewebe und im Liquor ist fast identisch. Mit Liquor wird die Hirnflüssigkeit bezeichnet, die alle Hohlräume im Gehirn und folglich auch das Ventrikelsystem ausfüllt.
Eine erste Bildverbesserung durch Bildverarbeitung bietet sich durch eine Bildglättung an. Die Mittelwertsglättung über eine 3x3-Umgebung unterdrückt zwar einen Teil des (statistischen) Rauschens, jedoch verwischen gleichzeitig die Kanten und Linien. Auch bei einer Glättung mittels eines Medianfilters, bei der eine solche Verschmierung weniger stark auftritt, ergeben sich bei nachfolgend angewandten Gradientenverfahren wie Roberts- bzw. Sobeloperator (1) keine brauchbaren Ergebnisse. Erst eine Filterung über eine 2D-Fourier-Transformation brachte eine Kontrastverstärkung an den Stellen minimaler Übergänge bei gleichzeitiger Rauschunterdrückung. Eine 2D-Fourier-Analyse transformiert dabei die Bilddaten in den Frequenzbereich. Es werden geeignet ausgewählte Frequenzen herausgefiltert bzw. stärker hervorgehoben und durch die Fourier-Synthese wieder in den Ortsbereich überführt. Die benutze Filterfunktion entspricht dabei nicht einem der Standardfilter (Tief-, Hoch- oder Bandpaßfilter), sondern wurde speziell für unsere Problemstellung empirisch entwickelt.

Kontrastverstärkung des Furchenmusters

Bei der räumlichen Anpassung unseres Modells an die 3D-rekonstruierte Gehirnoberfläche ist die Korrektur eines Rotationsfehlers um die Querachse erforderlich. Dieser Fehler ist verursacht durch unterschiedliche Bezugssysteme bei der CT-Aufnahme einerseits und der Modelle andererseits. Als geeignete Referenzstrukturen zur Kompensation des Rotationsfehlers bieten sich bestimmte Oberflächenstrukturen, die Hirnfurchen oder Sulci, an, insbesondere der Sulcus centralis. Vorteilhaft ist hier, daß dieser Sulcus relativ weit von der Rotationsachse entfernt liegt, die Streuung im Modell hinreichend gering und seine Lokalisation in den oberen CT-Schichten fast immer möglich ist. Eine Verbesserung der Lokalisation kann auch hier durch die oben beschriebene Filterung erzielt werden. Als effizienter - eine 2D-Fourier-Transformation ist trotz des Algorithmus von Cooley und Tukey (2) sehr rechenintensiv - und effektiver ist die Anwendung einer Kreuz-Korrelation mit einer Matrixgröße von 5x5. Da die Verlaufsrichtung des Sulcus centralis in verschiedenen menschlichen Gehirnen immer ähnlich ist, kann mit Hilfe dieses Vorwissens eine für das Sulcusmuster sehr typische Korrelationsmatrix definiert werden. Eine Anwendung ist in Abbildung 2 dargestellt.

Der Sulcus centralis wird nach vorne vom Sulcus praecentralis und nach hinten vom Sulcus postcentralis umgeben, welche beide in ähnlicher Weise wie der Sulcus centralis von der Umgebung abgrenzbar sind (siehe Abbildung 2). Die drei Sulci ergeben zusammen ein sehr charakteristisches Muster. Unter Berücksichtigung dieses Musters kann der Sulcus centralis auch dann automatisch lokalisiert werden, wenn er selbst nur schwer zu erkennen ist.

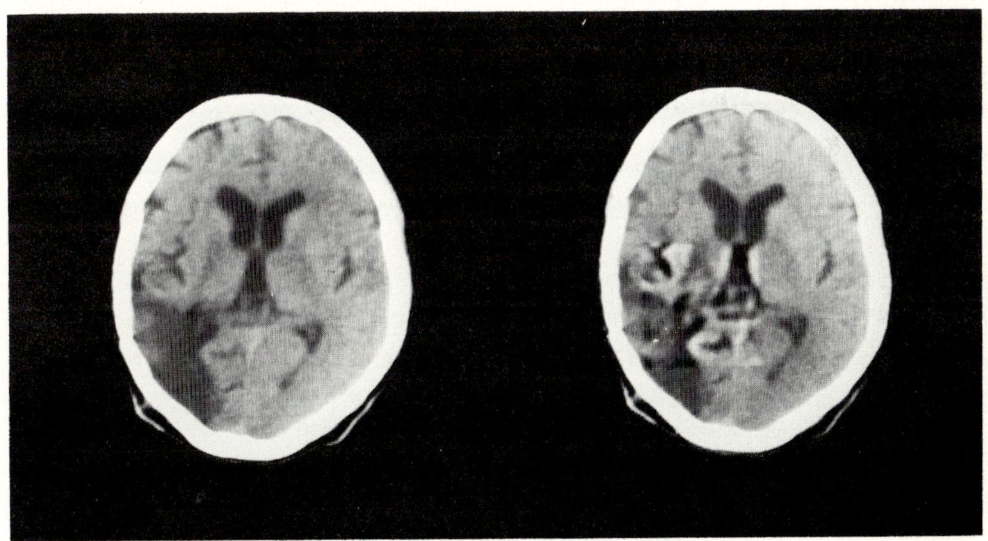

Abbildung 1: Läsionsabgrenzung vor (linkes Bild) und nach Filterung über eine 2D-Fourier-Transformation

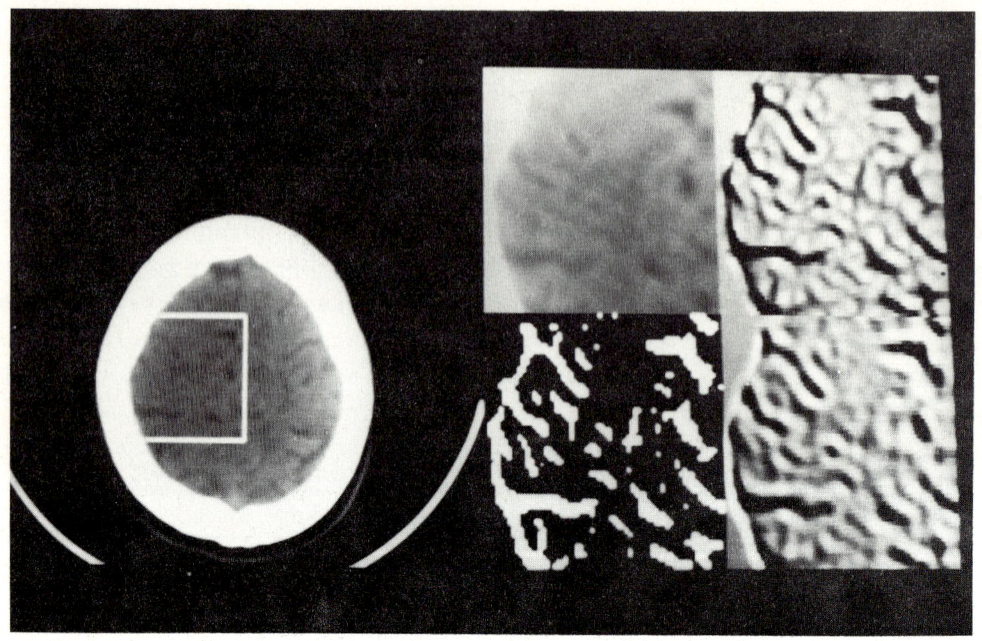

Abbildung 2: Kontrastverstärkung der zentralen Sulcusmuster in der linken Hemisphäre. Der mittlere Sulcus stellt den Sulcus centralis dar.

Literatur:

1) ROSENFELD/KAK "Digital Picture Processing". Vol. 1/2, Academic Press 1982
2) COOLEY, J.W., TUKEY, J.W. "An algorithm for machine calculation of complex Fourier series". Math. Computation Vol. 19, pp 297-301, April 1965

Computertomographische Rekonstruktion von Vektorfeldern

Dr.-Ing. Bernd Siemund
Universität Duisburg, Fachgebiet Nachrichtentechnik
Bismarkstraße 81, D-41 Duisburg

Einleitung

Zur Verifizierung von Modellvorstellungen zur Beschreibung von Entstehungsbränden ist es unumgänglich, Detailkenntnisse über die bei Brandversuchen real vorliegenden Verteilungen von Temperatur, Strömungsgeschwindigkeit und Aerosoldichte zu gewinnen. In /1/ wird ein computertomographisches Verfahren zur Bestimmung der Dichteverteilung der Partikelphase eines Aerosols in einer Meßebene beschrieben. Hierzu wird vom Rande des Meßgebietes die optische Extinktion zwischen einer Anzahl von Sende- und Empfangsmodulen bestimmt. Ein entsprechend realisiertes Meßsystem /2/ ist in der Lage, innerhalb von 1,5 Sekunden die notwendigen Messungen durchzuführen, die Dichteverteilung zu rekonstruieren und das Ergebnis der Rekonstruktion auf einem Farbmonitor darzustellen. Zusätzliche Informationen über Temperaturverteilung und Geschwindigkeitsvektorfeld der Gasphase des Aerosols können aus Ultraschall-Laufzeitmessungen gewonnen werden. Der Ausbreitungsweg der Schallwellen wird hierzu zunächst als geradlinig angesehen. Die Laufzeit der Schallwellen wird näherungsweise als die lineare Überlagerung des Einflusses der skalaren Verteilungsfunktion der Temperatur und der vektoriellen Verteilungsfunktion der Strömungsgeschwindigkeit der Gasphase beschrieben. Durch Messung in entgegengesetzten Richtungen und Mittelwertbildung kann der sich aus der Temperaturänderung ergebende Anteil des Meßwertes ermittelt, und die Temperaturverteilung in dem Meßgebiet mit den bekannten Rekonstruktionsalgorithmen für skalare Funktionen berechnet werden.

Radontransformation

Die Beschreibung einer skalaren Funktion durch ihre Integrale entlang von Geraden wird als Radontransformation (RT) bezeichnet /3/. Die Funktion sei auf einem Gebiet G definiert. Mit dem Ortsvektor $\underline{r}$ und dem Einheitsvektor $\underline{w}(\varphi)=(\cos\varphi,\sin\varphi)$ wird eine Projektionsgerade mit Abstand l vom Ursprung durch die Hessesche Normalform $\underline{r}\cdot\underline{w} = l$ beschrieben. Mit der Diracschen Delta-Distribution der Ebene und dem Operatorsymbol R lautet die RT:

$$H_R(l,\varphi) = [R\,h](l,\varphi) = \int_G \delta(l - \underline{r}\cdot\underline{w})\cdot h(\underline{r})\,d\underline{r} \tag{1}$$

Die inverse Radontransformation (IRT) wird durch den Operator R^{-1} repräsentiert. Ist die Funktion h stetig und beschränkt gilt für die Rücktransformierte h^*:

$$h^*(\underline{r}) = [R^{-1}R\,h](\underline{r}) = h(\underline{r}) \tag{2}$$

Neben der von Radon angegebenen Inversionsformel (IRT) gibt es eine Reihe von Rekonstruktionsalgorithmen die sich besser zur numerischen Berechnung der IRT eignen. Eine Zusammenstellung ist in /4/ und /5/ zu finden. Auf die Angabe einer konkreten Inversionsformel wird an dieser Stelle verzichtet.

Vektor-Radontransformation

Die Vektor-Radontransformation (VRT) mit dem Operator R_V wird als skalares Linienintegral über das Vektorfeld $\underline{h}$ längs der Projektionsgeraden g definiert. Das Vektorfeld $\underline{h}(\underline{r})$ nimmt nur in einem Gebiet G mit $\underline{r}$ R und $\underline{r}$ G Werte ungleich Null an.

$$H_V(l,\varphi) := [R_V\underline{h}](l,\varphi) = [R\,\underline{h}](l,\varphi)\cdot\underline{e}_\varphi \tag{3}$$

mit $\underline{h}(r,\varphi) = h_1(r,\varphi)\cdot\underline{e}_x + h_2(r,\varphi)\cdot\underline{e}_y$

$$\underline{e}_\varphi = (-\sin\varphi, \cos\varphi)$$

Der hierzu in gewissem Sinne orthogonale Spezialfall $\underline{e}$ = (cos ,sin) ergibt sich bei Elektronenstrahlmessungen an Magnetfeldern und wird in /6/ beschrieben.

Inverse Vektor-Radontransformation

Die inverse VRT (IVRT) wird analog zu /7/ definiert. Sie kann mit Hilfe der inversen RT beschrieben werden und wird durch den Operator $\underline{R}_V^{-1}$ gekennzeichnet.

$$\underline{h}^*(\underline{r}) := [\underline{R}_V^{-1}H_V](\underline{r}) = 2\cdot[\bar{R}^1\,(H_V(l,\varphi)\cdot\underline{e}_\varphi)](\underline{r}) \tag{4}$$

Diese IVRT kann also nach einer "Quadraturdemodulation" der Vektor-Radontransformierten komponentenweise mit den bekannten Rekonstruktionsalgorithmen durchgeführt werden. Mit Hilfe der Inversionsformel von Radon läßt sich die Transformation $\underline{h}^* = [\underline{R}_V^{-1}R_V\underline{h}]$ für eine harmonische Funktion $\underline{h}_h$, eine singuläre Quelle $\underline{h}_q$ und einen singulären

Wirbel $\underline{h}_w$ berechnen. Mit der Greenschen Funktion des Gebietes

$$G(\underline{r},\underline{r}_o) = -\ln(\underline{r}-\underline{r}_o) + W(\underline{r},\underline{r}_o) \quad (5)$$

$$\text{grad } W(\underline{r},\underline{r}_o) = \underline{w}(\underline{r},\underline{r}_o) = (w_1(\underline{r},\underline{r}_o),w_2(\underline{r},\underline{r}_o))$$

$$\underline{w}_j(\underline{r},\underline{r}_o) = (w_2(\underline{r},\underline{r}_o),-w_1(\underline{r},\underline{r}_o))$$

r_o: Ort der Singularität

ergeben sich für ein kreisförmiges Gebiet die Ergebnisse:

$$\underline{h}_h^*(\underline{r}) = \underline{h}_h(\underline{r}) \ ; \ \underline{h}_q^*(\underline{r}) = \underline{w}(\underline{r},\underline{r}_o) \ ; \ \underline{h}_w^*(\underline{r}) = 2\cdot\underline{h}_w(\underline{r}) + \underline{w}_j(\underline{r},\underline{r}_o) \quad (6)$$

Durch einen zusätzlichen iterativen Schritt (IIVRT) ist es möglich, $\underline{h}_w$ fehlerfrei zu rekonstruieren. Hierzu wird der Operator $\underline{R}_{IV}^{-1}$ mit

$$[\underline{R}_{IV}^{-1}H] := \frac{3}{2}\cdot[\underline{R}_V^{-1}H] - \frac{1}{2}\cdot[\underline{R}_V^{-1}R_V\underline{R}_V^{-1}H] = \frac{3}{2}\cdot h^* - \frac{1}{2}\cdot[\underline{R}_V^{-1}R_V\underline{h}^*] \quad (7)$$

definiert. Die elementare Berechnung von $\underline{h}^\# = [\underline{R}_{IV}^{-1}R_V\underline{h}]$ ergibt:

$$\underline{h}_h^\#(\underline{r}) = \underline{h}_h(\underline{r}) \ ; \ \underline{h}_q^\#(\underline{r}) = \underline{w}(\underline{r},\underline{r}_o) \ ; \ \underline{h}_w^\#(\underline{r}) = \underline{h}_w(\underline{r}) \quad (8)$$

Somit können auf einem kreisförmigen Gebiet quellenfreie Vektorfelder mit der IIVRT rekonstruiert werden.

Vektor-ART

Durch Diskretisierung (Operator disc) wird einer Funktion $h(\underline{r})$ ein Spaltenvektor $\underline{N}_h$ = [disc h] zugeordnet. Die diskretisierte Radontransformation (Gl. 1) lautet (/4/,/8/):

$$\underline{M}_H := \underline{\underline{A}}\cdot\underline{N}_h \quad (9)$$

Entsprechende Diskretisierung der Vektor-Radontransformation /9/ ergibt mit $\underline{h} = (h_1,h_2)$ und $\underline{\underline{N}}_h := (\underline{N}_{h1},\underline{N}_{h2})$:

$$\underline{M}_{HV} = \underline{\underline{C}}\cdot(\underline{\underline{A}}\cdot\underline{N}_{h1}) + \underline{\underline{S}}\cdot(\underline{\underline{A}}\cdot\underline{N}_{h2}) \quad (10)$$

$$\underline{M}_{HV} := [\text{disc } H_V] = (m_{HVk}) = (\ \underline{A}_k\cdot(\cos(\varkappa_k)\cdot\underline{N}_{h1} + \sin(\varkappa_k)\cdot\underline{N}_{h2})\ )$$

$$\underline{\underline{C}} := (\ \delta_{ik}\cdot\cos(\varkappa_k)) \ ; \quad \underline{\underline{S}} := (\ \delta_{ik}\cdot\sin(\varkappa_k))$$

$$\varkappa_k := \varphi_k + \frac{\pi}{2} \quad ; \ i,k = 1,K$$

Ausgehend von der Algebraischen Rekonstruktionstechnik (ART) wird ein entsprechendes vektorielles Verfahren (VART) zur Berechnung einer Lösung $\underline{\underline{N}} = (\underline{N}_x, \underline{N}_y)$ der diskretisierten VRT konstruiert.

$$\underline{\underline{N}}^{(1)} = (\text{[disc 0]} , \text{[disc 0]}) \tag{11}$$

$$d_k^{(i)} = - (\cos(\varkappa_k) \cdot \underline{A}_k \cdot \underline{N}_x^{(i)} + \sin(\varkappa_k) \cdot \underline{A}_k \cdot \underline{N}_y^{(i)} - m_{HVk})$$

$$\underline{\underline{N}}^{(i+1)} = \underline{\underline{N}}^{(i)} + \frac{\underline{B}_k^T}{\underline{A}_k \ \underline{B}_k^T} \cdot (\cos\varkappa_k, \ \sin\varkappa_k) \cdot d_k^{(i)}$$

$$k = (i \bmod K) + 1 \quad ; \quad i\text{: Iterationszähler}$$

Die Matrix $\underline{\underline{B}} = (\underline{B}_k)$ kann als das Ergebnis einer Tiefpaßfilterung von $\underline{\underline{A}}$ interpretiert werden /1/,/9/. Bei passender Wahl von $\underline{\underline{B}}$ ist $\text{disk}^{-1}\underline{\underline{N}}$ quellenfrei. Die Lösungen der VART haben unter dieser Voraussetzung die gleichen prinzipiellen Eigenschaften wie die der IIVRT (Gl. 10). Es ist jedoch kein kreisförmiges Rekonstruktionsgebiet erforderlich. Bei numerischen Simulationen zeigt zudem die VART bei stark unterbestimmten Gleichungssystemen ein besseres Verhalten als die IIVRT.

Diese Arbeit enstand im Rahmen des DFG Sonderforschungsbereiches 209 "Stoff- und Energietransport in Aerosolen".

Literaturverzeichnis:

/1/ I. Willms, Zur Anwendung computertomographischer Verfahren bei Aerosoldichtemessungen; Dissertation Universität Duisburg, 1983

/2/ H. Luck, B. Siemund, G. Lorbeer; The measurement of spatial aerosol distributions in enclosures by means of computed tomography; Part. Charact. 2(1985) p. 137-142, VCH-Verlagsg., Weinheim 1986

/3/ Johann Radon, Über die Bestimmung von Funktionen durch ihre Integralwerte längs gewisser Mannigfaltigkeiten; Ber. Verb. Sächs. Akad. Wiss., Leipzig, Math. Phys. KL. 69 (1917), Seiten 262-277

/4/ Gabor T. Herman, Image Reconstruction from Projections; Academic Press, London 1980

/5/ Gabor T. Herman, Image Reconstruction from Projections; Topics in Applied Physics Vol. 32; Springer-Verlag, Berlin 1979

/6/ J.B. Elsbrock, et al., Evaluation of three-dimensional micromagnetic strayfields by means of electron-beam tomography; IEEE Trans. on Magnetics 21 No. 5 (1985)

/7/ S. A. Johnson et al., Reconstructing three-dimensional fluid velocity vectorfields from acoustic transmission measurements; Acoust. Holo. 7, Kessler ed.; Plenum Press 1977

/8/ Hermann Schomberg, Nonlinear image reconstruction from projections of ultrasonics travel times and electric current densities; Mathematical Aspects of Computerized Tomography; G.T. Herman and F. Natterer ed.; Springer Verlag, Berlin 1981

/9/ Bernd Siemund, Ein Beitrag zur computertomographischen Rekonstruktion von skalaren und vektorwertigen Funktionen; Dissertation Universität Duisburg 1986

Some special applications of image processing in medicine

V. A. Pollak
Division of Biomedical Engineering
University of Saskatchewan
Saskatoon, Canada

Introduction

Digital image processing is rapidly becoming of age and finding more and more practical application in industry and other areas of the life of society. One of the disciplines which stands to profit significantly from the introduction of image processing is medicine and medical care. However, before turning to specific aspects of that field, it may be appropriate to view it from the much broader viewpoint of the impact of technology upon cost and performance of medicine in general and medical care in particular. The approach is justified by observing that new technology faces in medicine problems which are vastly different from those encountered in other fields of social activity.

First of all it is necessary to appreciate that medicine is the social activity which has until now been least transformed by the industrial revolution of the last two centuries. Most members of the medical profession are still regarding their discipline as an art rather than a science. At best they consider medicine as a "soft" science as opposed to the "hard" reasoning of the natural and engineering sciences. An immediate consequence of this attitude was a degree of tardiness in accepting organisational and technological changes, especially those involving high-tech arguments and transparent only to professionals with intricate knowledge in these areas. Also efforts to make the operation of the health care system more economical without sacrificing accessibility and performance carried much less weight than similar attempts in other fields.

Things began to change noticeably only in the last two decades in the aftermath of the oil shock of the seventies and the ensuing budgetary problems of all industrial nations. It was found that the cost of health care delivery had risen disproportionately and continued to do so at a rate much faster than the gross national product (GNP) of these countries. A variety of reasons is responsible for that; their analysis would far exceed the scope of this paper. Nor could it be overlooked that health care was consuming an increasingly large percentage of the GNP, which threatened to deprive other socially and politically sensitive areas of their financial basis. In consequence of that pressure began to rise to take steps to curb that increase and to make the system more economy conscious. Not surprisingly this new attitude ran into opposition from traditionally minded members of the system, who viewed this as an infringement upon their independence, authority and last not least their economic status. Initial expectations that introduction of advanced technology and administrative methods by itself could result in significant cost savings without impairing the quality of the care delivered proved soon to be erroneous. On the contrary, technology seemed to increase rather than to decrease cost. The immediate result of these experiences was that the acceptance of new technology became contingent not only upon the demonstration of significant patient benefits but also upon proof of an acceptable cost-benefit ratio with all its hidden fallacies. A completely new problem began to arise - how to express patient benefits in monetary terms. An all satisfying answer to this question is not in sight.

The indicated problems are found in all areas of medical technology, apply, however, especially to the new, highly sophisticated class of devices collectively labelled as "high tech". Unfortunately they are not always given adequate attention neither by the manufacturers of medical equipment nor by the personnel responsible for its introduction into clinical use. There is a strong tendency to concentrate upon technical performance parameters and upon capital cost, whilst disregarding operating expenses and the relation of the anticipated patient benefits to overall cost. Not adequately appreciated is also commonly the fact that full exploitation of advanced technology is more often than not contingent upon organisational changes and reeducation of the personnel involved.

Advanced pictorial methods are a significant part of the new developments in medical technology. Many of the new methods require processing of the acquired pictures before they can be interpreted. But processing proved soon to be a useful addition also to traditional imaging methods. New developments in computer technology have made picture processing faster, more powerful and economically accessible. The remainder of this presentation concentrates upon a brief survey of less well known applications of picture processing in medicine with apparently favourable cost-benefit ratios. The list of topics mentioned in the following is not intended to be complete nor exhaustive. They were selected to provide as much as possible an illustrative cross section of the field.

General facets of imaging methods in medicine

The desire of the physician to be able to view structures of the body not accessible to direct inspection of the intact surface is an old one, but one which long appeared to border on the impossible. Modern technology has not only opened the interior of the intact body to visual inspection, but made features visible which could not be seen, even if they had been accessible to direct observation with the help of visible light. It is thus not surprising that much effort has been spent on imaging devices which are not subject to these limitations. Some of them, e.g. the traditional X-ray machine, supply information directly in pictorial form, which can be read by the eye without further processing. In other cases the output signal of the machine is before transformation unintelligible by mere inspection. Computer assisted tomography and magnetic spin resonance are typical examples of that category.

The majority of the new sophisticated imaging devices are used primarily for diagnostic purposes. In therapy they are used largely in an auxiliary role.

Some less common methods of medical imaging and image processing

Conceptually related to computer assisted X-ray tomography is impedance tomography. The local absorbance of the illuminating X-rays in the former is here replaced by the impedance of the body at low to medium frequencies. Because of the long wave length of the illuminating radiation the resolution of the resulting image is poor. Nonetheless the method is useful for some special purposes, can, however, not be expected even to approach the significance of other tomographic techniques. It has the advantage that the equipment needed for implementation is inexpensive and with some caution the method does not pose any hazard to the patient.

Another uncommon imaging technique, the potential of which is as yet not fully explored, is thermography based upon measurement of high-frequency radiation with wave lengths beyond the far infrared. Much more common is, however, the same technique working in the intermediate infrared and using infrared imaging devices. Technological advances in the field of infrared imaging promise to improve the resolution and to decrease cost. Maps of the surface and subsurface temperature distribution can be produced in this way and serve as important aids for a number of pathological conditions, e.g. tumors of the female breast or some circulatory problems.

Stereo and pseudo-stereo techniques

Another unusual imaging method relies upon illumination of the body surface through a diffraction grating, which is in the simplest way formed by a grid of parallel wires. Illumination by collimated but non-coherent light is adequate. The grating produces an interference pattern on the body surface, which can be used for mapping the surface profile. Capital costs of the system are insignificant. The maps produced in this way are useful diagnostic aids for rehabilitation, forensic medicine, restorative and plastic surgery and others. The method is in a sense related to holography, which, however, requires illumination by coherent (laser) light. Holographic imaging is at present only little used in medical applications, but some of its modifications seem to have a large potential for the future. Many technical difficulties remain to be mastered before the method can gain the importance it seems to deserve.

Much interest is focused upon the production of synthetic holograms from a set of 2 D pictures to create a stereo view. The generation of 3 D pictures is attracting considerable attention in surgery and other fields, but centers mostly on other non-holographic approaches. Pseudo-stereoscopic pictures can be derived from stacks of tomographically obtained 2 D pictures using suitable, not necessarily plane, slice surfaces. Still more common is the use of shading and of axionometric displaying to produce an apparent 3 D impression. Conventional stereo viewers employing two laterally suitably displaced 2 D pictures are usable only for relatively few applications. Two displaced 2 D pictures can also be combined to a stereo display by differentiating them by colour or plane of polarization of the illuminating light. The advantage of the approach is that it can be easily applied to moving pictures. When electronic displays are the source, newly developed electronically controllable polarizing filters offer an advantageous option. Relatively new is the use of imaging mirrors or lenses with periodically varying focal distance. The same effect can be obtained by a revolving helical mirror. Which of these alternatives will find broadest acceptance is a question which still cannot be answered. Also the acquisition of the basic pictorial data for 3 D representation is a problem where many different techniques compete and the final outcome is far from being decided. In most cases are the original pictures acquired in planar, two-dimensional form. The lateral displacement of the source pictures, which for many medical images is difficult to implement directly, may have to be generated by suitable processing of the original(s). Also direct holographic picture acquisition has difficulties and must more often than not be replaced by synthetically generated holograms. Photogrametry and interferometry are for some purposes advantageous and in common use. Both approaches find obviously application also in medicine for other purposes than the collection of stereoscopic data. Summarizing it may be said that three-dimensional image representation has in medicine a large role to play, but much technical development work is still necessary before clinical use on a larger scale will become a practical possibility.

Pattern recognition

A rather seperate branch of image processing is machine recognition of (pictorial) patterns. Though the largest importance of this field lies in non-medical areas, it would be mistaken to underestimate the importance of this field for medicine. Unfortunately the discipline has not yet advanced to the stage where a reliable technique of pattern recognition suitable for the automatic or semi-automatic identification of a wide range of medical patterns was available, especially when the condition is added that the method should be suitable for clinical use at reasonable cost. If such a system was developed, it could not only improve the performance of many diagnostic methods, but also revolutionize certain fields of surgery and result in profound changes in many other medical disciplines.

Practical applications of picture processing

The first steps in this direction, some of which have already proven their clinical value, are efforts to read and interpret one-dimensional recordings of physiological parameters by machine. Noteworthy examples are computerized systems for the analysis of the electrocardiogram and electroencephalogram.

A promising addition to the already classical collection of interpretation methods in these fields is computerized mapping of bio-potentials on the surface of the body. Mapping can be stationary in time or result in a time variable picture; the latter places, of course, higher demands upon data acquisition techniques as well as the processing software and hardware. Another problem is the interpretation of the generated maps.

Imaging and image analysis also have an important part in sports medicine. A large role is played by simple recording and subsequent play back of the time course of the movements oth the athlete. The aim is to improve coordination of the activity of the muscles involved. The analysis is carried one step further by combining kinematic data with time-dependent electromyographic measurements. The approach is useful for the analysis of motor functions in general, e. g. in the diagnosis of motor disorders and in rehabilitation medicine. Digital techniques have proven useful, especially for gait analysis, and the interpretation of the measured trajectories of anatomical landmarks displayed in graphical form.

Digital picture processing has also a role to play in the design of aids for the visually impaired. Electronic reading aids, conversion of environmental features to forms which can be read by the tactile and/or acoustical sense are today the most prominent topics.

Also belonging into the arsenal of rehabilitation engineering but following a different philosophy with less emphasis upon the pattern recognition aspect is the production of prosthetic elements, which have to meet certain cosmetic requirements. Production of an artificial leg with shape, matching that of the remaining natural one, is a typical example. New and as yet untested is the exploitation of this technique in dentistry.

Digital picture analysis can be expected to become of importance in plastic surgery and even in some cases of restorative dentistry. The almost identical approach in reverse can be useful for the identification of the remains of victims of severe accidents, mainly of those involving fire and explosions, but also of otherwise unrecognizable naturally decomposed bodies.

Digial picture processing also holds the promise to become a valuable component in medical education, especially in fields like surgery and its specialized subdisciplines, in gynecology and obstetrics etc.. It allows the student to perform manipulations on a fantom and to illustrate immediately the consequences of mistakes or inappropriate approaches.

Therapeutic applications

One of the relatively few applications of picture processing in therapy is found in radiology, in the treatment (mostly of malignant tumors) by ionizing radiation. Careful planning of the irradiation pattern is here necessary the maximize the effect upon the diseased tissue and to minimize the damage to healthy tissue. Fantom experiments are the main tool for this purpose. Physical fantoms are costly and time consuming. A large part of the fantom work can, however, be replaced by computer simulation of the tissue and the pattern of the energy distribution of the applied radiation.

Another therapeutic application of picture processing involves the use of pseudo 3 D displays in surgery supplemented by indication of the instantaneous position of the instrument (e. g. scalpel) on the display. The technique should prove useful also for catherization, stone destruction and removal etc..

Picture processing for the laboratory

Working along different lines is the quantitative and qualitative evaluation of one-dimensional densitometric recordings, e. g. for thin layer chromatography and electrophoresis. Virtually all modern densitometers for this purpose are coupled to a computer which performs the analysis and not unfrequently controls also the operation of the instrument itself converting it to a degree to a semi-automatic device.

In principle similar but rather different in implementation is the analysis of two-dimensional separations obtained mostly by 2 D electrophoresis. The resolution of the 2 D approach is much higher than that of the 1 D technique and the amount of information to be assessed is consequently disproportionately larger. It may in fact be too large to make routine processing by eye practical. Data acquisition and processing methods are here in principle identical with those used in most standard branches of picture processing, but in detail they are modified to meet the special requirements of the purpose. Evaluation of 2 D separations (e. g. of proteins) is of importance in microbiology, genetics, toxiology, pharmacology and in many other fields where it is necessary to determine the composition of multicomponent solutions.

Another field where aspects of pattern recognition are of special importance is cytology and related branches of histology and pathology. The patterns to be recognized here are generally too complex and insufficiently well defined to be presently suitable for automatic analysis by computers. There are, however, subareas where attempts in this regard were successful. A typical example is caryotyping of chromosomes and the simpler task of cell counting in tissue fluids. Determination of the fertility of semen preserves and early detection of cancerous growth are other potentially important applications.

Computer consultation

Also quite different but still belonging to the field of pattern recognition is the area of machine supported consultation and decision making for diagnosis and therapy. At the first glance it might appear that the problem has little in common with picture processing, the core topic of this presentation. It is, however, in many cases possible and advantageous to arrange the principal diagnostic features of a particular case graphically in the form of a tree or a map. Image processing techniques can then be applied to compare the obtained map with a template one for a specific syndrom. The approach is in principle similar to the one used by the above mentioned technique of analysis of 2 D electrophoretic and occasionally chromatographic separations. In both cases the result is obtained as set of probability values which define the degree of similarity between the examined pattern and several selected template maps, which a priori would merit consideration. Interactive operation appears to be a must for methods of diagnostic support and the final decision will for a long time belong to the human operator, the physician.

Picture processing for the purpose of improving readability

Whilst automatic and even semi-automatic pattern recognition has still a long way to go, techniques for making the visual recognition of pictorial patterns easier have already reached a high level of efficiency. Many of the methods and algorithms available have general applicability, but some were developed specifically for medical purposes.

The field is much too broad and diversified to be covered here even in the form of a brief survey. Only a few characteristic applications are, therefore, quoted in the following to serve as illustrative examples.

The techniques involved amount primarily to operations of one kind or another on the three-dimensional coordinate system formed by the geometric coordinates of the pixels and their gray levels. Two more but generally less important coordinates have to be added for colour pictures. The most common operations involved are low pass filtering of one kind or another for noise reduction, high pass filtering for contour enhancement and interpolation to improve apparent resolution. Steepening or flattening of the gray scale is used for contrast modification. An extreme form of this technique is binarization of the picture. The gray scale resolution can also be enhanced by the introduction of false (artificially created) colours to enhance the perception of gray shade differences. Zooming is another frequently employed method. In some cases are the operations direction dependent or are employing non-linear transform laws. Histogram techniques are not unfrequently useful. Autocorrelation, crosscorrelation, convolution and deconvolution are also commonly employed, e. g. for feature enhancement or detection. Of the many available integral transforms, the Fourier transform is the most common one. Other transforms, e. g. the Walsh transform, the Haar and the Mellin transform are used to a lesser extent for special purposes. Fast and efficient execution of all these operations in two or three dimensions requires special algorithms, large storage capacity and is in most cases best carried out on specially designed hardware. The operations listed are standard not only for the enhancement of picture readability, but also for picture processing for other purposes. Image processing may sometimes not only help interpretation, but even salvage exposures, which otherwise would be lost because of wrongly chosen exposure time, excessive dynamic range of the source picture, fading etc.. The approach is particularly valuable for pictures which can not be repeated or where repetition of the exposure would be hazardous or costly.

Some methods in this class require special techniques of data acquisition. For example it is possible to improve the contrast of X-ray pictures and to make small details visible by addition or subtraction of two pictures obtained at different energies of the beam. In some cases, readability is improved by superimposing pictures obtained by completely different techniques, e. g. a potential map of the heart and an X-ray picture of the chest.

Electronic image display

The standard medium for displaying pictorial information for visual examination is today for the physician predominantly photographic film and sometimes paper. Resolution, dynamic range and sensitivity of todays photographic media are excellent. Still more important is that physicians are accustomed to the impression generated by this kind of display. There are, however, also disadvantages which in part could be overcome by replacing the photographic hard copy display by electronic viewing. There seems to be no objective difficulty which would prevent the substitution of photographic techniques by electronic methods in the majority of applications. Nonetheless the initial expectations that electronic displays with all their operational and economic advantages over photographic hard copy displays would soon replace the latter did not materialize. The most likely reason for that is the different visual character of the electronic display and the consequently necessary readjustment and the training of the examining physician, something most of them are reluctant to undergo.

PACS

Another important application of picture processing in medicine is the storage and retrieval of image information. In general it is desired that the location of both storage and retrieval points be flexible and not tied to the location of image aquisition nor to one another. The acronym PAC (picture acquisition and communication) is commonly used for this specialized field. The number of imaging methods for medical purposes has in the past decades substantially increased and this trend seems far from having terminated. Similarly the role image based information plays in medical practice is quantitatively and qualitatively on a significant upswing. It follows that also the demands upon the administration of this information and the importance of the latter have proportionately increased. The amount of information the system has to handle is, however, enormous and already now straining existing facilities to the limit. Despite that PAC systems have not yet found much favour in the Medical Records community. But this may well change in the near future. To achieve that, it will be necessary to give more attention to the clinical requirements of this service. From a purely technical point of view, a medium with much higher storage capacity than the present magnetic hard disk will be necessary. The most likely candidate for this role is the read-write optical disk. But even here data compression and subsequent expansion will be necessary, especially to keep transmission costs under control.The really necessary resolution and retrieval speed are also points in need of further study. The crucial point is, however, the organization of the system and the coding of the stored information. A medical PAC system operates inevitably as part of the Medical Records system, which in turn represents a gigantic data bank, containing besides pictorial information also verbal, numerical and graphical data. To achieve its full operational and economic potential (which is enormous) the contents of this data bank must be accessible according to a widely variable set of key codes. No present data bank system is as yet up to these requirements.

Similar though less stringent requirements are placed upon a medical advisory-reference system with remote access. Such a system should be valuable for sparsely populated areas where the density ot health care stations is by necessity low. Transmissions of pictorial information in a relativlely short time over low grade communication facilities is here an essential requirement. The transmission must be possible bidirectionally, from the data bank to the remote user and, reversely, from the user to the central station. A degree of preprocessing, coding and compression of the information to be transmitted may be needed in one or both directions. Processing on the receiving side, regardless of the direction of the flow of information, may be necessary to reduce the rate of in principle inevitable errors due to noise and distortion.

Conclusion

Summarizing it may be said that digital picture processing techniques with the related fields of data compression, storage and transmission can be expected to play an increasingly important role in clinical practice. If properly used, it should in most cases not be difficult to prove patient benefits as well as a favourable cost-benefit ratio. In the majority of cases picture processing is used to facilitate diagnosis. Real world diagnostic work proceeds always under constraints of time and space. These constraints not only affect cost but may even impair the depth to which diagnostic analysis can in actual practice be carried out. Modern methods of signal processing in general and picture processing in particular may then be able to reduce the equivocation of diagnostic results and remove the need for many additional tests and examinations. It is obvious that more precise diagnostics lead to improved therapy, a significant patient benefit. The reduced time required for examination and the removal of the need for some supplementary procedures result obviously in cost advantages at little investment expense. In some cases picture processing may even result in diagnostic findings, which would otherwise in all probability elude detection by traditional methods. In these cases, enormous patient benefits can be expected at little additional cost.

These considerations seem to be sufficient to make picture processing attractive with regard to both patient benefits and economics, especially when conventional picture acquisition technique can be used and no new sophisticated and expensive hardware for this purpose is required.

The reasoning above applies to PAC systems as well. Here, however, an additional time and cost saving feature should be considered, which already might by itself justify the introduction of PACS on a large scale. This feature is the fast availability of detailed epidemiological data, which can not unfrequently replace extensive clinical studies. The field is in rapid flux and the scenery changes almost day by day. Existing applications are modified and expanded as experience accumulates and hardware and software improve. At the same time new applications are certain to evolve. The list of those mentioned in this paper is even at todays level neither complete nor exhaustive. But returning to the key condition set out at the beginning of this presentation, permanent success and broad acceptance of a particular application is contingent upon a favourable cost-benefit ratio and adequate patient benefits over and above those presently available. Another prerequisite are changing attitudes of the medical community and modification of the organization and administrative structure of the health care system. In view of the sheer mass and inertia of that system, these changes will be slow to happen and resistance will be large. Attempts to proceed too fast might easily result not in progress, but in stiffening opposition and even temporary rejection.

BILDVERARBEITUNG IN DER PARAMETER-SELEKTIVEN KERNSPINTOMOGRAPHIE

Thomas Tolxdorff[+] und Klaus Gersonde[o]

[+]Abteilung Medizinische Statistik und Dokumentation
[o]Abteilung Physiologische Chemie
Klinikum der Rheinisch-Westfälischen-Technischen Hochschule (RWTH)
D-5100 Aachen

EINLEITUNG

Die Einführung der Kernspintomographie als ein neues bildgebendes Verfahren in der medizinischen Diagnostik hat völlig neue Perspektiven nicht nur in medizin-diagnostischer Hinsicht, sondern auch im Hinblick auf das Problem der digitalen Bilddatenverarbeitung eröffnet. Die in der Medizin bisher eingesetzten bildgebenden Verfahren machten die Verarbeitung nur eines physikalischen Parameters erforderlich; dies galt zunächst auch für die Kernspintomographie. Der bildgebende Meßwert der Kernspintomographie ist die Magnetisierung. Diese Größe wird durch verschiedene physikalische Prozesse beeinflußt. Somit ist die Magnetisierung eine vielschichtige Information, die ohne die Geschichte des experimentellen Vorgehens und der zugehörigen Datenanalyse keine vollständige Interpretation zuläßt. Erfaßt man nun diesen Meßwert nicht nur in der räumlichen, sondern auch in der zeitlichen Dimension, dann erhält man in einem einzigen Meßvorgang eine größere Zahl von Parametern (Relaxationszeiten T_1 und T_2, Partialvolumen α, Spindichte ρ), die quantitative Aussagen über die Beschaffenheit von Molekülen und über deren Umgebung erlauben. Die medizinische Diagnostik wird hierdurch erheblich bereichert. Das Besondere an dieser neuen Methode ist die Tatsache, daß sich nichtinvasiv nicht nur anatomische Strukturen im Körper darstellen lassen, sondern auch funktionelle Zustände, d.h. biochemische und biophysikalische Vorgänge bis hinunter zur Ebene der Moleküle und ihren Wechselwirkungen in der Zelle. Zugleich werden auch dynamische Vorgänge aller Art (Diffusion, Perfusion und Blutfluß im Gefäßsystem) für quantitative Messungen zugänglich. Die Beschreibung des Funktionszustandes des Gewebes, seine Unterscheidung von anderen Geweben und die Identifizierung von Krankheitsprozessen sind jedoch nur mit größerer Sicherheit möglich, wenn sich die Erkennung auf möglichst viele Eigenschaften eines Volumenelementes stützt. Die Bildverarbeitung in der NMR-Spektroskopie steht somit vor einem neuen Problem. Ein Bildelement enthält eine Vielzahl von Eigenschaften. Die Betrachtung eines Bildes kann aber immer nur auf einen ausgewählten Aspekt gerichtet sein. Die Auswahl muß nach bestimmten Regeln und Kriterien erfolgen. Man muß also die "richtige" Information aus einem Pool der Informationen heraussuchen. Dazu ist es unerläßlich, eine rechnergesteuerte Unterstützung bei der Bildgebung für den Arzt bereitzustellen.

SOFTWARESYSTEM

Die eben genannten erweiterten Aussagen der NMR-Bildgebung sind methodisch realisierbar mit Hilfe der parameter-selektiven Kernspintomographie [1,2]. An der RWTH Aachen wurde das Softwaresystem RAMSES (RWTH Aachen Magnetic Resonance Software System) entwickelt [3,4], mit dessen Hilfe die Vielschichtigkeit der NMR-Information aufgeschlüsselt und zugänglich gemacht werden kann. Unter dem Aspekt des Routineeinsatzes in der Medizin können mit diesem Softwaresystem Probleme der digitalen Bilderzeugung, der methoden-orientierten Bildverarbeitung, der Bildanalyse und der Erzeugung funktioneller Bilder bearbeitet werden. Der Schwerpunkt liegt in der digitalen Verarbeitung von Bildern und der Quantifizierung von Bilddaten. Beides bildet nun eine geeignete Basis für die Entwicklung von wissensbasierten Systemen für eine rechnerunterstützte Diagnostik.

MEDIZINISCHES ANWENDUNGSBEISPIEL

Die parameter-selektive Bilddarstellung mit gewebecharakterisierender Darstellung von Geweben soll hier an einem Beispiel die Beschreibung eines weiblichen Brusttumors demonstrieren. Es handelt sich um ein duktales Karzinom, das in axialer Schnittführung dargestellt wird. Fig. 1 zeigt in einer Darstellung mit 19 Grauwerten ein konventionelles Grauwertbild, das aus den 36 Echobildern der Untersuchung mit Hilfe der "Best-Echo-Bild" Technik erzeugt wurde [5]. Es ist durch größere Anteile von schnellen T_2-Relaxationsprozessen charakterisiert, da diese im Zeitbereich der frühen Echos noch nicht abgeklungen sind. Dieses Echobild zeigt anatomische Strukturen des Brustschnittes sowie einen raumfordernden Prozeß. Einen Hinweis auf die Tumorart oder dessen Homogenität gibt diese Darstellung jedoch nicht. Nichtsdestoweniger ist das Echobild bei Anwendung der Überlagerungstechnik von grauwertcodierten Echobildern mit farbcodierten parameter-selektiven Bildern [5] von hohem Wert, wenn die korrekte Lage der funktionellen Bildinformation von Interesse ist. Die Bildserie, die einen Tumor charakterisieren soll, wird mit Bildern der parameter-selektiven Bilddarstellung fortgesetzt. In Fig. 2 ist der raumfordernde Prozeß mit Hilfe von Wasserprotonen im T_2-selektiven Bild (A) sowie dessen Homogenität im α-selektiven Bild (B) gezeigt. Die Relaxationszeit T_2 ist die Repräsentationsvariable und bildet in der Selektionsbedingung als einziger Parameter eine Einschränkung durch Wahl des Selektionsfensters von 160 bis 240 ms. Nach der Klassifikation des Tumors durch T_2-Selektion soll die Homogenität des Tumorgewebes quantifiziert werden. Dazu wird eine zusätzliche Selektionsbedingung an das Partialvolumen α in Form des Selektionsfensters 80 bis 100 % gestellt. In Fig. 3B erhalten die Bildbereiche, die zu 100 % mit einer Substanz ausgefüllte Volumenelemente repräsentieren, eine helle Einfärbung. Die Bildbereiche mit absteigenden Partialvolumina werden durch absteigende Grauwerte codiert (aus drucktechnischen Gründen kann hier leider keine Darstellung in Farbe erfolgen). Fig. 3B zeigt, daß das Tumorgewebe vorwiegend aus Partialvolumina $\alpha > 80$ % besteht, also eine relativ homogene Gewebestruktur aufweist. Fig. 4A zeigt T_2 als Repräsentationsvariable im Fenster von 0.16 bis 0.24 Sekunden und Fig. 4B die Homogenität α im

Fenster von 80 bis 100 % mit den oben definierten Selektionsfenstern für T_2 und α sowie einem gegenüber Fig. 3A modifizierten Selektionsfenster für ρ von 45 bis 70 %. Man erhält hier in der parameter-selektiven Bilddarstellung der Relaxationszeit T_2 und der Homogenität α (Fig. 4A,B) die beste Separation des duktalen Karzinoms durch die gleichzeitige Wahl aller tumorspezifischen Selektionsfenster.

SCHLUSSFOLGERUNGEN

Die Technik der Parameter-Selektion erlaubt die Separierung biochemischer und biophysikalischer Prozesse und somit eine gezielte Gewebedarstellung. Die parameterselektive Kernspintomographie der weiblichen Brust erlaubt die Differenzierung maligner und benigner Brusttumore, so daß die Zahl unnötiger Biopsien erheblich eingeschränkt werden könnte. Eine Bildüberlagerungstechnik macht es möglich, funktionelle und anatomische Information in einem Bild gleichzeitig erscheinen zu lassen.

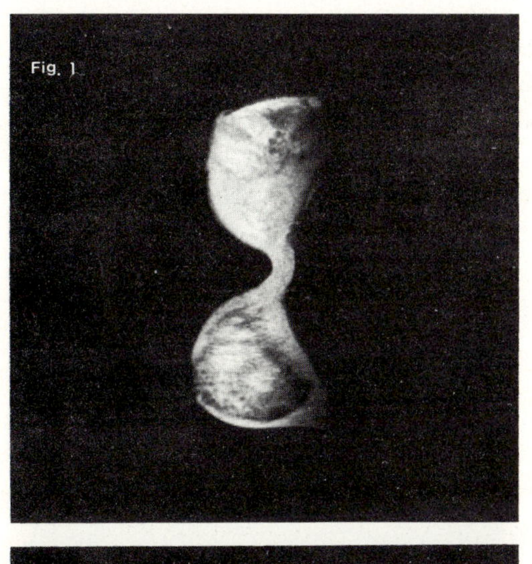

Fig. 1

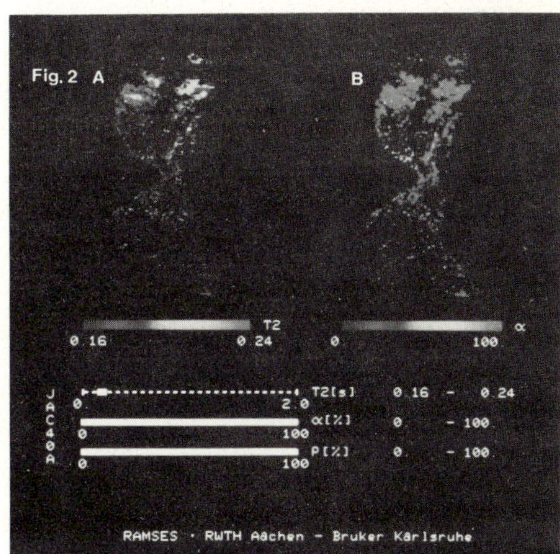

Fig. 2

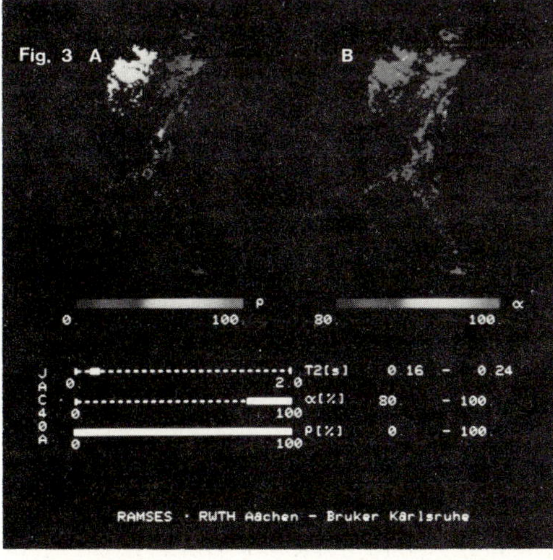

Fig. 3

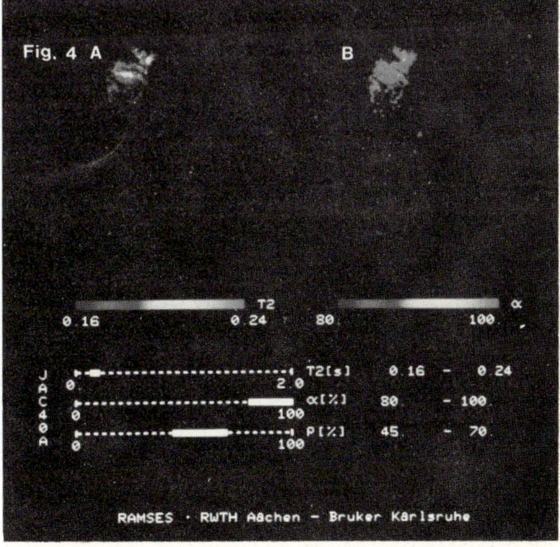

Fig. 4

LITERATUR

(1) Gersonde, K., Felsberg, L., Tolxdorff, T., Ratzel, D. and Ströbel, B.: Analysis of Multiple T_2 Proton Relaxation Processes in Human Head and Imaging on the Basis of Selective and Assigned T_2 Values. **Magn. Reson. Med. 1, 463-477, (1984).**

(2) Gersonde, K., Tolxdorff, T. and Felsberg, L.: Indentification and Characterization of Tissues by T_2-Selective Whole-Body Proton NMR-Imaging. **Magn. Reson. Med. 2, 390-401, (1985).**

(3) Tolxdorff, T., Felsberg, L., Mecking, B. und Gersonde, K.: RAMSES, ein universelles Verarbeitungs- und Informationssystem in der NMR-Diagnostischen Medizin. **Informatik-Fachberichte 127, 615-633, Springer-Verlag, Berlin (1986).**

(4) Tolxdorff,T.: Ein neues Software-System (RAMSES) zur Verarbeitung NMR-spektroskopischer Daten in der bildgebenden medizinischen Diagnostik. **Medizinische Informatik und Statistik 66, Springer-Verlag, Berlin, (1987).**

(5) Tolxdorff, T., Felsberg, L. and Gersonde, K.: Proton NMR Imaging: Contrast-Enhanced Images and Combination with Parameter-Selective Images by Overlay Display Technique. **Proceedings of the International Symposium Computer Assisted Radiology CAR 87, 3-11, Springer-Verlag, Berlin, (1987).**

Anwendungsbeispiele aus der Kunststofftechnik

- Qualitätskontrolle und Automatisierung durch Bildverarbeitungssysteme -

G. Menges, M. Haupt, K. Borgschulte
Institut für Kunststoffverarbeitung, RWTH Aachen
Pontstr. 49, 5100 Aachen

I) Prozeßüberwachung bei der Profilextrusion durch zweidimensionale Dimensionsvermessung mit optischen Sensoren

Die Herstellung maßhaltiger Profile gehört zu den schwierigsten Aufgaben in der Extrusionstechnik, da, insbesondere bei Fensterprofilen, eine Vielzahl von Geometriegrößen innerhalb enger Toleranzen eingehalten werden müssen. Neben der Überprüfung der Oberflächenbeschaffenheit, mechanischer Kennwerte und Schrumpf ist die Vermessung ein wichtiger Bestandteil der Qualitätskontrolle. Der Einsatz bildverarbeitender Systeme zur zweidimensionalen Dimensionsvermessung ermöglicht eine genaue, schnelle und objektive Geometrieüberwachung /1/.

Anforderungen

Von der Kunststoffindustrie werden folgende Forderungen an ein flexibles Profilmeßsystem gestellt :

- Vermessung verschiedener Profile mit einem Gerät
- Vergleich mit einem Musterprofil
- Variable Eingabe der Vermessungsposition
- Genauigkeit 0.1 mm bei max. 100 mm Profilgröße

Diese Anforderungen führten am Institut für Kunststoffverarbeitung zur Entwicklung eines

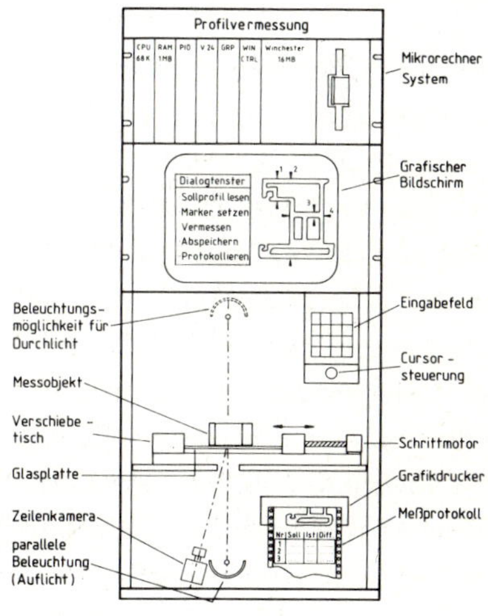

Bild 1 : Vermessungsaufbau

automatischen optoelektronischen Profilvermessungsgerätes /2/. Das System kann Profilkonturen erfassen, den Vergleich mit einem Musterprofil in Fremdfarbendarstellung graphisch anzeigen, Maße an frei definierten Konturpositionen errechnen und als Abstandswerte ausdrucken (Bild 1).

Bildverarbeitung

Das zweidimensionale Bild des Meßobjektes wird durch Abtasten des Profilquerschnittes mit einem hochauflösenden CCD-Zeilensensor und inkrementales Verschieben des Profils senkrecht zur Abtastrichtung erzeugt (Bild 2). Bei einfarbigen Profilen wird die benötigte Kanteninformation durch binäres Triggern, bei mehrfarbigen oder koextrudierten Profilen durch Kantendetektionsalgorithmen der Grauwertbildverarbeitung gewonnen.

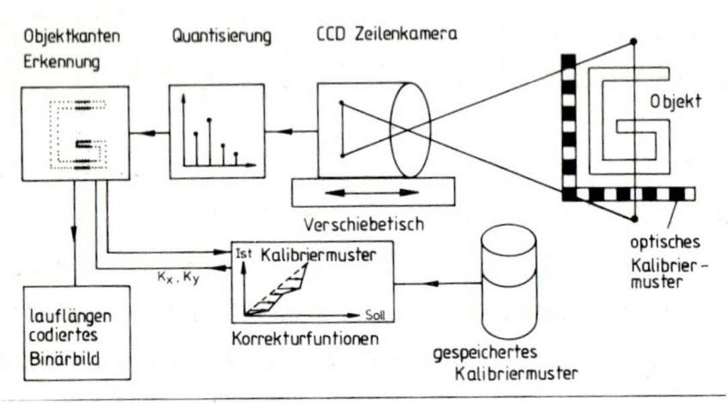

Bild 2 : Aufnahme eines Profilbildes

Während der Bildaufnahme ist die Lage der Profilprobe innerhalb des zugelassenen Bildbereiches frei wählbar. Daher muß vor der Vermessung das Profilbild mit dem Referenzbild in Deckung gebracht werden. Die Koeffizienten dieser zweidimensionalen Koordinatentransformation werden mit Hilfe eines modifizierten PSI-Algorithmus berechnet.

Ein Vermessungsbild besteht aus 4 Millionen Bildpunkten. Durch Binärisierung und Lauflängenkodierung der Bilder nach der Kantendetektion kann das Datenvolumen auf ca. 30 KByte reduziert werden. Zur Beschleunigung der rechenintensiven Prozeduren Kantendetektion und Bildkorrelation wird ein Image Pipeline-Prozessor verwendet.

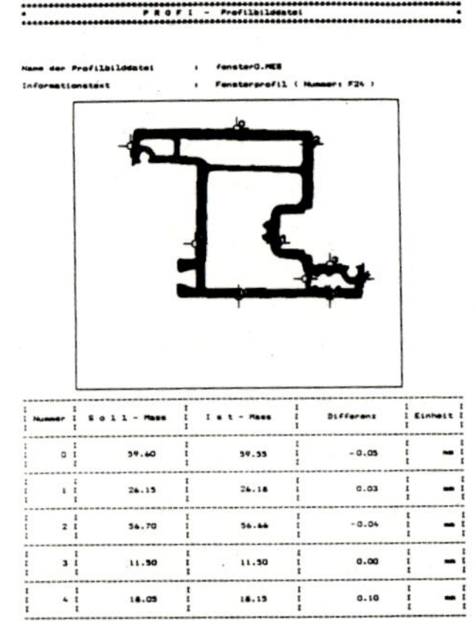

PROFI - Profilbilddatei

Name der Profilbilddatei : Fenster0.MES

Informationstext : Fensterprofil (Nummer: F24)

Nummer	Soll - Mass	Ist - Mass	Differenz	Einheit
0	59.60	59.55	-0.05	mm
1	26.15	26.18	0.03	mm
2	56.70	56.66	-0.04	mm
3	11.50	11.50	0.00	mm
4	18.05	18.15	0.10	mm

Bild 3 : Meßprotokoll

Ergebnisse

Mit diesem System kann ein Profilquerschnitt von 100*100 qmm auf 0.1 mm bei einer Verarbeitungszeit kleiner als zwei Minuten vermessen werden. Das Ergebniss wird in Fremdfarbendarstellung auf dem Kontrollmonitor und graphisch und tabellarisch auf einem Meßprotokoll ausgegeben (Bild 3).

II) Qualitätskontrolle von Gewebe aus Hochleistungsfasern

Die Verwendung von Gewebe aus Glas-, Kohle- und Aramidfasern oder anderer Materialien ermöglicht die Fertigung sehr leichter aber hochbeanspruchbarer Bauteile. Die hohen Anforderungen an die Festigkeit und Verarbeitung derartiger Bauteile erfordern einerseits eine vollautomatische Fertigung, andererseits muß aber auch eine optimale Qualität bei der Gewebeherstellung gewährleistet sein. Heutzutage erfolgt die Qualitätskontrolle ausschließlich durch Prüfpersonal, wobei ausreichend reproduzierbare, fehlerfreie und quantitative Ergebnisse meistens nicht erreicht werden.

Problemstellung und Prüfkomponenten

Die Probleme bei der automatischen Analyse des Gewebe sind in erster Linie :

- Wahl einer geeigneten Beleuchtungsquelle und -position, damit die Gewebestruktur deutlich sichtbar wird;
- Einsatz eines lichtempfindlichen und hochauflösenden optischen Sensors zur exakten Erfassung signifikanter Strukturmerkmale;
- Auswahl von Verfahren zur sicheren Erkennung und Klassifikation der verschiedenen Fehler;
- Verwendung einer leistungsfähigen Hardware zur Verarbeitung der hohen Datenmenge und des komplexen Algorithmus;

Neben der Bewältigung dieser Probleme sind beim praktischen Einsatz eines derartigen Prüfsystems in der Fertigung die Anpassung auf unterschiedliche Abzugsgeschwindigkeiten, Unempfindlichkeit gegen äußere Störeinflüsse und die Verarbeitung verschiedener Gewebematerialien zu berücksichtigen /5/.

Durch den Einsatz eines optischen Prüfsystems können diese Aufgaben automatisiert werden. Der Aufbau des Prüfsystems besteht aus drei wesentlichen Komponenten (Bild 4) :

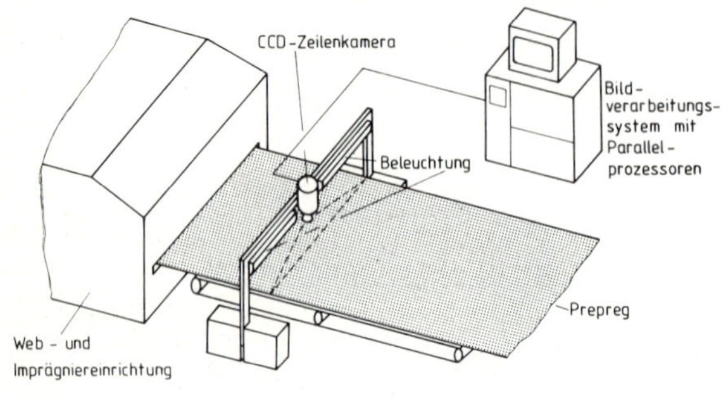

Bild 4 : Online Gewebekontrolle

hochauflösende Zeilenkamera (2048 Bildpunkte, 2 MHz Takt)
Bildaufnahmesystem zur Zwischenspeicherung der Bilddaten
Transputer-Parallelprozessorsystem zur Auswertung

Prüfablauf und Ergebnisse

Das Prüfsystem arbeitet derart, daß die Zeilenkamera kontinuierlich das bewegte Gewebe erfaßt und im Bildaufnahmesystem ein 2-dimensionales Bild aufgebaut wird. Dieses aufgebaute Bild wird dann in Blöcken zu 512 x 2048 Bildpunkten an das Transputersystem übergeben und anschließend ausgewertet. Während dieser Auswertung wird im Bildaufnahmesystem ein neues Bild gespeichert.
Diese Auswertung läuft dabei in mehreren Schritten ab :

- Für die Bildvorverarbeitung wurden Methoden der Kontraststeigerung und Datenreduzierung implementiert. Das Ziel war dabei, die zur Fehlererkennung relevanten Informationen hervorzuheben und die Bildgrauwerte auf diese zu reduzieren.

- Die Fehlererkennung basiert auf Verfahren zur Texturanalyse. Die Beleuchtungsfrage ist hierbei von großer Bedeutung, da durch eine bestimmte Beleuchtungsposition Reflexionen an den Schußfasern des Gewebes erzeugt weche letztendlich die auszuwertenden Informationen darstellen. Verschiedene Analyseverfahren - Gradientenoperatoren, Grauwertlauflängenstatistik und die Statistik der räumlichen Abhängigkeiten von Grauwerten - sind implementiert und können wahlweise ausgeführt werden.

- Im dritten Schritt, der Klassifizierung, werden verteilungsfreie und nichtparametrische Klassifikatoren eingesetzt. Sie ermöglichen die Einordnung der erkannten Fehler in zehn Fehlerklassen (z.B. Risse, Beulen, Falten, Harzanhäufungen, Trägerfolienreste oder Gewebeverschiebungen ; Bild 5).

Bild 5 : Fehlerfreies und fehlerhaftes Gewebe

Die Prüfgeschwindigkeit kann durch Erweiterung des Auswertungssystems nahezu unbegrenzt erhöht werden. In Abhängigkeit von der Gewebebreite und der Feinheit des Gewebes sind Abzugsgeschwindigkeiten von 10m/min bis 15m/min mit einer 100%- Kontrolle realisiert /6/.

Literatur

/1/ Breil,J. Automatisierungsmöglichkeiten an Rohr- und Profilextrusionsanlagen; Dissertation an der RWTH Aachen, 1986

/2/ Haupt,M. Aufbau und Programmierung einer optischen 2D-Vermessungsstation; Diplomarbeit am IKV/LfM, 1985 Betreuer : J.Breil; Prof. Meyer-Ebrecht

/3/ Paulsen,M. Untersuchung und Entwicklung von Algorithmen zur Korrelation zweier Binärbilder und Implementierung einer geeigneten Lösung in ein bestehendes Bildverarbeitungssystem; Studienarbeit am IKV/LfM,1986 Betreuer : M.Haupt; Prof. Meyer-Ebrecht

/4/ Gandelheidt,E. Entwicklung und Programmierung von Algorithmen der Bildverarbeitung für den Image Pipeline Prozessor uPD7281; Studienarbeit am IKV/LfM, 1986
Betreuer : M.Haupt; Prof. Meyer-Ebrecht

/5/ Indefrey,K. Automatisierte Qualitätskontrolle von CFK-Prepreg mit digitaler Bildverarbeitung; Diplomarbeit am IKV/LfM, 1985
Betreuer : H.Cherek; Prof. Meyer-Ebrecht

/6/ Faßbender,T. Optische Online-Qualitätskontrolle von Fasergewebe auf einem parallelverarbeitenden Bildauswertesystem; Diplomarbeit am IKV, 1987
Betreuer : K.Borgschulte

Pattern Recognition for Earthquake Detection

Manfred Joswig
Institut für Geophysik, Ruhr-Universität Bochum,
Postfach 10 21 48, D 4630 Bochum 1

Automatic detection of earthquakes in environmental noise is subject to research for more than 25 years. The traditional approaches describe the stationary noise level by some kind of median and standard deviation and determine a detection threshold [1],[2]. Each temporary signal with energy above the threshold triggers a detection. Bandpass- or ARMA-filtering is used to improve the signal to noise ratio. The main disadvantage of these detectors is their high false alarm rate, since they are not able to distinguish between small earthquakes and noise from traffic or industry, which has the same or even higher amplitude. This is due to their decision logic of negative kind: Every signal above the threshold is assumed to be caused by an earthquake.

Pattern recognition detectors are an application for positive decision logic: Only a sufficient similarity with one of some predefined patterns will trigger a detection. This approach equals the knowledge-based decision process of an human observer, when he examines the time-series or 'seismograms' of the earth motion. The application of selected seismograms as knowledge patterns however is not practical. This 'matched filter' approach fails, since too many different types of earthquakes are to detect and the exact waveform can't be predicted.

In syntactic pattern recognition [3],[4],[5], segments of the seismogram are characterized by a set of clusters. Their compilation forms sentences, which are compared by some grammar to predefined event types. Syntactic methods are based on the a priori assumption, that the human decision process is mainly a mental discurs [6].

Another approach starts from mental pictures as the key processing step between the impressions of sensing and the knowledge-based identification [7]. For earthquake detection these mental pictures may be any kind of simplified presentation of spectral energy versus time. So the pattern recognition detector consists of:

- ☐ The transformation of seismograms to mental pictures, for example by short time Fourier transforms and nonlinear scaling in the frequency and intensity axis (fig. 1).
- ☐ The knowledge base of typical earthquakes as a set of mental pictures with edge conversion (fig. 2).
- ☐ The pattern recognition process for mental pictures (fig. 3).
- ☐ The definition of some similarity measure for mental pictures, to speed up the consistency check of the knowledge base.

The pattern recognition process for mental pictures may be any algorithm, that is applicable to pattern recognition problems for 'optical' pictures like photo or tv.

The detector was first realized for processing local seismicity in the very noise environment of the Ruhrgebiet area in FRG. It's high detection probability and extremely low false alarm rate are superior to any other detector algorithm and comparable to the results of human analysis. Fig. 4 and 5 show seismograms and mental picture of some detected rockbursts of medium and very poor signal to noise ratio, fig. 6 is an example of rejected noise bursts.

[1] Freiberger, W. F., "An Approximate Method in Signal Detection", *Quarterly App. Math.*, vol. 20, pp. 373-378, 1963

[2] Blandford, Robert R., "Seismic Event Discrimination", *Bull. Seism. Soc. Am.*, vol. 72, pp. S69-S87, 1982

[3] Liu, Hsi-Ho and King-Sun Fu, "An Application of Syntactic Pattern Recognition to Seismic Discrimination", *IEEE Trans. Geosc. Rem. Sens.*, vol. GE-21, pp. 125-132, 1983

[4] Anderson, Kenneth R., "Syntactic Analysis of Seismic Waveforms using Augmented Transition Network Grammars", *Geoexploration*, vol. 20, pp. 161-182, 1982

[5] Anderson, Kenneth R. and James E. Gaby, "Dynamic Waveform Matching", Third International Symposium on Computer-aided Seismic Analysis and Discrimination, June 1983

[6] Haugeland, J., *Artificial Intelligence: The Very Idea*, MIT Press, Cambridge Mass., 1985

[7] Joswig, Manfred, "Methoden zur automatischen Erfassung und Auswertung von Erdbeben in seismischen Netzen und ihre Realisierung bei Aufbau des lokalen 'BOCHUM UNIVERSITY GERMANY'-Netzes", Dissertation, Ruhr-Universität Bochum, 1987

fig. 1

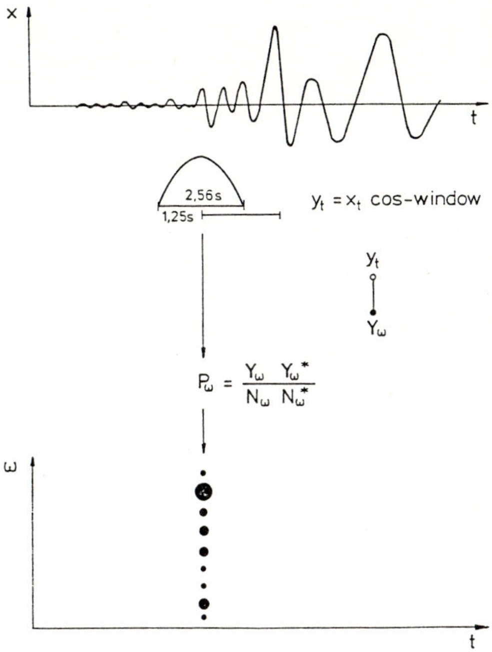

x
t
2,56s
1,25s
$y_t = x_t$ cos-window
y_t
Y_ω
$P_\omega = \frac{Y_\omega \; Y_\omega^*}{N_\omega \; N_\omega^*}$
ω
t

Calculation of Spectral Plots

fig. 2

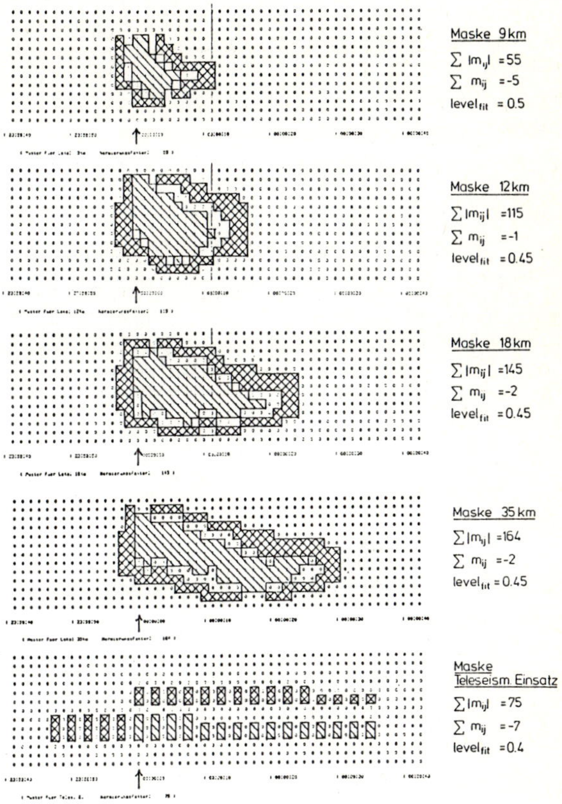

Maske 9km
$\sum |m_{ij}| = 55$
$\sum m_{ij} = -5$
$level_{fit} = 0.5$
Maske 12km
$\sum |m_{ij}| = 115$
$\sum m_{ij} = -1$
$level_{fit} = 0.45$
Maske 18km
$\sum |m_{ij}| = 145$
$\sum m_{ij} = -2$
$level_{fit} = 0.45$
Maske 35km
$\sum |m_{ij}| = 164$
$\sum m_{ij} = -2$
$level_{fit} = 0.45$
Maske
Teleseism. Einsatz
$\sum |m_{ij}| = 75$
$\sum m_{ij} = -7$
$level_{fit} = 0.4$

fig. 3

pattern 3
pattern 2
pattern 1
fit 3
fit 2
fit 1
event
+1
0
-1
t
transformed data

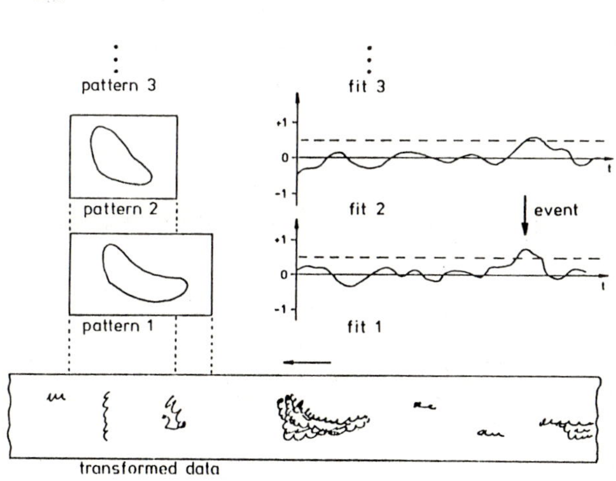

Pattern Recognition and Decision Process

fig. 4

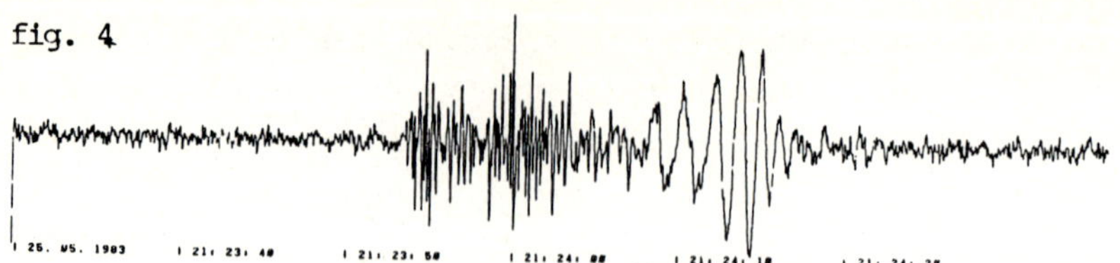

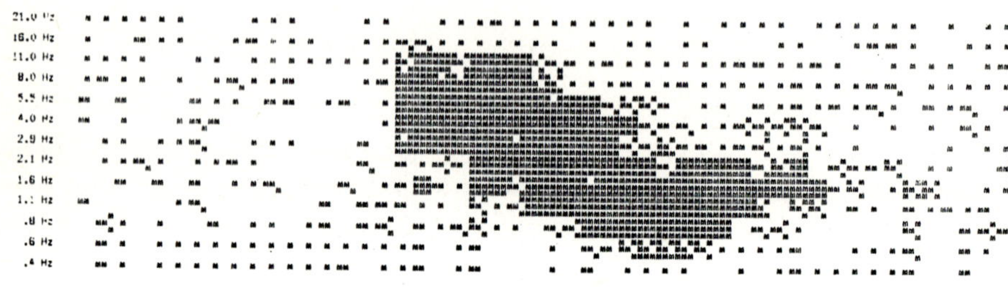

fig. 5

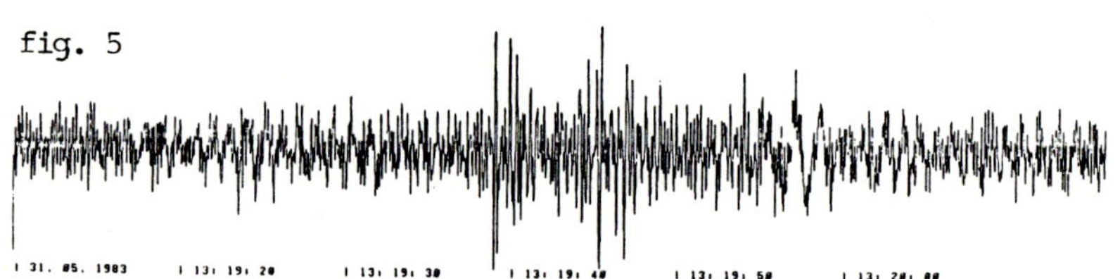

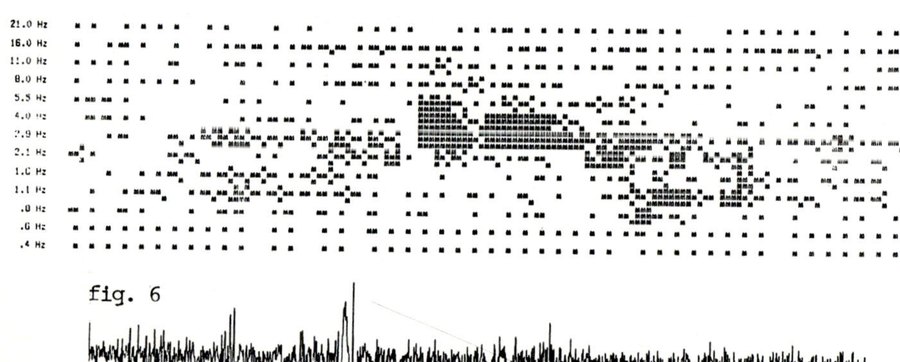

fig. 6

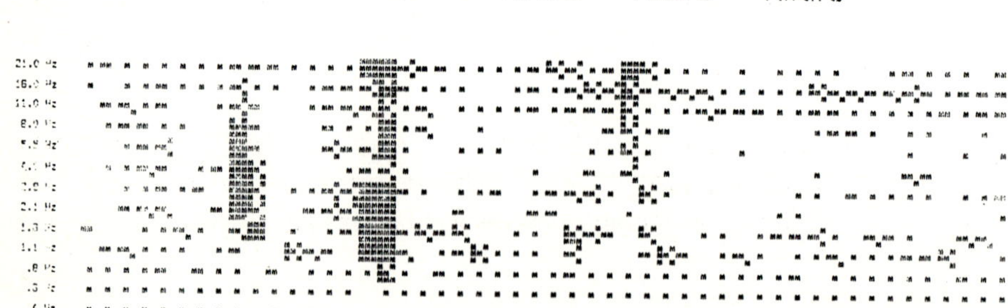

Navigation of an Airborne Vehicle by Model-Based Image Sequence Processing

R. Schmidt und H. Zinner
Messerschmitt-Bölkow-Blohm, Zentralbereich Technik
Postfach 801109, 8 München 80

Abstract

A method is proposed that enables navigation of an autonomous airborne vehicle by a passive imaging sensor, connected to the vehicle. Information necessary for the control of flight and for obstacle avoidance is derived from the changes of intensity in the image plane. The quality of the information can be improved by models incorporating knowledge of the environment and of the dynamics of the vehicle.

1. Introduction

Navigation by visual cues only, i. e. by processing the optical flow field across the retina, is an important principle used by animals. The optical flow provides propriospecific information to control egomotion as well as exterospecific information necessary for obstacle avoidance. In contrast, technical navigation systems rely mainly on inertial platforms. Because they yield only propriospecific information, additional sensors such as sonar or radar are necessary.

We describe a method that enables navigation of an autonomous airborne vehicle by a passive imaging sensor, connected to the vehicle. It provides propriospecific information to control the state of flight and exterospecific information for navigating through the environment. This task requires measuring and evaluating the changes of intensity in the image plane. Many methods have been proposed for this purpose, but none of them seem to be robust enough to be applied to the guidance and control of vehicles subject to general motion. We feel that the reason for this lack is that the tasks estimation, interpretation of optical flow, and control were always treated as separate entities.

The method described here aims to overcome these difficulties. Instead of first computing the optical flow and then interpreting the vector fields we relate the spatial and temporal changes of intensity in the image plane to the motion parameters of the vehicle directly. The algorithms are regularized and stabilized by two models: 1. the environment is described parametrically (exterospecific model); 2. the vehicle is portrayed by its dynamic equations (propriospecific model). The estimation of the motion parameters is embedded in a pixel-recursive algorithm that incorporates global knowledge derived from the models.

2. Optical flow and motion parameters

Let $\mathbf{V} = (u,v,w)^t$ be the translational velocity of the vehicle and let $\Omega = (p,q,r)^t$ be its angular velocity. Object points $\mathbf{X} = (X,Y,Z)^t$ are imaged by perspective projection: $(y,z)^t = f/X\,(Y,Z)^t$. The image plane is parallel to the Y-Z-plane at a focal distance f. Then the optical flow generated by the motion of the vehicle is given by (/1/)

$$\begin{aligned} \dot{y} &= \frac{1}{X}\,(-\,vf + uy) + pz + q\frac{yz}{f} - r\,(f + \frac{y^2}{f}) \\ \dot{z} &= \frac{1}{X}\,(-\,wf + uz) - py + q\,(f + \frac{z^2}{f}) - r\frac{yz}{f} \end{aligned} \tag{2.1}$$

X can be eliminated by introducing the quantities angle of attack $\alpha = w/u$ and angle of yaw $\beta = v/u$. Finally we obtain the equation (with $f = 1$)

$$\frac{z - \alpha}{y - \beta} = \frac{\dot{z} + py - q(1 + z^2) + ryz}{\dot{y} - pz - qyz + r(1 + y^2)} \,. \tag{2.2}$$

Instead of eliminating the dependence on the depth X, one can gain exterospecific information by modelling the environment parametrically, e. g. as a plane:

$$\frac{X}{A} + \frac{Y}{B} + \frac{Z}{C} = 1. \tag{2.3}$$

This assumption is adequate for vehicles flying at a certain height above ground. In this case the optical flow is given by

$$\begin{pmatrix} \dot{y} \\ \dot{z} \end{pmatrix} = (a + by + cz) \begin{pmatrix} y - \beta \\ z - \alpha \end{pmatrix} + p \begin{pmatrix} z \\ -y \end{pmatrix} + q \begin{pmatrix} yz \\ 1 + z^2 \end{pmatrix} - r \begin{pmatrix} 1 + y^2 \\ yz \end{pmatrix} , \tag{2.4}$$

where we used the abbreviations $a = u/A$, $b = u/B$, $c = u/C$. We relate these parameters to the change of intensity $I(y,z,t)$ in the image plane. If the change is due only to motion, then the gradient constraint equation is given by

$$I_y\,\dot{y} + I_z\,\dot{z} + I_t = 0. \tag{2.5}$$

Combine (2.4) and (2.5) and obtain after some simplifications /2/

$$\mathbf{I}^t\mathbf{p} + I_t = 0 \tag{2.6}$$

with the quantities

$$\mathbf{I} = (I_y,\ I_y y,\ I_y z,\ I_y y^2+I_z yz,\ I_y yz+I_z z^2, I_z,\ I_z y,\ I_z z)^t \text{ and}$$
$$\mathbf{p} = (-r-a\beta,\ a-b\beta,\ p-c\beta,\ -r+b,\ q+c,\ q-a\alpha,\ -p-b\alpha,\ a-c\alpha)^t. \tag{2.7}$$

The linear solutions **p** of (2.6) are called "pure" or "essential" parameters, /3/. They are nonlinear functions of the motion parameters. Two points should be emphasized:

1. The exterospecific model (2.3) provides information about the environment and permits navigation and obstacle avoidance because each of the quantities $1/a = A/u$, $1/b\ \beta = B/v$, $1/c\ \alpha = C/w$ is the so-called time to contact, i. e., the time that passes until the vehicle hits the surface. (Eq. (2.2) also contains some kind of depth information: By algebraic manipulations one can extract the quantity X/u. Clearly, A/u, B/v, C/u are more valuable for navigation.)

2. The assumption (2.3) regularizes the system and renders the solutions numerically stable. The kinematic quantities p, q, r, α, β and the orientation parameters a, b, c can be obtained analytically /4/. Our experiments have shown that the results obtained by solving (2.7) are much more accurate than by solving (2.2).

3. A pixel-recursive algorithm

The estimation of the motion parameters based on equ. (2.5) has several drawbacks: Firstly, the linear approximation may not be valid for real displacements; secondly, determining I_t may be subject to undersampling. These problems can be alleviated by a pixel-recursive algorithm developed recently /5/:

1. Estimate $(\dot{y}',\dot{z}')^t$ using (2.5) (or other information);
2. measure $DI_t = I(y+\dot{y}'t, z+\dot{z}'t, t+dt) - I(y,z,t)$;
3. evaluate $I_y\cdot(\dot{y}-\dot{y}') + I_z\cdot(\dot{z}-\dot{z}') + DI_t = 0$;
4. proceed to 1., using $\mathbf{p}(\dot{y},\dot{z})$ for a new estimate of the optical flow.

By this transformation the gradient algorithm (2.5) is applied to small displacements, reducing undersampling. A higher accuracy is achieved by less computational effort (I_y, I_z need be calculated only once). In contrast to other methods (e. g. /6/), our algorithm is based on an analytically correct description of <u>sensor</u> motion, incorporating global knowledge about the optical flow.

4. The dynamic model

The dynamic model relates the quantities obtained by image processing to the parameters that are necessary for the guidance and control of the vehicle. This is modelled as a six-degrees-of-freedom system with known aerodynamics, propulsion and mass characteristics. It can be described as a controlled stochastic system

$$\dot{\mathbf{x}} = \mathbf{F}(\mathbf{x},\mathbf{u},\kappa). \tag{4.1}$$

$\mathbf{x} = (p,q,r,u,v,w,z_g,\Phi,\Theta,\Psi,t)^t$ denotes the state vector with Euler angles Φ, Θ, Ψ ($\mathbf{F}$ is nonlinear in $\mathbf{x}$); $\mathbf{u} = (\eta,\zeta,\xi)^t$ is the vector of control commands (deflections of control surfaces) and $\kappa = (\Delta\alpha,\Delta\beta)^t$ is the stochastic input owing to wind. Details are described elsewhere /7/. The state vector is estimated in an Extended Kalman Filter by evaluating the information supplied by the imaging sensor. The measurement vector can be constructed using 1. the intensity changes in the image plane; 2. the pure parameters; 3. the motion parameters derived analytically from the pure parameters. We used the pure parameters:

$$\mathbf{h}(\mathbf{x}) = \mathbf{p}(\mathbf{x}) + \mathbf{v}, \tag{4.2}$$

because they are obtained by a linear least-squares-fit that estimates the covarince matrix of the error vector $\mathbf{v}$. By using (4.2) the full information provided by the imaging sensor is exploited for the guidance and control of the vehicle.

5. Experimental results

The method was tested in various simulations, where the vehicle had to navigate from one point to another (under the influence of wind and turbulence). Starting from real images, synthetic image sequences were generated using the simulated state of flight. As an example, fig. 1 shows two state parameters estimated by image processing (one iteration) and by Kalman filtering.

6. Conclusions

It has been shown, how an autonomous airborne vehicle can navigate by processing image sequences, provided by an imaging sensor connected to the vehicle. Computer simulations have proved that the information derived from the sensor is complete and sufficient to guide and control the vehicle.

Acknowledgements

The authors want to thank K.-H. Keil, U. Mackenroth and D. Wolf for valuable contributions. Part of this research was supported by the German Ministry of Defence.

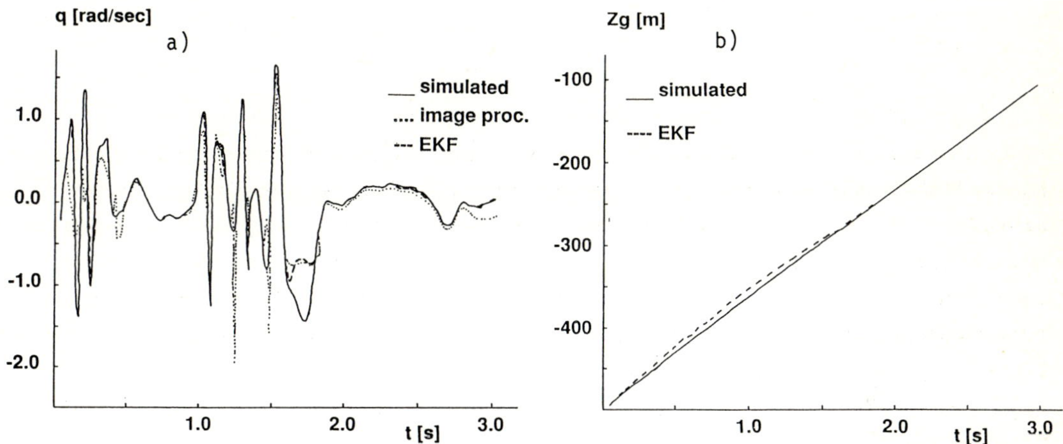

Fig. 1. Examples of state parameters. a) yaw rate; b) height above ground. V_o = 200 m/s, $\Theta_o = -45^o$.

References

/1/ Longuet-Higgins, H.C. and Prazdny, K.:The interpretation of moving retinal images. Proc. Roy. Soc. Lond. B**208**, 385-397 (1980).

/2/ Zinner, H.: Determining the kinematic parameters of a moving imaging sensor by processing spatial and temporal intensity changes. J. Opt. Soc. Am. A**3**, 1512-1517 (1986)

/3/ Tsai, R.Y. and Huang, T.S: Uniqueness and estimation of three-dimensional motion parameters of rigid objects with curved surfaces. IEEE **PAMI-6**, 13-27 (1984).

/4/ Longuet-Higgins, H.C.: The visual ambiguity of a moving plane. Proc. Roy. Soc. Lond. B**223**, 165-175 (1984).

/5/ Diehl, H. and Schmidt, R.: Ein pixel-rekursives Verfahren zur Bestimmung der Eigenbewegung von abbildenden Sensoren. In: Hartmann, G. (ed.): Mustererkennung 1986, p. 288, Springer, Berlin (1986).

/6/ Robbins, J.D. and Netravali, A.N.: Recursive motion estimation: A review. In: Huang, T.S. (ed.): Image Sequence Processing and Dynamic Scene Analysis, pp. 75-103, Springer, Berlin (1983)

/7/ Keil, K.-H. Mackenroth, U., Schmidt, R., Zinner,H.: Navigation von Fluggeräten mit starren abbildenden Sensoren. Proc. Jahrestagung der Deutschen Gesellschaft für Luft- und Raumfahrt, Berlin (1987) (in press)

IMAGES AND IMAGE PROCESSING FOR X-RAY SMALL ANGLE SCATTERING

H.W. Halling, H.G. Haubold
Zentrallabor für Elektronik, Institut für Festkörperforschung
Kernforschungsanlage Jülich GmbH, Postfach 1913, D-5170 Jülich

The advent of new powerful x-ray synchrotron radiation sources has enhanced the quality and speed of measurements for such methods as the x-ray small angle scattering and make new applications feasible. Two dimensional position sensitive detectors acquire scattering images with high resolution and large dynamic range. The high speed of image acquisition requires fast automatic preprocessing for corrections and calibration whereas the complexity of the monochromator, sample- and detector arrangements demand interactive data processing and finally archiving of measured- and processed data.

Introduction

For a variety of problem areas in biology, metallurgy and chemistry where shape and distribution of particles with sizes from 5 Å to 1000 Å are studied, x-ray small angle scattering (SAXS) is successfully applied [1]. X-ray energies in the range between 2-30 KeV are used for samples with thicknesses of .1 μm to 100 μm.

Typical application areas are studies concerning defects in crystals, concentration fluctuations in metallic alloys, microcrack and gas bubbles in sputtered thin films, colloids or polymers, dispersions and emulsions.

Fig. 1 shows the principle of small angle scattering measurements. A sample with a typical size of 1 mm^2 and a thickness of .1 μm to 100 μm is hit by a monochromatic x-ray beam. Scattering inhomogeneities with sizes of 5 to 1000 Å (1 Å = 10^{-8} cm) cause scattering of a certain amount of x-ray quants onto the detector plane. With scattering vectors /Q/ from 10^{-3} to 1 $Å^{-1}$ a two dimensional scattering image I(Q) builds up. Because no scattering amplitudes but only an intensity distribution is measured, the phase information is lost and a Fourier transformation cannot be applied to regain the precise structure of the sample but rather symmetries, size-and correlation information: Assuming a model of discrete particles, the intensity distribution can be written as a function of several terms [2]. The main terms are:

- the contrast, depending on the scattering density difference ϱf between particle p and matrix m (ϱ = number density of atoms with atomic scattering amplitude f),
- S(Q), is determined by the size and shape of the scattering regions (balls, discs, needles)
- Ψ(Q), describes the distribution of the particles in the matrix (interparticle correlation function).

The capability of the synchrotron radiation source to tune the x-ray energy continuously allowes to perform measurements on both sides of absorption edges of specific kind of atoms (use of anomalous scattering for contrast variation [3]). This can either be used to vary contrasts looking for special particles but also for finding out about the concentration distribution of certain atoms within the scattering particle. This contrast variation method is especially important for revealing the distribution of different components of multicomponent samples.

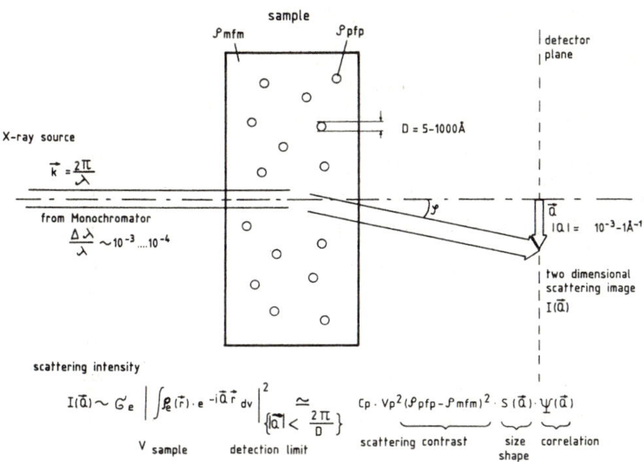

Fig. 1 SAS MEASUREMENT PRINCIPLE

Acquisition and preprocessing of measured data

Because of the high precision desired, real measurements are quite complicated. Instead of one measurement, a set of measurements is required. A typical set consists of two dimensional scattering images from the sample, a dummy sample without scattering regions (matrix only), a scattering image without sample (slit scattering background only), detector dark current and sensitivity. In addition to these two dimensional datasets, a one dimensional data set is acquired with data like the linear absorption coefficient as a function of x-ray energy or absolute calibration parameters. The two dimensional image set is used for performing automatic corrections for each x-ray energy. Typical correction are subtraction of dark current and slit scattering background. Other automatic calculations concern calculation of the Q vector for each detector resolution element. For the corrected data the statistical errors have to be calculated as well.

For a complete contrast variation measurement sequence an energy range has to be scanned with a resolution of about one eV. In order to achieve the required accuracy and to suppress unwanted harmonic x-ray energies a feedback loop in the monochromator set up is required and careful energy calibration must take place. Besides the necessity of an automatic data acquisition system a highly interactive behaviour of a small angle scattering system is essential. At the beginning of each measurement a complex interactive adjustment and calibration procedure of the monochromator and the detector system is required.

While acquiring data a monitor system shows the building up of the images. This is a first indication for the correct data acquisition procedure. Such a system must work with a short response time. After the data have been acquired and automatic corrections as mentioned above have been done, interactive actions are performed. One of them is to find the primary beam position in the image or the removal of inaccurate counting regions at the detector's edges.

The experienced user then must have the choice of proper averaging over selectable regions of interest. For certain problems linear cuts through images or radial averaging is required. Again, the width of these areas used for averaging and the directions of the cuts within the image etc. have to be chosen interactively. As an example Fig. 2 shows the determination of the radius of the scattering particle, which can be derived from the slope of the radial averaged intensity in a logarithmic Ivs-Q^2 plot in an interactive way.

With two dimensional data sets the user is able to regard the data with pseudo colours in two dimensions or in three dimensions from different views. In addition, certain aids like shadows or isoclines for the evaluation of certain structures are provided.

A set of procedures for data shaping are available like filtering (smoothing or edge detection) or inversion, scaling, addition and substraction. For most of these, the user has the choice between several alternatives and it is up to him to choose the most appropriate and immediately check the results via three dimensional views (Fig. 3).

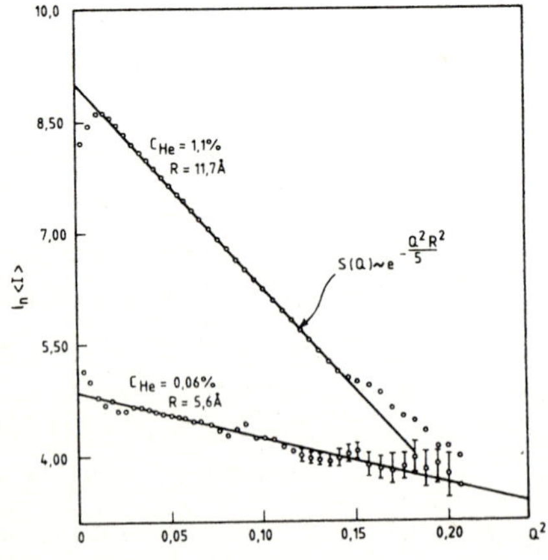

Fig. 2 INTERACTIVE PROCESSING - RADIAL AVERAGING

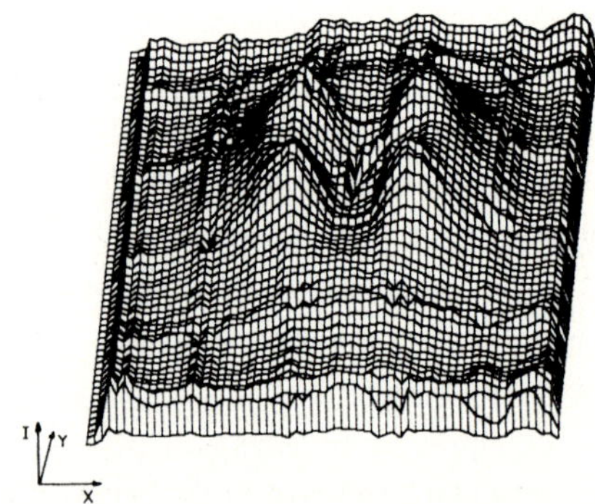

Fig 3 SAS IMAGE AFTER CORRECTION AND SMOOTHING

In order to solve more ambitions physical problems like to find out the concentration of certain components inside scattering regions, more sophisticated procedures must be applied. An example is the contrast variation method where sequences of images have to be processed.

Fig. 4 shows the structure of a computer based SAS system which is capable of processing 512 x 512 pixel images, controlling the adjustment and calibration of the monochromator and the detector system. Interactive and automatic data processing, graphic representation in two or three dimensions are provided. Finally archiving of the acquired and processed data sets and communication with larger processing systems is supported.

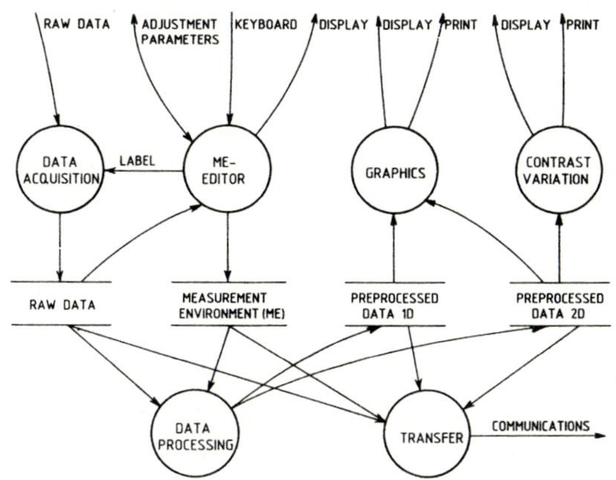

Fig. 4 SAS SOFTWARE SYSTEM STRUCTURE

An implemented SAXS system for measurements at the synchrotron source at Hasylab (DESY, Hamburg) is shown in Fig. 5. It consists of the monochromator control responsible for DC- and steppermotors, a piezo positioned feedback loop, acquisition of x-ray spectra etc.

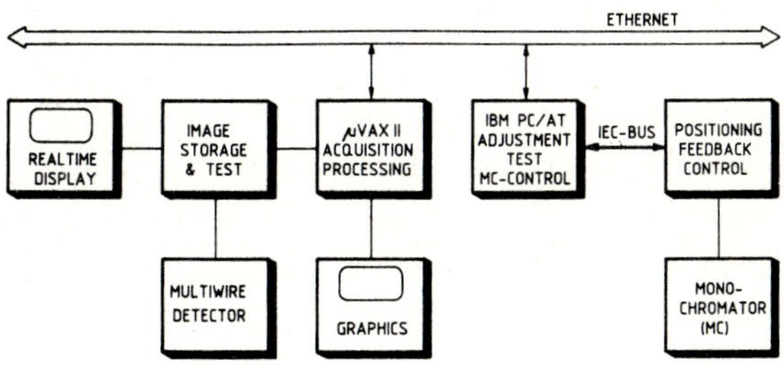

Fig. 5 VAX BASED SAS SYSTEM

The second part is the detector system which is based on a two dimensional multiwire gas proportional counter with useful resolution elements (200 x 200), a CAMAC data acquisition and monitoring system and a fast link to the µVAX II based central data acquisition and processing system. While the control software is mainly written in Turbo Pascal the software of the mainframe is written in FORTRAN comprising graphics- and communication packages. All of the software is menudriven and does extensivly use graphics.

Future developments will concentrate on the detector system. For the high intensities which are available at synchrotron sources, counting multiwire detectors are not an optimal choice. In the future these will be replaced by integrating detectors like CCD devices or special storage tubes similar to vidicon principles and standard video techniques which facilitate interfacing.

The processing power of the front end computers must be raised in order to cope with the processing speed required for image processing. Finally larger storage devices based on optical discs will be used.

Conclusion

X-ray small angle scattering is becoming a routine method for solving a variety of problems in biology, metallurgy and chemistry. Computer based systems are the tools which enable scientists to solve problems interactively. The proper processing and display of images plays an important role within such systems.

Literature references

1 Synchrotronstrahlung in der Festkörperforschung
Vorlesungsmanuskripte 18. IFF Ferienkurs 16.-27. März 1987

2 The Optical Principles of the Diffraction of X-rays
R.W. James, London, G. Bell & Sons LTD 1958

3 A Comparison of the Merits of Isotopic Substitution in Neutron Small-Angle Scattering and Anomalous X-ray Scattering for the Evaluation of Partial Structure Functions in a Ternary Alloy
J.P. Simon, J.Appl. Cryst. (1985) 18, 230-236

A New 'Switched Kalman Filter' For 3D-Contourmeasuring Problems With A Laser Diode Range Finder

Otmar Loffeld
Institut für Nachrichtenverarbeitung, Universität-GH-Siegen
Hölderlinstraße 3, D-5900 Siegen

Abstract

The paper presents a new approach to Kalman filtering of measurement data which result from three dimensional objects, that are scanned with a Laser Diode Range Finder (LDRF). In order to cope with the measurement 'jumps' resulting from the edges of a scanned object, a Kalman filter is combined with a recursive maximum likelihood - 'jump'-detecting and -estimating algorithm. Proper incorporation of this additional information in the Kalman filter yields the 'Switched Kalman filter' with superior properties concerning the tracking of edges as well as stationary precision. Some preliminary test results are given.

I. Introduction

During the past several years much effort has been spent in the development of three dimensional image processing techniques based on either stereo camera or on greytone images. Yet the problems of shading and varying illumination of the objects, that are to be measured, have not been solved completely up to now, not to speak of the computer load, that is caused by those techniques. Recent approaches to these problems utilize direct distance measuring techniques by means of either microwaves, ultrasonic waves or laser light, which are self-illuminating and therefore independent of secondary illumination of industrial scenes. The latter technique seems to yield superior results concerning the diameter of the measuring beam, which is very crucial for the angular resolution /2,3/. The 3-dimensional image can then be gathered directly (distance versus xy-plane) by distance measuring along scanning lines. These lines can be organized in form of classical rows or more sophisticatedly along two dimensional contours in the xy-plane. In this special case the processing of the measurements only along that contour would not discard any information between neighbouring 'rows' as there is no information between two neighbouring points on the two sides of a contour edge. Earlier results concerning two-dimensional contour tracking have been reported in /10/ and more recent investigations

/12/ show very promisng results. These directly measuring techniques, however, have in common that they deliver more or less erroneous measurements, that have to be processed before utilizing them any further. The means at hand for this task is to apply Kalman filters, which are optimal for a wide class of applications. Even if based on relatively simple models of the 'truth', they operate quite satisfactorily, despite of being somewhat suboptimal in this case. Here we deal with Kalman filtering of measurement data resulting from objects, which are scanned line by line and therefore can contain 'jumps' due to the edges.

II. Linear Stochastic Models with Unknown Jumps

We will now be seeking for an appropriate model of a technical height profile along a scanning line. As there is no à-priori information about jumps available, we model a jump in a linear stochastic system by an additional deterministic control input $\underline{u}(k)$, in this work referred to as an 'update'-process. For the sake of simplicity we will consider 'time'-invariant matrices throughout this work, but the derivation as well as the results are easily extended to 'time'-varying systems. Let the model be described by:

$$\underline{x}(k+1) = A\,\underline{x}(k) + \underline{w}(k) + \underline{u}(k) \tag{1}$$

$$\underline{y}(k) = C\,\underline{x}(k) + \underline{v}(k) \tag{2}$$

where:

$\underline{x}(k)$ is the [nxl] vector of states at point kS

S is the distance between two sampling points along the sampling axis, and will be referred to as the 'time' for convenience.

$\underline{w}(k)$ is the driving noise vector which consists of white Gaussian noise, which is uncorrelated with all other noises in the system

$\underline{y}(k)$ is the [mxl] vector of observations of the state

$\underline{v}(k)$ is the [mxl] measurement noise vector of white Gaussian noise

Let the statistical parameters of the noises given by:

$$E\{\underline{w}(k)\}=\underline{0}\ ;\ E\{\underline{v}(k)\} = \underline{0} \tag{3}$$

$$E\{(\underline{w}(k) - E\{\underline{w}(k)\})\ (\underline{w}(l) - E\{\underline{w}(l)\})^T\} = Q(k)\ \delta_{k,l} \tag{4}$$

$$E\{(\underline{v}(k) - E\{\underline{v}(k)\})\ (\underline{v}(l) - E\{\underline{v}(l)\})^T\} = R(k)\ \delta_{k,l} \tag{5}$$

$$E\{(\underline{w}(k) - E\{\underline{w}(k)\})\ (\underline{v}(l) - E\{\underline{v}(l)\})^T\} = 0 \text{ for arbitrary } k,l \tag{6}$$

Let us assume that the edges are rather infrequent, that means there is only one 'jump' for at least N sampling points. Then $\underline{u}(k)$ can be modeled in the following way:

$$\underline{u}(k) = \underline{u}_\theta\ \delta_{k,\theta} \tag{7}$$

that means a discrete jump of height $\underline{u}_\theta$ occurs from point θ to $\theta+1$ in the sequence of states. The Kalman filter based on exact knowledge of the model would be:

$$\hat{\underline{x}}^-(k+1) = A\,\hat{\underline{x}}^+(k) + \underline{u}(k) \tag{8}$$
$$\underline{r}\,(k+1) = \underline{y}(k+1) - C\,\hat{\underline{x}}^-(k+1) \tag{9}$$
$$\hat{\underline{x}}^+(k+1) = \hat{\underline{x}}^-(k+1) + K(k+1)\underline{r}(k+1) \tag{10}$$
$$P^-(k+1) = A\,P^+(k)\,A^T + Q(k) \tag{11}$$
$$K(k+1) = P^-(k+1)\,C^T\,[C\,P^-(k+1)\,C^T + R(k+1)]^{-1} \tag{12}$$
$$P^+(k+1) = P^-(k+1) - K(k+1)\,C\,P^-(k+1) \tag{13}$$

For the filter application regarded here, $\underline{u}(k)$ must be considered as a deterministic, yet unknown parameter vector. Now, since neither the height of the jump nor the sampling point where it occurs are known in advance, the Kalman filter can use only an estimate $\hat{\underline{u}}(k)$ of the input $\underline{u}(k)$ at an estimated 'time' $\hat{\theta}$, thus in cases of 'jumps', equations (8) and (11) will have to be modified to:

$$\hat{\underline{x}}^-(k+1) = A\,\hat{\underline{x}}^+(k) + \hat{\underline{u}}(k) \tag{8a}$$

$$P^-(k+1) = F(P^+(k), P_{\hat{\underline{u}e}}(k), P_{\hat{\underline{x}},\hat{\underline{u}}}(k), Q(k)) \tag{11a}$$

where (11a) indicates that the new prediction error covariance, based on the estimated update $\hat{\underline{u}}(k)$, will be a function of the filter error covariance $P^+(k)$, the error covariance of the 'jump'estimation, denoted by $P_{\hat{\underline{u}e}}(k)$, the crosscovariance $P_{\hat{\underline{x}},\hat{\underline{u}}}(k)$ between state estimation and 'jump'estimation error, and the driving noise covariance. The incorporation of the jump-information switches the Kalman filter, so it is termed a 'Switched Kalman filter' /5,6/.

III. Combined edge detection and jump estimation

To derive a combined edge detection and estimation algorithm, we compute the state $\underline{x}(k+i)$ in relation to the state $\underline{x}(k)$ and a deterministic input $\underline{u}_\theta$ at time θ, where: $k \leq \theta < k+i$ and $1 \leq i \leq N$. (Note that N denotes the length of an observation window). Straight forward calculations yield:

$$\underline{x}(k+i) = A^i\underline{x}(k) + \sum_{j=0}^{i-1} A^{(i-1-j)}\underline{w}(k+j) + A^{(k+i-\theta-1)}\underline{u}_\theta \quad \text{for } i \geq 1 \text{ and } k+i \geq \theta+1 \tag{12}$$

If we introduce the augmented observation vector $\underline{y}_a(k)$, where:

$$\underline{y}_a(k) = [\underline{y}(k+1)^T|\underline{y}(k+2)^T| \;\dots\; |\underline{y}(k+N)^T]^T , \tag{13}$$

the augmented driving noise vector $\underline{w}_a(k)$:

$$\underline{w}_a(k) = [\underline{w}(k)^T|\underline{w}(k+1)^T| \;\dots\; |\underline{w}(k+N-1)^T]^T \tag{14}$$

with covariance $Q_a(k)$, and the augmented measurement noise vector $\underline{v}_a(k)$:

$$\underline{v}_a(k) = [\underline{v}(k+1)^T|\underline{v}(k+2)^T| \;\dots\; |\underline{v}(k+N)^T]^T \tag{15}$$

with covariance $R_a(k)$, we get by straight forward calculations from (12,2,13,14,15):

$$\underline{y}_a(k) = C_a\,\underline{x}(k) + G_a\,\underline{w}_a(k) + B_{a\theta}\underline{u}_\theta + \underline{v}_a(k) \tag{16}$$

where the partitioned matrices C_a, G_a, $B_{a\theta}$ are given by:

$$C_a = \begin{bmatrix} C A \\ C A^2 \\ C A^3 \\ \cdot \\ \cdot \\ C A^N \end{bmatrix} ; \qquad G_a = \begin{bmatrix} C & 0 & 0 & 0 & \cdot & \cdot & 0 \\ CA & C & 0 & 0 & \cdot & \cdot & 0 \\ CA^2 & CA & C & 0 & \cdot & \cdot & 0 \\ \cdot & \cdot & \cdot & \cdot & \cdot & \cdot & \cdot \\ \cdot & \cdot & \cdot & \cdot & \cdot & \cdot & \cdot \\ \cdot & \cdot & \cdot & \cdot & & C & 0 \\ CA^{N-1} & CA^{N-2} & CA^{N-3} & CA^{N-4} & & CA & C \end{bmatrix};$$

(17) (18)

The augmented observation matrix $B_{a\theta}$ for the jump $\underline{u}_\theta$ is clearly a function of the offset $\nu' = \theta - k$, where the jump occurs:

$$B_a = \begin{bmatrix} 0 \\ \cdot \\ 0 \\ C \\ CA \\ \cdot \\ CA^{(N-1-\nu')} \end{bmatrix} \begin{matrix} \text{<-- point } k+1 \\ \\ \\ \text{<-- At this point } \theta = k+\nu'+1 \text{ can } \underline{u}_\theta \text{ be observed for} \\ \text{the first time !} \\ \\ \text{<-- point } k+N \end{matrix} \tag{19}$$

$\underline{y}_a(k)$ may be regarded as a 'noisy' observation of the jump at point θ, so we introduce the conditional density of the augmented observation vector $\underline{y}_a(k)$ based on the complete measurement history $\underline{Y}(k) = [\underline{y}(0)^T|\underline{y}(1)^T| \ldots |\underline{y}(k)^T]^T$ up to point k and based on the fact that a jump $\underline{u}(\theta) = \underline{u}_\theta$ occurs at time $\theta = \tilde{\theta}$, which can be shown to be Gaussian. We have:

$$f_{\underline{y}_a(k)/\underline{Y}(k),\underline{u}(\theta),\theta}(\xi_{ak}/\underline{Y}_k,\tilde{\underline{u}}_\theta,\tilde{\theta}) = [(2\pi)^{Nm/2} \det \{P_{y_a y_a}(k)\}^{1/2}]^{-1} \exp [\%]$$

where: $$\% = -\frac{1}{2} [\xi_{ak} - \hat{\underline{y}}_a(k)]^T P_{y_a y_a}(k)^{-1} [\xi_{ak} - \hat{\underline{y}}_a(k)] \tag{20}$$

with conditional mean:

$$\hat{\underline{y}}_a(k) = E\{\underline{y}_a(k)/\underline{Y}(k)=\underline{Y}_k,\ \underline{u}(\theta)=\tilde{\underline{u}}_\theta,\ \theta=\tilde{\theta}\} = C_a \hat{\underline{x}}^+(k) + B_{a\tilde{\theta}}\, \tilde{\underline{u}}_\theta \tag{21}$$

and conditional covariance:

$$\begin{aligned} P_{y_a y_a}(k) &= E\{(\underline{y}_a(k) - \hat{\underline{y}}_a(k))\ (\underline{y}_a(k) - \hat{\underline{y}}_a(k))^T/\underline{Y}(k)=\underline{Y}_k,\ \underline{u}(\theta)=\tilde{\underline{u}}_\theta,\ \theta=\tilde{\theta}\} \\ &= C_a P^+(k) C_a^T + G_a Q_a(k) G_a^T + R_a(k) \end{aligned} \tag{22}$$

In order to introduce the concepts of statistical hypothesis testing, we formulate the two hypotheses to be tested against each other:

H_0: No jump in the window [k,k+N], that means: $\tilde{\theta} > k+N$

H_1: Jump in the observation window, that means: $k \leq \tilde{\theta} < k+N$

These two hypotheses correspond to different conditional mean values of observations:

$$H_0: \quad \hat{\underline{y}}_a(k) = C_a \hat{\underline{x}}^+(k) \tag{23}$$

$$H_1: \quad \hat{\underline{y}}_a(k) = C_a \hat{\underline{x}}^+(k) + B_{a\tilde{\theta}}\, \tilde{\underline{u}}_\theta \tag{24}$$

Now the maximum likelihood estimates $\hat{\underline{u}}(k)$, $\hat{\theta}(k)$ of the unknown values of $\underline{u}(\theta)$, $\theta(k)$ are, assuming that H_1 is true, given by:

$$\hat{\underline{u}}(k),\ \hat{\theta}(k) = \arg\max_{\tilde{u},\ \tilde{\theta}}\ f_{\underline{Y}_a(k)/\underline{Y}(k),\underline{u},\theta}\,(\xi_{ak}/\underline{Y}_k,\tilde{\underline{u}},\tilde{\theta})/H_1 \tag{25}$$

By maximizing equation (25) we get for $\hat{\underline{u}}(k)$, given that we already know $\hat{\theta}(k)$:

$$\hat{\underline{u}}(k)_{|\hat{\theta}} = [\ B_{a_{\hat{\theta}}}^{T}\ P_{Y_aY_a}(k)^{-1}\ B_{a_{\hat{\theta}}}\ ]^{-1}\ B_{a_{\hat{\theta}}}^{T}\ P_{Y_aY_a}(k)^{-1}\ \underline{r}_a(k)$$

where:

$$\underline{r}_a(k) = \xi_{ak} - C_a\ \hat{\underline{x}}^{+}(k) \tag{26}$$

ξ_{ak} are the realisations of the augmented observation vector. The term $\underline{r}_a(k)$ can be interpreted as a 'homogenous' residual vector, as it consists of the difference between the actual measurements $\underline{y}(k+1),\underline{y}(k),\ \ldots\underline{y}(k+N)$ and their 'predictions', computed by the propagation of the actual state estimate $\hat{\underline{x}}^{+}(k)$ into the future. Furthermore it should be noticed, that $\hat{\underline{u}}(k)$ is an explicit function of $\hat{\theta}(k)$. In order to decide, whether a jump occurs in the window [k,k+N], we consider the generalized likelihood ratio (GLR) /7,8/, which is defined by:

$$\Lambda(k) = \frac{f_{\underline{Y}_a(k)/\underline{Y}(k),\underline{u}(k),\theta}\,(\xi_{ak}/\underline{Y}_k,\hat{\underline{u}},\hat{\theta})/H_1}{f_{\underline{Y}_a(k)/\underline{Y}(k)}\,(\xi_{ak}/\underline{Y}_k)/H_0} \tag{27}$$

To select either H_0 or H_1 we use the decision rule:

$$\Lambda(k) \underset{H_0}{\overset{H_1}{\gtrless}} \eta\ . \tag{28}$$

Taking the natural logarithm of (27), as both densities are Gaussian, we get by straight forward calculations:

$$l(k,\hat{\theta}(k),\hat{\underline{u}}(k)_{|\hat{\theta}}) = 2\ \ln(\Lambda(k)) = \underline{r}_a(k)^{T}B_{a\hat{\theta}}\ \hat{\underline{u}}(k)_{|\ \hat{\theta}} \tag{28}$$

which has to be maximized by $\hat{\underline{u}}(k)$ and $\hat{\theta}(k)$. Thus the MLE $\hat{\theta}(k)$ is the value of $k \leq \hat{\theta}(k) < k+N$ that maximizes :

$$l(k,\ \hat{\theta},\hat{\underline{u}}_{|\hat{\theta}}) = \max_{\tilde{\theta}}\ \underline{r}_a(k)\ B_{a\tilde{\theta}}\ \hat{\underline{u}}_{|\ \tilde{\theta}} \tag{29}$$

Then our decision rule is:

$$l(k,\ \hat{\theta},\hat{\underline{u}}_{|\hat{\theta}}) \underset{H_0}{\overset{H_1}{\gtrless}} \varepsilon = 2\ \ln\ \eta \tag{30}$$

where ε is a design parameter which is chosen to provide an acceptable tradeoff between false alarms and miss alarms. Some guidelines for the choice of ε are given in /7,8/. The principle of cooperation between Switched Kalman filter and jump estimator is depicted in figure 1.

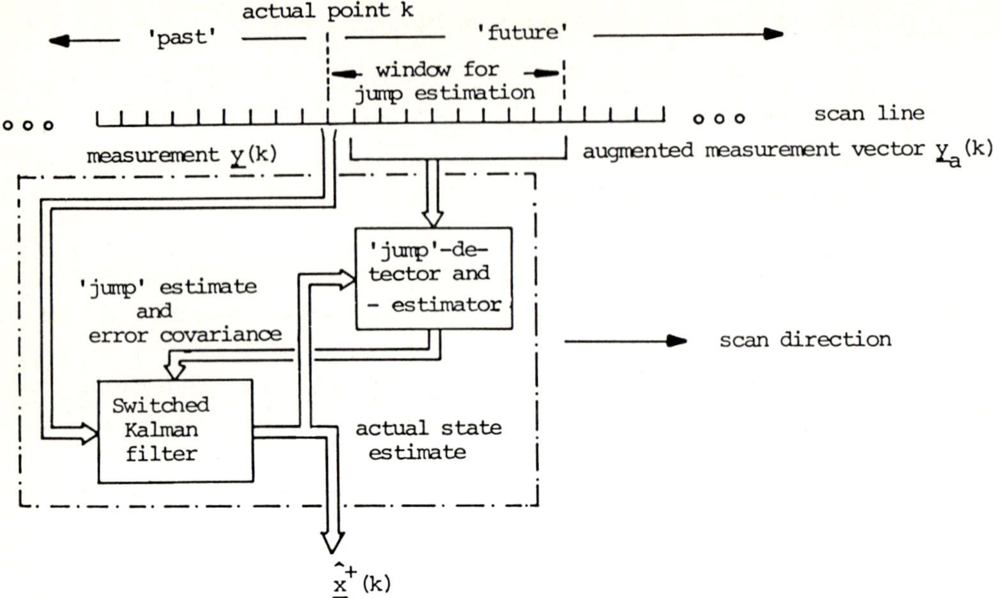

Figure 1: Cooperation of Switched Kalman filter and jump estimator

IV. Recursive Computation of the Maximum Likelihood Estimates

Equation (26) can be considered as a weighted least squares algorithm for the jump estimation, where $P_{\underline{y}_a\underline{y}_a}^{-1}$ is the weighting matrix of the squared error. Solely the matrix inversions needed to evaluate equation (26) are tedious, so we intend to derive a recursive algorithm, that is equivalent to (26), but which processes the measurement vectors $\underline{y}(k+i)$ sequentially. It can be shown, that any weighted least squares algorithm can be brought into a recursive form without loss of performance, if the weighting matrix is diagonal (blockdiagonal). Now as $P_{\underline{y}_a\underline{y}_a}(k)$ is not diagonal in this case, a diagonalisation procedure should be applied. First we note that equation (26) is the weighted least squares solution of the following observing problem:

$$\underline{r}_a(k) = B_{a\theta}\tilde{\underline{u}}(k) + \underline{n}_a(k) \tag{31}$$

where: $E\{\underline{n}_a(k)\} = \underline{0}$; $E\{(\underline{n}_a(k)-E\{\underline{n}_a(k)\})(\underline{n}_a(k)-E\{\underline{n}_a(k)\})^T\} = P_{\underline{y}_a\underline{y}_a}(k)$

We might now apply any invertible transformation to this equation, without changing the solution, so consider:

$$\underline{r}_a^*(k) = T\,\underline{r}_a(k) = T\,(B_{a\theta}\tilde{\underline{u}}(k) + \underline{n}_a(k)) = B_{a\theta}^{\sim *}\underline{u}(k) + \underline{n}_a^*(k) \tag{32}$$

where: $E\{\underline{n}_a^*(k)\} = \underline{0}$;

$$P_{Y_aY_a}(k) = T^{-1}E\{(\underline{n}_a^*(k)-E\{\underline{n}_a^*(k)\})(\underline{n}_a^*(k)-E\{\underline{n}_a^*(k)\})^T\}\ (T^{-1})^T \tag{33}$$

If we let: $\text{cov}\{\underline{r}_a(k),\underline{r}_a(k)\} = E\{(\underline{n}_a^*(k)-E\{\underline{n}_a^*(k)\})(\underline{n}_a^*(k)-E\{\underline{n}_a^*(k)\})^T\} = I$, which is diagonal, we get for the transforming matrix T^{-1}:

$$T^{-1}(T^{-1})^T = S\ S^T = P_{Y_aY_a}(k) \tag{34}$$

One particular useful choice for S is the Cholesky lower triangular composition of $P_{Y_aY_a}(k)$, that is:

$$S = T^{-1} = \sqrt[c]{P_{Y_aY_a}(k)} \tag{34}$$

which can be computed recursively from $P_{Y_aY_a}(k)$. Then for the parameters and matrices of equation (32) we have:

$$E\{(\underline{n}_a^*(k)-E\{\underline{n}_a^*(k)\})(\underline{n}_a^*(k)-E\{\underline{n}_a^*(k)\})^T\} = I \tag{35}$$

$$B_{a\tilde{\theta}} = S\ B_{a\tilde{\theta}}^* \tag{36}$$

$$\underline{r}_a(k) = S\ \underline{r}_a^*(k) \tag{37}$$

It should be noted, that it is not necessary to invert S to compute $\underline{r}_a^*(k)$ and $B_{a\tilde{\theta}}^*$ these values may be computed recursively, beginning with the first component, due to the lower triangular structure of S.

The transformed augmented observation matrix $B_{a\tilde{\theta}}^*$ may be written in form of a partitioned matrix, analogously to equation (18):

$$B_a^{\sim *} = \begin{bmatrix} 0 \\ \cdot \\ 0 \\ B_{\tilde{\theta},\tilde{\theta}+1} \\ B_{\tilde{\theta},\tilde{\theta}+2} \\ \vdots \\ B_{\tilde{\theta},N} \end{bmatrix} \begin{matrix} \text{<-- point } k+1 \\ \\ \\ \text{<-- At this point } \tilde{\theta}=k+\tilde{\nu}+1\text{, the observation begins!} \\ \\ \\ \text{<-- point } k+N \end{matrix} \tag{38}$$

The augmented transformed observation vector may be partitioned in the same way:

$$\underline{r}_a^*(k) = [\underline{r}_{a01}^{*\,T} \mid \underline{r}_{a02}^{*\,T} \mid \ldots\ \underline{r}_{a0N}^{*\,T}]^T \tag{39}$$

With the transformed observation vector $\underline{r}_a^*(k)$, which is disturbed by uncorrelated measurement noise of unit covariance, it is now possible to formulate the recursive ML-estimating algorithm for $\hat{\underline{u}}(k)_{|\tilde{\theta}}$, given that $\theta = \tilde{\theta}$. This condition is omitted for convenience in the following. If the index "i" denotes the number of the iteration cycle, we can write:

$$\hat{\underline{u}}_{i+1}(k) = \hat{\underline{u}}_i(k) + K_{i+1}^u\ [\ \underline{r}_{a0i+1}^* - B_{\tilde{\theta},i+1}\ \hat{\underline{u}}_i(k)] \tag{40}$$

$$K_{i+1}^u = P_i^u\ B_{\tilde{\theta},i+1}^T\ [B_{\tilde{\theta},i+1}\ P_i^u\ B_{\tilde{\theta},i+1} + I\ ]^{-1} \tag{41}$$

$$P_{i+1}^u = P_i^u - K_{i+1}\ B_{\tilde{\theta},i+1}\ P_i^u \tag{42}$$

The algorithm is started with initial conditions $\hat{\underline{u}}_0 = \underline{0}$ and $(P_0^u)^{-1} = 0$, since there is no a-priori information about a jump available. Thus an 'inverse covariance' formulation of (40)-(42) could be applied advantageously, allowing also a viable start up procedure for the edge detecting and estimating filter. A comment should be made, concerning the minimum iteration count, that is needed to attain a 'full rank' estimate of $\underline{u}(k)$. For each problem, there is a minimum number of measurements, that have to be processed in order to ensure complete observability of the jump. The number of iterations must be chosen large enough to ensure this complete observability. A further remark should be made on the operation of the filter algorithm above: The matrix partitions $B_{\tilde{\theta},i+1}$ are zero, unless the iteration count i is not larger than the offset $\tilde{\nu}' = \tilde{\theta} - k$. This causes the filter algorithm to start with the first measurement vector at point $\tilde{\theta}$ +1 and then continue up to point k+N in order to compute the estimate $\hat{\underline{u}}(k)_{|\tilde{\theta}}$, based on the assumption that the jump occurs at point $\theta = \tilde{\theta}$. With the estimate based on that assumption, it is now possible to compute the likelihood ratio for that point $\tilde{\theta} = k + \tilde{\nu}'$. Following from equation (27) and taking into account that $\mathrm{cov}\{\underline{r}_a^*(k), \underline{r}_a^*\} = I$, we identify the MLE $\hat{\theta}(k) = k + \hat{\nu}'(k)$ to be that value $\tilde{\theta}$, which maximizes:

$$l(k, \hat{\theta}, \hat{\underline{u}}_{|\hat{\theta}}) = \max_{\tilde{\theta}} \underline{r}_a^*(k)^T B_{a\tilde{\theta}} \hat{\underline{u}}(k)_{|\tilde{\theta}} \tag{43}$$

V. Computation of the Prediction Error Covariance $P^-(k+1)$

Once the jump has been detected and estimated, the Kalman filter is updated according to equation (8a) and the prediction error covariance matrix has to be corrected due to the incorporation of the jump estimate. The prediction error covariance matrix can be evaluated by straight forward considerations. Remembering that : $P^-(k+1) = E\{(\underline{x}(k+1) - \hat{\underline{x}}^-(k+1))(\underline{x}(k+1) - \hat{\underline{x}}^-(k+1))^T \ / \ \underline{Y}(k) = \underline{Y}_k, \hat{\underline{u}}(k) = \hat{\underline{u}}_k\}$, we can find after some tedious conversions and reductions:

$$P^-(k+1) = P_N^u - Q(k) - A\,P^+(k)\,A^T \tag{44}$$

where: P_N^u is evaluated by (42) after the last cycle

VI. Simulations

The Switched Kalman filter was tested together with two different jump detecting and estimating algorithms: 1.) the algorithm outlined in this work, and 2.) the 'Willsky' algorithm /7,8/, which uses the residuals of a Kalman filter to detect and estimate a state jump. The simulation modeled a typical laser scanning problem /5/ of a three dimensional object. The Kalman filter was a two state filter, with the state variables $x_1(k)$ and $x_2(k)$ denoting the 'target' distance and the derivative of the target distance with respect to the scanning line. The simulated height profile

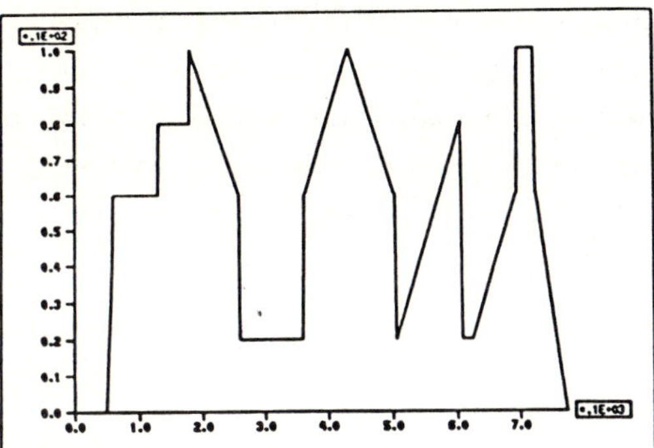

Figure 2: Simulated height profile

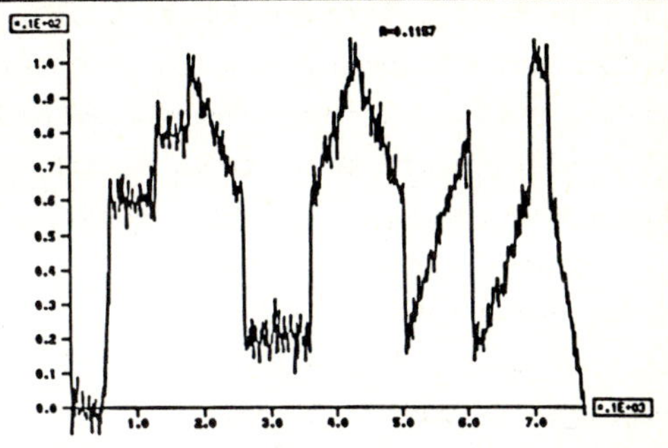

Figure 3: Disturbed measurements

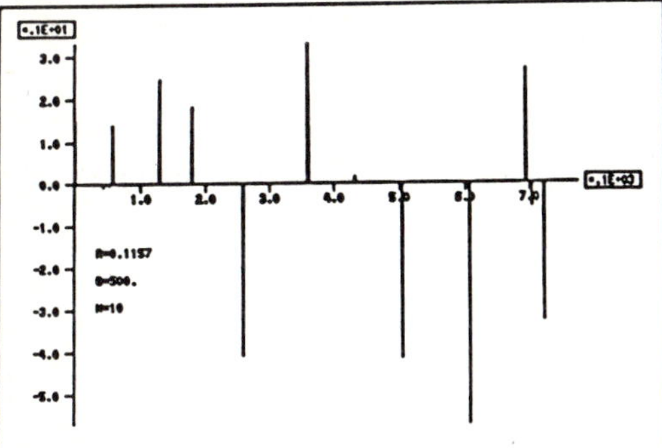

Figure 4: Estimated jumps $\hat{u}_1(k)$

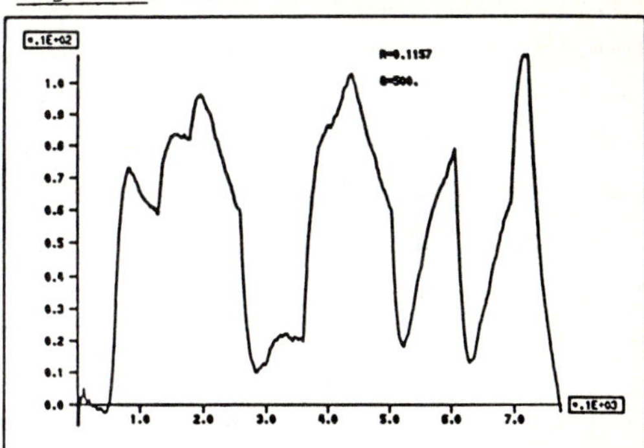

Figure 6: Distance estimates of Standard Kalman filter

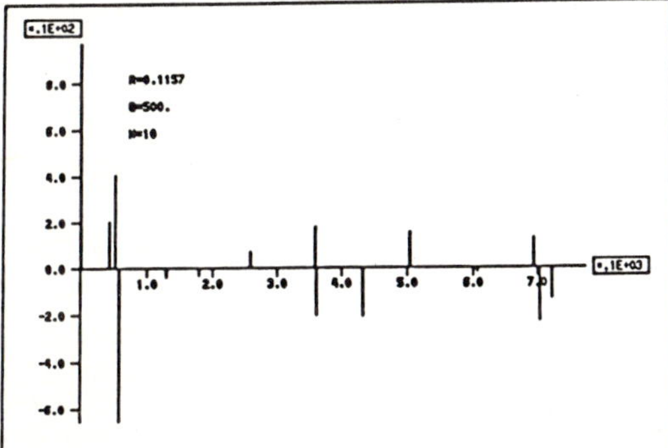

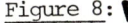

◄ Figure 5: Estimated jumps $\hat{u}_2(k)$

Figure 7: Distance estimates of Sw. Kalman filter applying the recursive ML-jump estimator

Figure 8: Distance estimates of Sw. Kalman filter applying the 'Willsky'algorithm

Figure 7: ▼ ▲ Figure 5: Figure 8: ▼

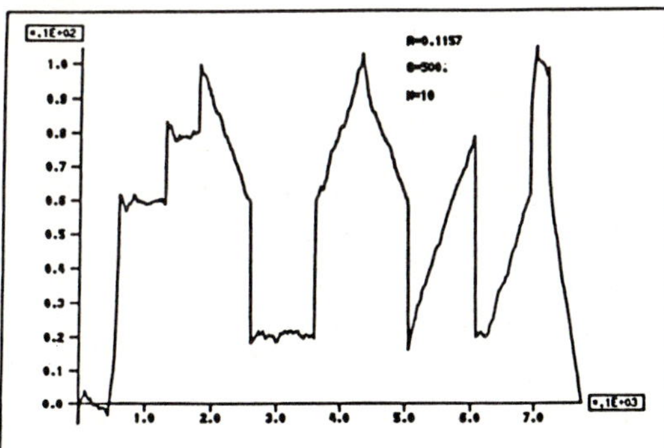

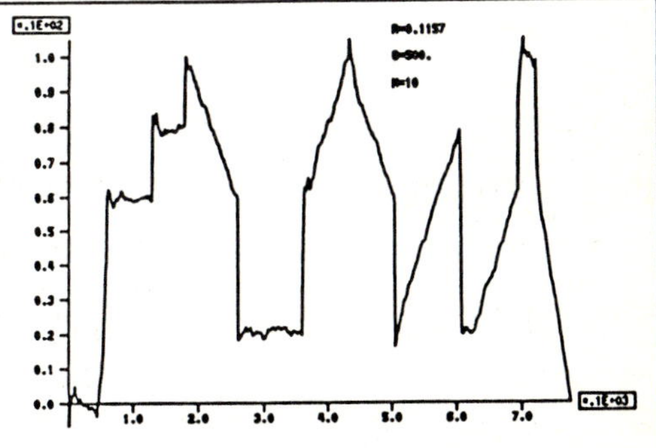

is given in figure 2. The height profile contaminated by white Gaussian noise is depicted in figure 3. Figure 4 and 5 present the estimated jumps $\hat{u}_1(k)$ and $\hat{u}_2(k)$, whereas figure 6 gives an impression of the standard Kalman filter operation. Finally the figures 7 and 8 show the filter estimates of the Switched Kalman filter applying the jump estimation algorithm formulated in this work (fig. 7) and the 'Willsky'algorithm (fig. 8).

VII. Results and Conclusions

For the applications regarded herein, the Switched Kalman filter in conjunction with the ML-jump detector and estimator formulated in this work seems to be slightly better than in conjunction with the 'Willsky'filter. This may be caused by the fact that the ML-estimator, derived in this paper, allows an exact computation of the prediction error covariance after a state update, whereas the Willsky approach offers only an approximated formula for this covariance. Yet, the results of a detailed analysis and comparison of both approaches in form of Monte Carlo simulations have to be waited for. There is also a point of criticism of the approach taken in this paper: The incorporation of 'future' measurement data for the jump estimate yields statistical dependencies between the new state estimate and the future measurement data. This is a contradiction to the classical Kalman filter theory, since the actual state estimates are not only based on the 'past', but also on the future. Therefore smoother theory should be applied to describe the whole problem. The neglect of these facts, however, does not seem to degrade the filter performance seriously, so it may be justified.

Acknowledgment

The author would like to thank his colleagues Mr. K. Hartmann and Mr. L. Tran duc for the valuable discussions and Mr. K.D. Menk and Mr. P. Kreutz for the computations and their work pertaining to the results reported herein. Finally I should like to thank Prof. R. Schwarte for the support of and his interest in my work.

References

1. Schwarte, R. "Performance Capabilities of Laser Ranging Sensors", Proc. of ESA Workshop Splat, Les Diablerets, March 1984.
2. Schwarte, R. "Laser-Radar für hochauflösende Entfernungsmessung und räumliches Sehen", ET-Kolloquium der Technischen Universität Stuttgart, 11.12.84.
3. Schwarte, R. "Intelligentes Laserradar für Roboter", Tagungsbericht "Zukunftsweisende Robotertechnik", Universität-GH-Paderborn, 15.03.85.
4. Schwarte, R. " Anwendung eines laseroptischen Sensorsystems in der fertigungstechnischen Automatisierung", Tagung "Zukunftsweisende Robotertechnik", Technische Universität Bochum, Sep. 1985.
5. Loffeld, O. "Ein neuartiges 'Switched Kalman-Filter' mit geringer Wortbreite für die hochauflösende Entfernungsmessung nach den Laserpuls-Laufzeitverfahren" Vom

Fachbereich Elektrotechnik der Universität-GH-Siegen genehmigte Dissertation, Siegen, Sep. 1986.

6. Loffeld, O. "A Switched Kalman Filter for the Implementation on Microprocessor Units for On-line Applications" Proc. IASTED International Symposium Applied Signal Processing and Digital Filtering, Paris, June 1985
7. Willsky, A. S. and Jones, H. L. "A Generalized Likelihood Ratio Approach to State Estimation in Linear Systems Subject to Abrupt Changes" IEEE Proc. Decision and Control, Nov. 1974.
8. Willsky, A. S. and Jones, H. L. "A Generalized Likelihood Ration Approach to the Detection and Estimation of Jumps in Linear Systems" IEEE Transactions on Automatic Control, Feb. 1976.
9. Basseville, M. and Beneviste, A. (ed.), Detection of Abrupt Changes in Signals and Dynamical Systems, Lecture Notes in Control and Information Sciences, Springer Verlag, 1986.
10. Raubenheimer, H.R., "Konturverfolgung und -segmentation mit Kalman-Filtern," Dissertation an der Fakultät für Bergbau, Hüttenwesen und Maschinenbau der Technischen Hochschule Clausthal, 1984.
11. Menk, K.D., "Entwicklung von Kantenestimations und Detektionsalgorithmen für Kalman-Konturverarbeitungsfilter", Diplomarbeit am Institut für Nachrichtenverarbeitung der Universität Siegen, 1987.
12 Kreutz, P., "Entwicklung und Untersuchung von Kalman-Filtern für die zweidimensionale Konturverfolgung mit einem Laserscanner", Diplomarbeit am Institut für Nachrichtenverarbeitung der Universität Siegen, To appear.

EIN GERÄUSCHUNTERDRÜCKUNGSSYSTEM MIT ZWEIDIMENSIONALER MIKROFONGRUPPE UND NACHGESCHALTETER ADAPTIVER WIENER FILTERUNG

Rainer Zelinski
Forschungsinstitut der Deutschen Bundespost, Außenstelle Berlin
Ringbahnstraße 130, 1000 Berlin 42

Bei vielen Aufgaben der Sprachverarbeitung (z.B. Spracherkennung, Freisprechen ohne Handapparat) treten Probleme auf, wenn Umgebungsgeräusche in das Aufnahmemikrofon gelangen. Die bisher untersuchten Verfahren zur Geräuschreduktion sind im wesentlichen eindimensionale Verfahren, die sich in zwei Kategorien einteilen lassen: Geräuschfilterung mit einem einzigen Aufnahmemikrofon [1] (erfordert in der Regel einen Sprachpausendetektor) oder Geräuschkompensation mit zusätzlichem Referenzmikrofon und Kompensationsfilter [2]. Beide Verfahren sind in der Praxis oft nicht einsetzbar, weil entweder das Störsignal instationär ist (bewirkt fehlangepaßte Störstatistik bei der Geräuschfilterung) oder das Störsignal nicht als einzelne punktförmige Störquelle modelliert werden kann (Versagen der Geräuschkompensation [3,4]).

Es wird ein neues Geräuschunterdrückungssystem mit mehrdimensionaler Signalverarbeitung vorgestellt (Bild 1), das auf einer zweidimensionalen Mikrofongruppe mit 4 Mikrofonen (M1 ... M4) aufbaut.

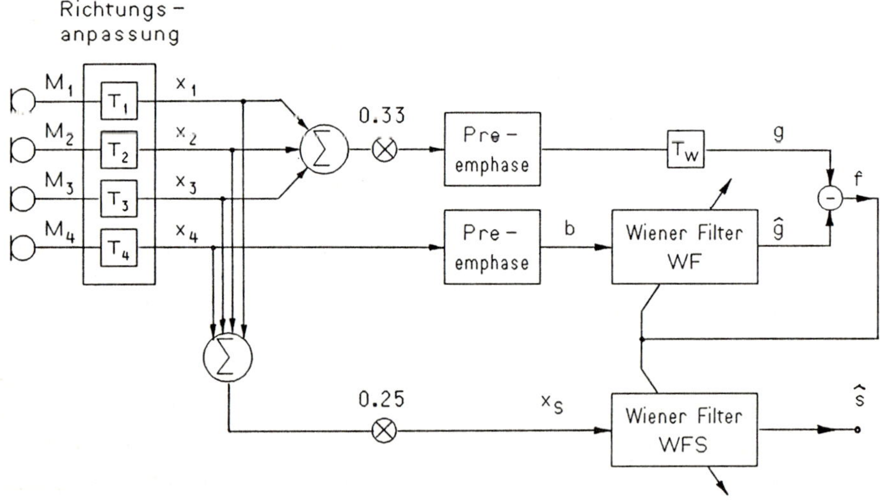

Bild 1: Blockbild des Geräuschunterdrückungssystems

Die erste Systemstufe ("Richtungsanpassung") besteht aus 4 Zeitverzögerungselementen T_1 ... T_4, die so eingestellt werden, daß die Signalkomponente aus der gewünschten Richtung - also hier das Sprachsignal s -

in den Signalen $x_1 \ldots x_4$ gleichzeitig eintrifft:

$$x_i = s + n_i \quad ; \quad i = 1 \ldots 4$$

Die Störsignalkomponenten n_i werden durch das Umgebungsgeräusch verursacht. Die geometrische Anordnung der Mikrofongruppe und der umgebende Raum seien dabei so beschaffen, daß die Störsignalkomponenten n_i untereinander näherungsweise als unkorreliert betrachtet werden können.

Die Reduktion des Umgebungsgeräusches wird in zwei Schritten durchgeführt. Im ersten Schritt wird die Richtwirkung der Mikrofongruppe ausgenutzt: Das Summensignal x_S enthält einen Störsignalanteil, der - im Vergleich zu einem Einzelkanal x_i - um etwa 6 dB gedämpft ist. Im zweiten Schritt erfolgt eine adaptive Wiener Filterung des Signals x_S mit dem Wiener Filter WFS. Der verbliebene Störanteil in x_S wird dabei zusätzlich abgeschwächt; das Ausgangssignal $\hat{s}$ präsentiert das geschätzte Sprachsignal.

Die Adaption des Wiener Filters wird allein durch die aktuell auftretenden Signale $x_1 \ldots x_4$ gesteuert: Aus den Signalen $x_1 \ldots x_3$ wird ein gemitteltes Signal g berechnet, das - im Vergleich zum Einzelkanal x_i - weniger Störungen enthält und daher als Näherung für das gewünschte Sprachsignal s dient. Das Signal x_4 dient als "beobachtetes Signal" b am Eingang des Filters WF und führt auf die Signalschätzung $\hat{g}$. Wenn die Störsignalkomponenten $n_1 \ldots n_4$ tatsächlich unkorreliert sind, so kann das Wiener Filter nur die gemeinsame Signalkomponente in b und g - das Sprachsignal s - schätzen. Das Filter WF führt also eine optimale Abschwächung des Geräuschanteiles durch. Das Summensignal x_S wird nun mit dem Filter WFS gefiltert, das als direkte Kopie von WF arbeitet ("Sklaven-Filter").

Das Zeitverzögerungsglied T_W in Bild 1 ist auf die halbe Filterlaufzeit von WF eingestellt. Dies ermöglicht eine nichtkausale Filterung und damit eine erhöhte Geräuschunterdrückung. Die beiden Systemstufen "Preemphase" führen eine Höhenanhebung um 6 dB/Oktave durch und bewirken damit eine genauere Reproduktion der höheren Sprachfrequenzen.

Der Abgleich des Wiener Filters WF wird durch das Fehlersignal

$$f(j) = g(j) - \hat{g}(j) = g(j) - \underline{w}^T(j) \cdot \underline{b}(j)$$

gesteuert; dabei ist j der Zeittakt, $\underline{w}$ der Vektor der Filterkoeffizienten in WF und $\underline{b}$ der Vektor aus den zeitlichen Abtastwerten des Signals b . Die Filteradaption erfolgt gemäß dem LMS-Algorithmus [2] mit einer zusätzlichen Leistungsnormierung des Adaptionsfaktors μ :

$$\underline{w}(j+1) = \underline{w}(j) + 2\,\mu(j) \cdot f(j) \cdot \underline{b}(j) \qquad \text{mit } \mu(j) = \alpha \,/\, [\underline{b}^T(j) \cdot \underline{b}(j)]$$

Das Geräuschreduktionssystem nach Bild 1 wurde in einer Reihe von Experimenten untersucht. Als Aufnahmeraum dient ein typisch möblierter Laborraum mit 25 m^2 Grundfläche, in dem vier verschiedene räumlich ausgedehnte Geräuschquellen (mit elektromotorischem Antrieb) aufgestellt sind. Die 4 Mikrofone sind in den Ecken eines Quadrats der Kantenlänge 60 cm angeordnet (zweidimensionale Mikrofongruppe). Der Abstand zum Sprecher (Nutzsignal) beträgt etwa 60 cm . Die adaptive Steuerung der Systemstufe "Richtungsanpassung" ist derzeit noch nicht realisiert; die Zeitverzögerungen $T_1 \ldots T_4$ sind fest auf den Zielsprecher eingestellt.

Zunächst wurde untersucht, wie weit die Annahme von untereinander unkorrelierten Störsignalkomponenten n_i zutrifft. Die Komponenten n_i sind zwar zunächst nicht unabhängig voneinander, da sie ja von gemeinsamen Störquellen verursacht werden. Mit wachsendem Mikrofonabstand nimmt jedoch - bedingt durch die hohe Zahl von unterschiedlichen Schallrückwürfen an den Wänden - die Korrelation zwischen zwei Störsignalkomponenten ab. Bild 2 zeigt dazu die Meßergebnisse (Kohärenzfunktion).

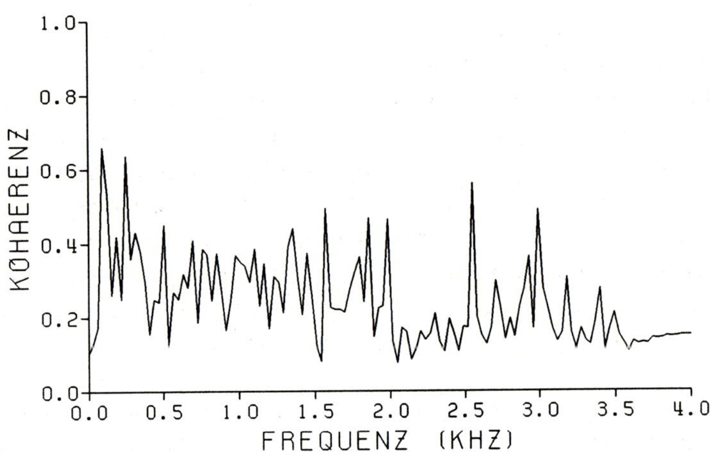

Bild 2: Gemessene Kohärenzfunktion zu den Störsignalen in zwei benachbarten Mikrofonen (Mikrofonabstand 60 cm)

Es wird deutlich, daß nur noch eine geringe Restkorrelation vorliegt; eine Ausnahme bilden tiefe Frequenzen und einzelne diskrete Frequenzen, die den Oberwellen der Motordrehungen zugeordnet sind.

Zur Verarbeitung von gestörter Sprache muß das adaptive Wiener Filter hinreichend genau und schnell dem zeitvariablen Kurzzeit-Leistungsdichtespektrum des Sprachsignals folgen können. Bei der schließlich gewählten Einstellung von $\alpha = 0.1$ und der Filterlänge von 33 Koeffizienten ergibt sich eine effiziente Geräuschunterdrückung, wobei die Sprachsignalkomponente in x_S nach subjektiven Maßstäben praktisch nicht verzerrt wird. Für Frequenzen oberhalb von 0.3 kHz ist - im Vergleich zu einer

1-Mikrofon-Aufnahme - eine Geräuschdämpfung von insgesamt 17...20 dB erreichbar (Bild 3).

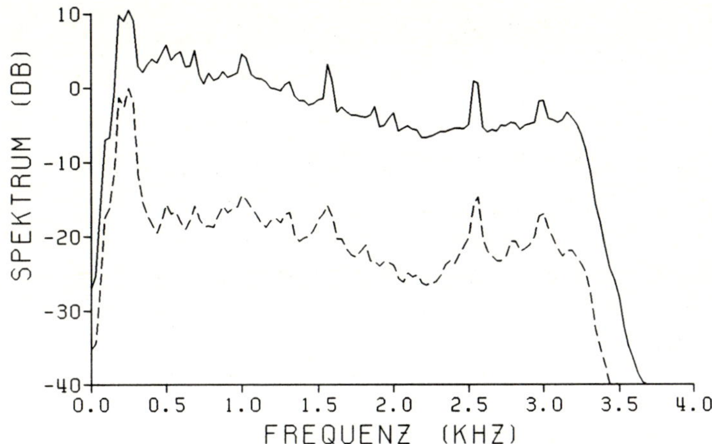

Bild 3: Gemessene Leistungsdichtespektren des Geräusches (Sprache s=0)
durchgezogene Linie: Spektrum für Einzelkanal x_i
gestrichelte Linie : Spektrum für Ausgangssignal $\hat{s}$

Dieser Wert ist zusammengesetzt aus einem Anteil von 6 dB für den Richtwirkungsgewinn der Mikrofongruppe und einem Anteil von 11...14 dB für den Zusatzgewinn durch das Wiener Filter WFS.

Die subjektive Bewertung der verarbeiteten Sprachproben zeigt, daß auch der Einfluß des Sprach-Nachhalls deutlich reduziert wird. Dies ist darauf zurückzuführen, daß die Nachhallsignale ebenso wie die Störsignalkomponenten untereinander näherungsweise unkorreliert sind und daher vom Geräuschunterdrückungssystem zusammen mit dem Störsignal abgeschwächt werden.

Das vorgestellte Geräuschunterdrückungssystem mit räumlicher Auswertung des Schallfeldes arbeitet auch dann sehr wirksam, wenn die Störquellen räumlich ausgedehnt sind oder in großer Zahl auftreten. Auch instationäre Störquellen führen nicht zu Problemen, da keinerlei Vorabkenntnisse über die Statistik von Nutz- und Störsignal nötig sind.

Literatur:

[1] Lim, J.S.; Oppenheim, A.V. : Enhancement and Bandwidth Compression of Noisy Speech. Proc. IEEE 67 (1979), S. 1586-1604.

[2] Widrow, B. et al.: Adaptive Noise Cancelling: Principles and Applications. Proc. IEEE 63 (1975), S.1692-1716.

[3] Armbrüster, W.; Czarnach, R.; Vary,P. : Adaptive Noise Cancellation with Reference Input - Possible Applications and Theoretical Limits. European Signal Process. Conf. EUSIPCO-86, The Hague, S. 391-394.

[4] Rodriguez, J.J.; Lim, J.S. : Adaptive Noise Reduction in Aircraft Communication Systems. IEEE Int. Conf. Acoustics, Speech, and Signal Processing ICASSP87, Dallas, Paper 6.1, S.169-172.

The Feature Locating for Fingerprint Recognition

Wang Houshu Yu Shenglin Wu Yiquan

Nanjing Aeronautical Institute
The People's Republic of China

This paper describes a procedure for feature locating. The determining of reference coordinate is very important for fingerprint recognition, so it is described in detail in this paper.

1. Introduction

The feature locating in fingerprint image is to determine a reference coordinate in which the positions of minutiae may be fixed up, depending on the textures of the core of the print. The coordinates must be not changeable owing to the variation of the input of the print.

2. The block of the fingerprint recognition system

Fig.1 is an automatic fingerprint recognition system.

3. Fringe flow mode extracting and smoothing

A figure of a bilevel picture of a practical fingerprint and its figure matrix are shown in fig.2. The processed bilevel picture is divided into 20X20 subdivisions, and the angle of the ridge of every

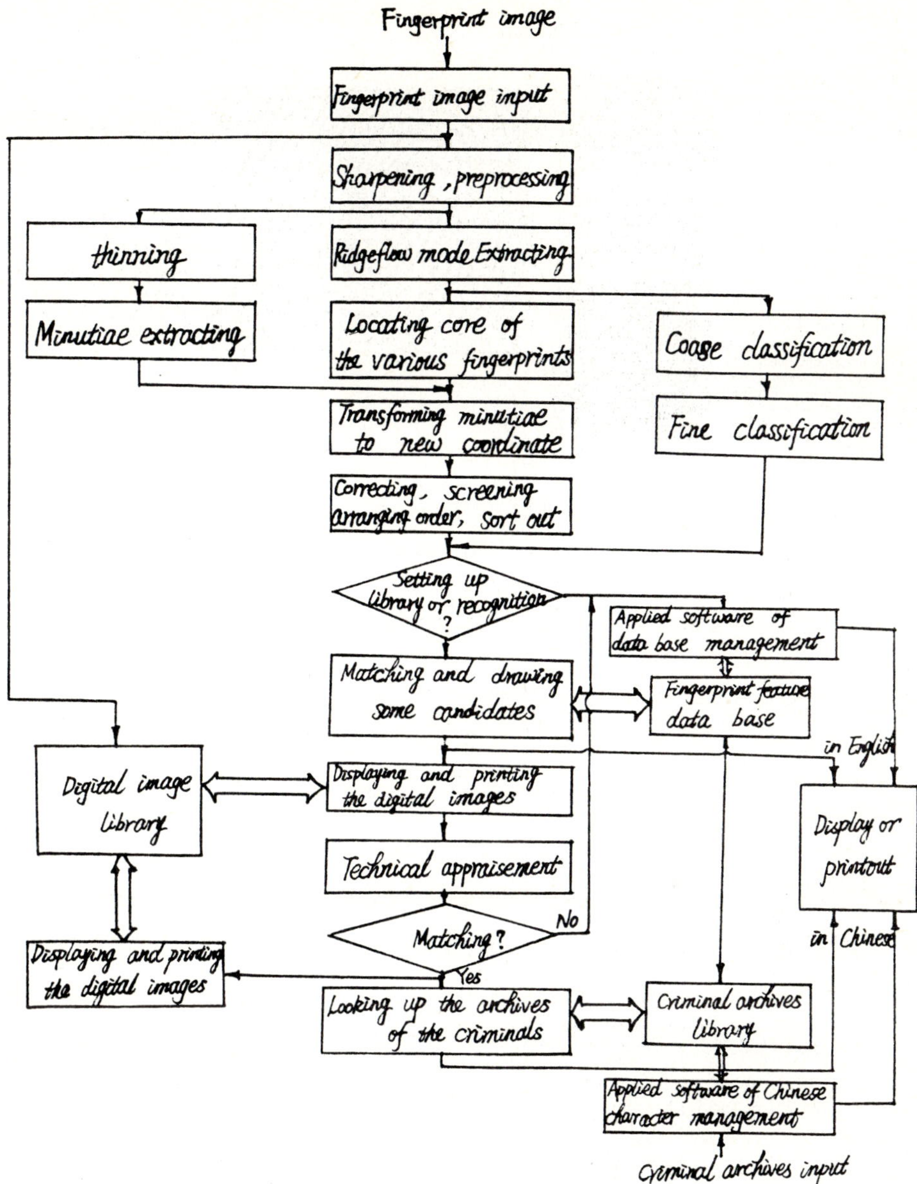

Fig.1 An Automatic Fingerprint Recognition System

subdivision is determined by the 2X2 mask. The value of the angle is approximately estimated by the following formulae

$$AVDIR = \begin{cases} \dfrac{0^\circ \times N_{0^\circ} + 45^\circ \times N_{45^\circ} + 90^\circ \times N_{90^\circ} + 135^\circ \times N_{135^\circ}}{N_{0^\circ} + N_{45^\circ} + N_{90^\circ} + N_{135^\circ}} & N_{45^\circ} \geq N_{135^\circ} \\[2ex] \dfrac{180^\circ \times N_{180^\circ} + 45^\circ \times N_{45^\circ} + 90^\circ \times N_{90^\circ} + 135^\circ \times N_{135^\circ}}{N_{180^\circ} + N_{45^\circ} + N_{90^\circ} + N_{135^\circ}} & N_{45^\circ} < N_{135^\circ} \end{cases}$$

Fig.3 is a bilevel picture of a whorl and its ridge flow in experiment.

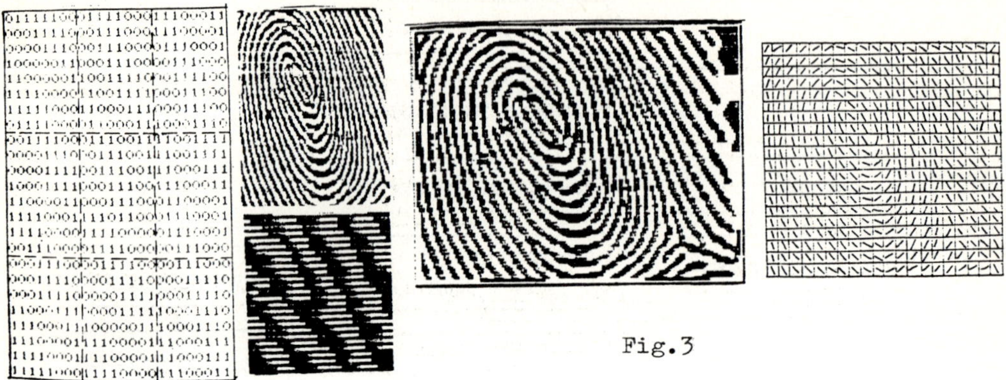

Fig.3

Fig.2

For rejecting the noise it is necessary to smooth the ridge flow having been extracted from the original fingerprint image. The weighting was proceed by the nonlinear mode of probablity in relaxation method, by the weighting of the reciprocal of the absolute value of difference angle, by the mean value and the standard variation of angulars of the 8-neighbors.

4. Computation of the core

At first we found out the center point in ridge flow matrix of the fingerprint by means of the certain procedure and then the more precise center point is computated by a series of formulae. At last, the center point for loop, whorl and tent :

$$YCP=YC2-\frac{XC*(YC2-YC1)}{LSR} \quad ; \quad XCP=(I-1)*LSR+\frac{LSR}{2}+LSR-XC$$

The center point for plain arch:

$$YCP=YC1; \quad XCP=\frac{SA(L)*X(L-1)+SA(L)*X(L)+SA(L+1)X(L+1)}{SA(L-1)+SA(L)+SA(L+1)}$$

5. Computation of angle

The feature data will be rotated about this point through a angle AROTA to the standard orientation for recognition or library. The rotation angle can be computated as follows:

$$AROTA=\frac{\sum_{ROTAK\alpha} (AVDIR(I,J3)+AVDIR(I,J4+1)/2}{K}$$

6. Correction, transformation, screening, rearranging

1)It is possible to appeare errors between the angles of the minutiae and the neighbor ridge flow. In this case, we use the four neighbor ridge flow angles to correct the minutie angle. The weight is selected as:

$$AW(K)=\frac{1}{((XR(K)-A(I,1))^2+(YR(K)-A(I,2))^2)^{1/2}}$$

or $$AW(K)=\sqrt{2}LSR-\sqrt{(XR(K)-A(I,1))^2+(YRCK)-A(I,2))^2} \quad (K=1,2,3,4)$$

Then the difference of minutiae angles and the weight average angles of neighbor ridge divided into 16 subdivisions, and according to correcting criteria we correct the minutia angles.

2)We build a new coordinate. The original point of the new coordinate is the (XCP,YCP), and the X-axis is fixed up by the rotation angles AROTA.

3)We preserved the minutiae in a defining circular area or rectangular area.

4)For fastening the operating velocity, we selected a rapid method of arranging minutiae orders.Because of the defection of fingerprint image, there are existing some called pseudominutiae during extracting the minutiae. We rejected these poseudominutiae.

The results of experiment of some fingerprint recognition are quite efficient and satisfactory. Three types of feature locations are shown on Fig.4 .

Fig.4

Poster

PHYSIOLOGISCHE MEßSIGNALE UND DEREN AUSWERTUNG ZUR UNTERSTÜTZUNG MEDIZINISCHER ENTSCHEIDUNGEN BEI OPERATIONEN

A.JANITZKI
Fachgebiet Nachrichtentechnik, FB14, Universität-GH-Paderborn,
Pohlweg 47-49, 4790 Paderborn

Einleitung

Mit Hilfe eines Multimikroprozessorsystems lassen sich Meßsignale mit nicht allzu hohen Frequenzanteilen, wie sie häufig bei einer Reihe von physiologischen Messungen auftreten, dokumentieren und weiterverarbeiten. Im realisierten, mobil einsetzbaren Gerät werden die Sensorsignale analog durch Filter und Nichtlinearitäten, im digitalen Bereich durch Programmauswahl von wenig aussagekräftigen Signalanteilen befreit /1/. Der physiologische Meßparameter Hautwiderstand gestattet es, indirekt Aussagen über die Informationsverarbeitung im menschlichen Zentralnervensystem zu machen /2/. Er wurde beispielsweise auch zum Verhaltensstudium von Autofahrern erfolgreich verwendet /3-6/.

Verfahren

Bei Operationen, insbesondere an Risikopatienten, ist es für die Anaesthesieführung wichtig, laufend und möglichst gut über die Situation und den Zustand des Zentralnervensystems des Patienten informiert zu sein. Dies geschieht herkömmlich mit einer Reihe von Messungen wie z.B. Elektrokardiagramm mit Pulsfrequenzmessung, Blutdruckmessung, sowie besonders auch durch visuelle Beobachtung des Patienten. Erst neuerdings werden in der Forschung auch Hirnstromaktivitäten (Elektroencephalogramm EEG und evozierte Potentiale) mit zur Anaesthesieführung herangezogen. Diese Messungen sind jedoch wegen unzureichender Kenntnisse über die zentralnervöse Informationsverarbeitung sowie wegen sehr schwacher Meßsignale, die Meßspannungen liegen im Bereich von Nanovolt, sehr schwierig aufzunehmen und zu interpretieren und werden daher kontrovers diskutiert. Sie können deshalb im klinischen Bereich routinemäßig noch nicht verwendet werden. Demgegenüber ist der Hautwiderstand ein Parameter, der, anders als die meisten EEG-Signale, ausschließlich effektorische Komponenten enthält und somit wesentlich einfacher zu interpretieren ist. Effektorische Signale sind Aktionen und Reaktionen des Gehirns, welche die Erfolgsorgane aktivieren, und das sind im Falle der Hautwiderstandsmessung peripher die eccrinen Schweißdrüsen. Sie werden mit Hilfe des sympathischen Nervensystems angesteuert und besitzen - unterschiedlich zu den meisten Effektoren - keinen Antagonisten (Parasympathicus). Daher läßt sich mit Hilfe der Hautwiderstandsaktivität unwillkürlich und direkt die sympathische Aktivität, die vom Gehirn gesendet wurde, messen; und diese ist mit ausschlaggebend für die Regelung der Herztätigkeit, die Wärmeregulierung der Haut, für den Blutdruck und eine Reihe weiterer durch das autonome Nervensystem angesteuerte Funktionen. Für die Anaesthesieführung bedeutet daher die Verfolgung des Hautwiderstandsverhaltens einen sehr einfach und unblutig zu messenden, aussagekräftigen, schnellen Parameter, der beispielsweise einen gefährlichen Blutdruckanstieg (Ursache dafür ist oft die Zunahme sympathischer Aktivität) während einer Operation bereits im Entstehen anzeigen kann.

Daneben eignet sich die Auswertung der Hautwiderstandsaktivität eines Patienten besonders auch zur aktuellen Einschätzung der Narkosetiefe und -wirkung. Dies ist besonders gut erkennbar an unterschiedlichen Narkoseformen /7,8/, wo z.B. eine sehr tiefe Narkose die effektorische Aktivität des Gehirns ganz unterdrücken kann, mit allen organisch daraus resultierenden Folgen.

Letztlich stellen sich die Schwankungen des Hautwiderstandes vor, während und nach operativen Eingriffen als eine gute Meßgröße zur Einschätzung der Beanspruchung und Belastung des Patienten heraus. Dadurch läßt sich die Dosierung von Medikamenten nicht nur von der Erfahrung der Mediziner her festlegen, sondern sie kann mit Hilfe der Messung und Auswertung streßrelevanter Aktivitäten des sympathischen Nervensystems sehr schnell individuell bestimmt werden.

Beispiel

Ein Beispiel für die Nützlichkeit von Hautwiderstandsmessungen aus der interdisziplinären Zusammenarbeit mit Medizinern im klinischen Bereich zeigt exemplarisch Fig. 1, wo das Hautwiderstandsverhalten eines Patienten kurz vor und nach Anästhesiebeginn aufgezeichnet wurde. Die Hautwiderstandsaktivität ist vor Beginn eines operativen Eingriffs meist recht hoch, wie hier, und kann individuell Werte von 10 - 20 diskriminierbaren Reaktionen pro Minute erreichen, wenn nicht entsprechende Medikamente dem entgegenwirken. Bei der Anaesthesie handelt es sich um eine Leitungsanaesthesie. Sie unterbricht die Leitung von Nervenimpulsen vom und zum Gehirn. Dadurch sinkt die Amplitude der Hautwiderstandsreaktionen innerhalb weniger Minuten ab. Die Häufigkeit der Reaktionen nimmt ebenfalls ab. Im Idealfall bei vollständiger Blockade verschwinden alle Ausschläge, so daß die Auswertung der Messung einmal die Wirksamkeit der Blockade quantisiert und zum anderen auch eine Aussage über die Streßsituation des Patienten ermöglicht. Bei fehlenden Hautwiderstandsreaktionen ist auch keine Belastungsituation für das Zentralnervensystem des Patienten vorhanden. Der Nachweis hierüber gelingt besonders gut, wenn noch ein zweiter Meßkanal zum Vergleich herangezogen wird /7/. Läßt die Blockadewirkung der Anästhesie nach, so wächst auch die Amplitude der Hautwiderstandsreaktionen und der Anästhesist kann mit diesem Prozeßparameter die Wirksamkeit der Narkose optimieren. Zusammen mit weiteren Meßsignalen wie z.B. der Pulsfrequenz und dem Blutdruck läßt sich so eine sehr präzise multifunktionelle Darstellung aufbereiten, die Patienten und behandelnden Ärzten von großem Nutzen sein kann.

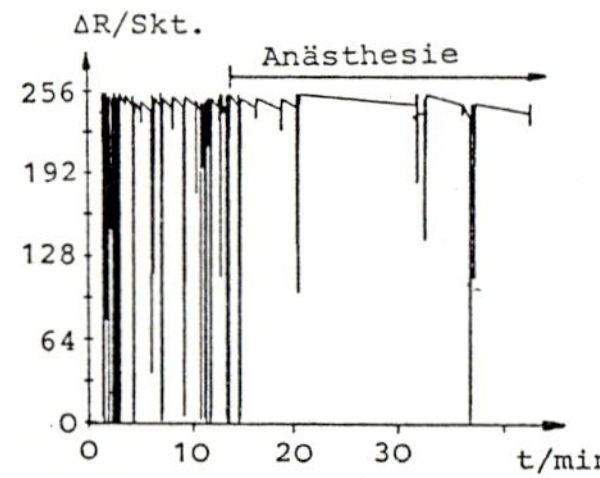

Fig. 1 Hautwiderstandsaktivität eines Patienten beim Beginn der Anästhesie, aufgenommen mit dem mobilen Meßsystem nach /1/.

Literatur

/1/Janitzki A., Föckeler W. A mobile system for signal adaptive data storage - application in physiological measurements. Measurement, 1987, accepted for publ.

/2/Janitzki A., Vedder N. Mehrkanal-Hautwiderstandsmessungen. Biomedizinische Technik, 32, 98-107, 1987

/3/Janitzki A. Ein Ansatz zur Reduzierung der Unfallquote infolge menschlichen Versagens. Straßenverkehrstechnik, 6, 174-179, 1982

/4/Janitzki A., Kumm W. Zur Eignung von induktiven Entropiewerten bei der Beurteilung des individuellen Leistungsverhaltens von Autofahrern. Frequenz, 41, 76-82,1987

/5/Janitzki A. A method to distinguish safe from less safe driving. Proc. 4th IFAC/IFIP/IFORS Int. Conf. on Control in Transportation systems, Baden-Baden, 1983

/6/Föckeler W. Analyse des individuellen Hautwiderstandsverhaltens von Autofahrern zur Untersuchung des aktuellen Zustandes im System Mensch-Maschine. Zwischenbericht zum DFG Projekt Ku 300/3 1, Paderborn, März 1986

/7/Janitzki A., Götte A. Hautwiderstandsmessungen zum Aktivitätsnachweis des Sympathicus bei der Spinalanästhesie Regional-Anaesthesie, 9,49-53, 1986

/8/Janitzki A., Götte A., Nolte H., Meyer J. Monitoring von sympathischer Aktivität nach Blockaden des Ganglion Stellatum. Zentraleuropäischer Anaesthesiekongreß, München, 1987

Nachtrag zur Theorie

HAUSDORFF DISTANCE AND DIGITAL FILTERS

A. Andreev

One of the main steps for designing digital filters is the problem for approximation of the amplitude-frequency characteristic of the filter. As a rule it is necessary to approximate a functions with jumps (step function) by means of polynomials or rational functions. There exist many approaches for solving that problem under different restrictions and the nature of the problem shows that the Hausdorff distance between the functions is a suitable metric as a criterion.

In accordance with the Hausdorff metric it is possible to determine the polynomial or the rational function avoiding some defects connected with the uniform metric, Fourier series and some other approaches. The Hausdorff distance between the functions was introduced by Bl. Sendov in 1960 [1] and its most important properties can be found in the book [2].

1. One-dimensional case. Let us denote $R = \{x:-\infty<x<\infty\}$, $S^M(R) = \{[a,b]: a\in R, b\in R, -M \leq a < b \leq M, M > 0\}$ and $A_\Delta^M = \{f:\Delta\to S^M(R),\ \Delta = [a,b]\}$ be the set of all segment-valued bounded functions in the interval Δ. For $f\in A_\Delta^M$ we denote by $F(f)$ the complemented graph of f in the following way:

$$F(f) = \bigcap_{g\in F_\Delta, \bar{f}\subset g} g$$

where F_Δ is the set of all sets in the plane which are closed, and bounded, convex with respect to y and their projection on the real axis is Δ, $\bar{f}$ is a graph of the function f.

Let $H_n(T_n)$ be the set of all algebraical (trigonometrical) polynomials of degree $\leq n$ and $R_{n,m} = \{f:f = p/q,\ p\in H_n(T_n),\ q\in H_m(T_m)\}$ is the set of all rational functions.

For $f,g\in A_\Delta^M$ the one-sided distance is defined as

$$h(\Delta,\alpha;f,g) = \max_{(x,y)\in F(f)} \ \min_{(\xi,\eta)\in F(g)} \ \max\{\alpha(x)^{-1}|x-\xi|, |y-\eta|\},$$

and $r(\Delta,\alpha;f,g) = \max\{h(\Delta,\alpha;f,g), h(\Delta,\alpha;g,f)\}$ is the Hausdorff distance between f and g, where $\alpha(x) > 0$ is a parameter. For a given $f\in A_\Delta^M$ we define $E_h(R_{m,k},\Delta,\alpha;f) = \inf\{h(\Delta,\alpha;q,f):q\in R_{m,k}\}$, $E_r(R_{m,k},\Delta,\alpha;f) = \inf\{r(\Delta,\alpha;q,f):q\in R_{m,k}\}$ - the best approximation according to one-sided and Hausdorff distance. Such $p\in R_{m,k}$ and $q\in R_{m,k}$ for which $E_h(R_{m,k},\Delta,\alpha,f) = h(\Delta,\alpha,p,f)$, $E_r(R_{m,k},\Delta,\alpha,f) = r(\Delta,\alpha,q,f)$ are called rational functions of the best h and r approximation (such p and q exist for every $f\in A_\Delta^M$, k, m and $\alpha(x) > 0$).

Theorem 1. If the function f A_Δ^M is λ-monotonous and $R_r(R_{m,k},\Delta,\alpha;f) \leq \lambda/\alpha$ then there exists only one function $p \in R_{m,k}$ such that $r(\Delta,\alpha;p,f) = E_r(R_{m,k},\Delta,\alpha;f) = E_h(R_{m,k},\Delta,\alpha;f) = h(\Delta,\alpha;p,f)$.

The function $f \in A_\Delta^M$ is called λ-monotonous if f is monotonous in every subinterval $\Delta' \subset \Delta$ such that the length of Δ' is less than λ and f is a constant in $[a, a-\lambda]$ and $[b-\lambda,b]$.

Theorem 2. Let the function $f \in A_\Delta^M$ be λ-monotonous and $E_h(R_{m,k},\Delta,\alpha;f) \leq \lambda/\alpha$. A necessary and sufficient condition for the function $p \in R_{m,k}$ to satisfy $r(\Delta,\alpha;p,f) = E_r(R_{m,k},\Delta,\alpha;f)$ is the existance of L points $\{x_i\}_{i=1}^{L}$, $L = m+k+2-d$, $d = \min(\mu,\nu)$, $p(x) = \sum_{j=0}^{m-\nu} a_j x^j / \sum_{j=0}^{k-\mu} b_j x^j$,

$x_i \in \Delta$, such that $\Psi(\Delta,\alpha,p,f;x_i) = h(\Delta,\alpha;p,f)$, $\operatorname{sign}(f(x_i)-p(x_i)) = (-1)^i\varepsilon$, $\varepsilon=\pm 1$, $i = 1,2,\ldots,L$, where $\Psi(\Delta,\alpha,p,f;x) = \min_{(\xi,\eta)\in F(f)} \max\{\alpha^{-1}|\chi-\xi|,|p(x)-\eta|\}$.

Theorem 2 allows us to modify the second Remez' algorithm and to find an algorithm for numerical determination of the rational function of the best Hausdorff approximation for λ-monotonous functions.Let us remember that if the bounded function f is not λ-monotonous then the rational function of the best Hausdorff approximation is not unique.If $f \in A_\Delta^M$ is λ-monotonous it is possible to prove the convergence theorem for the mentioned algorithm (given in [3]),i.e.the sequence of rational functions $p^{(k)} \in R_{n,m}$ generated by the algorithm converges uniformly.

The suggested algorithm was used for numerical finding of the rational functions of the best Hausdorff approximation of the following functions:

$$f_1(x)=\operatorname{sign} x,\quad f_2(x) = \begin{cases} 0, & -1\leq x<-0.5, \\ 1, & -0.5\leq x<0, \\ 1-x, & 0\leq x<0.5, \\ 0.5, & 0.5\leq x\leq 1, \end{cases}$$

in the interval $[-1, 1]$. On the Fig. 2 the function f_1 and its elements of the best Hausdorff approximation $p \in R_{5,5}$ and $r \in R_{10.0}$ are given. On the Fig. 1 the function f_2 and its rational functions $q \in R_{7,6}$ and $s \in R_{14,0}$ are given.

It is possible to modify Theorem 2 and by its help to determine the polynomial or the rational function of the best Hausdorff approximation of a given bounded function under some restrictions. For exmaple for the function f

$$f(x) = \begin{cases} -1, & -1\leq x\leq 0, \\ 0, & 0<x<0.5, \\ 1, & 0.5\leq x\leq 1. \end{cases}\quad u(x) = \begin{cases} +\infty, & -1\leq x<-0.1 \\ 0.04, & -0.1\leq x\leq 0.4, \\ +\infty, & 0.4<x\leq 1 \end{cases}\quad l(x) = \begin{cases} -1.02, & -1\leq x<0.1 \\ -0.04, & 0.1<x\leq 0.6 \\ -\infty, & 0.6<x\leq 1 \end{cases}$$

we want to obtain the polynomial $p \in H$ of the best Hausdorff approximation, such that $l(x) \leq p(x) \leq u(x)$.

We have $0.117... \leq r([-1,1],1;f,p) \leq 0.126...$ On Fig. 3 the graphs of f, p, u and l are given

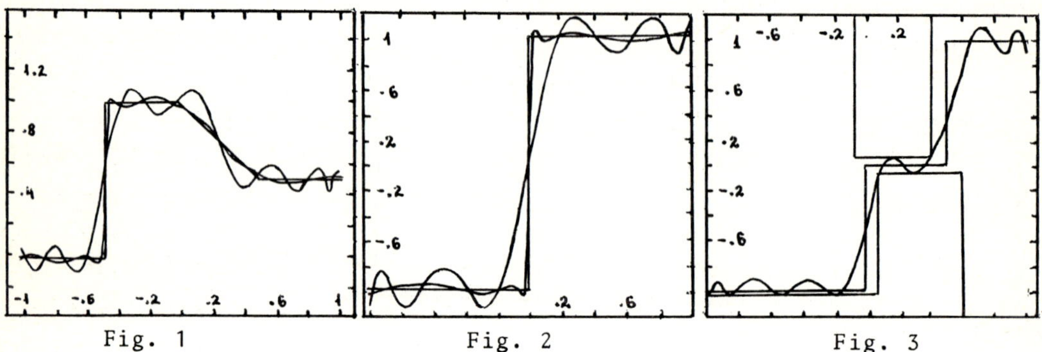

Fig. 1 Fig. 2 Fig. 3

The next four tables show the order of the approximation of the functions f_1 and f_2 at different m and k.

m	k	$E_r(R_{m,k},[-1,1],1,f_1)$	m	k	$E_r(R_{m,k},[-1,1],1,f_1)$
8	0	0.171...	4	4	0.0470...
10	0	0.147...	5	5	0.0278...
12	0	0.130...	6	6	0.0175...
14	0	0.117...	7	7	0.0112...
16	0	0.106...	8	8	0.0077...
18	0	0.098...	9	9	0.0054...
20	0	0.091...	10	10	0.0036...
70	0	0.032...			
-m	k	$E_r(R_{m,k},[-1,1],1,f_2)$	m	k	$E_r(R_{m,k},[-1,1],1,f_2)$
13	0	0.0842...	6	7	0.0170
14	0	0.0805...	7	6	0.0145
70	0	0.0270...	7	7	0.0137

2. Multidimensional case. Let us denote by E^m the m-dimensional Enclidean space. The Hausdorff distance between two sets $A \subset E^m$, and $B \subset E^m$ is defined as

$$r(A,B) = \max\{\max_{a \in A} \min_{b \in B} \rho(a,b), \max_{b \in B} \min_{a \in A} \rho(a,b)\}$$

where

$\rho(a,b) = \max\{|a_1-b_1|,|a_2-b_2|,...,|a_m-b_m|\}$, $a = (a_1,a_2,...,a_m)$, $b = (b_1,b_2,...,b_m)$.

Let $K \subset E^m$ is a compact set. We denote by F_k the set of all bounded and closed point set $F \subset E^{m+1}$ such that:

1. If $x(x_1,x_2,...,x_m,x_{m+1}) \in F$, then $x^*(x_1,x_2,...,x_m) \in K$ and for every $x^*(x_1,x_2,...,x_m) \in K$ there exists a point $x(x_1,x_2,...,x_m,x_{m+1}) \in F$;

2. F is convex with respect to the axis x_{m+1}, that is if

$x'(x_1',x_2',\ldots,x_{m+1}')\in F$ and $x''(x_1'',x_2'',\ldots,x_{m+1}'')\in F$ then $\alpha x' + (1-\alpha)x''\in F$ for every $\alpha\in[0,1]$.

If the function f is defined in K we shall define again by F(f) the complemented graph of f as

$$F(f) = \bigcap_{g\in F_k,\bar{f}\subset g} g$$

where $\bar{f}$ is the graph of the function f, i.e. $(x_1,x_2,\ldots,x_m,(f(x))\in E^{m+1}$. It is obvious that $F(f)\in F_k$ and if f is a continous function then $F(f) = \bar{f}$. For the bounded functions f and g, defined in $K\subset E^m$, we denote by $r(f,g) = r(F(f),F(g))$ the Hausdorff distance and by $H_n^m(T_n^m)$ the set of all algebraical (trigonometrical) polynomials of m variables of degree $\le n$. Let F_K^M be the space of all $f\in F_K$ for which for every point $(x_1,x_2,\ldots,x_{m+1})\in f$ is is fulfiled that $|x_{m+1}| \le M$. Let us consider the best approximation of a given function $f\in F_K^M$ by means of algebraical polynomials

$$E_{n,r}(f) = \inf_{p\in H_n^m} r(f,p).$$

Theorem (see [4]). If $f\in F_K$ (here K is a unit cube in E^m) then

$$E_{n,r}(f) = m\,\frac{\ln n}{n} + O(\frac{1}{n}).$$

For the filter synthesis in multidimensional case we have to approximate regular bodies like cylinder or parallelepiped and that fact gives possibility to use the well-working method in one-dimensional case when we approximate by means of polynomials and rational functions.

Institute of Mathematics,
Bulgarian Academy of Sciences,
Sofia 1090, Bulgaria

REFERENCES

1. Bl. Sendov. Approximation of functions in Hausdorff metric by means of algebraic polynomials. Ann. de l'Univer. de Sofia, Fac. des Math., vol. 55, 1960/1961, 1-39 (Bulgarian).
2. Bl. Sendov. Hausdorff Approximation, Bulgarian Academy of Sciences, Sofia, 1979 (Russian).
3. P.G. Marinov, A.S. Andreev. A modified Remez algorithm for approximate determination of the rational function of the best approximation in Hausdorff metric, Compt. rend. de l'Acad. bul. des Sci., T40, No 3, 1987, 13-16.
4. Bl. Sendov, V.A. Popov, Approximation of functions of several variables in Hausdorff metric by means of algebraic polynomials. Ann. de Univ. de Sofia, Fac. des Math. Vol. 63, 1968/69, 61-76 (Russian).

Autorenindex

Affeld, K. 155
Ameling, W. 37 163 278
Andreev, A. 384
Appel, U. 88
Arp, F. 187

Bamler, R. 73
Baston, J. 265
Behrens, K. 176
Bohlmann, H. 233
Boie, W. 128
Borgschulte, K. 341
v. Brandt, A. 203
Brennecke, R. 314
Bruck, G. 124
Bunke, H. 143
Butzer, P. L. 45

Chen, S.-C. 295
Coy, D. 116

Dallas, W. J. 302
Danielsson, P. E. 239
Dekker, S. T. 269
Du, Y. 191

Ebert, A. 181
Eckhardt, U. 131
Eggers, H. 52
Engelhorn, M. W. 16 155
Erbel, R. 314
Eulenberg, H. 282

Föhr, R. 278
Franke, U. 135 195

Garnett, E. S. 310
Gebhardt, R. 37
Gersonde, K. 337
Gilge, M. 199 286
Glünder, H 84
Graf v. Keyserlingk 159 319
Griesinger, M. 120
Groen, F. C. A. 269
Gütschow, K. 255

Guse, W. 199 286

Halling, H. W. 356
Hammer, B. 203
Haubold, H. G. 356
Haupt, M. 341
Higgins, J. R. 49
Hinsen, G. 45
Huang, T. S. 64
Huisman, W. C. 207

Janitzki, A. 381
Janke, P. 290
Jensch, P. 163
Jilin, L. 155
Jonker, P. P. 269
Joswig, M. 347
Jung, D. 314

Kauff, P. 295
Kikuschi, M. 172
Kroschel, K. 168

Loffeld, O. 361
Loken, M. 310

Maderlechner, G. 131
Makhamreh, I. 229
Manstetten, P. 212
Maqusi, M. 229
Martin, R. 37
Marvasti, F. A. 97
Massen, R. 290
Mauer, E. 176 181
Meissner, P. 233
Menges, G. 341
Mester, R. 135 195
Meyer, J. 314
Meyer-Ebrecht, D. 263
Mugler, D. H. 41

Nahmias, C. 310
Natterer, F. 29
Niemann, K. 159
Nitezki, P. 273

Paulus, E. 56
Penner, E. 37

Peters, L. E. 163
Pollak, V. A. 327
Ptacek, W. 88

Radig, B. 102
Rösch, J. 290
Rosenblatt-Roth, M. 77
Rumatowski, K. 60

Saito, M. 172
Sawicki, J. 80
Schäfer, R. 295
Schartl, M. 155
Schilling, T. 263
Schmidt, R. 139 351
Schomberg, H. 2
Schorer, B. 120
Schwarz, H. 259
Schwarzmann, P. 120
Sekiya, T. 172
Siemund, B. 323
Silverberg, M. 128
Simnacher, M. 290
Splettstößer, W. 41
Steinmetz, E. 314

Tolxdorff, T. 337

Wade, W. R. 93
Wang, H. 376
Wasel, J. 319
Watanabe, A. 172
Wendler, T. 217
Winkler, W. 263
Wittlich, N. 314
Wu, Y. 376

Yu, S. 376

Zelinski, R. 372
Zinner, H. 351